Modeling Water Quality

in Distribution Systems

Modeling Water Quality in Distribution Systems

Second Edition

Robert M. Clark

American Water Works Association

Modeling Water Quality in Distribution Systems, Second Edition

AWWA Publications Manager: Gay Porter De Nileon
Project Manager/Technical Editor: Martha Ripley Gray
Production: Cheryl Armstrong
Cover Art: Karen Staab

Library of Congress Cataloging-in-Publication Data

Clark, Robert Maurice.
Modeling water quality in distribution systems / Robert M. Clark. -- 2nd ed.
p. cm.
Rev. ed. of: Modeling water quality in drinking water distribution systems / Robert M. Clark, Walter M. Grayman. 1998.
Includes bibliographical references and index.
ISBN 978-1-58321-816-7
1. Water quality--Computer simulation. 2. Water--Distribution--Computer simulation. I. Clark, Robert Maurice. Modeling water quality in drinking water distribution systems. II. Title.
TD482.C58 2011
628.1'44--dc23

2011022334

Printed in the United States of America

American Water Works Association

6666 West Quincy Avenue
Denver, CO 80235-3098
303.794.7711
www.awwa.org

Dedication

To my supportive and understanding family, especially my grandchildren, Thomas, Nicholas, Benjamin, and Eleanor. They have helped me to appreciate and understand the real meaning of life.

Contents

4 Modeling and Monitoring Chemical and Microbiological Contaminants 45

5 Hydraulic Modeling Principles 61

6 Water Quality Modeling Principles 85

11 Planning and Designing a Water Quality Simulation Study 223

12 Limitations of Modeling 255

Appendix A Calibration/Validation and Tracer Studies 275

Appendix B Water Quality Modeling Examples 311

Appendix C Water Quality Effects of Tanks and Storage 345

Appendix D Conversion Charts 365

Appendix E Color Figures 375

Index 391

About the Author 399

List of Figures

NOTE: *Figures on pages 375–390 appear in color in Appendix E.*

List of Tables

Preface

When I initiated USEPA's research studies in an attempt to understand the factors that influence water quality in drinking water distribution systems, I had no way of anticipating how the studies would eventually impact our field. While my colleagues and I were pursuing this research, we were frequently warned that our efforts were a waste of time and resources and were told repeatedly that drinking water networks were much too complicated to lend themselves to systematic scientific study, let alone be amenable to predictive modeling. Nevertheless, we committed ourselves to conducting carefully constructed field studies as the basis for developing and validating our hydraulic/water quality models, and to developing models on a logical, incremental basis.

As our work progressed, it became obvious that drinking water distribution systems were much more dynamic then we originally thought, and we moved from the development of steady-state models to dynamic models. An important principle that developed from our research was the need to be meticulous in developing properly calibrated hydraulic models to provide the basis for our water quality predictions. The progress of this research is discussed in chapter 8.

A problem we encountered early in our work was the large amount of data being generated from our network analysis. Because of limitations in available software, we were forced to run our hydraulic and water quality models separately, but it became clear that we needed to integrate these two aspects of our efforts. As a result, EPANET was developed to provide this increased efficiency. The immediate, widespread response to the availability of this tool was phenomenal. Two decisions that I believe eventually shaped the future of hydraulic/water quality modeling and its applications were the decision to make EPANET a totally public sector model and the decision not to provide technical support for its implementation.

At one point, we contemplated a licensing arrangement with AWWA and distributing EPANET in a manner similar to some of the software developed by other agencies of the federal government. However, we decided not to follow that path but to make EPANET freely available in the public sector. Consequently, it has become widely used throughout the world and has also become the basis for many proprietary models. The decision not to provide technical support for EPANET, I believe, provided the basis for the development and application of many of the proprietary models in use today. In addition to the development of many proprietary models, a large number of special-purpose modifications have been made to EPANET, primarily by academics. As discussed in chapter 10, USEPA is planning to eliminate all of its technical and development support for EPANET. This is most likely because the EPANET code is widely available and easy to modify, and the original code and many modifications can be obtained from various sources with minimal effort.

As discussed in chapter 12, water quality modeling is an extremely powerful tool for analyzing the fate and effect of contaminant transport in drinking water distribution systems. However, it is important for modeling practitioners and decision makers to recognize that modeling has limitations and that models are only representations of reality, not reality itself. This caution is repeated throughout the book. It has been, however, highly gratifying to see that hydraulic/water quality modeling tools have progressed from being a curiosity, as they were when we started our work, to the situation today in which they have become widely accepted and incorporated as a routine tool for application in our field. I hope this book helps contribute to and continue this progress.

This second edition extends and expands the first edition and addresses the many developments that have taken place in the field over the last 14 years.

Chapter 1 discusses the general characteristics of water supply in the United States. It addresses issues related to water scarcity, water security, the effects of climate change, and the general organization of water supplies in the United States. It also reviews the technical elements common to most US water supplies and discusses basic design and operational philosophies that are predominant in the industry.

Chapter 2 reviews public health and regulatory issues related to water supply. This chapter discusses the distribution system as a barrier to waterborne disease, water quality deterioration in distribution systems, the Safe Drinking Water Act, and recent recommendations from the National Research Council related to drinking water distribution systems.

Chapter 3 addresses the fundamentals of water quality modeling and monitoring in drinking water distribution systems. Various types of distribution system models are discussed, and issues such as the importance of adequate hydraulic representation and model calibration are reviewed. The important interactions between water quality modeling and monitoring are highlighted, and current commercial water quality models are discussed.

Chapter 4 discusses typical chemical and microbiological contaminants found in drinking water and the integration of modeling and monitoring. Important issues include the following monitoring issues: regulatory monitoring, security, process control, routine and compliance monitoring, automated/online monitoring, monitoring equipment, establishing requirements, monitoring parameters, number and location of monitors, monitoring characteristics, amenability to remote monitoring and SCADA integration, and online sampling/control devices. Several monitoring case studies are presented.

Chapter 5 reviews the important hydraulic principles that underlie water quality modeling. In addition, the chapter discusses the application of hydraulic models, including modeling temporal variations in systems, model calibration, fire-flow issues, and flushing water mains.

Chapter 6 provides an overview of fundamental water quality principles. EPANET as well as other water quality models are reviewed, and commercial hydraulic water quality software is discussed.

Chapter 7 presents various models that predict disinfectant decay and the formation of trihalomethanes. Exposure modeling is also discussed.

Chapter 8 provides an overview of some of the first water quality modeling studies in the United States, including studies conducted in collaboration with the North Penn Water Authority, the South Central Connecticut Regional Water Authority, and the North Marin Water District. Results from an important study, conducted by the American Water Works Association Research Foundation (now Water Research Foundation) in collaboration with the USEPA, are presented.

Chapter 9 presents several case studies in which water quality modeling has played a critical role. They range from waterborne outbreak studies conducted in Cabool and Gideon, Mo., and in Walkerton, Ont., to chronic-exposure examples as demonstrated in Dover Township, N.J., and Camp Lejeune, N.C. The use of water quality modeling to assess the vulnerability of a utility to security threats is illustrated by a case study conducted by East Bay MUD in California.

Chapter 10 addresses advanced concepts in water quality modeling. These concepts include the application of geospatial tools, the integration of geographic information systems and water quality modeling, multispecies modeling, distribution systems security, water quality and pressure transients, current USEPA research activities, and several newly emerging research areas.

Chapter 11 provides guidelines on how to get started in modeling and in planning and designing a water quality simulation study. The topics covered include the need to define the scope and the problem definition associated with a modeling study. Issues addressed include: model selection and setup, network representation, calibration/validation model application analysis and display of data, and typical model applications.

Chapter 12 is intended to remind practitioners of some of the limitations associated with water quality modeling. To illustrate these cautions, the inconsistencies regarding our understanding of the nature of bulk and wall reactions associated with chlorine decay are discussed.

Appendix A provides an in-depth discussion of the potential associated with the calibration and validation of models using tracer techniques. Appendix B contains several example applications of EPANET. Appendix C reviews the very important and frequently unappreciated aspect of water quality modeling associated with proper tank design and operation. Appendix D contains typical conversion factors that apply to water quality modeling, and Appendix E contains color figures that support the previous chapters and appendices.

Acknowledgments

This is the second edition of *Modeling Water Quality in Distribution Systems*. The first, which I coauthored with Walter M. Grayman, was published in 1998. The research and engineering studies described in both editions took place over a span of more than 20 years and could not have been completed without the advice and encouragement of many colleagues and associates. I would like to acknowledge Dr. Walter Grayman, of course, as well as Dr. Richard Males, Mr. Benjamin Lykins, Dr. James Goodrich, and Dr. Rolf Deininger, who were involved in all phases of our early research. I would also like to acknowledge the following individuals: Dr. Lewis Rossman, for his outstanding efforts in developing and validating EPANET; Mr. Harry Borchers, Ms. Judy Coyle, and Mr. David Milan, from the North Penn Water Authority, who were instrumental in implementing the first field studies ever conducted to demonstrate the of potential water modeling; Mr. Alan Hess, Mr. Ken Skov, and Mr. Darrell Smith of the South Central Connecticut Regional Water Authority, who provided many useful ideas and were deeply involved in validating many concepts described in both editions; Ms. Gayle Smalley, of the North Marin Water District, who assisted in the first application of water quality modeling to study total trihalomethane propagation; Dr. John Vasconcelos, formerly of MWH Global, who field tested, validated, and applied EPANET to a broad cross section of water utilities in the United States; and Mr. John Hill, of the Missouri Department of Natural Resources, and Dr. Fred Angulo, from the Centers for Disease Control and Prevention, for their help in applying EPANET, for the first time, to a waterborne outbreak situation. Dr. Paul Boulos, president and chief operating officer, Innovyze (formerly MWH Soft), was an early participant in our studies and provided many useful comments related to this second edition.

I am very grateful to Ms. Lucille Garner, formerly with the US Environmental Protection Agency, and Mr. Sri Panguluri, senior engineer for the Shaw Group (Cincinnati, Ohio, office), for their assistance in preparing this second edition of the book. I would also like to express my appreciation to Dr. Steven Buchberger, professor of civil and environmental engineering at the University of Cincinnati, and his graduate students Joseph Wright, Chandler McCoy-Simandle, and Nestor Alonso Mancipe Munoz; and Matthew D. McCutcheon, a water resources engineer with Camp Dresser McKee (Columbus, Ohio, office), for allowing me to use their work in Appendix B.

I would like to acknowledge Ms. Megan G. Roberts, senior principal engineer, Hazen and Sawyer, P.C., and Ms. Pat Kline, former AWWA small systems engineer, for their many useful and helpful comments in preparing this second edition. Finally, I would like to express my deepest appreciation to Martha Ripley Gray, Gay Porter De Nileon, and Cheryl Armstrong of the American Water Works Association editorial staff and Deborah J. Lynes for their assistance in producing and publishing this book.

Chapter 1

Characteristics of US Water Supply Systems

Introduction

The quest for pure water is well documented even in earliest times. One of the earliest recorded works on medicine advises "to heat foul water by boiling and exposing to sunlight and by dipping seven times into it a piece of hot copper and then to filter and cool in an earthen vessel. This direction is given by the god who is the incarnation of medical science." This statement is attributed to the "Naghrund Bhuson," a collection of medical maxims from the *Ayura Veda*, the earliest known Sanskrit work on medicine (Baker 1949). Many references are made to the need to acquire and purify water throughout history. The earliest known apparatus for obtaining clarified liquids was pictured on Egyptian walls in the fifteenth and thirteenth centuries BCE. In the Bible, Moses is said to have sweetened bitter waters by casting them into a tree shown him by the Lord. During his forty years wandering in the wilderness, Moses is said to have brought forth water by smiting a rock (Baker 1949).

The Romans demonstrated the importance of an adequate water supply to the development of organized cities and communities. Vitruvius (84 BCE) discussed the various elements of water supply in Book VIII of *De Architectura* including a source of information on springs and the uses and quality of water. One of the most famous water engineers in history was Sextus Julius Frontinus (AD 40–103), the water commissioner of Rome, who in AD 97 wrote a two-volume treatise describing the Roman water system. Frontinus declared that the Roman aqueducts were the real mark of civilization and wrote of settling reservoirs at the head of one of the aqueducts supplying Rome and of pebble catchers built into most of the aqueducts. As with water systems in modern times, the Roman water systems were constructed to support a number of uses, such as supplying water for the public baths and fountains, and not solely for drinking water (Mays 2002).

In the United States, one of the first water systems was built in 1754 by the Moravian settlement in Bethlehem, Pa. It consisted of spring water forced by a pump through bored logs. The city of Philadelphia was also developing a water system at that time. Cast iron has been the material used by most older cities in the United States since its introduction in the early 1800s. The first cast-iron main was laid in Philadelphia in 1817. In 1948, ductile-iron pipe, which has the favorable characteristics of both steel and cast iron, was invented. Boston started using ductile iron in 1968 and has used it exclusively since 1970

(Mays 2000). The estimated 980,000 mi (1.6×10^6 km) of water mains in the United States are being replaced at an estimated rate of once every 200 years (Grigg 2005).

A large range in type and age of pipes makes up today's water distribution systems. The oldest cast-iron pipes from the late nineteenth century typically have an average useful life-span of about 120 years because of the pipe wall thickness (AWWA 2001, AWWSC 2002). In the 1920s, the manufacture of iron pipes changed to improve pipe strength but with a thinner wall and an average life of about 100 years. Pipe manufacturing continued to evolve in the 1950s and 1960s with the introduction of ductile-iron pipe. Polyvinyl chloride (PVC) pipes were introduced in the 1970s and high-density polyethylene in the 1990s, both of which are very resistant to corrosion but with lower strength than ductile iron.

Water quality can be enhanced by source water protection and by treatment, including disinfection. Some pathogens have a relatively short life in water and will not survive storage for even a few days. Filtration will remove bacteria, cysts, and larger particles, but viruses will pass through the filter. Also, the filters need to be cleaned or replaced regularly so that they do not become a source of pollution themselves. Chlorination is probably the most effective disinfectant and should be maintained at 0.2–0.5 mg/L of water. Water can be boiled in an emergency to eliminate pathogens, and municipalities may issue orders to households to boil water when pathogens enter the water supply.

In the United States and throughout the world, drinking water utilities face the challenge of providing potable water to their consumers. However, many factors can result in the degradation of water quality before it is delivered to the customers. For example, treated water may be transmitted through a network of corroded or deteriorated pipes.

Water Scarcity Issues

It is estimated that for the period 1990–2025 there will be water scarcity problems in 118 countries. Approximately 1 billion people face absolute water scarcity because they live in arid regions that lack sufficient water resources. Another 348 million people face severe water scarcity because they live in regions where existing water resources require high-cost water-acquisition developments such as dams, wells, reservoirs, and aqueducts to tap their potential (Bruins 2000).

Global water use has more than tripled since 1950 mainly based on the availability of surface water resources such as rivers, lakes, and groundwater. Water use is generally categorized as follows:

- Domestic or urban water use
- Agricultural water use
- Industrial water use

Irrigation is the largest consumer of freshwater, using over 70 percent of all developed water supplies in the world. Most often, increasing water demands are met by large engineering projects that involve building dams and storing surface water. Some very large surface water projects include the Sardar Sarovar Dam in India and the Three Gorges Project in China. The number of large dams in the world increased from 5,000 in 1950 to about 38,000 by 1995. There has also been increased drilling of groundwater wells. For example, in India the area irrigated from groundwater is greater than all other surface water irrigation combined. Throughout the world, groundwater resources are being overutilized. For example, in northern China, groundwater tables are declining at approximately 1 m (3.28 ft) each year; in the Punjab region of India, groundwater tables are declining 20 cm (7.87 in.) each year. Rivers and lakes will eventually dry up as groundwater disappears (Bruins 2000).

The United States is generally blessed with adequate water resources. The contiguous United States receives an average of approximately 75 cm (30 in.) of rainfall per year, although a great deal of variability is found across the country (Clark and Sethi 2002). The Northwestern part of the United States has the most rainy days in the contiguous United States, while the Southwestern United States has the fewest (NRC 1999). However, the Southwest often receives great amounts of precipitation over a short period of time, while the Northeast has more precipitation in smaller amounts over a longer period of time. Evaporation is important to water supply because it represents the natural loss of otherwise available water. Annual average values of evaporation range from 50 cm (20 in.) in northern Maine to 215 cm (86 in.) in southwestern California.

Representative soil water budgets for the United States show that the magnitude of surpluses and deficits varies greatly by region and season. The largest deficits occur in the Southwest, while the greatest surpluses occur in the Pacific Northwest and the East. The average annual runoff for the contiguous United States is approximately 30 cm (12 in.) but varies widely. The rates range from less than 0.6 cm (0.25 in.) in some Western areas to more than 50 cm (20 in.) in some parts of the East. Year-to-year variations in precipitation and runoff can also be significant. Changes may occur over a span of decades or centuries, and climatologists have only begun to understand how large an impact these changes might have on local and regional weather patterns (NRC 1999).

Ninety-seven percent of all freshwater on earth is groundwater and only two percent of freshwater is contained in lakes and rivers. The United States obtains at least 23 percent of its domestic and industrial water supply from local and regional aquifers, many of which are restored at rates often less than 0.2 percent per year by volume. Almost all rural homes and many cities depend heavily on groundwater. Throughout the West and Midwest, large aquifers have declined substantially because extraction rates have exceeded recharge rates (NRC 1999).

Significant spatial differences often separate areas of high water demand from areas of water surplus, and therefore water transfers play an important role in the United States. Between 1935 and 1965, water storage in western US reservoirs increased from about 5 million acre-ft to 270 million acre-ft (6.2 billion–332.9 billion m^3). For example, the Colorado River is fed from the mountains and once flowed through the deserts of the Southwest and emptied into the Gulf of California. By compact the entire flow was apportioned to seven states, including California, Nevada, and Arizona, and the river's water has been transported to regions outside the river's watershed. Because of erroneous long-term stream flow measurements, more water was apportioned than would normally be available. A series of reservoirs made the water readily available for consumption, so that little flow now reaches the ocean. Southern California is the nation's largest importer of water; its imported water is used for both urban water supply and agriculture, especially irrigation. Water is supplied to Southern California not only from the Colorado River but also from Northern California and the Owens River Valley on the east side of the Sierra Nevada Mountains (NRC 1999).

Another form of water transfer that may have multiple impacts is groundwater extraction. First, as the water table drops, pumping of groundwater becomes increasingly difficult and expensive. Second, declining groundwater can often cause depleted stream flow and depleted local surface water supply. Finally, excessive groundwater extraction can cause the ground to subside; subsidence of more than 9 m (29.5 ft) has been recorded in Southern California, and some formerly inhabited areas along the Gulf Coast are now inundated. Because of heavy irrigation, some areas also have excessive groundwater recharge. As water tables rise, the slightly saline groundwater can move to the surface by capillary action and evaporate, concentrating salt and other minerals at the surface, often damaging the soil, and polluting local groundwater (NRC 1999).

Water Security Issues

Once water supplies and systems are established, they may be vulnerable to drought, earthquakes, and terrorism. Drought affects more people than any other natural hazard, earthquakes can affect water security in modern urban communities, and terrorism is always a problem. Some are concerned that water scarcity might become the basis for future wars (Bruins 2000).

Droughts have led to crises and disasters throughout history and in many parts of the world. Drought may affect both developing and developed countries, and according to the United Nations' Office of Foreign Disaster Assistance, no other natural disaster has resulted in as many displaced persons in the twentieth century. For example, a drought in the Great Plains in the United States in the 1930s caused serve economic hardship in Missouri, Kansas, Nebraska, Oklahoma, South Dakota, and Arkansas. The Great Plains also experienced droughts in the 1950s, 1970s, and 1990s. One drought in Ethiopia affected more then 40 million people (Bruins 2000).

Unlike droughts, which are described as a creeping phenomenon, the damage associated with earthquakes is concentrated in time and space. In 1906, an earthquake in San Francisco caused numerous pipes to rupture and caused drowning of dozens of residents when broken water pipes flooded the Valencia hotel. It was impossible to control the firestorms that spread through the area, engulfing many buildings with temperatures reported to have reached 2000°F (1093.2°C). In 1995, a major earthquake directly hit the city of Kobe, Japan (Great Hanshin Earthquake). The quake lasted 20 seconds; 6,434 people died, 14,679 were injured, and 300, 000 people were displaced. Of 67,421 fully collapsed structures, 6,985 were burned to the ground, and a citywide power failure and a near-citywide water supply failure occurred (Clark and Deininger 2001).

Even greater damage resulted from the The 2011 Tōhoku earthquake, also known as the Great East Japan Earthquake. The resulting tsunami caused even greater damage then the Great Hanshin Earthquake. It is estimated that 45,700 buildings were destroyed and 144,300 were damaged. National police have reported 15,538 deaths, 5,685 injuries, and 7,060 people missing. At least 1.5 million households were reported to have lost access to water supplies (Wikipedia 2011).

Terrorism is also a very real, ongoing issue in our time. The President's Commission on Critical Infrastructure Protection concluded that the nation's water supply systems might be vulnerable to attack (Clark and Deininger 2000). According to Bruins (2000), Israel included Arab villages in the National Carrier System in order to limit the potential posed by terrorists. Water played an important role in the peace treaty that Israel and Jordan signed on Oct. 26, 1994. To this point, the worst-case scenarios have not materialized over water disputes in the Middle East.

Effects of Climate Change

Climate change may adversely affect water quality in streams and other raw water sources in the United States and consequently impact the design and operation of existing and future drinking water treatment plants. Increases in precipitation or precipitation intensity and resulting increases in runoff and stream flows can result in problematic turbidity levels, increased levels of organic matter, potentially high levels of pathogens, and increased levels of pesticides in lakes, rivers, and streams. Climate change may affect both surface water and groundwater quality. Although some areas may experience increases in runoff caused by shifts in the spatial and temporal distribution of precipitation, some areas may experience droughts resulting in elevated levels of potentially toxic algae and high concentrations of organic matter, bacteria, and so on (Whitehead et al. 2009, Interlandi and Crockett 2003, Jacobs et al. 2001). Some of these changes may have an adverse impact on the ability of water treatment plants (WTPs) to meet Safe Drinking Water Act (SDWA) regulations, thereby requiring treatment systems to make changes, sometimes major changes, in their operations.

Organization of Water Supplies in the United States

Water supplies in the United States were initially constructed in cities primarily for fire prevention, but most were eventually adapted to serve commercial and residential properties with water. The first US municipal water utility was established in Boston in 1652 to provide domestic water and fire protection (Hanke 1972). By 1860, 136 water systems operated in the United States, most of which supplied water from springs low in turbidity and relatively free from pollution (Baker 1949). Today, more than 162,000 water systems in the United States meet the federal definition of a public water system (USEPA 2005b).

Water utilities in the United States vary greatly in size, ownership, and type of operation. The SDWA defines public water systems as consisting of community water supply systems; transient noncommunity water supply systems; and nontransient noncommunity water supply systems. A community water supply system serves year-round residents and ranges in size from those that serve as few as 25 people to those that serve several million. A transient noncommunity water supply system serves areas such as campgrounds or gas stations where people do not remain for long periods of time. A nontransient noncommunity water supply system serves primarily nonresidential customers but must serve at least 25 of the same people for at least six months of the year (such as schools, hospitals, or factories that have their own water supply). Thirty-three percent (52,838) of these systems are categorized as community water supply systems, 55 percent are categorized as transient noncommunity water supplies, and 12 percent (19,375) are nontransient noncommunity water systems (USEPA 2005b).

Overall, public water systems serve 297 million residential and commercial customers. Most systems (98 percent) serve less than 10,000 people, but almost three quarters of all Americans get their water from community water supplies serving more than 10,000 people (USEPA 2005b). Not all water supplies deliver water directly to consumers, but rather deliver water to other supplies. Community water supply systems are defined as "consecutive systems" if they receive their water from another community water supply through one or more interconnections (Fujiwara et al. 1995).

Some utilities rely primarily on surface water supplies while others rely primarily on groundwater. Surface water is the primary source of 22 percent of the community water supply systems, while groundwater is used by 78 percent of community water supply systems. Of the noncommunity water supply systems (both transient and nontransient), 97 percent are served by groundwater. Many systems serve communities using multiple sources of supply such as a combination of groundwater and/or surface water sources. In a grid/looped system, the mixing of water from different sources can have a detrimental influence on water quality, including taste and odor, in the distribution system (Clark et al. 1988; 1991a,b).

Some utilities, like the one operating in New York City, own large areas of the watersheds from which their water source is derived, while other utilities depend on water pumped directly from major rivers like the Mississippi River or the Ohio River, and therefore own little if any watershed land. The SDWA was amended in 1986 and again in 1996 to emphasize source water protection in order to prevent microbial contaminants from entering drinking water supplies (Borst et al. 2001). Owning or controlling its watershed provides an opportunity for a drinking water utility to exercise increased control of its source water quality (Peckenham et al. 2005).

The water supply industry in the United States has a long history of local government control over operation and financial management, with varying degrees of oversight and regulation by state and federal government. Water supply systems serving cities and towns are generally administered by departments of municipalities or counties (public systems) or by investor-owned companies (private systems). Public systems are predominantly owned by local municipal governments, and they serve approximately 78 percent of the total population that uses community water supplies. Approximately 82 percent of urban water systems (those serving more than 50,000 persons) are publicly owned. About 33,000 privately owned water systems serve the remaining 22 percent of people served by community water systems. Private systems are usually investor-owned in the larger population size categories but can include many small systems as part of one large organization. These investor-owned utilities are in business to generate profits for their shareholders. In the small- and medium-sized categories, the privately owned systems tend to be owned by homeowners, associations, or developers, and they simply provide water to their clientele with little regard for profit. Finally, several classifications of state-chartered public corporations, quasi-governmental units, and municipally owned systems exist that operate differently than traditional public and private systems. These systems include special districts, independent nonpolitical boards, and state-chartered corporations (NRC 2006). Table 1-1 provides a snapshot of the size and the population served for public water systems in the United States.

Common Elements in Water Systems

Distribution system infrastructure is a major asset of a water utility, even though most of the components are either buried or located inconspicuously. Drinking water distribution systems are designed to deliver water from a source (usually a treatment facility) in the required quantity and quality and at satisfactory pressure to individual consumers in a utility's service area. In general, to continuously and reliably move water between a source and a customer, the system would require storage reservoirs/tanks and a network of pipes, pumps, valves, and other appurtenances. This infrastructure is collectively referred to as the drinking water distribution system (Walski et al. 2003).

Table 1-1. Size and the population served for public water systems in the United States in 2003 (USEPA 2005a)

System Size by Population Served		Very Small 500 or less	Small 501–3,300	Medium 3,301–10,000	Large 10,001–100,000	Very Large >100,000	Total
CWS	No. systems	30,417	14,394	4,686	3,505	361	53,363
	Pop. served	5,010,834	20,261,508	27,201,137	98,706,485	122,149,436	273,329,400
	% of systems	57%	27%	9%	7%	1%	100%
	% of pop.	2%	7%	10%	36%	45%	100%
NTNCWS	No. systems	16,785	2,786	97	16	2	19,686
	Pop. served	2,327,575	2,772,334	506,124	412,463	279,846	6,298,342
	% of systems	85%	14%	0%	0%	0%	100%
	% of pop.	37%	44%	8%	7%	4%	100%
TNCWS	No. systems	85,366	2,657	96	29	4	88,152
	Pop. served	7,315,647	2,602,706	528,624	619,248	12,269,000	23,335,225
	% of systems	97%	3%	0%	0%	0%	100%
	% of pop.	31%	11%	2%	3%	53%	100%
Total No. of systems		132,566	19,837	4,879	3,550	367	161,201

CWS = community water supply
NTNCWS = nontransient noncommunity water supply
TNCWS = transient noncommunity water supply

Impact of System Design and Operation on Water Quality

Based on the design and configuration of a particular system, many opportunities arise for water quality to change as water moves between the treatment plant and the consumer. These unwanted changes may occur for various reasons including: failures at the treatment barrier, transformations in the bulk phase, corrosion and leaching of pipe material, biofilm formation, and mixing between different sources of water. Many researchers have investigated the factors that influence water quality deterioration once it enters the distribution system. It has been well documented that bacteriological growth can cause taste-and-odor problems, discoloration, slime buildup, and economic problems, including corrosion of pipes and biodeterioration of materials. Bacterial numbers tend to increase during distribution and are influenced by several factors, including bacterial quality of the finished water entering the system, temperature, residence time, presence or absence of a disinfectant residual, construction materials, and availability of nutrients for growth (Geldreich et al. 1972, LeChevallier et al. 1987, Maul et al. 1985a and b, Zhang and DiGiano 2002, Camper et al. 2003).

The branch and grid/loop are the two basic configurations for most water distribution systems. A branch system is similar to that of a tree branch with smaller pipes branching off larger pipes throughout the service area. This type of system is most frequently used in rural areas, and the water has only one possible pathway from the source to the consumer. A grid/loop system is the most widely used configuration in large municipal systems and consists of interconnected pipe loops throughout the area to be served. In this type of system, the water can follow several pathways from the source to the consumer. Transmission mains are typically 20 to 24 in. in diameter or larger. Dual-service mains that serve both transmission and distribution purposes are normally 12 to 20 in. in diameter. Distribution mains are usually 6 to 12 in. in diameter in every street. Service lines are typically 1 in. in diameter. Specific pipe sizes can vary depending on the extent of the distribution system and the magnitude of demand. Looped systems provide a high degree of reliability should a line break occur, because the break can be isolated with little impact on consumers outside the immediate area (Clark and Tippen 1990, Clark et al. 2004).

Storage Tanks/Reservoirs

Tanks and reservoirs are used to provide storage capacity to meet fluctuations in demand, to provide reserves for firefighting use and other emergency situations, and to equalize pressures in the distribution system (Boulos et al. 2006). The most frequently used type of storage facility is the elevated tank, but other types of tanks and reservoirs include in-ground tanks and open or closed reservoirs. Construction materials include concrete and steel. An issue that has drawn a great deal of interest is the problem of water turnover within storage facilities. Much of the water volume in storage tanks is dedicated to fire protection. Unless utilities make a deliberate effort to exercise (fill and draw) their tanks or to downsize the tanks when the opportunity presents itself, both water aging and water mixing problems can occur. The latter can lead to stratification and/or large stagnant zones within the water volume. Some of these issues will be discussed later in this book.

Pipe Network

The systems of pipes or *mains* that carry water from the source (such as a treatment plant) to the consumer are often categorized as transmission/trunk, distribution, and service mains. Transmission/trunk mains are usually designed to convey large amounts of water over long distances, such as from a treatment facility to a storage tank within the distribution system. Distribution mains are typically smaller in diameter than the transmission mains and generally follow city streets. Service mains are pipes that carry water from the distribution main to the building or property being served. Service lines can be of any size, depending on how much water is required to serve a particular customer, and are sized so that the utility's design pressure is maintained at the customer's property for the desired flows.

The most commonly used pipes today for water mains are ductile iron, prestressed concrete, polyvinyl chloride (PVC), reinforced plastic, and steel. In the past, unlined cast-iron pipe and asbestos–cement pipes were frequently used. Even a medium-sized water utility may have hundreds of miles of pipes constructed from various types of materials, ranging from new, lined, or plastic pipes to unlined pipes that are more than 50 years old. Over time, biofilms and tubercles attached to pipe walls can result in both loss of carrying capacity and a significant loss of disinfectant residual, thereby adversely affecting water quality (Clark and Tippen 1990).

The mains should be placed in areas along the public right of way, which provides for ease of access, installation, repair, and maintenance. Broken or leaking water mains should be repaired as soon as possible to minimize property damage and loss of water. In the past, it has been standard practice to maintain the carrying capacity of the pipe in the distribution system as high as possible to provide the design flow and keep pumping costs as low as possible. However, there has been recent concern that excess capacity can lead to long residence times and thus contribute to deterioration in water quality.

Valves

Two general types of valves are found in a distribution system: isolation valves and control valves. Isolation valves are used in the distribution system to isolate sections for maintenance and repair and are typically located in a system so that the areas isolated will cause a minimum of inconvenience to other service areas. Maintenance of the valves is one of the major activities carried out by a utility. Many utilities have a regular valve-turning program in which a percentage of the valves are opened and closed on a regular basis. It is desirable to turn each valve in the system at least once per year. In large systems, this may or may not be practical, but periodic exercise and checking of valve operations should occur. This practice minimizes the likelihood that valves will become inoperable due to corrosion. The implementation of such a program ensures that, especially during an emergency, water can be shut off or diverted and that valves have not been inadvertently closed.

Control valves are used to regulate the flow or pressure in a distribution system. Typical types of control valves include pressure-reducing valves, pressure-sustaining valves, flow control valves, throttling valves, and check valves.

Pumps

Pumps are used to impart energy to the water in order to boost it to higher elevations or to increase pressure. Routine maintenance, proper design and operation, and testing are required to ensure that they will meet their specific objectives. Pump tests are typically run every 5 to 10 years to check the head-discharge relationship for the pump. Many system designers recommend two smaller pumps instead of one large pump to ensure redundancy.

Hydrants and Other Appurtenances

Hydrants are primarily a part of the firefighting infrastructure of a water system. Although water utilities usually have no legal responsibility for fire flow, developmental requirements often include fire flows, and thus distribution systems are designed to support needed fire flows where practical (AWWA 2008). Proper design, spacing, and maintenance are needed to ensure an adequate flow to satisfy firefighting requirements. Fire hydrants are typically exercised and tested periodically by water utility or fire department personnel. Fire-flow tests are conducted periodically to satisfy the requirements of the Insurance Services Office (ISO 2003) or as part of a water distribution system calibration program. Other appurtenances in a water distribution system include blow-off valves and air-release valves.

Basic Design and Operational Philosophies

A detailed understanding of *how water is used* is critical to understanding water distribution system design and operation. Almost universally, the manner in which industrial and residential customers use water drives the overall design and operation of a water distribution system. Generally, water use varies both spatially and temporally. Besides customer consumption, a major function of most distribution systems is to provide adequate standby fire-flow capacity (Fair and Geyer 1971). For this purpose, fire hydrants are installed in areas that are easily accessible by firefighters and are not obstacles to pedestrians and vehicles. The ready-to-serve requirements for firefighting are governed by the National Fire Protection Association (NFPA), which establishes standards for firefighting capacity of distribution systems (Cote et al. 2008). In order to satisfy this need for adequate standby capacity and pressure (as mentioned earlier), most distribution systems use standpipes, elevated tanks, and large storage reservoirs. Additionally, most distribution systems are *zoned*. Zones are areas or sections of a distribution system of relatively constant elevation. Zones can be used to maintain relatively constant pressures in the system over a range of ground elevations. Sometimes, zone development occurs as a result of the manner in which the system has expanded.

Conservative design philosophies, aging water supply infrastructure, and increasingly stringent drinking water standards have resulted in concerns over the viability of drinking water systems in the United States. Questions have been raised over the structural integrity of these systems as well as their ability to maintain water quality from the treatment plant to the consumer. The *Clean Water and Drinking Water Infrastructure Gap Analysis* (USEPA 2002), which identified potential funding gaps between projected needs and spending from 2000 through 2019, estimated a potential 20-year funding gap for drinking water capital and operations and maintenance, ranging from $45 billion to $263 billion, depending on spending levels. Based on current spending levels, the United States faces a shortfall of $11 billion annually to replace aging facilities and comply with safe drinking water regulations. Federal funding for drinking water in 2005 remained level at $850 million—less than 10 percent of the total national requirement (ASCE 2005). Parts of many systems are approaching or exceed 100 years old, and an estimated 26 percent of the distribution system pipe in this country is unlined cast iron and steel in poor condition. At current replacement rates for distribution system components, it is projected that a utility will replace a given pipe every 200 years (Kirmeyer et al. 1994). Grigg (2005) provides comprehensive guidance to utilities on how to assess options for distribution system renewal. Grigg's (2005) report contains a knowledge base on condition assessment, planning and prioritization, and renewal methods.

The effect of designing and operating a system to maintain adequate fire flow and redundant capacity can result in long travel times for water between the treatment plant and the consumer. These long travel times and low velocities may be detrimental to meeting the drinking water maximum contaminant levels (MCLs). Long residence times may lead to formation of disinfection by-products (DBPs), loss of disinfectant residuals, bacterial growth, and formation of biofilm.

Nonpotable waters (e.g., sea, river, and lake water) without adequate treatment have been used for fire protection for many years, often with disastrous results. However, reclaimed wastewater (in cases where its quality is better managed than the aforementioned unregulated sources) has been effectively used for providing fire protection. St. Petersburg, Fla., has been operating such a system to bolster fire-protection capacity since 1976. The reclaimed water hydrants are distinguished from potable water hydrants by color and their special valves. If the reclaimed water system is designed for fire protection, the potable water piping can have a very small diameter and investments can be made in higher quality pipe materials, which, with much shorter residence time in the system, would vastly improve the quality of the water at the tap. With this in mind, where retrofitting one of the two systems is necessary, it might be wiser to use the existing potable water system for the reclaimed water and retrofit with new, high-quality, smaller potable water lines (Okun 1996).

Many systems serve communities with multiple sources of supply, such as a combination of wells and/or surface sources. In a grid/looped system, this configuration will influence water quality in a distribution system because of the effect of mixing of water from these different sources. These interactions are a function of complex system hydraulics (Clark et al. 1988, Clark et al. 1991a). Water quality models can be very useful in defining mixing and blending zones within water utility distribution networks. Mixing of water in a network can result in taste and odor problems or other water quality problems and can influence maintenance, repair, and rehabilitation procedures.

References

American Society of Civil Engineers (ASCE). 2005. *Report Card for America's Infrastructure: Drinking Water.* Reston, Va.: ASCE.

American Water Works Association (AWWA). 2001. *Reinvesting in Drinking Water Structure: Dawn of the Replacement Era.* Denver, Colo.: AWWA.

AWWA. 2008. *AWWA Manual M31, Distribution System Requirements for Fire Protection.* 4th ed. Denver, Colo. AWWA.

American Water Works Service Co. (AWWSC). 2002. Deteriorating Buried Infrastructure Management Challenges and Strategies. http://www.epa.gov/safewater/tcr/pdf/infrastructure.pdf.

Baker, M.M. 1949. *The Quest for Pure Water.* New York: AWWA, 1–7.

Borst, M., M. Krudner, L. O'Shea, J.M. Perdek, D. Reasoner, and M.D. Royer. 2001. Source Water Protection: Its Role in Controlling Disinfection By-Products (DBPs) and Microbial Contaminants. In: *Controlling Disinfection By-Products and Microbial Contaminants in Drinking Water*, ed. R.M. Clark and B.K. Boutin. EPA/600/R-01/110. Washington, D.C.: USEPA Office of Research and Development.

Boulos, P.F., K.E. Lansey, and B.W. Karney. 2006. *Comprehensive Water Distribution Systems Analysis Handbook for Engineers and Planners.* 2nd ed. Broomfield, Colo.: MWH Soft, 660 pp.

Bruins, H.J. 2000. Proactive Contingency Planning vis-à-vis Declining Water Security in the 21st Century. *Journal of Contingencies and Crisis Management*, 8(2):63–72.

Camper, A.K., K. Brastrup, A. Sandvig, J. Clement, C. Spencer, and A.J. Capuzzi. 2003. Impact of Distribution System Materials on Bacterial Regrowth. *Journal AWWA*, 95(7):107–121.

Clark, R.M., W.M. Grayman, and R.M. Males. 1988. Contaminant Propagation in Distribution Systems. *Journal of Environmental Engineering*, 114(2).

Clark, R.M., and D.L. Tippen. 1990. Water Supply. In *Standard Handbook of Environmental Engineering*, ed. Robert A. Corbitt. New York: McGraw-Hill, 5.173–5.220.

Clark, R.M., W.M. Grayman, and J.A. Goodrich. 1991a. Water Quality Modeling: Its Regulatory Implications. In *Proc. AwwaRF/EPA Conf. on Water Quality Modeling in Distribution Systems, Cincinnati, Ohio.* Denver, Colo.: AwwaRF.

Clark, R.M., W.M. Grayman, J.A. Goodrich, R.A. Deininger, and A.F. Hess. 1991b. Field Testing of Distribution Water Quality Models. *Journal AWWA*, 83(7):67–75.

Clark, R.M., and R.A. Deininger, 2000. Protecting the Nation's Critical Infrastructure: The Vulnerability of US Water Supply. *Journal of Contingencies and Crisis Management*, 8(2):73–80.

Clark, R.M., and R.A. Deininger, 2001. Minimizing the Vulnerability of Water Supplies to Natural and Terrorist Threats. In *Proc. AWWA IM Tech Conf., Atlanta, Ga., April 8–11.* Denver, Colo.: AWWA, 1–20.

Clark, R.M., and V. Sethi. 2002. Water Conservation Practices and Policies in the U.S. In *Regional Water System Management: Water Conservation, Water Supply and System Integration*, ed. E. Cabera, R. Cobacho, and J.R. Lund. Lisse, Netherlands: Swets & Zeitlinger B.V., 131–148.

Clark, R.M., W.M. Grayman, S.G. Buchberger, Y. Lee, and D.J. Hartman. 2004. Drinking Water Distribution Systems: An Overview. In *Water Supply Systems Security*, ed. Larry W. Mays. New York: McGraw-Hill, 4.1–4.49.

Cote, A.E., J.R. Hall, P. Powell, C.C. Grant, and R.E. Solomon. 2008. *Fire Protection Handbook 2008.* Quincy, Mass: National Fire Protection Association.

Fair, G.M., and J.C. Geyer. 1971. *Water Supply and Waste-Water Disposal.* New York: John Wiley and Sons.

Fujiwara, M., J. M. Manwaring, and R. M. Clark. 1995. Drinking Water in Japan and the United States: Conference Objectives. In *Drinking Water Quality Management*, ed. R.M. Clark and D.A. Clark. Lancaster, Pa.: Technomic.

Geldreich, E.E., H.D. Nash, D.J. Reasoner, and R.H. Taylor. 1972. The Necessity of Controlling Bacterial Populations in Potable Water: Community Water Supply. *Journal AWWA*, 64:596–602.

Grigg, N.S. 2005. *Assessment and Renewal of Water Distribution Systems.* AwwaRF Report No. 91025F. Denver, Colo.: AwwaRF.

Hanke, S.H. 1972. Pricing Urban Water. In *Public Prices for Public Products*, ed. Selma Mushkin. Washington, D.C.: Urban Institute, 283–306.

Insurance Services Office (ISO). 2003. *Fire Suppression Rating Schedule.* New York: Insurance Services Office.

Interlandi, S.J., and C.S. Crockett. 2003. Recent Water Quality Trends in the Schuylkill River, Pennsylvania, USA: A Preliminary Assessment of the Relative Influences of Climate, River Discharge, and Suburban Development. *Water Resources*, 37:1737–1748.

Jacobs, K., D.B. Adams, and P. Gleick. 2001. Potential Consequences of Climate Variability and Change for the Water Resources of the United States. In *Climate Change Impacts on the United States: The Potential Consequences of Climate Variability and Change*, ed. J. Meilillo, A. Janetos, and T. Karl. New York: Cambridge University Press.

Kirmeyer, G.J., W. Richards, and C.D. Smith. 1994. *An Assessment of the Condition of North American Water Distribution Systems and Associated Research Needs.* Denver, Colo.: AwwaRF and AWWA.

LeChevallier, M.W., T.M. Babcock, and R.G. Lee. 1987. Examination and Characterization of Distribution System Biofilms. *Applied and Environmental Microbiology*, 53:2,714–2,724.

Maul, A., A.H. El-Shaarawi, and J.C. Block. 1985a. Heterotrophic Bacteria in Water Distribution Systems. I. Spatial and Temporal Variation. *The Science of the Total Environment*, 44:201–214.

Maul, A., A.H. El-Shaarawi, and J.C. Block. 1985b. Heterotrophic Bacteria in Water Distribution Systems. II. Sampling Design for Monitoring. *The Science of the Total Environment*, 44:215–224.

Mays, L.W. 2000. Introduction. In *Water Distribution Systems Handbook*, ed. L.W. Mays, New York: McGraw-Hill, 1.1–1.30.

Mays, L.W. 2002. Urban Water Infrastructure: A Historical Perspective. In *Urban Water Supply Handbook*, ed. L.W. Mays, New York: McGraw-Hill, 1.3–1.66.

National Research Council (NRC). 1999. *New Strategies for American's Watersheds.* Committee on Water Science Management, National Academy of Sciences. Washington, D.C.: National Academies Press, 56–101.

National Research Council. 2006. *Drinking Water Distribution Systems: Assessing and Reducing Risks.* Washington, D.C.: National Academies Press.

Okun, D. 1996. Distributing Reclaimed Water Through Dual Systems. *Journal AWWA*, 89 (11):52–64.

Peckenham, J.M., C.V. Schmitt, J.L. McNelly, and A.L. Tolman. 2005. Linking Water Quality to the Watershed: Developing Tools for Source Water Protection. *Journal AWWA*, 97(9):62–69.

US Environmental Protection Agency (USEPA). 2002. *The Clean Water and Drinking Water Infrastructure Gap Analysis.* Washington, D.C.: USEPA Office of Water.

USEPA. 2005a. *Drinking Water Infrastructure Needs Survey and Assessment Third Report to Congress.* EPA 816-R-05-001. Washington, D.C.: USEPA Office of Water.

USEPA. 2005b. *Factoids: Drinking Water and Ground Water Statistics for 2003.* EPA 816-K-05-01. Washington, D.C.: USEPA Office of Water.

Walski, T.M, D.V. Chase, D.A. Savic, W.M. Grayman, S. Beckwith, and E. Koelle. 2003. *Advanced Water Distribution Modeling and Management.* Waterbury, Conn.: Haestad Press, 1–4.

Whitehead, P.G., R.L. Wilby, R.W. Battarbee, M. Kernan, and A.J. Wade. 2009. A Review of the Potential Impacts of Climate Change on Surface Water Quality. *Hydrological Sciences*, 54(1):101–123.

Wikipedia. 2011. 2011 Tōhoku earthquake and tsunami. http://en.wikipedia.org/wiki/Japanese_earthquake_2011.

Zhang, W., and F.A. DiGiano. 2002. Comparison of Bacterial Regrowth in Distribution Systems Using Free Chlorine and Chloramine: A Statistical Study of Causative Factors. *Water Research*, 36(6):1469–1482.

Chapter 2

Public Health and Regulatory Issues in Water Supply

Introduction

Water is absolutely essential for human existence. The human body consists of 65 to 70 percent water, and an active person loses approximately 2 L (0.53 gal) of water every day through exhaling, urine, stools, and perspiration. These lost fluids must be replaced, and in hot climates the amount of replacement liquid may be as high as 5 to 10 L (1.32–2.64 gal) of water per day. The number of days a healthy person can survive without water is limited and is temperature dependent. In a moderate temperature range of 16–21°C (61–70°F), the survival period is approximately 18 days, but in a more extreme environment of 39–49°C (102.2–120.2°F), the survival period is about 2 days. The stages of dehydration for an 80-kg (176-lb) person are as follows (Bruins 2000):

- A loss of 3.2 L (0.85 gal) of water (4 percent loss in body weight) leads to a feeling of apathy.
- A loss of 6.4 L (1.70 gal) of water (8 percent loss in body weight) leads to a condition in which it becomes very difficult to speak.
- A loss of 9.6 L (2.54 gal) of water (12 percent loss of body weight) leads to situation where it becomes impossible to swallow and a person can only be saved by the infusion of intravenous fluids.
- A loss of 14.4 L (3.80 gal) of water (18 percent loss in body weight) usually results in death.

Obviously the quantity of water consumed is important, but water quality is equally significant, and water should be free from chemical and biological contaminants. Each year, millions of people, particularly children, die from diarrhea caused by contaminated drinking water. Water-related diseases include waterborne infections such as cholera, typhoid, infective hepatitis, schistosomiasis, and guinea worm.

In the United States, drinking water treatment has played a major role in protecting public health by reducing waterborne disease. For example, the typhoid death rate for one year in the 1880s was 158 per 100,000 in Pittsburgh, Pa. (Fujiwara et al. 1995). Similarly, both typhoid case and death rates for the city of Cincinnati declined more than tenfold during the period 1898 to 1928 because of the use of sand filtration, disinfection via chlorination, and the application of drinking water standards (Clark et al. 1985). By 1935, the typhoid death rate had declined to 5 per 100,000. Such reductions in waterborne disease outbreaks were brought about by the use of sand filtration, disinfection, and the application of drinking water standards (Clark et al. 1991b). By the end of the nineteenth century, however, waterborne disease had become recognized as a serious problem in industrialized river valleys. This led to the more routine use of water treatment prior to its distribution. Figure 2-1 shows the decline in typhoid incidence rate for the Cincinnati Water Works, and Figure 2-2 shows a similar decline for the death rate during this period.

It is generally accepted that water treatment in the United States has been a major contributor to safeguarding the nation's public health. Such dramatic reductions in waterborne disease outbreaks were brought about by the application of drinking water standards and "multiple barriers" of protection. The multiple-barrier concept includes the use of conventional treatment (e.g., sand filtration) in combination with disinfection to provide safe and aesthetically acceptable drinking water. Residual disinfectant levels serve to protect water quality within a distribution system prior to its delivery (Clark et al. 1991a).

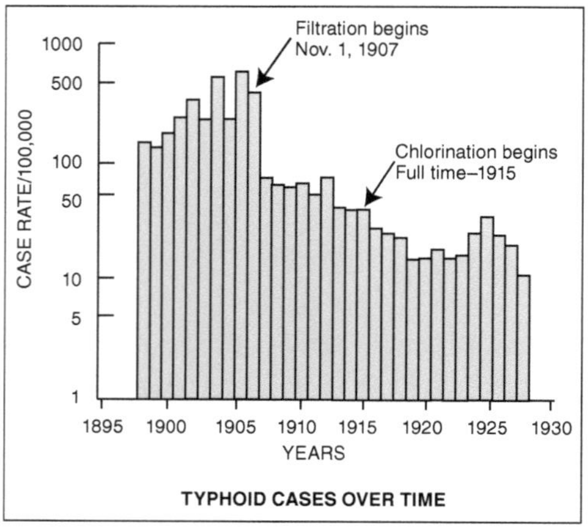

Figure 2-1. Typhoid case rate for Cincinnati Water Works (Clark et al. 1985)

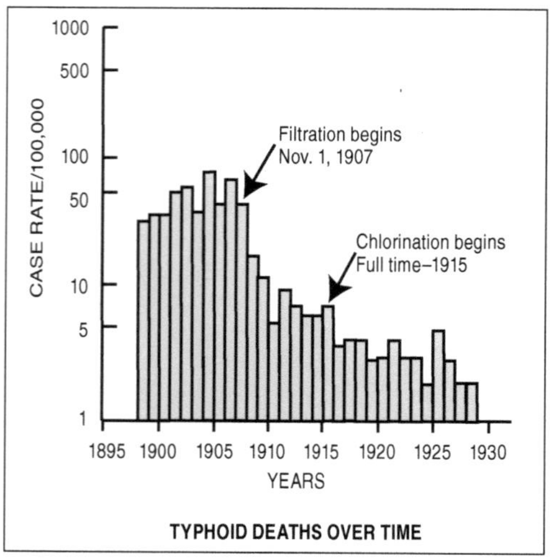

Figure 2-2. Typhoid death rate for Cincinnati Water Works (Clark et al. 1985)

However, periodically water supply problems have continued to surface. Some of the more notable water supply problems that occurred during the 1980s and 1990s in the United States are summarized in Clark and Grayman (1998). For example, in 1993 in Milwaukee, Wis., an estimated 400,000 water consumers were infected during a cryptosporidiosis outbreak. The number of deaths associated with this outbreak is estimated at nearly 100 immunocompromised individuals. Boil-water orders in the Manhattan borough of New York City and Washington, D.C., drew attention to drinking water problems in major metropolitan areas. In November 1993 in Gideon, Mo., out of a population of 1,000, 400–500 people contracted *Salmonella typhimurium* (Clark et al. 1996). The *Salmonella* outbreak contributed to the death of 7 of these individuals. It is believed that bird droppings contaminated storage tanks. During the period Dec. 15, 1989, to Jan. 20, 1990, residents and visitors to Cabool, Mo, population 2,090, experienced 240 cases of diarrhea and 6 deaths (Geldreich et al. 1992). The organism *Escherichia coli* serotype 0157:H7, associated with the feces of healthy dairy cattle, was isolated in many of the stool samples of ill people. As in the case of Gideon, Cabool used a nondisinfected groundwater. In both situations, deaths were among the elderly in nursing homes. One consequence of these investigations is the finding that immunocompromised people (elderly and young) are often less capable of surviving waterborne illness than younger, healthier people (Clark et al. 1994a).

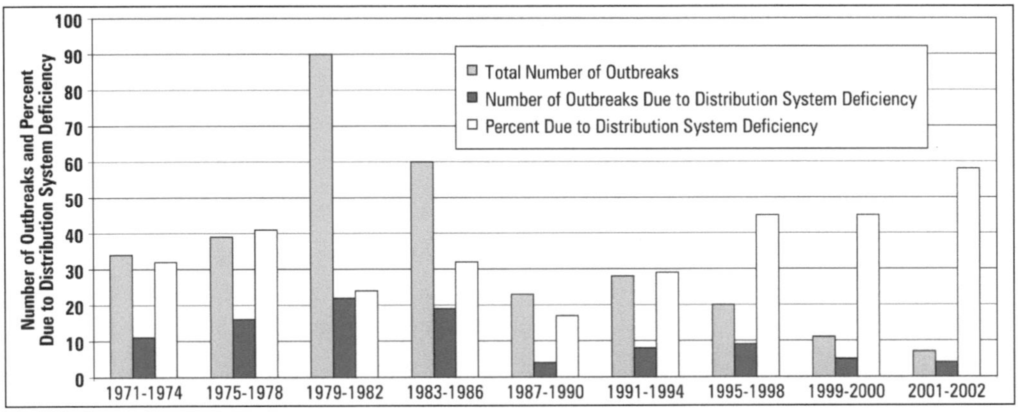

Figure 2-3. Waterborne disease outbreaks in community water systems (CWS) associated with distribution system deficiencies (Panguluri et al. 2005)

In addition, a rising number of giardiasis cases were reported especially in small communities in the 1980s and 1990s.

These types of problems are certainly not confined to the United States. A USEPA team was invited by the Centers for Disease Control and Prevention (CDC) to assist in evaluating the waterborne disease potential associated with a cholera outbreak in Peru in 1991. The investigative team concluded that one of the major factors in this outbreak was the marginal condition of the distribution systems in the areas visited. It was observed that water leaving a treatment plant had adequate disinfectant residuals that quickly disappeared within the distribution system. Systems experienced intermittent operation, fluctuating pressures, frequent pipe breaks, high water losses, and unplanned cross-connections. Little repair and maintenance were practiced by the utilities visited. It was the opinion of the investigative team that many of the waterborne disease problems associated with cholera were related to improperly operated and poorly maintained distribution systems (Clark et al. 1993a).

The National Research Council (NRC 2006) examined the causes of waterborne outbreaks in the United States reported by various investigators between 1971 and 2004. Figure 2-3 presents the total number and proportion of waterborne diseases associated with distribution system deficiencies.

The Distribution System as a Barrier to Waterborne Disease

Developing an effective drinking water infrastructure is essential to public health, spurs productivity and profitability, increases the tax base, and creates skilled jobs in construction, engineering, and manufacturing (Clark et al. 1988, Clark et al. 1993b). On the one hand, it has been estimated that 57,000 jobs are created for every $1 billion invested in distribution infrastructure (Kirmeyer et al. 1994). On the other hand, improperly designed and maintained systems can be an economic liability.

An infrequently considered factor that may influence water quality in a distribution system is the effect of mixing water from different sources. Water distribution systems frequently draw water from multiple sources, such as a combination of wells and/or surface sources. The mixing of waters from different sources that takes place within a distribution system is a function of complex system hydraulics (Clark and Grayman 1998, Clark et al. 1991a, Clark et al. 1991c).

Much has been written about the multiple-barrier concept and its application in providing protection against waterborne disease (Clark et al. 1993a). The multiple-barrier concept has traditionally referred to the use of conventional treatment in combination with disinfection to provide safe and aesthetically acceptable drinking water. However, the distribution system may be the final barrier against waterborne disease.

Water Quality Deterioration in Distribution Systems

Even medium-sized water utilities may have hundreds of miles of pipes constructed from various types of materials, ranging from new lined pipes to unlined pipes that are more than 50 years old. Over time, biofilms and tubercles attached to pipe walls can result in a significant loss of disinfectant residual and, thereby, adversely affect water quality.

Many opportunities arise for water quality to change as it moves between the treatment plant and the consumer. Figure 2-4 illustrates some of the transformations that take place in the water flow through the pipe (bulk water phase) and at the pipe wall (Clark et al. 1994b). Cross-connections, failures at the treatment barrier, and transformations in the bulk phase can all degrade water quality. Corrosion, leaching of pipe material, and biofilm formation and scour can occur at the pipe wall to degrade water quality.

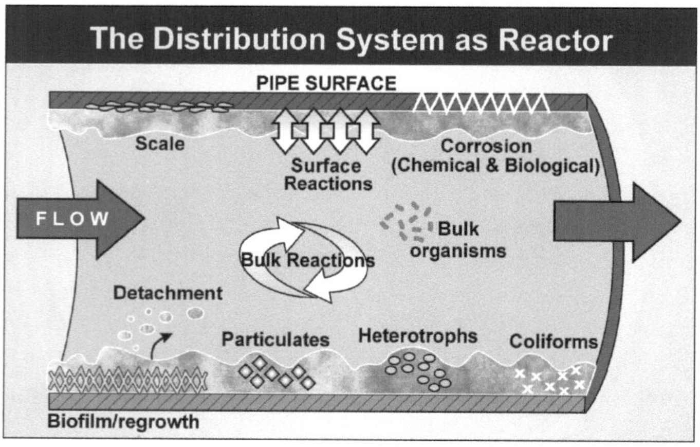

Figure 2-4. Transformations in the bulk phase and at pipe wall (Adapted from MSU–CBE, P. Dirckx)

Unlike hydraulic modeling of distribution systems, water quality modeling is a relatively recent development dating back only to the early 1980s. As will be discussed, water quality models can be used to trace the movement of contaminants in the water distribution system. These models may also be used to calculate the age of the water within the distribution system or to determine the source(s) of water throughout the system. Water quality models generally use the results of hydraulic models and, like hydraulic models, may be based on either steady-state or extended-period simulation (EPS). Contaminants may be treated as either conservative (do not degrade) or as experiencing decay or growth as they move through the system.

Many researchers have investigated how water deteriorates once it enters the distribution system. It has been documented that bacteriological quality changes can cause taste-and-odor problems, discoloration, slime growths, and economic problems, including corrosion of pipes and biodeterioration of materials. Bacterial numbers tend to increase during distribution and are influenced by a number of factors, including bacterial quality of the finished water entering the system, temperature, residence time, presence or absence of a disinfectant residual, construction materials, and availability of nutrients for growth (Geldreich et al. 1972, LeChevallier et al. 1987, Maul et al. 1985a and 1985b).

The relationship of bacteriological quality to turbidity and particle counts in distribution water was studied by McCoy and Olson (1986). One upstream and one downstream sampling site in each of three distribution systems (two surface water supplies and one groundwater supply) were sampled twice per month over a one-year period. Degradation of bacterial water quality was shown to be the result of unpredictable intermittent events that occurred within the system.

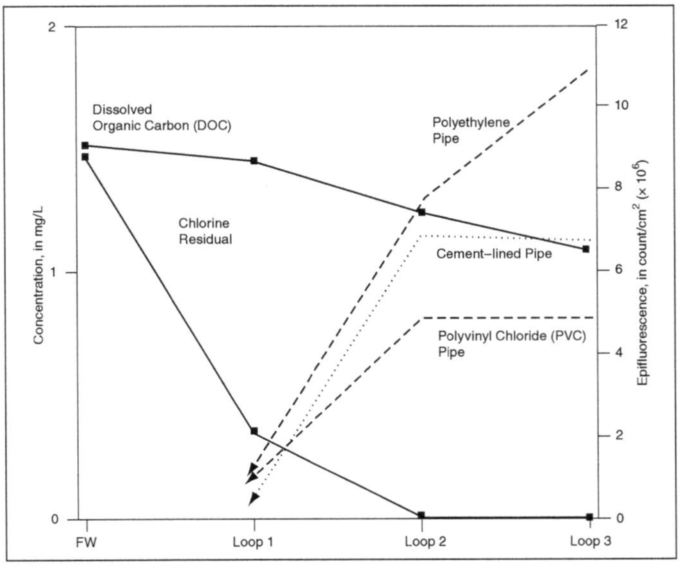

Figure 2-5. Loss of chlorine residuals and DOC and growth in bacterial populations for three different pipe materials with time in a distribution simulator

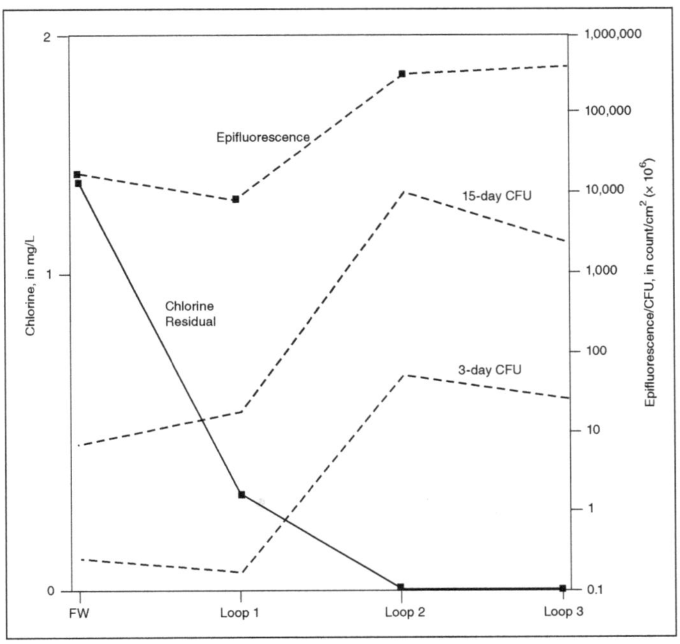

Figure 2-6. Loss of chlorine residuals and growth in bacterial populations with time in a distribution simulator

Figures 2-5, 2-6, and 2-7 illustrate some of the complex chemical and microbiological changes that take place in distribution networks (Clark et al. 1994b). These data are from a unique pilot facility in Nancy, France. The pilot facility consisted of two pilot plants operating in parallel, each feeding a separate simulated distribution system. The source of water for the pilot plant was a nondisinfected raw surface water. Each simulated system consisted of three pipe loops in series. Each of the loops had a one-day residence time.

It is difficult at this time, given our current level of knowledge, to predict changes in water quality during transport in a distribution network. However, as can be seen from these data, as the water moves through the loops, there is a tendency for the disinfectant residual to drop and for various measures of biological activity in the bulk phase to increase.

Based on the study in Nancy, Figure 2-5 illustrates the decrease in dissolved organic carbon (DOC) that took place as water traveled through the pilot network. This supports the assumption that

DOC was used for biological growth. The biofilm buildup increased with residence time in the loop and increased dramatically after disinfectant residuals disappeared. This is consistent with results obtained by LeChevallier et al. (1987). Polyethylene coupons sustained the highest biofilm buildup, perhaps caused by materials in the pipe that encourage biofilm growth. Clearly, differences in biofilm formation occur based on the type of pipe material.

Figure 2-5 shows the steady loss of disinfectant residual through the pipe loops. The chlorine residual dropped most rapidly because of its reaction with organic material in the bulk phase of the water and with the pipe wall material. As the residuals decreased, the organisms in the bulk water increased dramatically.

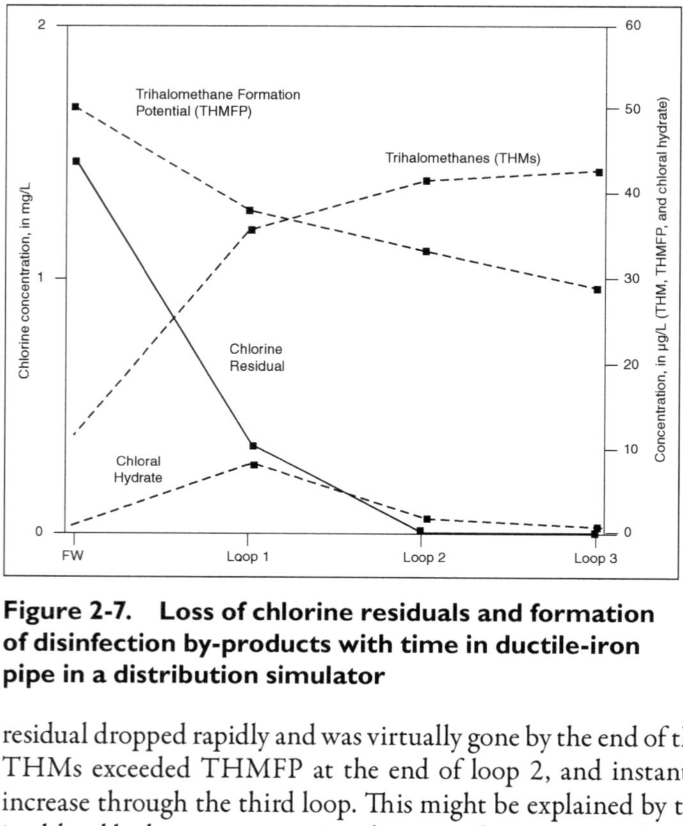

Figure 2-7. Loss of chlorine residuals and formation of disinfection by-products with time in ductile-iron pipe in a distribution simulator

As shown in Figure 2-6, bacterial populations increase over time in the distribution simulator as chlorine residual decreases. Based on the data presented. the bacterial populations seem to drop as they consume the available nutrients.

Figure 2-7 shows trihalomethane (THM) changes that took place across the pipe loops for postdisinfection with chlorine. The drop in trihalomethane formation potential (THMFP) was no doubt related to the consumption of DOC.

As shown in Figure 2-7, the disinfectant residual dropped rapidly and was virtually gone by the end of the second loop. Instantaneous THMs exceeded THMFP at the end of loop 2, and instantaneous THMs continued to increase through the third loop. This might be explained by the increase and then decrease in chloral hydrate concentrations between the treatment plant and the end of the third loop. Chloral hydrate may be converted to THMs in the absence of chlorine. These results are consistent with other research that has shown this same phenomenon (Stevens et al. 1989).

Safe Drinking Water Act

Concerns about waterborne disease and uncontrolled water pollution resulted in a dramatic increase in federal water quality legislation between 1890 and 1970. Even though significant advances were made to eliminate waterborne disease outbreaks during that period, the focus of drinking water concerns began to change. By the 1970s, more than 12,000 chemical compounds were known to be in commercial use, and many more were being added each year. Many of these chemicals cause contamination of groundwater and surface water and are known to be carcinogenic and/or toxic. The passage of the Safe Drinking Water Act (SDWA) of 1974 was a reflection of concerns about chemical contamination (SDWA 1974).

On Dec. 16, 1974, the US Congress passed the SDWA, which authorized the US Environmental Protection Agency (USEPA) to promulgate the first set of federal regulations which would "protect health to the extent feasible, using technology, treatment techniques, and other means, which the Administrator determines are generally available (taking costs into consideration)" (SDWA 1974). As a result, a set of regulations was promulgated in 1975, which became effective June 24, 1977. These were known as the National Interim Primary Drinking Water Regulations (NIPDWR). The NIPDWR established enforceable maximum contaminant levels (MCLs) for 10 inorganic contaminants, 6 organic contaminants, turbidity, coliform, radium-226, radium-228, gross alpha activity, and man-made radionuclides. The NIPDWR also established monitoring and analytical requirements for determining compliance.

Amendments to the SDWA promulgated in 1986 required the US Environmental Protection Agency to establish maximum contaminant level goals (MCLGs) for each contaminant that may have an adverse health effect. Each goal is required to be set at a level at which no known or anticipated adverse effects on health occur, thereby allowing an adequate margin of safety (Clark at al. 1987). Maximum contaminant levels or enforceable drinking water standards must be set as close to MCLGs as feasible.

The 1996 amendments to the SDWA nearly doubled the text of the act and provided for several new programs and initiatives, including the following: encouraging capacity development on the part of drinking water utilities, guidelines for water conservation, a requirement for utilities to increase efforts at informing their customers as to their status in complying with drinking water regulations, establishing an infrastructure grants program, emphasizing the need to bring small systems into compliance, and creating a whole new emphasis on source water protection.

Originally, the emphasis on understanding, interpreting, and implementing the SDWA and its regulations was primarily focused on treated water. However, substantial evidence indicates that many factors can cause water quality to deteriorate between the treatment plant and the point of consumption. Some of these factors include the following: chemical and biological quality of source water; effectiveness and efficiency of treatment processes; adequacy of the treatment facility, storage facilities, and distribution system; age, type, design, and maintenance of the distribution network; and quality of treated water (Clark and Coyle 1990).

Most of the regulations established under the SDWA were promulgated with little understanding of the effect that the distribution system can have on water quality. However, the SDWA has been interpreted as meaning that some MCLs shall be met at the consumers' tap, and thus involves the entire distribution system.

Safe Drinking Water Act regulations that emphasize system monitoring include the Surface Water Treatment Rule (SWTR), the Total Coliform Rule (TCR), the Lead and Copper Rule (LCR), and the Stage 1 Disinfectants and Disinfection By-Products Rule (D/DBPR1) and the Stage 2 Disinfectants and Disinfection By-Products Rule (D/DBPR2) (originally the total trihalomethane regulation), which addresses trihalomethanes and haleocetic acids. The D/DBPR, SWTR, and TCR specify treatment and monitoring requirements that must be met by all public water suppliers.

The SWTR requires that a detectable disinfectant residual be maintained at representative locations in the distribution system to provide protection from microbial contamination. The TCR regulates coliform bacteria that are used as "surrogate organisms" to indicate whether or not system contamination is occurring. Monitoring for compliance with the Lead and Copper Rule is based entirely on samples taken at the consumer's tap. The current standard for trihalomethanes (THMs) is 0.08 mg/L for systems serving more than 10,000 people, but the D/DBPR imposes the current (or a reduced) THM level on all systems. This regulation also requires monitoring and compliance at selected monitoring points in the distribution system. Some of these regulations may, however, provide contradictory guidance. For example, the SWTR and TCR recommend the use of chlorine to minimize risk from microbiological contamination. However, chlorine or other disinfectants interact with natural organic matter (NOM) in treated water to form disinfection by-products (DBPs). Raising the pH of treated water will assist in controlling corrosion but will increase the formation of THMs.

A summary of the evolution of federal drinking water regulation since the passage of the SDWA in 1974 is presented in Figure 2-8.

In addition to the rules and regulations promulgated under the SDWA, security has recently become an issue for the water utility industry. Security of water systems is not a new issue. The potential for natural, accidental, and purposeful contamination of water supplies has been the subject of many studies. For example, in May 1998, President Bill Clinton issued Presidential Decision Directive (PDD) 63 that outlined a policy on critical infrastructure protection, including US water supplies. However, it was not until after Sept. 11, 2001, that the water industry focused on the vulnerability of the nation's water supplies to security threats. In recognition of these issues, President George W. Bush signed the Public Health Security and Bioterrorism Preparedness and Response Act of 2002 (Bioterrorism Act) into law in June 2002 (PL107-188). Under the requirements of the Bioterrorism Act, community water systems (CWSs) serving more than 3,300 people are required to prepare vulnerability assessments and emergency response plans. CWSs are public water systems (PWSs) that supply water to the same population throughout the year.

Table 2-1 summarizes the key requirements of the regulations presented in Figure 2-8 from a distribution system compliance perspective. However, no evidence suggests that current regulations have resulted in a reduction in the proportion of outbreaks attributable to distribution system–related factors. In 2000, the Federal Advisory Committee for the Microbial/Disinfection By-Products Rule recommended that USEPA evaluate available data and research on aspects of distribution systems that may create risks to public health. In 2003, USEPA committed to revising the TCR and is currently exploring the possibility of revising the TCR to provide a comprehensive approach for addressing water quality in the distribution system environment. To aid in this process, USEPA requested that the National Academies' Water Science and Technology Board conduct a study of water quality issues associated with public water supply distribution systems and their potential risks to consumers (NRC 2006). The revised TCR has been proposed and is now in review and will be discussed later. Recommendations from this study will be discussed later in this volume.

Some important distribution system water quality concerns are: maintenance of proper disinfectant levels; minimization of DBP formation; turbidity, taste, color, and odor issues; distribution tank mixing and utilization; main repair and pressure stabilization; flow management; cross-connection control; and backflow prevention. It should be noted that water quality goals can be difficult to achieve and can be contradictory. For example, an important goal is to maintain a positive disinfectant residual in order to protect against microbial contamination. However, DBPs (TTHMs) will increase as water moves through the network as long as disinfectant residual and NOM are available. Other DBPs (haloacetic acids [HAA5]) are degraded biologically when free chlorine or chloramines are nearly absent.

Table 2-1. Selected rules and regulations dealing with distribution systems (not inclusive) (Panguluri et al. 2005)

Regulation	Key Distribution System Requirements
Safe Drinking Water Act (SDWA)	Gives USEPA the authority to establish national primary and secondary drinking water regulations (MCLs and MCLGs).
National Interim Primary Drinking Water Regulations (NIPDWR)	The NIPDWR adopted at the passage of the SDWA required that representative coliform samples be collected throughout the distribution system.
Total Trihalomethane (TTHM) Rule	Established a standard for TTHMs as 0.08 mg/L.
1986 Amendments to the Safe Drinking Water Act (86SDWAA)	Established the MCLG concept.
Total Coliform Rule (TCR)	Regulates coliform bacteria, which are used as surrogate organisms to indicate whether or not treatment is effective and system contamination is occurring.
Surface Water Treatment Rule (SWTR)	Requires using chlorine or some other disinfectant.
Lead and Copper Rule (LCR)	Monitoring for compliance with the LCR is based entirely on samples taken at the consumer's tap.
Information Collection Rule (ICR)	Provided data to support the interim and long-term enhanced SWTR, and D/DBPR2.
1996 Amendments to the Safe Drinking Water Act (96SDWAA)	Has many provisions dealing with distribution systems, including the role that surface water quality can play in influencing the quality of distributed water.
Interim Enhanced Surface Water Treatment Rule (IESWTR)	Provisions to enhance protection from pathogens, including *Cryptosporidium*, and intended to prevent increases in microbial risk while large systems comply with the D/DBPR1.
Stage 1 Disinfectants and Disinfection By-Products (D/DBPR1)	This standard applies to all haloacetic acids (HAAs) and lowered the standard for TTHMs from 0.1 mg/L to 0.08 mg/L for community water supplies in the United States and requires monitoring and compliance at selected points in the distribution system.
Long Term Interim Enhanced Surface Water Rule 1 (LT1ESWTR)	Provisions to enhance protection from pathogens, including *Cryptosporidium*, and prevent increases in microbial risk for systems serving less than 10,000 people while they comply with the D/DBPR1.
Stage 2 Disinfectants and Disinfection By-Products Rule (D/DBPR2)	Requiring systems delivering disinfected water to meet maximum contaminant levels as an average at each compliance monitoring location (instead of as a systemwide average as in previous rules) for total trihalomethanes (TTHMs) and five haloacetic acids (HAA5). The rule targets systems with the greatest risk and builds incrementally on existing rules. Intended to reduce DBP exposure and related potential health risks and provide more equitable public health protection.

(continued)

Table 2-1. Selected rules and regulations dealing with distribution systems (not inclusive) (Panguluri et al. 2005) (continued)

Long Term Enhanced Surface Water Treatment Rule 2 (LT2ESWTR)	The LT2 Enhanced Surface Water Treatment Rule is intended to reduce incidence of disease associated with *Cryptosporidium* and other pathogenic microorganisms in drinking water. The rule applies to all public water systems that use surface water or groundwater that is under the direct influence of surface water. The rule will bolster existing regulations and provide a higher level of protection for drinking water supplies by: • Targeting additional *Cryptosporidium* treatment requirements to higher-risk systems • Requiring provisions to reduce risks from uncovered finished water storage facilities • Providing provisions to ensure that systems maintain microbial protection as they take steps to reduce the formation of disinfection by-products.
Revised Total Coliform Rule (RTCR)	The proposed RTCR shifts the regulatory focus to assessment of potential problems in the distribution system and fixing those problems, as appropriate. It proposes: • Removing the maximum contaminant level goal (MCLG) and maximum contaminant level (MCL) of zero for total coliforms and the associated public notice for total coliforms • Setting triggers for assessment and corrections that parallel the previous MCL for total coliforms • Maintaining a zero MCLG and MCL for *E. coli* • Introducing a tiered approach for assessment and corrections

Recommendations From the NRC Report

As mentioned previously, the National Research Council of the National Academy of Sciences (NAS) convened a committee on Public Water Supply Distribution Systems: Assessing and Reducing Risks in 2004. It released a report of its findings on Sept. 8, 2006, entitled *Drinking Water Distribution Systems: Assessing and Reducing Risks* (NRC 2006). Risk was defined as the possibility of suffering harm from a hazard. The report suggests that the distribution system is the remaining component of public water supplies yet to be adequately addressed in national efforts to eradicate waterborne disease. The report recommends that the USEPA should focus on working with representatives from states, water systems, and local jurisdictions to establish the elements that constitute acceptable cross-connection control programs. Existing plumbing codes should be consolidated into one uniform national code that utilities develop. The report recommends that:

- Storage facilities should be inspected on a regular basis.
- Better sanitary practices are needed during installation, repair, replacement, and rehabilitation of distribution system infrastructure.
- Water residence times in pipes, storage facilities, and premise plumbing should be minimized.
- Positive water pressure should be maintained.

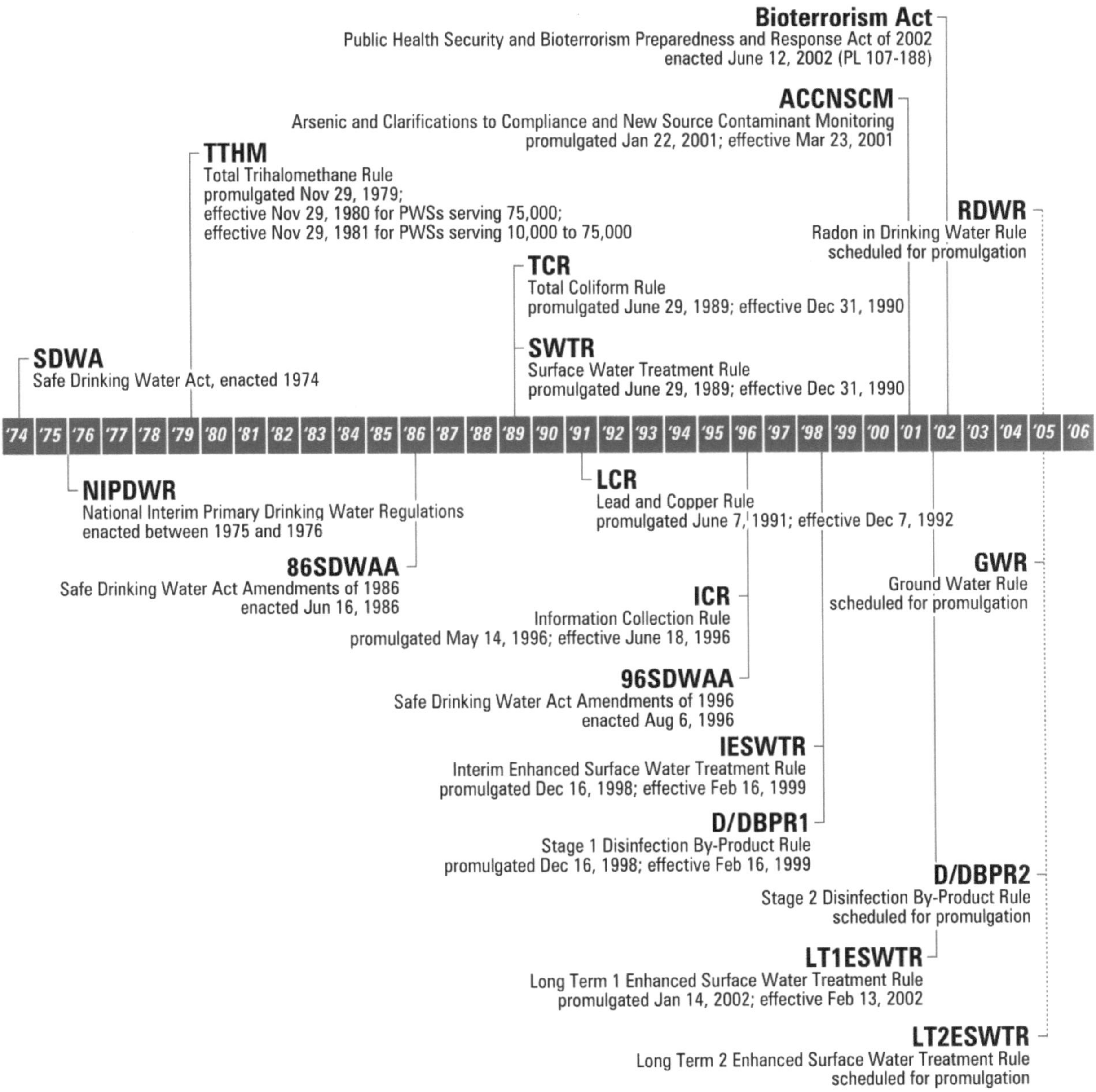

Figure 2-8. Evolution of safe drinking water regulations in the United States (Panguluri et al. 2005)

- Distribution system monitoring and modeling are critical to maintaining hydraulic integrity.
- Microbial growth and biofilm development in distribution systems should be minimized.
- Residual disinfectant choices should be balanced to meet the overall goal of protecting public health.
- Standards for materials used in distribution systems should be updated to address their impact on water quality, and research is needed to develop new materials that will have minimal impacts.
- Although it is difficult and costly to perform, condition assessment of buried infrastructure should be a top priority for utilities.
- Cross-connection control should be in place for all water utilities.
- Where feasible, surge protection devices should be installed.
- Prior to distribution, the quality of treated water should be adjusted to minimize deterioration of water quality.

The NRC devoted a considerable amount of the report to the role of premise plumbing and its role in protecting water quality. The report recommends that:

- Communities squarely address the problem of *Legionella*, via both changes to the plumbing code and new technologies.
- Communities should better assess cross-connections in the premise plumbing of privately owned buildings, and inspections for cross-connections and other code violations at the time of property sale should be required.
- USEPA should create a homeowner's guide and Web site that highlights the nature of the health threat associated with premise plumbing and mitigation strategies that can be implemented to reduce the magnitude of the risk.

The report also identifies the following research gaps:

- Distribution system ecology is poorly understood, making risk assessment via pathogen occurrence measurements difficult.
- There has been inadequate investigation of waterborne disease outbreaks associated with distribution systems, especially in premise plumbing.
- Epidemiology studies that specifically target the distribution system component of waterborne disease are needed.
- External and internal corrosion should be better researched and controlled in standardized ways.
- Research is needed to better understand how to analyze data from online, real-time monitors in a distribution system.
- Research is needed that specifically addresses potential problems arising from premise plumbing.
- Current microbial monitoring is limited in its ability to indicate distribution system contamination events, such that new methods and strategies are needed.
- A rigorous standardized set of network model development and calibration protocols should be developed.
- Additional research, development, and experimental applications in data integration are needed so that distribution system models can be used in real-time operation.

Current Activities

On June 17, 2010, the USEPA administrator proposed revisions to the 1989 Total Coliform Rule that require systems that have an indication of coliform contamination in the distribution system to assess the problem and take corrective action that may reduce cases of illnesses and deaths caused by potential fecal contamination and waterborne pathogen exposure. This proposal also updates provisions in other rules that reference analytical methods and other requirements in the current TCR.

As with the current Total Coliform Rule, the proposed Revised Total Coliform Rule applies to all public water systems. Under the current TCR, each total coliform–positive sample is assayed for fecal coliforms and/or *E. coli.* Fecal coliform bacteria are a subgroup of total coliforms that traditionally have been associated with fecal contamination. Since the promulgation of the TCR, more information and understanding of the suitability of fecal coliform and *E. coli* as indicators have become available. Study has shown that the fecal coliform assay is imprecise and too often captures bacteria that do not originate in the human or mammal gut (Edberg et al. 2000). *E. coli* is a more restricted group of coliform bacteria that almost always originate in the human or animal gut (Edberg et al. 2000). Thus, *E. coli* is a better indicator of fecal contamination than fecal coliforms. The proposed RTCR establishes a maximum contaminant level goal (MCLG) and maximum contaminant level (MCL) for *E. coli.* It takes a preventive approach to protecting public health by establishing a coliform treatment technique for protection against potential fecal contamination. The treatment technique uses both total coliforms and *E. coli* monitoring results to start an evaluation process that, where necessary, will require the PWS to conduct follow-up corrective action that could prevent future incidences of contamination and exposure to fecal contamination and/or waterborne pathogens. Its proposal includes:

- *Monitoring.* As with the current TCR, PWSs will continue to monitor for total coliforms and *E. coli* according to a sample siting plan and schedule specific to the system. Sample siting plans under the proposed RTCR must continue to be representative of the water throughout the distribution system.
- *Seasonal systems.* The proposed RTCR establishes monitoring requirements for seasonal systems for the first time. Seasonal systems represent a special case in that the shutdown and startup of these water systems present additional opportunities for contamination to enter or spread through the distribution system. Seasonal systems must demonstrate completion of a state-approved startup procedure.
- *Assessment and corrective action.* As part of a treatment technique, all PWSs are required to assess their systems when monitoring results show that the system may be vulnerable to contamination. Systems must conduct a simple self-assessment (Level 1) or a more detailed assessment (Level 2) depending on the severity and frequency of contamination. The system is responsible for correcting any sanitary defect(s) found through either a Level-1 or Level-2 violation.

References

Bruins, H.J. 2000. Proactive Contingency Planning vis-à-vis Declining Water Security in the 21st Century. *Journal of Contingencies and Crisis Management*, 8(2):63–72.

Clark, R.M., J.A. Goodrich, and J.C. Ireland. 1984. Cost and Benefits of Drinking Water Treatment. *Journal of Environmental Systems,* 14(1):1–30.

Clark, R.M., J.Q. Adams, and R.M. Miltner. 1987. Cost and Performance Modeling for Regulatory Decision Making. *Water*, 28(3):20–27.

Clark, R.M., W.M. Grayman, and R.M. Males. 1988. Contaminant Propagation in Distribution Systems. *Jour. Environ. Eng.*, 114(2).

Clark, R.M., and J.A. Coyle. 1990. Measuring and Modeling Variations in Distribution System Water Quality. *Journal AWWA*, 82(8):46.

Clark, R.M., W.M. Grayman, and J.A. Goodrich. 1991a. Water Quality Modeling: Its Regulatory Implications. In *Proc. AWWARF/EPA Conf. on Water Quality Modeling in Dist. Systems.* Cincinnati, Ohio: USEPA.

Clark, R.M., D.J. Ehreth, and J.J. Convery. 1991b. Water Legislation in the U.S.: An Overview of the Safe Drinking Water Act. *Toxicology and Industrial Health*, 7(516):43–52.

Clark, R.M., W.M. Grayman, J.A. Goodrich, R.A. Deininger, and A.F. Hess. 1991c. Field Testing Distribution Water Quality Models. *Journal AWWA*, 83(7):67.

Clark, R.M., J.A. Goodrich, and L.J. Wymer. 1993a. Effect of the Distribution System on Drinking Water Quality. *Jour. Water Supply: Research and Technology–AQUA*, 42(1):30–38.

Clark, R.M., W.M. Grayman, R.M. Males, and A.F. Hess. 1993b. Modeling Contaminant Propagation in Drinking Water Distribution Systems. *Jour. Environ. Eng.*, 119(2):349–364.

Clark, R.M., J.E. Hill, J.A. Goodrich, J.A. Barnick, and F. Abdesaken. 1994a. Impact of Tanks and Reservoirs on Water Quality in Drinking Water Distribution Systems: Regulatory Concerns. In *Proc. AWWA Annual Conference.* Denver, Colo.: AWWA.

Clark, R.M., B.W. Lykins Jr., J.C. Block, L.J. Wymer, and D.J. Reasoner. 1994b. Water Quality Changes in a Simulated Distribution System. *Journal of Water Supply: Research and Technology–AQUA*, 43(6):263–277.

Clark, R.M., E.E. Geldreich, K.R. Fox, E.W. Rice, C.H. Johnson, J.A. Goodrich, J.A. Barnick, and F. Abdesaken. 1996. Tracking a *Salmonella* Serovar *Typhimurium* Outbreak in Gideon, Missouri: Role of Contaminant Propagation Modeling. *Journal of Water Supply: Research and Technology–AQUA,* 45(4):171–183.

Clark, R.M., and W.M. Grayman. 1998. *Modeling Water Quality in Drinking Water Distribution Systems.* Denver, Colo.: AWWA.

Edberg, S.C., E.W. Rice, R.J. Karlin, and M.J Allen. 2000. *Escherichia coli*: The Best Biological Drinking Water Indicator for Public Health Protection. *Journal of Applied Microbiology,* 88:106S–116S.

Fujiwara, M., J.M. Manwaring, and R.M. Clark. 1995. Drinking Water in Japan and the United States: Conference Objectives. In *Drinking Water Quality Management*, ed. R.M. Clark and D.A. Clark. Lancaster, Pa.: Technomic.

Geldreich, E.E., H.D. Nash, D.J. Reasoner, and R.H. Taylor. 1972. The Necessity of Controlling Bacterial Populations in Potable Water: Community Water Supply. *Journal AWWA*, 64(9):596–602.

Geldreich, E.E., K.R. Fox, J.A. Goodrich, E.W. Rice, R.M. Clark, and D.L. Swerdlow. 1992. Searching for a Water Supply Connection in the Cabool, Missouri, Disease Outbreak of *Escherichia Coli* 0157:H7. *Water Research*, 26(8):1127–1137.

Kirmeyer, G.J., W. Richards, and C.D. Smith. 1994. *An Assessment of the Condition of North American Water Distribution Systems and Associated Research Needs.* Denver, Colo.: AwwaRF and AWWA.

LeChevallier, M.W., T.M. Babcock, and R.G. Lee. 1987. Examination and Characterization of Distribution System Biofilms. *Applied and Environmental Microbiology*, 53:2714–2724.

Maul, A., A.H. El-Shaarawi, and J.C. Block. 1985a. Heterotrophic Bacteria in Water Distribution Systems. I. Spatial and Temporal Variation. *The Science of the Total Environment*, 44:201–214.

Maul, A., A.H. El-Shaarawi, and J.C. Block. 1985b. Heterotrophic Bacteria in Water Distribution Systems. II. Sampling Design for Monitoring. *The Science of the Total Environment*, 44:215–224.

McCoy, W.F., and B.H. Olson. 1986. Relationship Among Turbidity Particle Counts and Bacteriological Quality Within Water Distribution Lines. *Water Research*, 20:1023–1029.

National Research Council (NRC). 2006. *Drinking Water Distribution Systems: Assessing and Reducing Risks.* Washington, D.C.: National Academies Press, 15–46.

Panguluri, S., W.M. Grayman, and R.M. Clark. 2005. *Distribution System Water Quality Report: A Guide to the Assessment and Management of Drinking Water Quality in Distribution Systems.* Cincinnati, Ohio: USEPA Office of Research and Development.

Presidential Decision Directive (PDD) 63. 1998. *Critical Infrastructure Protection.* The White House, Washington, D.C.

Public Health Security and Bioterrorism Preparedness and Response Act of 2002. 2002. Public Law (PL) 107-188.

Safe Drinking Water Act (SDWA). 1974. Public Law 93–523.

Stevens, A.A., L.A. Moore, and R.J. Miltner. 1989. Formation and Control of Non-Trihalomethane Disinfection By-Products. *Journal AWWA,* 84:54–60.

Chapter 3

Modeling and Monitoring Water Quality: An Overview

Introduction

The use of mathematical models for the analysis of water distribution systems was first proposed in the 1930s by Hardy Cross (1936). Since that time, as illustrated in Figure 3-1, models have evolved from the early manual Hardy Cross network flow analysis, to the development and wide availability of computer-based hydraulic network models in the 1970s and 1980s, to the emergence of distribution system water quality models later in the 1980s. Modeling and mapping systems are now being integrated into comprehensive, user-friendly systems for analyzing and displaying hydraulic and water quality parameters in a distribution system. As will be discussed later in this book, water quality models have reached operational status, but research and development continues to further the understanding of the processes taking place in the distribution system and to translate this understanding into usable tools.

In applying models to a water distribution system, the degree of temporal (over time) variation and the specific issues that are being studied determine the types of models that are most applicable. For example, *steady-state modeling* represents external forces as constant in time (static) and determines solutions that would occur if the system were allowed to reach equilibrium (Wood 1980a). In *dynamic modeling*, demands and supplies are allowed to vary with time and the resulting temporal solution is determined (Clark et al. 1988a, Clark et al. 1988b).

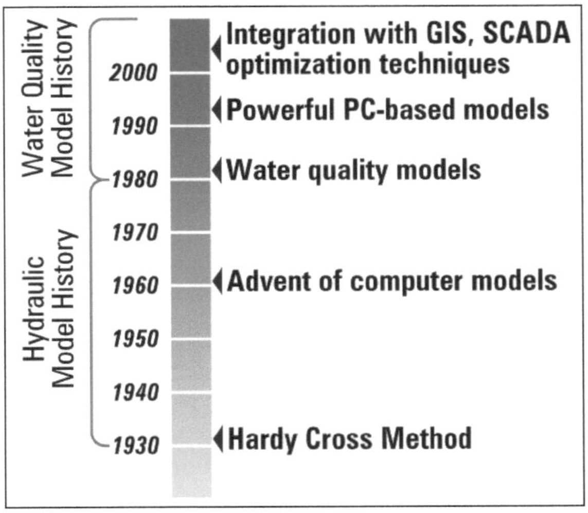

Figure 3-1. Illustration of the evolution of hydraulic and water quality models (Panguluri et al. 2005)

In both steady-state and dynamic modeling, a distribution system is represented by a link–node network (i.e., pipes are represented as links, and junctions of pipes, wells, tanks, or starts and ends of pipes as nodes). Hydraulic models are used to determine flows and velocities in links and pressures at nodes.

Water quality models are used to determine variations in the concentrations of a contaminant throughout the distribution system. The hydraulic and water quality models may be an integrated package, or the results from a hydraulic model may be input to a water quality model for subsequent analysis.

Water distribution system models have become increasingly accepted within the water utility industry as a mechanism for simulating the hydraulic and water quality behavior in water distribution system networks. Current water distribution modeling software is powerful, sophisticated, and user-friendly. Many software packages are integrated with geographic information system (GIS) and computer-aided design (CAD) technology in order to facilitate model construction and storage and display of model results. Early network models simulated only steady-state hydraulic behavior. In the 1970s, modeling capability was expanded to include extended-period simulation (EPS) models that could accommodate time-varying demand and operations. Subsequently, in the early 1980s, investigators began introducing the concept of water quality modeling. Most water distribution system modeling software packages now routinely incorporate water quality simulation capability. Also, in the early 1980s, the concept of modeling water quality in distribution system networks was developed based on steady-state formulations (Clark et al. 1986). By the mid-1980s, water quality models were developed that incorporated the dynamic behavior of water networks (Grayman et al. 1988a). The usability of these models was greatly improved in the 1990s with the introduction of the public domain EPANET model (Rossman 2000) and other Windows-based commercial water distribution system models.

Modeling Applications

In 1991, the USEPA and American Water Works Association Research Foundation (AwwaRF, now the Water Research Foundation [WaterRF]) cosponsored a conference called Water Quality Modeling in Distribution Systems to identify the state-of-the-art in water quality contaminant propagation modeling. The symposium identified the following 11 categories for water quality modeling research and application (AwwaRF/USEPA 1991):

- Model development
- Chemical, physical, and biological processes
- Modeling application
- Water quality studies
- Calibration of models
- Regulatory implications
- Monitoring strategies
- Legal proceedings
- Management and operational cases
- Operation and design of storage tanks
- Health effects

Although this agenda was established in 1991, it became the template for research in the field of distribution system water quality modeling as followed by WaterRF and the USEPA.

More recently, transient models for simulating water hammer (a transient phenomenon) and tank mixing/aging models have either been incorporated into or integrated with water distribution system models. Algorithms have been developed that enable users to optimize water system design and operation, assist in model calibration, and perform probabilistic analyses. Each of these model types is briefly described later in this chapter.

Types of Distribution System Models

A variety of computer-based mathematical models of water distribution systems has been developed and used by the water utility industry to assess the movement and fate of contaminants within the distribution system. Such models may be divided into three general categories:

- *Hydraulic models*, which simulate the flow quantity, flow direction, and pressure in the system.
- *Steady-state water quality models and flow-tracing models*, which determine the movement of contaminants, including their flow paths and travel times, through the network under steady-state operational and demand conditions.
- *Dynamic water quality models*, which simulate the movement and transformation of substances in the water under conditions that vary over time.

Two related technologies, which are briefly discussed in this book, are

- *Optimization models incorporating water quality*, which examine a wide range of operational and/or design variables and select the best solution based on a stated objective function and specified constraints.
- *Analysis/display methods*, which perform various analyses and/or display the results of water quality sampling or modeling.

Each type of model serves a particular purpose in assessing distribution system water quality and is essential in investigating water quality issues in a distribution system.

Hydraulic Models

As mentioned previously, mathematical methods for analyzing the flow in networks have been in use for over half a century (Cross 1936). Computer-based models for performing this type of analysis were first developed in the 1950s and 1960s and were greatly expanded and made more available in the 1970s and 1980s (Wood 1980b, Sarikelle et al. 1989).

Hydraulic models were developed to simulate flow and pressures in a distribution system, either under steady-state conditions or under time-varying demand and operational conditions. These are generally referred to as *extended-period simulation* (EPS) models. Hydraulic models may also incorporate optimization components to aid the user in selecting system parameters that result in the best match between observed system performance and model results (Gessler and Walski 1985). The theory and application of hydraulic models are thoroughly explained in many widely available references (Walski et al. 2003, AWWA 2004, Larock et al. 2000).

Hydraulic models also provide flow information used in distribution system water quality models. Hydraulic models and water quality models can be tightly bundled into a single entity, or a stand-alone hydraulic model can be used to generate a file containing hydraulic flow conditions that can be used in a stand-alone water quality model. Both techniques are in common use today.

Water Quality Models

Water quality models use the output of hydraulic models in conjunction with additional inputs (described later in this section) to predict the temporal and spatial distribution of a variety of constituents within a distribution system. These constituents include:

- The fraction of water originating from a particular source.
- The age of water (e.g., duration since leaving the source).

- The concentration of a nonreactive constituent or tracer compound either added to or removed from the system (e.g., chloride or fluoride).
- The concentration of a reactive compound including the concentration of a secondary disinfectant with additional input of its loss rate (e.g., chlorine or chloramines) and the concentration of disinfection by-products with their growth rate (e.g., THMs).

For convenience, we can classify water quality models into steady-state and dynamic water quality models.

Steady-State Models

Wood (1980b) suggested the use of models to study the spatial distribution of slurry flow in a pipe network. Males et al. (1985) used simultaneous mass balance equations to calculate the spatial distribution of variables that could be associated with links and nodes such as concentration, travel times, costs, and other variables in drinking water distribution systems. This model, called SOLVER, was a component of the Water Supply Simulation Model (WSSM), an integrated database management, modeling, and display system that was used to model steady-state water quality in networks (Clark and Males 1986). A similar formulation was later used in a 166-link representation of the Alameda County, Calif., Water District with three sources of water of differing hardness (Chun and Selznick 1985, Metzger 1985).

A more general "marching out" solution was proposed by Males et al. (1988). An alternative steady-state "incremental" solution was introduced by Clark et al. (1988b) for calculating spatial patterns of concentrations, travel times, and the percentage of flow from sources. In this approach, links are hydraulically ordered starting with source nodes and progressing through the network until all nodes and links are addressed.

Alternative methodologies for predicting water quality and determining the source of delivered flow under steady-state conditions were investigated by Wood and Ormsbee (1989). An iterative cyclic procedure was found to be both an effective and efficient method. This procedure was similar to the incremental solutions described previously for networks that they term as *source dependent* (i.e., networks in which the nodes can be hydraulically sequenced starting from sources). However, for nonsource-dependent networks, which are rare, their algorithm iterated until a unique solution was found.

Dynamic Models

Though steady-state water quality models proved to be useful tools for investigating the movement of a contaminant under constant conditions, the need for models that represented the dynamics of contaminant movement was recognized. In the mid-1980s, several models that simulate the movement and transformation of contaminants in a distribution system under time-varying conditions were developed and applied. Three such models were initially introduced at the American Water Works Association (AWWA) Distribution Systems Symposium in 1986 (Clark et al. 1986, Liou and Kroon 1986, Hart et al. 1986).

Grayman et al. (1988b) developed and applied a dynamic water quality simulation model that used flow results previously generated by a hydraulic model and a numerical scheme to route conservative (i.e., concentration does not degrade with time) and nonconservative (i.e., concentration changes with time) contaminants through a network. In this model, each pipe link was represented as a series of "sublinks" and "subnodes," with the length of each sublink selected to approximate the distance that a contaminant will travel during each time step. The number of sublinks varies with the velocity of flow in a link (Grayman et al. 1988a). Kroon and Hunt (1989) developed a similar numerical model originally implemented on a minicomputer and originally used on a personal computer–based (PC based) workstation

(Liou and Kroon 1986). This model was directly tied to a hydraulic model and generated both tabular and graphical output displaying the spread of contaminants through a network. Hart et al. (1986) developed a model using the GASP IV simulation language.

EPANET was initially developed in 1993 as a distribution system hydraulic–water quality model to support research efforts at USEPA (Rossman et al. 1994). The development of the EPANET software has also satisfied the need for a comprehensive public-sector model and has served as the hydraulic and water quality "engine" for many commercial models.

EPANET can be used for both steady-state and EPS hydraulic simulations. In addition, it is designed to be a research tool for modeling the movement and fate of drinking water constituents within distribution systems. EPANET can be operated in the SI (metric or Système International d'Unités) or British systems of measurement.

The water quality routines in EPANET can be used to model concentrations of reactive and conservative substances, changes in age of water and travel time to a node, and the percentage of water reaching any node from any other node. Outputs from EPANET include:

- Color-coded network maps
- Time series plots
- Tabular reports

Example outputs from EPANET are presented in Figures 3-2 (see appendix E), 3-3, and 3-4.

The history and proliferation of water quality modeling in distribution systems can be traced back to the workshop held in 1991 and another held in 2003. The results of these workshops are presented in AwwaRF/USEPA (1991) and Powell et al. (2004).

Model Applications

The types of models described were applied in several research studies. Some of these studies will be described in more detail elsewhere in this book. In a cooperative agreement between the USEPA and the North Penn Water Authority, a series of water quality models were developed and applied in order to study the movement of water through the distribution system (Clark and Coyle 1990). In this system, trihalomethanes (THMs) were higher in the surface water source, while hardness was higher in the groundwater sources. Modeling was used to trace the movement of these constituents and, ultimately, to alter the operation in order to minimize water quality variation. The temporal and spatial variability of these constituents was quite significant and, as has been confirmed in many other studies, the dynamics of the contaminant movement in the system exceeded the general expectations of the water utility (Clark et al. 1988c).

In a cooperative agreement involving the American Water Works Association Research Foundation (AwwaRF), now called the Water Research Foundation (WaterRF), the South Central Connecticut Regional Water Authority (SCCRWA), Montana State University, and Stoner Associates, water quality models were developed and applied to the Saltonstall service area of the SCCRWA to study the chlorine residual in the system and its impact on bacterial regrowth (Characklis 1988). These models were later applied to the Philadelphia Suburban Water Company's Great Valley Division to determine the percentage of water from various groundwater and surface sources reaching various parts of the system (Kroon 1990).

In another cooperative agreement, USEPA and the University of Michigan applied water quality models to the Cheshire service area of the SCCRWA. A highly skeletonized representation (i.e., network with only major transmission lines) of the full SCCRWA system was developed, and the effects of skeletonization on water quality were studied (Clark et al. 1990). Both

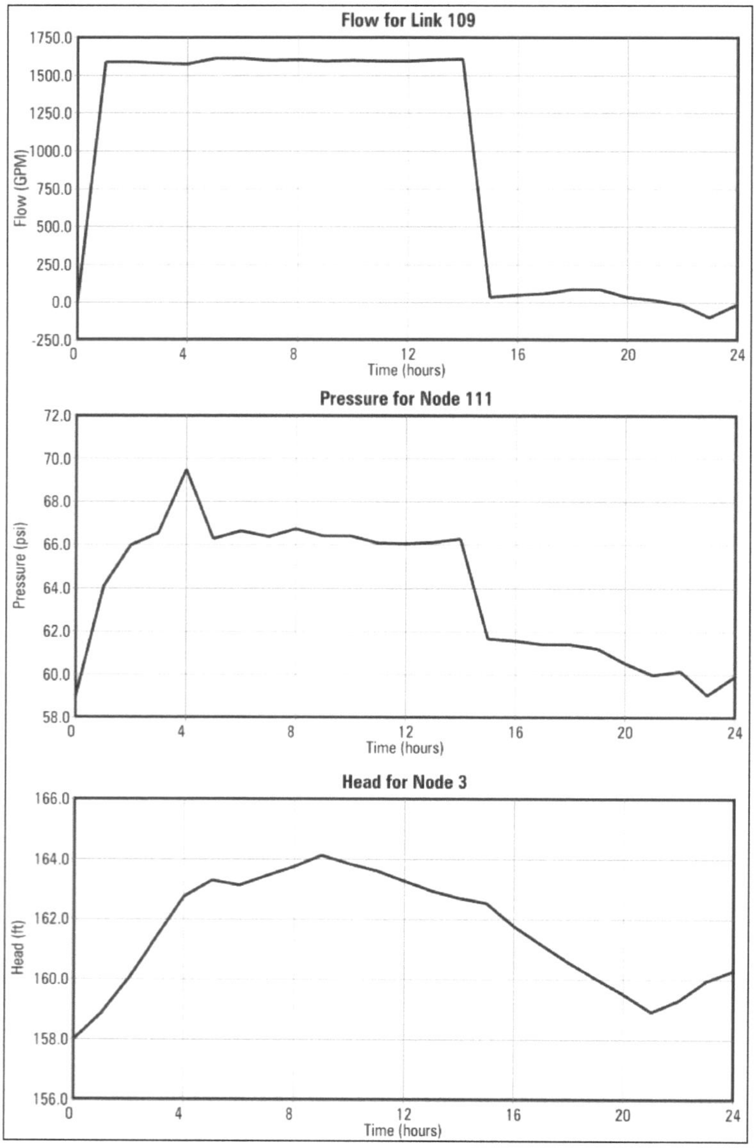

Figure 3-3. Sample EPANET time series plots of flow, pressure, and tank water level (head) (Panguluri et al. 2005)

fluoride (used as a tracer) and chlorine residual were modeled in this study. This study reinforced many of the previous studies in illustrating the significant dynamics in water quality within a distribution system resulting in large variations over the network and over time in concentrations (Clark et al. 1990).

Model Calibration

Calibration is an important part of the "art" of modeling drinking water distribution systems. *Model calibration* is the process of adjusting model input data (or, in some cases, model structure) so that the simulated hydraulic and water quality outputs sufficiently mirror observed field data. One way of viewing calibration is to think of a television screen showing observed and predicted values with knobs available to adjust the predicted values. Calibration is the process of adjusting the knobs so that the predicted value provides the closest estimate to the actual (observed) values. Calibration can be difficult, costly, and time-consuming.

The extent and difficulty of calibration are minimized by developing an accurate representation of the network and its components. Accurate system performance and operations data should be collected to validate the network simulations. A traditional technique for calibration is to use fire-flow pressure measurements. Pressures and flow in isolated pipe sections are measured in the field, and the pipe friction factors are adjusted to reflect the data.

Another method is to use water quality tracers. Naturally occurring or added chemical tracers may be measured in the field and the results used to calibrate hydraulic and water quality models. Calibration should be an ongoing process in which the degree of accuracy of the simulations is continually being improved (Cesario 1995).

A popular method for calibrating water quality models is the use of fluoride as a tracer. Typically, for systems that routinely fluoridate their water and whose background level of fluoride is negligible, fluoride injection is discontinued for a period during which distribution system sampling occurs. Further sampling can then take place during the transient period when fluoride injection begins again. A large-scale calibration study of this type was performed in the Cheshire Service Area of the SCCRWA distribution system and will be discussed in detail later. In this study, fluoride was turned off for a 7-day period and then restarted for another 7 days. Intensive sampling occurred during this 14-day period. Before the study, water quality models were applied to the area to predict the fluoride movement to establish sites and timing for sampling. Over 2,000 grab samples were taken during the study. These samples were augmented by samples taken using an automated sampler receiving information from a fluoride probe. Following the sampling study, a water quality model was applied under the hydraulic conditions existing during the sampling period. The model was found to give good agreement with the sample results. Fluoride concentrations measured at the inlet and outlet of a storage tank in the Cheshire system were used to calibrate a model of the tank and to demonstrate the existence of short-circuiting of flow in the tank.

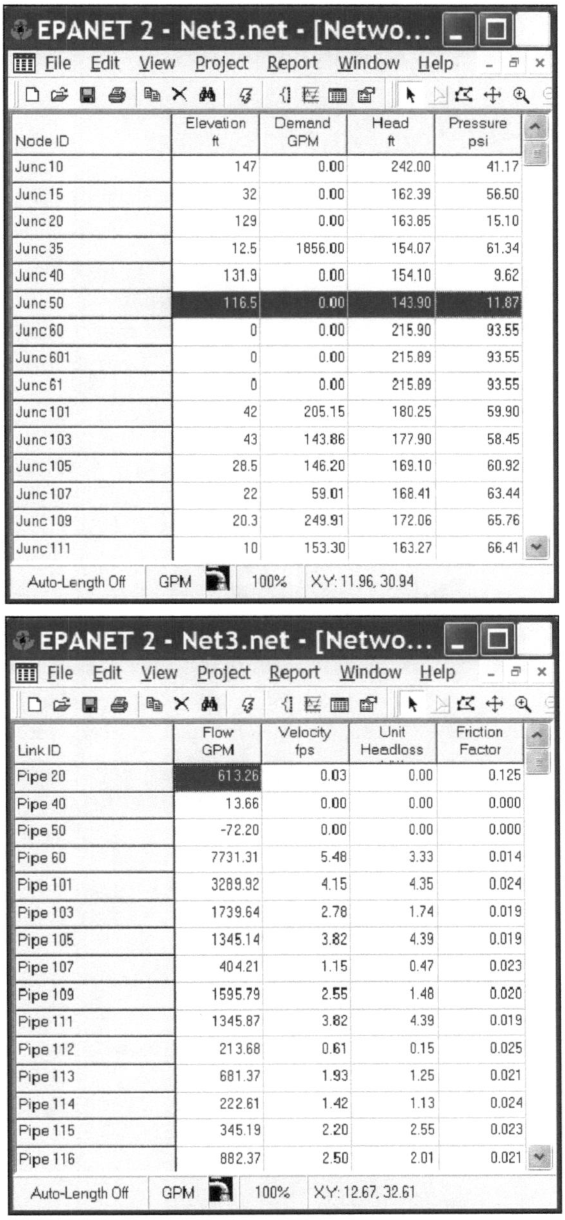

EPANET 2 - Net3.net - [Netwo...

File Edit View Project Report Window Help

Node ID	Elevation ft	Demand GPM	Head ft	Pressure psi
Junc 10	147	0.00	242.00	41.17
Junc 15	32	0.00	162.39	56.50
Junc 20	129	0.00	163.85	15.10
Junc 35	12.5	1856.00	154.07	61.34
Junc 40	131.9	0.00	154.10	9.62
Junc 50	116.5	0.00	143.90	11.87
Junc 60	0	0.00	215.90	93.55
Junc 601	0	0.00	215.89	93.55
Junc 61	0	0.00	215.89	93.55
Junc 101	42	205.15	180.25	59.90
Junc 103	43	143.86	177.90	58.45
Junc 105	28.5	146.20	169.10	60.92
Junc 107	22	59.01	168.41	63.44
Junc 109	20.3	249.91	172.06	65.76
Junc 111	10	153.30	163.27	66.41

Auto-Length Off GPM 100% XY: 11.96, 30.94

EPANET 2 - Net3.net - [Netwo...

File Edit View Project Report Window Help

Link ID	Flow GPM	Velocity fps	Unit Headloss	Friction Factor
Pipe 20	613.26	0.03	0.00	0.125
Pipe 40	13.66	0.00	0.00	0.000
Pipe 50	-72.20	0.00	0.00	0.000
Pipe 60	7731.31	5.48	3.33	0.014
Pipe 101	3289.92	4.15	4.35	0.024
Pipe 103	1739.64	2.78	1.74	0.019
Pipe 105	1345.14	3.82	4.39	0.019
Pipe 107	404.21	1.15	0.47	0.023
Pipe 109	1595.79	2.55	1.48	0.020
Pipe 111	1345.87	3.82	4.39	0.019
Pipe 112	213.68	0.61	0.15	0.025
Pipe 113	681.37	1.93	1.25	0.021
Pipe 114	222.61	1.42	1.13	0.024
Pipe 115	345.19	2.20	2.55	0.023
Pipe 116	882.37	2.50	2.01	0.021

Auto-Length Off GPM 100% XY: 12.67, 32.61

Figure 3-4. EPANET sample tabular outputs (at time 10.00 hr) (Panguluri et al. 2005)

The effects of skeletonization of networks on the accuracy of hydraulic and water quality modeling have also been studied. These studies have concluded that skeletonization based on sound engineering judgment is a cost-effective means of representing complex networks but that water quality modeling results are generally more sensitive to skeletonization than are hydraulic model results.

Regulatory Implications

A synergistic relationship appears to exist between regulation of water quality in the distribution system and water quality modeling. Much of the impetus and interest in water quality modeling has been spurred by the passage of the Safe Drinking Water Act and its amendments. The SDWA specifically requires that detectable levels of chlorine residual (0.1 mg/L or greater) be maintained and that heterotrophic plate counts (HPCs) not be over 500 cfu/100 mL. In particular, new regulations require that water quality meet specified standards at the tap.

Water quality complaints fall into several categories, including taste-and-odor and red water complaints. Tracing flows from the source, for example, can be helpful in responding to taste-and-odor complaints. Customers do not usually complain of lack of chlorine residuals or increasing HPCs, but modeling can also be used to solve these problems. Under D/DBPR2, an initial distribution system evaluation (IDSE) is required for each community water supply to determine where the highest levels of total trihalomethanes (TTHMs) and five haloacetic acids (HAA5) can be found in a water system's distribution system. Models can be used to select maximum age sites under the IDSE.

Water quality modeling provides a tool that water utilities can use to estimate the effects at the tap of various treatment and delivery scenarios. Additionally, regulations regarding disinfection may cause significant changes in the disinfection processes. Water quality modeling provides a mechanism for determining the effect of changes in disinfection practices on delivered water.

Monitoring Strategies

In the future, water quality models may be used to select monitoring station locations to maximize the informational content of sampling results. For example, models can be used to identify low or stagnant flow areas that might in turn be candidate sites for distribution network monitoring stations.

Water quality modeling has been used extensively in legal proceedings involving the movement of a contaminant in a distribution system. However, because of the closed nature of the legal process, information on only a few studies has been made available, thus, the lessons that could be learned from such applications cannot be shared. Typically, water quality models are used in legal proceedings as a means of showing the movement of water from a contaminated source to some point and to prove or disprove that the plaintiff has been damaged by the defendant. In theory, utilities would like to have water quality models, but the time and effort involved make them a low priority until legal problems arise. In such cases, the utility has to backtrack and recreate the incident.

Commercial Hydraulic–Water Quality Modeling Software

In addition to EPANET, several commercial software packages are widely used in the United States and internationally. Most of these packages are based on the EPANET formulation and include value-added components as parts of a graphical user interface (GUI) that increase the capability of the software. Examples of such value-added components that are part of one or more of the commercially available software packages include:

- *Scenario manager:* Manage inputs and outputs of a group of model runs.
- *Calibration optimization:* Utilize genetic algorithm optimization technique to determine model parameters that best fit a set of field data.
- *Design optimization:* Utilize genetic algorithm (or other) optimization techniques to select pipe sizes that minimize costs or other selected objectives.
- *Integration with GIS or CAD:* Water distribution model directly integrates with GIS or CAD to assist in constructing or updating model and display results.
- *Flexible output graphics:* Provide convenient ways to modify parameters for graphical displays of output data.
- *Energy management:* Calculate energy use for a selected alternative.
- *Automated fire-flow analysis:* Assess the availability of fire flow at a range of nodes and determine whether a system meets fire-flow requirements.
- *Water security and vulnerability assessment methods, skeletonization, and demand allocation tools.*

Table 11-1 (see p. 231) provides a summary some of the major commercial software and Web links where additional details may be obtained on specific features and current version availability/pricing.

Additional Modeling Tools

In addition to standard hydraulic and water quality modeling of distribution systems, several other related types of models can be used to assess hydraulic and water quality behavior in distribution systems. These include: storage modeling tools, transient (water hammer) modeling tools, optimization tools, and probabilistic models. Each of these types of models are briefly described and discussed in the following sections.

Storage Modeling Tools

An important aspect of water quality and contaminant propagation in drinking water distribution systems is the effect of system storage. Most utilities use some type of ground or elevated storage system to process water during time periods when treatment facilities would otherwise be idle. It is then possible to distribute and store water at one or more locations in the service area closest to the user.

The principal advantage of distribution storage is that it equalizes demands on supply sources, production works, and transmission and distribution mains. As a result, the sizes or capacities of these elements may be minimized and peak power tariff periods for pumping can often be avoided. Additionally, system flows and pressures are improved and stabilized to better serve the customers throughout the service area. Finally, reserve supplies are provided in the distribution system for emergencies, such as firefighting and power outages.

In most municipal water systems, less than 25 percent of the volume of the storage in tanks is actively used (on a daily basis) under routine conditions. As the water level drops, tank controls require high-service pumps to start in order to satisfy demand and refilling of the tanks. The remaining water in the tanks (70 to 75 percent) is normally held in reserve as dedicated fire or emergency storage. This water tends to be stagnant and may cause water quality problems.

Storage tanks and reservoirs are the most visible components of a water distribution system but are often the least understood in terms of their effects on water quality. Although these facilities can play a major role in providing hydraulic reliability for firefighting needs and in providing reliable service, they may also serve as vessels for unwanted complex chemical and biological changes that may result in the deterioration of water quality. These storage tanks and reservoirs also contribute to increased residence time in drinking water systems. This increased residence time can contribute to the loss of disinfectant residuals and cause subsequent growth of microorganisms. Modeling can provide information on what will happen in existing, modified, or proposed distribution system tanks and reservoirs under a range of operating situations (Grayman et al. 2004a).

Three primary types of models are used for representing storage tanks and reservoirs: computational fluid dynamics (CFD) models, compartment models, and physical-scale models. In mathematical models, equations are written to simulate the behavior of water in a tank or reservoir. These models range from detailed representations of the hydraulic mixing phenomena in the facility, called CFD models, to simplified conceptual representations of the mixing behavior, called compartment or systems models. Physical-scale models are constructed from materials such as wood or plastic. Dyes or chemicals are used to trace the movement of water through the model. Storage tank modeling is discussed in more detail in appendix C.

CFD Models

CFD models use mathematical equations to simulate flow patterns, heat transfer, and chemical reactions. Partial differential equations representing conservation of mass, momentum, and energy are solved numerically for a two- or three-dimensional grid that approximates the geometry of the tank. CFD modeling has been used widely in the chemical, nuclear, and mechanical engineering fields, and in recent years has emerged as a modeling tool in the drinking water industry (Grayman and Arnold 2003). CFD models can be used to simulate temperature variations, unsteady hydraulic and water quality conditions, and decay of constituents in storage facilities. Significant experience is required to apply CFD models, and model run times of many hours, days, or even weeks are required for complex situations. Figure 3-5 (see appendix E) depicts a graphical output from a CFD model showing the concentration throughout a tank at a snapshot in time resulting from a tracer that has been injected into the inflow.

Many generalized CFD software packages are available that can be used to construct CFD models of tanks. Examples of such packages are listed in Table 3-1. These packages vary in terms of capabilities, solution methods, ease of use, and support. Prior to selection of a package, the specific needs and capabilities of the user should be carefully evaluated.

Generally the purchase or lease of these packages is significant (typically on the order of $25,000 per year), and significant training/expertise is required to effectively apply them (Panguluri et al. 2005).

Compartment Models

Compartment models are a class of models in which physical processes (e.g., the mixing phenomena in the tank or reservoir) are represented by highly conceptual, empirical relationships. This type of model is also referred to as a black box model or input–output model. Since such models do not use detailed mathematical equations to describe the movement of water within the tank, they rely on engineering judgment or upon field data and past experience to define the parameters that control the behavior of the model. Compartment models are used in water distribution network models to represent mixing in tanks and reservoirs. Various assumptions can be made in these models about the mixing behavior

Table 3-1. Example CFD modeling software packages (Panguluri et al. 2005)

CFD Package	Company	Web Site
CFD-ACE	CFD Research Corp.	www.cfdrc.com
Cfdesign	Blue Ridge Numerics	www.brni.com
CFX	Ansys, Inc.	www.software.aeat.com/cfx
FLOW-3D	Flow Science, Inc.	www.flow3d.com
Fluent	Fluent, Inc.	www.fluent.com
PAB3D	Analytical Services and Materials	www.asm-usa.com
Phoenics	CHAM	www.cham.co.uk
Sinda/Fluint	C&R Technologies	www.crtech.com
SWIFT	AVL	www.avl.com

in tanks including complete and instantaneous mixing, plug flow, last-in/first-out (LIFO) behavior, and multicompartment models. Both conservative substances and substances that decay according to a first-order decay function may be simulated in addition to simulation of water age. Compartment models are relatively easy to use and run in seconds as opposed to the long run times of CFD models.

Compartment models of tanks are available as part of most water distribution system models. EPANET and several of its derivative commercial models allow the user to select from four options: a complete mix model, a plug flow first-in/first-out (FIFO) model, a LIFO (short circuiting) model, and a two-compartment model. A stand-alone model called CompTank provides a wide range of alternatives and allows the user to model water age and reactive or conservative substances over a long period of time (Grayman et al. 2000). This model uses tank inflow and outflow information that is generally available from supervisory control and data acquisition (SCADA) records as its primary input.

Physical-Scale Models

Physical-scale models provide a relatively inexpensive mechanism for studying the mixing characteristics of tanks. In a physical-scale model, a tracer chemical is added to the inflow (or internally within the model) and the movement of the tracer is monitored during the experiment (Grayman et al. 2000). Tracer substances include visible dyes, which are appropriate for developing a qualitative understanding of mixing behavior, and chemicals (e.g., calcium chloride) that can be measured by sensors in the tanks and used for quantitative assessments. Use of tracers of different density or careful control of temperature of the tracer can be used to study the impacts of thermal variations on mixing. Laws of similitude in hydraulics must be followed in order to account for the scaling effects. Scale models can vary in size and complexity from small tabletop models to large-scale models built in hydraulics laboratories.

Figure 3-6 (see appendix E) depicts such a large-scale model. Roberts and Tian (2002) have applied three-dimensional laser-induced fluorescence to provide detailed measurements of mixing in tanks. They report the results of extensive experiments on jet-induced mixing in water storage tanks in two papers. The experiments were conducted on three styles of storage tanks using a newly developed three-dimensional laser-induced fluorescence (3DLIF) system that can measure the movement of tracer concentrations in the tanks and its temporal evolution through the mixing process. Various inlet geometries were tested for each tank style. According to Tian and Roberts (2008a), without buoyancy effects the 3DLIF technique enabled complex flow patterns to be observed. For cylindrical tanks, it was found that the dimensionless mixing time increases with increasing depth-to-diameter ratio. Vertical nozzles at the bottom mixed most efficiently. If a single vertical nozzle is used, placing it near a sidewall is preferred. Mixing becomes somewhat more rapid as the number of nozzles increases. Mixing times for rectangular tanks are generally similar to cylindrical tanks. The use of a tube intended to enhance mixing actually inhibited mixing in standpipe tanks.

Tian and Roberts (2008b) also observed that small density differences can prevent mixing. The criteria to predict whether mixing occurs depend primarily on the total inflow momentum flux, density difference, water depth, and nozzle locations and orientations. Good mixing can be achieved by injecting positively buoyant inflows horizontally at the bottom or negatively buoyant inflows horizontally near the water surface, resulting in long jet entrainment path lengths. Mixing can usually be accomplished by relatively simple nozzle configurations, provided they are suitably configured, and overly elaborate mixing devices are probably unnecessary.

Transient Analysis

Transient pressures in pressurized drinking water supply distribution systems may be caused by the sudden shutdown of pumps or by operational events such as flushing, hydrant use, and main breaks (Lansey and Boulos 2005). A hydraulic transient is a rapid change in pressure associated with a pressure wave that moves rapidly through a piping system. Figure 3-7 illustrates an event that results in a negative pressure transient for 22 sec caused by a power outage associated with a lightning strike. If the magnitude of the resulting pressure wave is large enough and adequate transient control measures are not in place, a transient can cause a water hammer, leading to failure of hydraulic components (Boulos et al. 2005). It can also lead to instantaneous low or negative pressures that can result in intrusion of untreated water into the pipe, potentially resulting in contamination. Transient events are highly dynamic and sophisticated, and mathematical models are required to analyze their movement in a distribution system.

In a series of research projects (LeChevallier et al. 2003, Gullick et al. 2004), the frequency and location of low and negative pressures in representative distribution systems were measured under normal operating conditions and during specific operational events. These investigators confirmed that fecal indicators and culturable human viruses were present in the soil and water exterior to the distribution system pipes. Their research shows that a well-calibrated hydraulic surge model can be used to simulate the occurrence of pressure transients under a variety of operational scenarios, and a model can also be used to determine optimal mitigation measures.

Although data are insufficient to indicate whether pressure transients pose a substantial risk to water quality in the distribution system, mitigation techniques can be implemented. These techniques include the maintenance of an effective disinfectant residual throughout the distribution system, leak control, redesign of air relief venting, installation of hydropneumatic tanks, and more rigorous application of existing engineering standards.

Many surge analysis and design rules have evolved over time to help utilities cope with the complexity of transient phenomena and have been applied to simplify analysis by restricting both the number and difficulty of the transient cases that need to be evaluated. However, many published guidelines are so misleading and so frequently false that they should only be used with extreme caution, if at all. Jung et al. 2007 found that many accepted rules are often

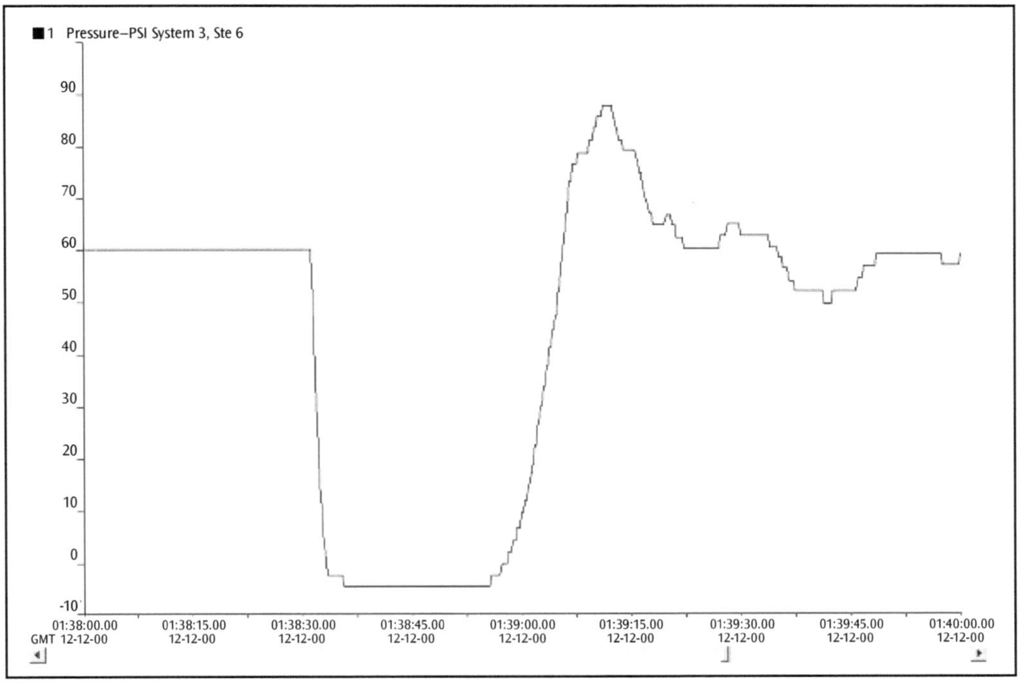

Figure 3-7. Negative pressure transient associated with a power outage (Panguluri et al. 2005)

questionable and that only systematic and informed water hammer analysis can be expected to resolve complex transient characterizations and adequately protect distribution systems from rapid transients.

Covas and Ramos (2010) have proposed the use of inverse transient analysis as a means of detecting leaks in pipes. These investigators found that they could successfully identify leaks of reasonable size provided that the physical and hydraulic characteristics were established. Several commercial software packages for performing transient analysis in water distribution systems are available. Examples of such software are listed in Table 3-2. The technical capabilities, user interface, solution methods, graphical display, and technical support and training vary considerably among the packages.

Optimization Tools

Optimization tools allow the user to evaluate a large number of options and to select the specific alternative that gives the best results in terms of predefined objective functions. In the area of water distribution system analysis, optimization models are used for calibration, design, and operational purposes. These applications are briefly described in the following subsections.

Optimizing Calibration

Calibration of a water distribution system model involves adjustments in various model parameters so that the model agrees with field measurements of flow, pressure, and tank water level. Such a tool is used most frequently with flow and pressure measurements taken during flow (hydrant) tests to stress the system or from SCADA data. Parameters that are typically adjusted include roughness factors, demands, and status of isolation valves.

Table 3-2. Transient modeling software packages (adapted from Panguluri et al. 2005)

Transient Modeling Software	Company	Web Site
AQUIS Surge	7 Technologies	www.7t.dk/aquis
HAMMER	Bentley/Haestad Methods	www.bentley.com
Hytran v3.0	Hytran Solutions	www.hytran.net
Impulse	Applied Flow Technology	www.aft.com/products/impulse
InfoWorks WS	Innovyze	www.innovyze.com
Surge	KYPIPE LLC	www.kypipe.com

Use of manual adjustment techniques may involve many runs of a distribution system model until the resulting predicted flows, pressures, and tank levels approximate the values observed in the field. When an optimization model is applied, the user defines an objective function, such as minimizing the square of the difference between observed and predicted values (for pressure, flow, and tank level). The optimization algorithm then uses some type of controlled search method to identify the set of model parameters that will result in the best results (i.e., minimize the error). The user will generally set constraints on parameters so that the resulting values are reasonable. For example, the user may specify that the allowable range for the roughness factor for a certain set of ductile-iron pipes range between 90 and 120.

Over the past 40 years, various techniques have been applied as part of automated calibration methods (Rahal et al. 1980, Walski et al. 2003). The most common optimization technique in use today couples a hydraulic model with an optimization routine using genetic algorithms. Genetic algorithms are based on the theory of genetics in which successive population trials are generated with the "fittest" ones surviving to breed and evolve into increasingly desirable offspring solutions. The fitness of a solution is based on the objective functions that were previously described. Genetic algorithm–based calibration tools are available as optional components of several water distribution system analysis software packages.

Design Optimization

In a manner analogous to the calibration optimization technique described above, design optimization techniques evaluate a large number of distribution system design options and select the one that provides the best solution (Lansey 2000). Schaake and Lai (1969) first proposed such an approach and applied it to the design of major transmission lines providing water to New York City. Since that time, numerous papers have been written on the subject (Walski et al. 2003) and have included a variety of techniques such as linear programming, dynamic programming, mixed integer programming, metaheuristic algorithms, gradient search methods, and enumeration methods. In recent years, metaheuristic (e.g., genetic algorithm) methods have been favored for this problem and have been widely used in a variety of situations and are included in several commercial software packages (Wu et al. 2001). The user should, however, be aware that, metaheuristic algorithms do not guarantee optimality. These algorithms must be run several times to ensure near optimal solutions.

Typically, design optimization tools limit a user to choose from designated piping options and to size the pipes to meet present and future demands. Cost minimization is the most common objective function. Additionally, some researchers have incorporated reliability and capacity considerations (Mays 1989).

Optimization of Operation

Models can also be used to optimize operations of a distribution system (Goldman et al. 2000). The most common areas of operation where such models have been applied are in energy management and water quality (Hannoun and Boulos 1997). Chase et al. (1994) describe a computer program to control energy costs that incorporates a hydraulic model, a pump optimization program, and an interface. In the water quality area, Uber et al. (2003) used optimization techniques to determine optimal location and operation of chlorine booster stations. Jentgen et al. (2003) implemented a prototype energy and water quality management system at Colorado Springs Utilities. This system combines a simplified distribution system model and an optimization routine to adjust operation of the water system and power generation system in near real-time.

Parks and VanBriesen (2009) have proposed rules for locating booster disinfection to protect a drinking water distribution system against contamination events. Their analysis shows that it is important to properly locate these stations for maximum effectiveness.

Kang and Lansey (2010) have combined optimal valve operation and booster disinfection to improve system water quality. They have shown that this approach results in reduction of total chlorine applied while maintaining minimum chlorine residual levels.

Huang and McBean (2008) have proposed the use of Bayesian analysis for estimating chlorine wall decay coefficients in water supply systems. Their approach provides an efficient approach for estimating EPANET water quality modeling.

Probabilistic Models

Hydraulic and water quality models of distribution systems are deterministic models. For a set of network parameters and specific operations and demands, the model produces a single set of resulting flows and pressures. However, uncertainty exists in many of the aspects of these models including parameters such as roughness, demands, actual inside diameter of pipes, valve settings, and system controls. This uncertainty is generally caused by both imperfect knowledge and natural variability. An emerging procedure is to embed a deterministic network model within a probabilistic framework and to examine the effect of uncertainty on the results.

The most common approach to incorporating uncertainty in models is the use of a Monte Carlo simulation (Vose 2000). In this method, probability distributions are assigned to model parameters to represent the uncertainty associated with each parameter. The distribution system model is then run many times with parameter values being randomly drawn from the probability distributions. The results of many simulation runs are combined to determine the most likely result and a distribution of results. This approach has been used in legal cases where historical contamination events have been reconstructed (Grayman et al. 2004b), in evaluation of the impacts of purposeful contamination (Murray et al. 2004), and in modeling bacterial regrowth in distribution systems (DiGiano and Zhang 2004).

References

American Water Works Association (AWWA). 2004. *AWWA Manual M32, Computer Modeling of Water Distribution Systems*. 2nd ed. Denver, Colo.: AWWA.

AwwaRF/USEPA. 1991. *Water Quality Modeling in Distribution Systems*. W.M. Grayman, R.M. Clark, and R.M. Males, eds. Denver, Colo.: AwwaRF and AWWA.

Boulos, P.F., B.W. Karney, D.J. Wood, and S. Lingireddy. 2005. Hydraulic Transient Guidelines for Protecting Water Distribution Systems. *Journal AWWA*, 97(5), 111–124.

Cesario, L. 1995. *Modeling, Analysis, and Design of Water Distribution Systems*. Denver, Colo.: AWWA.

Characklis, W.G. 1988. *Bacterial Regrowth in Distribution Systems*. Denver, Colo.: AwwaRF and AWWA.

Chase, D.V., D. Guastella, K.L Lai, and L.E. Ormsbee. 1994. Energy Management for Large Scale Water Supply Systems. In *Proc. AWWA Computer Conference, Los Angeles, Calif.* Denver, Colo.: AWWA.

Chun, D.G., and H.L. Selznick. 1985. Computer Modeling of Distribution System Water Quality. In *ASCE Spec. Conf. on Comp. Applic. in Water Res*. New York: American Society of Civil Engineers.

Clark, R.M., and R.M. Males. 1986. Developing and Applying the Water Supply Simulation Model. *Journal AWWA*, 78(8):61.

Clark, R.M., W.M. Grayman, R.M. Males, and J.A. Coyle. 1986. Predicting Water Quality in Distribution Systems. In *Proc. 1986 AWWA Distribution System Symp*. Denver, Colo.: AWWA.

Clark, R.M., W.M. Grayman, R.M. Males, and J.A. Coyle. 1988a. Contaminant Propagation in Distribution Systems. *Jour. Environ. Engg.*, 114(2).

Clark, R.M., W.M. Grayman, R.M. Males, and J.A. Coyle. 1988b. Development, Application, and Calibration of Models for Predicting Water Quality in Distribution Systems. In *Proc. 1988 AWWA Water Quality and Treatment Conference*. Denver, Colo.: AWWA.

Clark, R.M., W.M. Grayman, R.M. Males, and J.A. Coyle. 1988c. Modeling Contaminant Propagation in Drinking Water Distribution Systems. *Jour. Water Supply: Research and Technology—AQUA*, 3:137–151.

Clark, R.M., and J.A. Coyle. 1990. Measuring and Modeling Variations in Distribution System Water Quality. *Journal AWWA*, 82(8):46.

Clark, R.M., W.M. Grayman, J.A. Goodrich, R.A. Deininger, and A.F. Hess. 1990. Water Quality Modeling and Sampling Study in a Distribution System. In *Proc. 1990 AWWA Distribution System Symposium*. Denver, Colo.: AWWA.

Covas, D., and H. Ramos. 2010. Case Studies of Leak Detection and Location in Water Pipe Systems by Inverse Transient Analysis. *Journal of Water Resources Planning and Management*, 136(2):248–257.

Cross, H. 1936. *Analysis of Flow in Networks of Conduits or Conductors*. Univ. of Ill. Eng. Experiment Station Bulletin 286. Urbana: University of Illinois Press.

DiGiano, F.A., and W. Zhang. 2004. Uncertainty Analysis in a Mechanistic Model of Bacterial Regrowth in Distribution System. *Environmental Science and Technology*, 38(22):5925–5931.

Gessler, J., and T.M. Walski. 1985. *Water Distribution System Optimization*, TREL-85-11, WES. Vicksburg, Miss.: US Army Corps of Engineers.

Goldman, F.E., A.B.A. Sakarya, L.E. Ormsbee, J.G. Uber, and L.W. Mays. 2000. Optimization Models for Operation. In *Water Distribution System Handbook*, ed. Larry Mays. New York: McGraw-Hill.

Grayman, W.M., R.M. Clark, and R.M. Males. 1988a. Modeling Distribution-System Water Quality: Dynamic Approach. *Jour. Water Resources Planning and Management*, 114(3):295–312.

Grayman, W.M., R.M. Clark, and R.M. Males. 1988b. A Set of Models to Predict Water Quality in Distribution Systems. In *Proc. Intern. Symp. on Computer Modeling of Water Dist. Systems*. Lexington: University of Kentucky.

Grayman, W.M., L.A. Rossman, C.N. Arnold, R.A. Deininger, C.D. Smith, J.F. Smith, and R. Schnipke. 2000. *Water Quality Modeling of Distribution System Storage Facilities*. Denver, Colo.: AwwaRF and AWWA.

Grayman, W.M., and C. N. Arnold. 2003. Overview of CFD Methods in Analysis of Distribution System Tanks and Reservoirs. In *Proc. AWWA Annual Conference*. Denver, Colo.: AWWA.

Grayman, W.M., L.A. Rossman, R.A. Deininger, C.D. Smith., C.N. Arnold., and J.F. Smith. 2004a. Mixing and Aging of Water in Distribution System Storage Facilities. *Journal AWWA*, 96(9):70–80.

Grayman W.M., R.M. Clark, B.L. Harding, M. Maslia, and J. Aramini. 2004b. Reconstructing Historical Contamination Events. In *Water Supply Systems Security*, ed. Larry Mays. New York: McGraw-Hill.

Gullick, R.W., M.W. LeChevallier, R.C. Svindland, and M. Friedman. 2004. Occurrence of Transient Low and Negative Pressures in Distribution Systems. *Journal AWWA*, 96(11):52–66.

Hannoun, I.A., and P.F. Boulos. 1997. Optimizing Distribution Storage Water Quality: A Hydrodynamic Approach. *Jour. Applied Mathematical Modeling*, 21(8):495–502.

Hart, F.L., J.L. Meader, and S.N. Chiang. 1986. CLNET—A Simulation Model for Tracing Chlorine Residuals in a Potable Water Distribution Network. In *Proc. 1986 AWWA Distribution System Symposium*. Denver, Colo.: AWWA.

Huang, J.J., and E.A. McBean. 2008. Using Bayesian Statistics to Estimate Chlorine Wall Demand Coefficients for Water Supply Systems. *Journal of Water Resources Planning and Management*, 134(2).

Jentgen, L., S. Conrad, R. Riddle, E. Von Sacken, K. Stone, W.M. Grayman, and S. Ranade. 2003. *Implementing a Prototype Energy and Water Quality Management System*. Denver, Colo.: AwwaRF.

Jung, B. B.W. Karney, P.F. Boulos, and D.J. Wood. 2007. The Need for Comprehensive Transient Analysis of Distribution Systems. *Journal AWWA*, 99(1):112–123.

Kang, D., and K. Lansey. 2010. Real-Time Optimal Operation and Booster Disinfection for Water Quality in Water Distribution Systems. *Journal of Water Resources Planning and Management,* 136(4).

Kroon, J.R., and W.A. Hunt. 1989. Modeling Water Quality in the Distribution Network. In *Proc. 1989 AWWA Water Quality Technology Conference*. Denver, Colo.: AWWA.

Kroon, J.R. 1990. An Innovation in Distribution System Water Quality Modeling. *Waterworld News*, 6(4).

LeChevallier, M.W., R.W. Gullick, M.R. Karim, M. Friedman and J.E. Funk. 2003. The Potential for Health Risks from Intrusion of Contaminants Into the Distribution System From Pressure Transients, *Journal of Water and Health*, 01:3–14.

Lansey, K.E. 2000. Optimal Design of Water Distribution Systems. In *Water Distribution System Handbook,* ed. L.W. Mays. New York: McGraw-Hill.

Lansey, K.E., and Boulos P.F. 2005. *Comprehensive Handbook on Water Quality Analysis for Water Distribution Systems*. Broomfield, Colo.: MWH Soft.

Larock, B.E., R.W. Jeppson, and G.Z. Watters. 2000. *Hydraulics of Pipeline Systems*. Boca Raton, Fla.: CRC Press, 12–26.

Liou, C.P., and J.R. Kroon. 1986. Propagation and Distribution of Waterborne Substances in Networks. In *Proc. 1986 AWWA Distribution System Symposium*. Denver, Colo.: AWWA.

Males, R.M., R.M. Clark, P.J. Wehrman, and W.E. Gates. 1985. Algorithm for Mixing Problems in Water Systems. *Jour. of Hyd. Eng.,* 111(2).

Males, R.M., W.M. Grayman, and R.M. Clark. 1988. Modeling Water Quality in Distribution Systems. *Jour. Water Resources Planning and Management*, 114(2).

Mays, L.W., ed. 1989. *Reliability Analysis of Water Distribution Systems*. New York: ASCE.

Metzger, I. 1985. Water Quality Modeling of Distribution Systems. In *ASCE Spec. Conf. on Comp. Applic. in Water Res*. New York: ASCE.

Murray, R., R. Janke, and J. Uber. 2004. The Threat Ensemble Vulnerability Assessment (TEVA) Program for Drinking Water Distribution System Security. In *Proc. ASCE-EWRI World Water & Environmental Resources Congress, Salt Lake City, Utah*. Reston, Va.: ASCE.

Panguluri, S., W.M. Grayman, and R.M. Clark. 2005. *Distribution System Water Quality Report: A Guide to the Assessment and Management of Drinking Water Quality in Distribution Systems.* Cincinnati, Ohio: USEPA Office of Research and Development.

Parks, S.L.I., and J.M. VanBriesen. 2009. Booster Disinfection for Response to Contamination in a Drinking Water Distribution System. *Journal of Water Resources Planning and Management,* 134(6).

Powell, J., J. Clement, M. Brandt, R. Casey, D. Holt, W. Grayman, and M. LeChevallier. 2004. *Predictive Models for Water Quality in Distribution Systems.* Denver, Colo.: AwwaRF.

Rahal, C.M., M.J.H. Sterling, and B. Coulbeck. 1980. Parameter Tuning for Simulation Methods of Water Distribution Networks. In *Proc. Institute of Civil Engineers*, Part 2, 69, 751.

Roberts, P.J.W., and X. Tian. 2002. Application of Three-Dimensional Laser-Induced Fuorescence to Stratified Turbulent Mixing Processes. In *Proc., ASCE Hydraulic Measurements and Experiments Conference*. Reston, Va.: ASCE.

Rossman, L.A., R.M. Clark, and W.M. Grayman. 1994. Modeling Chlorine Residuals in Drinking Water Distribution Systems. *Journal Environmental Engineering*, 120(4):803–820.

Rossman, L.A. 2000. *EPANET Version 2 User's Manual.* EPA/600/R-00/057. Cincinnati, Ohio: USEPA Water Supply and Water Resources Division.

Sarikelle, S., Y. Chuang, and G.A. Loesch. 1989. Analysis of Water Distribution Systems on a Supercomputer. In *Proc. AWWA Comp. Spec. Conf.* Denver, Colo.: AWWA.

Schaake, J., and D. Lai. 1969. *Linear Programming and Dynamic Programming Applications to Water Distribution Network Design*. Report 116. Cambridge, Mass.: MIT Department of Civil Engineering.

Tian, X., and P.J.W. Roberts. 2008a. Mixing in Water Storage Tanks. I: No Buoyancy Effects. *Journal of Environmental Engineering*, 134(12):974–985.

Tian, X., and P.J.W. Roberts. 2008b. Mixing in Water Storage Tanks. II: With Buoyancy Effects. *Journal of Environmental Engineering*, 134(12):986–995.

Uber, J., F. Shang, M. Polycarpou, and Z.Wang. 2003. Feedback Control of Booster Chlorination Systems. Denver, Colo.: AwwaRF.

Uber, J., F. Shang, and L. Rossman. 2004. Extensions to EPANET for Fate and Transport of Multiple Interacting Chemical or Biological Components. In *Proc. ASCE-EWRI World Water & Environmental Resources Congress, Salt Lake City, Utah*. Reston, Va.: ASCE.

Vose, D. 2000. *Risk Analysis: A Quantitative Guide*. Chichester, UK: John Wiley and Sons.

Walski, T.M., D.V. Chase, D.A. Savic, W. Grayman, S. Beckwith, and E. Koelle. 2003. *Advanced Water Distribution Modeling and Management*. Waterbury, Conn.: Haestad Press.

Wood, D.J. 1980a. *Computer Analysis of Flow in Pipe Networks*. Lexington.: University of Kentucky Press.

Wood, D.J.1980b. Slurry Flow in Pipe Networks. *Jour. Hydraulics*, 106(1).

Wood, D.J., and L.E. Ormsbee. 1989. Supply Identification for Water Distribution Systems. *Journal AWWA*, 81(7):74.

Wu, Z.Y., Boulos, P.F., Orr, C.H., Ro, J.J. 2001. Using Genetic Algorithms to Rehabilitate Distribution Systems. *Journal AWWA*, 93(11):74–85.

Chapter 4

Modeling and Monitoring Chemical and Microbiological Contaminants

Introduction

In the United States and throughout the world, drinking water utilities face the challenge of providing potable water to their consumers despite the many factors that can result in the degradation of water quality before it is delivered to the customers. Frequently, raw water is derived from surface or groundwater sources that may be subject to naturally occurring or accidental contamination (ILSI 1999, Gullick et al. 2003). Treated water may be transmitted through a network of corroded or deteriorating pipes.

The goal of drinking water utilities in the United States is to provide potable water when and where it is needed in sufficient quantity, at appropriate pressure, and at an acceptable level of quality. Two techniques that can be instrumental in helping drinking water utilities achieve this goal are routine monitoring and distribution system modeling. Drinking water of "acceptable quality" is defined by the Safe Drinking Water Act (SDWA) of 1974 and is framed in terms of the maximum contaminant levels (MCLs), rules, and regulations promulgated under the act. Regulatory requirements are set by the USEPA or by the individual states that have been granted primacy by the USEPA, and states may institute more stringent requirements if they deem it necessary.

Based on the design and configuration of a particular system, many opportunities exist for water quality to deteriorate as water moves between the treatment plant and the consumer. These unwanted changes may occur for various reasons, including failures at the treatment barrier, transformations in the bulk phase, corrosion and leaching of pipe material, biofilm formation, and mixing between different sources of water. Many researchers have investigated the factors that influence water quality deterioration once it enters the distribution system. It has been well documented that bacteriological growth can cause taste-and-odor problems, discoloration, slime buildup, and economic problems, including corrosion of pipes and biodeterioration of materials (Water Research Centre 1976). Bacterial numbers tend to increase during distribution and are influenced by several factors, including bacterial quality of the finished water entering the system, temperature, residence time, presence or absence of a disinfectant residual, construction materials, and availability of nutrients for growth (Clark and Grayman 1998, Zhang and DiGiano 2002, Camper et al. 2003).

In pipes, it has been found that chlorine can be lost both through the interaction with natural organic material (NOM) in the bulk phase and with pipe walls themselves in transporting finished water. This mechanism for loss of chlorine may be even more serious than long residence times in tanks. The pipe wall demand, possibly caused by biofilm and tubercles, may use up the chlorine very rapidly in a distribution system. Maintaining adequate levels of disinfectant residual may require routine cleaning/replacement of pipes and intensive treatment (Clark et al. 1993a and 1993b).

Some important distribution system water quality concerns are maintenance of proper disinfectant levels; minimization of DBP formation; turbidity, taste, color, and odor issues; distribution tank mixing and utilization; main repair and pressure stabilization; flow management; cross-connection control; and backflow prevention.

It should be noted that water quality goals can be difficult to achieve and can be contradictory. For example, an important goal is to maintain a positive disinfectant residual in order to protect against microbial contamination. However, disinfection by-products (DBPs) such as total trihalomethanes (TTHMs) will increase as water moves through the network as long as disinfectant residual and NOM are available. Other DBPs (HAA5) are degraded biologically when free chlorine or chloramines are nearly absent (Liang and Singer 2003).

Water distribution system models have become increasingly accepted within the water industry as a mechanism for "simulating" the behavior of real water distribution networks. They are intended to replicate the behavior of a real or proposed system and are routinely used for a variety of purposes including: capital investment decisions, development of master plans, estimation of fire protection capacity, design of new systems and extension of existing systems, energy management, water quality studies, and daily operational decisions. Currently available water distribution models have become very sophisticated, and many incorporate geographic information system (GIS) capability. Early models simulated hydraulic behavior only and were steady state in nature. But with the advent of more powerful computers, extended-period simulation (EPS) models were developed. In the early 1980s, investigators began introducing the concept of water quality modeling, and now most water distribution system models routinely incorporate water quality simulation capability. In addition to hydraulics and water quality, many distribution system models are capable of simulating water hammer (surge) and tank mixing. Algorithms have been developed, and incorporated into distribution system models, to "optimize" water system design.

Monitoring of a water supply system serves as a means of gathering data about the physical or water quality characteristics of a system. If the system does not experience major temporal or spatial changes in quality, then only a minimal monitoring system would be required in order to characterize these changes. However, if variation or the potential for variation exists in the system's water quality, the monitoring program must be designed to capture that variability. An effective monitoring program can capture the variation or lack of variation at a specific point in a water distribution system but cannot provide an overall understanding of changes throughout the system. Modeling water distribution system water quality is emerging as a tool that can effectively complement systemwide monitoring, as will be discussed later in this chapter.

This chapter will discuss the basic principles underlying routine water quality modeling in drinking water distribution systems and will discuss new developments in monitoring techniques. In addition, the potential for the integration of monitoring and modeling into a decision support system will be explored.

Distribution System Water Quality

The US Environmental Protection Agency (USEPA) has promulgated many rules and regulations as a result of the SDWA that require drinking water utilities to meet specific guidelines and numeric standards for water quality. Some of the rules that specifically target water quality within the distribution system are the Lead and Copper Rule (LCR), the Surface Water Treatment Rule (SWTR), the Total Coliform Rule (TCR), and the Disinfectants and Disinfection By-Products Rule (D/DBPR). The LCR establishes monitoring requirements for lead and copper within tap water samples. The SWTR establishes the minimum required detectable disinfectant residual and the maximum allowed heterotrophic bacterial plate count, within the distribution system. The TCR requires monitoring of the distribution systems for total coliforms, fecal coliforms, and/or *E. coli*. The D/DBPR addresses the maximum disinfectant residual and concentration of disinfection b-products such as total trihalomethanes and haloacetic acids in the distribution systems (Panguluri et al. 2005).

In the United States, disinfection has become an essential part of a drinking water treatment train and is considered to be one of the major public health advances of the twentieth century. Chlorine and chloramines are most often used because they are very effective disinfectants, and residual concentrations can be maintained in the water distribution system. Some utilities (in the United States and Europe) use ozone and chlorine dioxide as oxidizing agents for primary disinfection prior to the addition of chlorine or chloramines for residual disinfection. For example, the Netherlands identifies ozone as the primary disinfectant, as well as common use of chlorine dioxide, but typically uses no chlorine or other disinfectant residual in the distribution system (Connell 1998). While disinfectants are effective in controlling many microorganisms, they can react with naturally occurring organic (and/or inorganic) matter (NOM) in the treated and/or distributed water to form potentially harmful disinfection by-products (DBPs). To minimize the formation of DBPs, the USEPA has promulgated regulations that specify maximum residual disinfectant level goals (MRDLGs) for chlorine (4 mg/L as chlorine), chloramines (4 mg/L as chlorine), and chlorine dioxide (0.8 mg/L as chlorine dioxide). In addition, MCLs for the DBPs total trihalomethanes (TTHMs) and haloacetic acids (HAA5) have been established as 0.080 and 0.060 mg/L, respectively. The TTHMs include chloroform, bromodichloromethane, dibromochloromethane, and bromoform. The HAA5 include monochloroacetic acid, dichloroacetic acid, trichloroacetic acid, monobromoacetic acid, and dibromoacetic acid. In order to meet these requirements, utilities may need to remove DBP precursor material from water prior to disinfection by applying appropriate treatment techniques or modify their disinfection process (Panguluri et al. 2005). A summary of the evolution of federal drinking water regulation since the passage of the SDWA in 1974 is presented in Figure 2-8, and Table 2-1 summarizes the current SDWA rules and regulations governing distributed drinking water.

Prior to the passage of the SDWA, most drinking water utilities concentrated on meeting drinking water standards at the treatment plant, even though it had long been recognized that water quality could deteriorate in the distribution system. After its passage, the SDWA was interpreted by the USEPA as meaning that some federal water quality standards should be met at various points within the distribution system rather than at the water treatment plant discharge. Consequently, water quality in the distribution system became a focus of regulatory action and a major interest to drinking water utilities.

Despite the existence of the rules and regulations promulgated under the SDWA, no monitoring requirements are in place for many contaminants that have the potential to degrade distribution system water quality. Furthermore, some epidemiological and outbreak investigations conducted in the last five years suggest that a substantial proportion of waterborne disease outbreaks, both microbial and chemical, are attributable to problems within distribution systems (Craun and Calderon 2001, Blackburn et al. 2004, USEPA 1996). However, no evidence indicates that current regulations have resulted in a reduction in the proportion of outbreaks attributable to distribution system–related factors. In 2000, the Federal Advisory Committee for the Microbial/Disinfection By-Products Rule recommended that USEPA evaluate available data and research on aspects of distribution systems that may create risks to public health. In 2003, USEPA committed to revising the TCR and is currently exploring the possibility of revising the TCR to provide a comprehensive approach for addressing water quality in the distribution system environment. To aid in this process, USEPA requested that the National Academies' Water Science and Technology Board conduct a study of water quality issues associated with public water supply distribution systems and their potential risks to consumers (NRC 2006).

Under the SDWA, monitoring is required for all contaminants regulated under the act. It is required both at the entry point to a distribution system and, in some cases, at various points within the distribution system. Routine monitoring will satisfy regulatory requirements and is a useful tool for assessing water quality throughout a system. The TCR, SWTR, D/DBPR, and LCR explicitly require monitoring in the distribution system. Some of these requirements are summarized in Table 4-1 (Owens 2001).

Monitoring Principles

Monitoring a water supply system and its various components facilitates the gathering of data about the state of the system (physical, operational, and water quality). If the state of the system has minimal changes in time or space, a simple monitoring system may be sufficient to define and manage the system characteristics. However, if there is potential for significant variation in the state of the system, the monitoring system must be adequately designed to capture that variability. Thereafter, depending on the type and magnitude of variability, an appropriate response can be provided to restore the "normal" system state.

Table 4-1. Federal distribution system water quality monitoring requirements (USEPA 1989a, 1989b, 2003)

Regulation	Monitoring Requirement
Total Coliform Rule	Samples must be collected at sites that are representative of the water throughout the distribution system based on a site plan that is subject to review by the primacy regulatory agency.
	The minimum number of samples that must be collected per month depends on the population served by the system.
	For each positive total coliform sample, there are various repeat sampling requirements.
Surface Water Treatment Rule; Interim Enhanced Surface Water Treatment Rule (IESWTR)	Disinfectant residuals must be measured at TCR monitoring sites.
Lead and Copper Rule	All systems serving a population > 50,000 people must do water quality parameter (WQP) monitoring.
	The number of sample sites for Pb/Cu and WQP monitoring is based on system size.
Disinfectants/Disinfection By-Products Rule – Stage 2	Standard monitoring program (SMP) will require one year of data on THMs and HAA5s. Number of sampling locations based on utility size and source characteristics. Modeling can reduce sampling requirement.
Possible Future Distribution System Rule	Monitoring requirements unknown.

In a distribution system, water quality may vary because of factors such as normal patterns in water consumption, seasonal variations, source water quality, components of the distribution system, operation of the system, retention time in storage, travel time in the piping system, or the condition of the system itself. Variability may also result from unusual occurrences, such as intentional/accidental intrusions of contaminants, or chemical processes such as nitrification. Design of a water quality monitoring program must take into account both the nature of the variability and the manner in which monitoring data will be used. In other words, the objective of the monitoring program must be defined along with appropriate output or reporting requirements.

In general, monitoring systems can be defined based on the uses or needs of the monitoring program, the general type of monitoring to be performed (manual grab sampling and/or continuous automated online monitoring), or the specific monitoring equipment characteristics. It is important to first establish a clear objective(s) for monitoring. Thereafter, depending on the availability of funding, need, and expertise, one should select the appropriate sampling technique(s) and monitoring equipment. Once an appropriate monitoring system has been selected and implemented, it is important to operate and maintain the program to achieve optimal results and benefits. However, the system should be flexible enough so that it can be modified in case it does not meet the original objective(s).

Online continuous monitoring systems have resulted in large streams of data that document minute-by-minute changes in water quality that exist at various points in a water network. The application of this technology has the potential for providing new insights as to how water distribution systems may be operated and designed to improve water quality. However, these systems will require a relatively high level of sophistication in terms of data management, including the capability to generate real-time reports, graphical and visual representation of information, and compliance reports for meeting drinking water standards. Some of these data streams may well reveal excursions in water quality that constitute violations of current or future drinking water standards, or a security-related incident. This type of information may put pressure on drinking water utilities and regulatory agencies to take remedial action, possibly on an emergency basis, even when such actions may not be fully justified (or warranted). However, careful planning and negotiations with appropriate regulatory authorities to define these potential "excursions" and the proper corrective action to be taken would prevent any misunderstandings and minimize or eliminate the potential for unjustified enforcement or response actions.

Currently most routine water quality monitoring in distribution systems is carried out through manual grab samples followed by analysis in the field or in the laboratory. Essentially all regulatory monitoring is still carried out by this method. For example, samples required under the SWTR are manually collected at sites within the distribution system and manually tested for disinfectant levels in the field. Samples taken to satisfy the requirements of the TCR are manually collected in the field and subsequently analyzed in the laboratory. Manual sampling is labor intensive, and the number of samples that can be collected is limited by personnel and analysis costs. In addition, important events that occur between samples are lost (e.g., nighttime events), or unusual results may be dismissed (Premazzi and Hargesheimer 2002).

Interest is growing within the drinking water industry in moving from grab-sampling to online monitoring. This trend is being reinforced because of increased emphasis among water utilities on automating data processing and control functions (Premazzi and Hargesheimer 2002). Online monitoring requires a mechanism for moving the sampled water from the distribution system to an instrument, instrumentation for analyzing the water, a mechanism for communicating the results, and a means of assessing the results of the monitoring. Additionally, the instrumentation must be periodically calibrated and maintained for quality control/quality assurance.

In the past, distribution system online monitors were typically housed in a controlled environment with takeoffs from the distribution system to the instrument. This resulted in most instrumentation being located at facilities such as treatment plants, tanks, and pump stations. The instrumentation was sometimes connected to a supervisory control and data acquisition (SCADA) system so that results could be communicated to a central office. Recently, some instrumentation has been designed for installation in access holes (or junction chambers) or for direct insertion into water mains.

Monitors can be categorized by the types of parameters (contaminants, agents, characteristics) that the monitor is used to measure. The three broad categories usually used are physical parameters, chemical parameters, and biological parameters (Premazzi and Hargesheimer 2002, Clark et al. 2004b) as discussed below:

- Physical monitors, which are used to measure the physical characteristics of water
- Chemical monitors, which are used to measure inorganic or organic chemicals in the water
- Biological monitors, which use biological organisms as sentinels to determine the likely presence of some form of toxic contaminant in a water sample

Establishing Monitoring Objectives

In order to define and implement an effective monitoring plan, clear objectives must be established. Collecting data just for the sake of accumulating information is not cost-effective. In drinking water systems, there are several specific reasons to collect data, and, typically, the monitoring system is tailored to meet one or more of these needs. The objectives of monitoring distribution systems can be broadly classified as discussed below.

Regulatory-Driven Monitoring

Various federal, state, or other governmental agencies have regulations that specify distribution system monitoring requirements. The specific federal distribution system monitoring requirements (existing and proposed) are summarized in Table 4-1. In some cases, states have imposed more stringent criteria and monitoring requirements.

Security-Related Monitoring

Assessments performed by utilities and various research studies have identified that water distribution systems are vulnerable to intentional (or accidental) contamination. In addition to "hardening" systems in order to deter intentional contamination, monitoring as part of an environmental warning system (EWS) has emerged as a logical approach to cope with potential contamination events. No existing or proposed standards are in place for such monitoring. However, it is well recognized that monitors will need to be sufficiently sensitive to a broad range of potential contaminants and appropriately located to detect a contamination event within a reasonable time. Additionally, as detailed in USEPA's *Response Protocol Toolbox* (USEPA 2003/2004), monitors must be an integral part of an emergency response management plan in order to be effective. Extensive research and development are under way on monitor development, calibration, and placement in response to the perceived security monitoring needs.

Currently, USEPA has an ongoing test program to evaluate the potential of sensors monitoring routine online water quality parameters, such as pH, oxidation reduction potential (ORP), free chlorine, total organic carbon (TOC), conductivity, and turbidity, to serve as rapid detection devices for detecting contamination events in distribution systems. Online monitors were selected because response time is critical for achieving the objective of providing early warning. Both bench- and pilot-scale studies are being conducted at the Water Awareness Technology Evaluation Research and Security (WATERS) Center within USEPA's Test and Evaluation (T&E) Facility in Cincinnati, Ohio. The bench-scale runs are designed to identify the detection threshold of each sensor for specific contaminants. The pilot-scale runs are designed to evaluate overall response of the selected sensors by injecting known quantities of potential contaminants into the distribution system simulator (DSS). For this purpose, several pilot-scale DSSs have been fabricated and used for these test runs. The sensor data are collected continuously and archived electronically to establish stable baseline conditions and to also record sensor responses to injected contaminants. Grab samples are collected periodically before and after injection of contaminants to confirm the sensor results.

Process Control–Related Monitoring

Monitors can also be used in a distribution system to provide real-time or near real-time information on water quality that can then be used to control treatment processes at a treatment plant or in the distribution system. The use of continuous chlorine monitors in the distribution system to control disinfectant feed rates at the plant or their use at in-distribution system booster chlorination stations are examples of this type of monitoring (Uber et al. 2003).

The American Society of Civil Engineers (ASCE), in concert with other leading organizations, entered into a cooperative agreement with the USEPA to develop standards documents and guidance aimed at enhancing the physical security of the nation's water and wastewater/stormwater systems. Under this agreement, ASCE is leading the effort to develop guidelines for designing an online contaminant monitoring system (OCMS). The *Interim Voluntary Guidelines for Designing an OCMS* were published in December 2004 (ASCE 2004). This document provides comprehensive information on several topics, including rationale for OCMS and system design basics, selection and siting of instruments, data analysis, and use of distribution system models.

Monitoring Equipment Overview

In general, monitors can be categorized by the types of parameters (contaminants, agents, and characteristics) that the monitor is used to measure. For establishing water quality, the monitors are designed to measure one or more parameters that represent physical, chemical, and/or biological characteristics of the system. Typically, in manual grab sampling programs, handheld physical and/or chemical parameter measuring devices are used. These handheld devices are carried to the sampling location along with appropriate containers to collect water samples for performing more complex chemical and biological analyses in a laboratory. The online sampling devices are more complex devices that are designed to automatically measure, record, and display specific physical, chemical, or biological parameters.

Establishing Monitoring Requirements

Selection of the types, numbers, and locations of monitors is dependent on the nature of the monitoring program desired. These requirements depend upon the overall monitoring objectives and the distribution system site-specific requirements. For example, a monitor used for regulatory purposes may need to monitor different constituents than one used as part of a process control or security system. Similarly, a different monitor may be needed for a utility that uses chlorine as the disinfectant compared to one that uses chloramine. The site-specific monitoring requirements can be evaluated and represented in the following terms:

- Monitoring parameters
- Number and location of monitors
- Monitor characteristics (e.g., detection limits, sampling frequency, cost, false negatives/false positives), and amenability to remote monitoring and SCADA integration

These requirements are further discussed in the following subsections.

Monitoring Parameters

The parameters to be monitored depend strongly upon the specific use of the monitor and upon utility-specific situations. For regulatory purposes, the regulations typically specify the minimum set of parameters that must be sampled.

Number and Location of Monitors

For selecting monitoring locations in distribution systems, two related decisions must be made: (1) how many monitors to place in the system, and (2) where to place them. The number of monitors is generally controlled by the monitoring objective (e.g., regulatory requirement) or by budgetary factors, while the location of monitors is a more complex issue that can be addressed in many ways. For example, for compliance with the TCR and the SWTR, specific requirements describe the number of samples that must be taken.

Watson et al. (2004) use mixed-integer linear programming models for sensor placement over a range of design objectives. Bahadur et al. (2003) describes an approach using PipelineNet in which GIS data and hydraulic model results are used to guide the manual placement of monitors in order to fulfill some general criteria. In a case study conducted with personnel at a water utility, 25 potential monitoring sites were identified and subsequently reduced to 2 best sites using the GIS/PipelineNet framework. Other researchers have investigated the optimal location and the cost-benefit tradeoffs with selecting monitoring systems in drinking water distribution systems (Berry et al. 2004, Johansson 1991, Kessler et al. 1998, Ostfeld 2004, Ostfeld and Salomons 2004).

Monitoring Characteristics

The following characteristics of monitors must be evaluated prior to selecting an appropriate device:

- Minimum detection limit (MDL)
- False negatives/false positives
- Sampling frequency
- Amenability to SCADA integration
- Operation and maintenance requirements
- Combinations of monitors
- Costs

The heightened level of concern over the need to protect water distribution systems has led to the initiation of research into the development of CWSs for both source and finished waters (Clark et al. 2004a). CWSs are intended to reliably identify low probability/high impact contamination events in source or distributed water. The International Life Sciences Institute (ILSI) developed a report (ILSI 1999) focused on the development of environmental warning systems (EWSs) for source water. The same development principles apply to distribution systems. EWSs applied to distribution systems are commonly referred to as CWSs. The following design requirements for EWSs were identified by ILSI in its report:

- Provides warning in sufficient time to respond to a contamination event and prevent exposure of the public to the contaminant
- Capable of detecting all potential contamination threats
- Remotely operable
- Identifies the point at which the contaminant was introduced
- Generates a low rate of false positive and false negative results
- Provides continuous, year-round surveillance
- Produces results with acceptable accuracy and precision
- Requires low skill and training
- Is affordable to the majority of public water systems

A key aspect of an effective EWS will be the need for it to operate in a remote monitoring and reporting mode.

Amenability to Remote Monitoring and SCADA Integration

For a comprehensive, networkwide water quality remote monitoring program, it is essential to ensure that the system and its monitored components are amenable to remote monitoring and SCADA integration (Haught and Panguluri 1998). The SCADA component adds the element of control to the monitored network. Most utilities have some sort of SCADA functionality to automate and monitor the key water treatment and/or distribution operations. The control logic is typically triggered based on a specified time and/or event. For example, the pumps may be set to fill a distribution system tank at midnight and when the tank level monitor detects that the tank is full (an event reported through the SCADA system), the control logic to turn off the pumps is initiated. This type of control logic can be enhanced to perform control functions based on detection of water quality change in the distribution system. However, to achieve this functionality, one needs to understand the following three major components of a remote monitoring and/or control system (or SCADA):

- Online sampling instruments (e.g., pH, oxidation reduction potential [ORP]) and/or control devices (e.g., pump, valves).
- SCADA or remote monitoring network, and field wiring and communications media.

Electric power is generally required for operating online sampling and control devices. If electric power is not readily available at the desired location where a monitor is to be installed, consider the costs for installing a suitable power apparatus (e.g., a solar panel, battery pack).

Online Sampling/Control Devices

Online sampling/control devices can be the most expensive component of a SCADA system. The sensors, switches, monitors, and controllers used in a SCADA system may vary widely, depending on the parameters that need to be controlled and/or observed. The cost for online sampling devices can range from a few hundred dollars to over $100,000. Control units such as sample feed pumps or shutoff valves are less expensive (Panguluri et al. 1999). Smart systems have higher initial costs, but overall costs are reduced because the communication costs (e.g., long-distance phone costs) are negligible as most of the burden is transferred from the main computer to the individual SCADA unit at the remote site (Panguluri et al. 1999).

Monitoring Case Studies

USEPA has conducted research into the use of remote monitoring and control technology alternatives for many years. These projects have involved both water treatment systems and water distribution (Clark et al. 2004b).

Washington, D.C., Remote Monitoring Network

Following a number of coliform violations, USEPA's Region 3 office directed the Washington, D.C., District of Columbia Water and Sewer Authority (WASA) to implement a number of corrective actions for its water distribution system (Clark et al. 1999). Remote monitoring of water quality parameters within the distribution system was identified as being one possible method for identifying water quality problems. Consequently in 1997, USEPA initiated a study to install a remote network at various locations in Washington, D.C., to monitor water quality within the distribution system (Meckes et al. 1998). The WASA staff teamed with USEPA to select appropriate locations within the distribution system for installation of online sampling stations. Following are some of the study objectives:

- Development of methods to monitor real-time water quality at various locations within the distribution system.
- Field evaluation of sensors and remote monitoring technologies for inclusion in the network.
- Development of effective methods to publish real-time data that enhanced consumer confidence.
- Evaluation of costs associated with implementing such systems, and identification of potential problems and suggestions for remedial actions when implementing remote monitoring networks.

Free chlorine, pH, temperature, and turbidity were selected as the monitored parameters based on the availability of online sensor technologies. The selection was based on the premise that these were parameters that could be reliably monitored continuously, and the selected instruments required limited maintenance. Additionally, WASA used its SCADA system to track various operating parameters within the distribution system. During the evaluation, it was clear that use of the existing SCADA system to manage the monitored data provided clear advantages over other available systems. Using the existing SCADA system minimized long-term on-site support costs.

After suitable location(s) were identified, customized sampling and monitoring systems were built. The remote monitoring system in Washington, D.C., was implemented in three phases. In the first phase, a remote monitoring system was installed at the Fort Reno #2 tank (Figure 4-1), which provided security and easy access to the distribution system. Subsequently, based on initial success at this location, two other sites (Bryant Street and Blue Plains) were selected and added to the remote monitoring network in the second phase. The third phase involved the development of a Web-based application to publish the real-time data in order to enhance consumer confidence.

Figure 4-1 shows the Fort Reno Tank #2 remote sampling system, and Figure 4-2 shows the WASA remote monitoring system layout and data transmission scheme. Although WASA's SCADA system used a proprietary operating system, it provided a personal computer (PC) link that was used to "dump" data into a regular PC for further processing. The hardware-based feature enabled tight security; an authorized end user could only copy the relevant data published on the PC and could not directly access the SCADA system. This feature also eliminated any potential interference between the sampling system data and other distribution system operations data. Unfortunately, the USEPA funding for this study was terminated and, as a result, the systems and the Web site are currently not operational. The overall project, however, did demonstrate that such systems could be developed and operated.

Tucson Water Monitoring Network

Based on a grant received from the USEPA's Environmental Monitoring for Public Access and Community Tracking (EMPACT) Program, the city of Tucson implemented a comprehensive water quality monitoring program (Panguluri et al. 2005). The city's EMPACT goals included the following: implementing enhanced monitoring of the utility's potable distribution system, providing the community with near real-time water quality information on Tucson Water's Web site (www.cityoftucson.org/water), and creating community partnerships to better inform water consumers about water quality and resource issues. The water quality monitoring and data collection tools provided through EMPACT also enable the utility to track and respond to real-time changes in system water quality.

Tucson Water's distribution system consists of 1 central drinking water distribution system that serves the majority of the customers and 10 isolated drinking water distribution systems. All 11 drinking water distribution systems cover a service area of 300 mi^2 and serve 680,000 customers in the Tucson metropolitan area. The two types of source water that supply the central distribution system are native groundwater and renewable recharged surface water from the Colorado River. The source water that supplies the 10 isolated distribution systems is groundwater.

For the purposes of monitoring, the central distribution system is divided into 10 water quality zones, and each isolated distribution system is considered an individual water quality zone. A water quality zone is defined as an area of the distribution system that is similar in water quality characteristics, water pressure, and geographical and political boundaries. Each water quality zone has a set number of dedicated sampling stations and points-of-entry (POEs). The dedicated sampling stations monitor the quality of the drinking water in the distribution system before delivery to the customer. The POEs are usually individual wells that represent the water quality of a single well or, in a few cases, combined POE systems that represent the collective blended water quality from a group of wells that directly supply Tucson's drinking water.

In total, 262 dedicated sampling stations and approximately 154 active POEs are located within the multiple distribution systems. In addition, 22 online water quality stations (for monitoring chlorine residual, total dissolved solids, pH, and temperature) are located throughout the central distribution system at strategic locations, such as reservoirs, well sites, and booster stations, as one of the primary objectives of the EMPACT program.

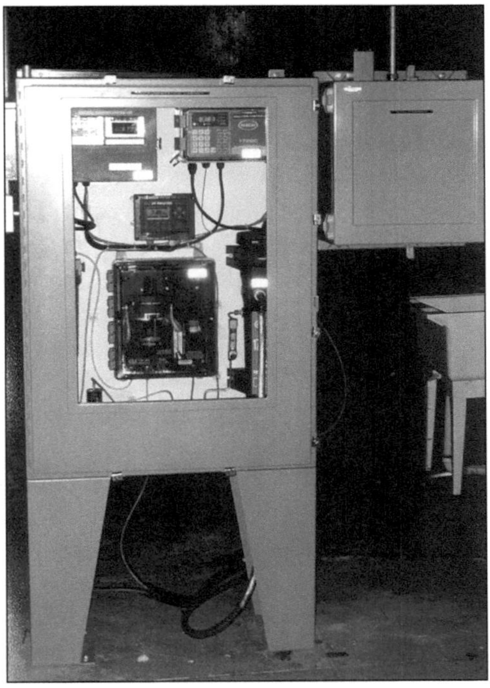

Figure 4-1. Remote monitoring system installed at the Fort Reno #2 tank in WASA

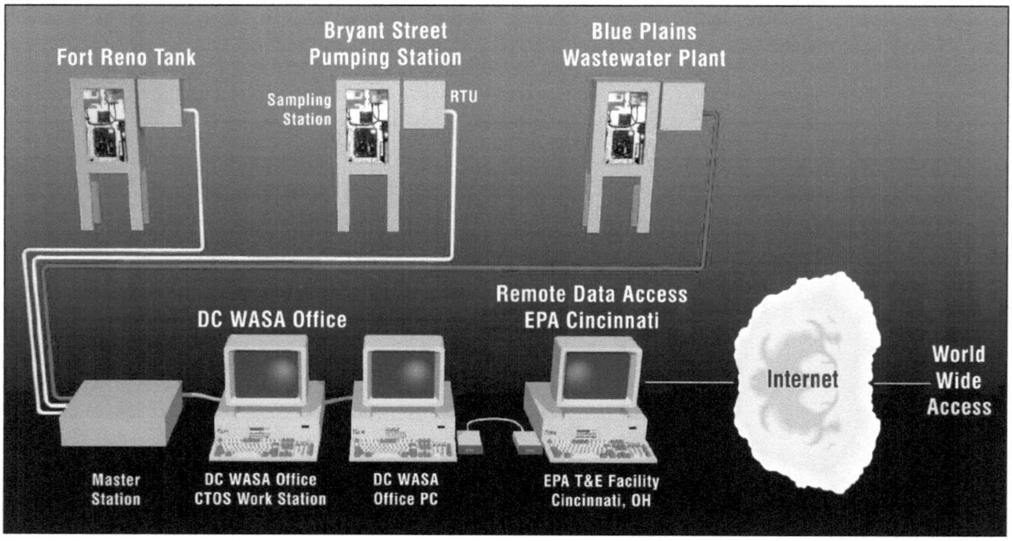

Figure 4-2. WASA remote monitoring system layout and data transmission scheme

Figure 4-3 depicts a continuous water quality monitoring station. The monitoring frequency ranges from triannually (for grab sample locations) to every 60 sec (for continuous monitoring stations), depending on the location and specific monitoring program that is being utilized for that location.

The comprehensive water quality monitoring program encompasses the entire distribution system. Source waters are monitored and sampled according to the Arizona Department of Water Resources and the Arizona Department of Environmental Quality (ADEQ) regulations, while the drinking water is monitored according to USEPA and ADEQ regulations and drinking water standards. Drinking water is also evaluated against a set of consumer-established water quality goals. Special-purpose samples are taken to characterize and track changing trends in water quality for both source water and drinking water. All data sets are utilized to track and monitor changes in water quality to learn the baseline water quality operating parameter levels and also to be able to identify and react appropriately when a contamination event occurs. Most of the analysis is conducted by the utility's water quality laboratory and all the results are tracked through the Water Quality Management Division.

All 262 dedicated sampling locations are monitored at least once each month for total coliform and chlorine residual, while 26 other parameters are monitored once every three months. Based on the water quality measurements collected each month from these 262 sampling locations, the trends in water quality conditions are determined for each water quality zone and for the distribution system as a whole. This information can be found on the aforementioned Web site in the Water Quality section under Tucson's Water Quality and Water Quality in My Neighborhood links. The water quality information displayed on two interactive maps shows data charts and tables for each location that is sampled under the Water Quality program. In addition, the information provided to all Tucson water customers in the annual water quality report or consumer confidence report is based on POE monitoring data.

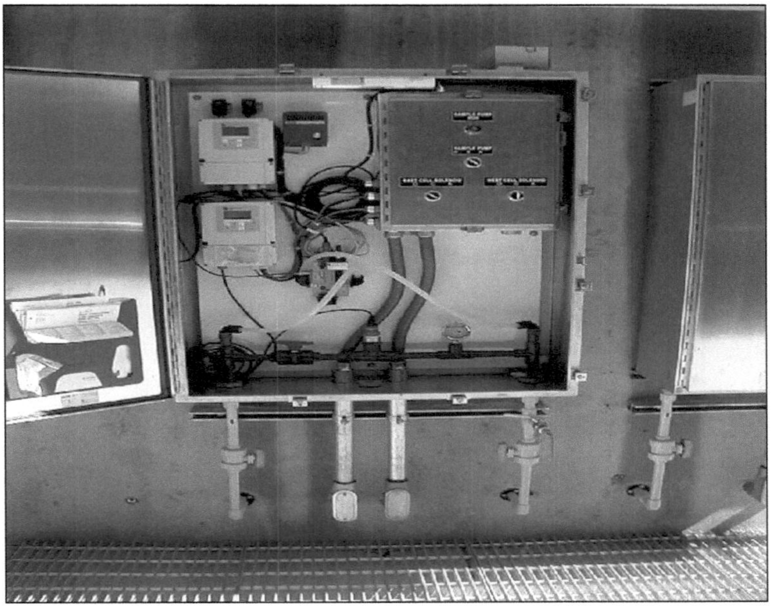

Figure 4-3. Continuous water quality monitoring station in Tucson

North Marin Water District

The North Marin Water District (NMWD) provides an excellent example of the power of monitoring integrated with water quality modeling. As will be discussed in chapter 8, the NMWD study was one of USEPA's first full-scale water quality modeling studies. The NMWD is located Novato, Calif., and at the time of the study used two sources of water—Stafford Lake and the North Marin Aqueduct. The North Marin Aqueduct is a year-round source, but Stafford Lake was used only during the warm summer months when precipitation is virtually nonexistent and demand is high (Clark et al. 1994).

The water quality of the two sources differs greatly. Stafford Lake water has a high humic content and is treated with conventional treatment and prechlorination doses of between 5.5 and 6.0 mg/L. The treated water has a residual of 0.5 mg/L when it leaves the treatment plant clearwell. Total trihalomethane formation potential (TTHMFP) levels in the Stafford Lake water were very high. The North Marin Aqueduct water was derived from the Raney Well Field along the Russian River with a very low TTHM potential. The modeling study revealed dramatic differences in water quality at selected sampling stations over a 48-hr period. Figure 8-24 shows the results of the modeling predictions superimposed on the TTHM data collected during the study period. Examining the data alone resulted in very confusing results because of the dramatic variation in TTHM levels at these sites over the course of the study. However, the modeling results showed that a definite cycle occurred in which the water from the high TTHM source dominated at one point during the day and water from the low TTHM sources dominated during the rest of the day. This NMWD study will be discussed in more detail in chapter 8.

Summary

Distribution system monitoring is intended to identify the spatial and temporal variations in water quality that take place in a drinking water system. Monitoring data can be used to satisfy various objectives, such as regulatory requirements, security requirements, or process control requirements. The costs of implementing such a system can best be justified if the resulting data can be used for more than one of the aforementioned objectives.

A monitoring program can implement either routine grab sampling or continuous monitoring. A combined approach, utilizing both continuous and grab sampling data, may prove to be very effective as the basis for a comprehensive systemwide monitoring plan. In the past, distribution system online monitors were typically housed in a controlled environment with sample lines from the distribution system to the instrument. This resulted in most instrumentation being located at facilities such as treatment plants, tanks, and pump stations. The instrumentation was sometimes connected to a SCADA system, so that results could be communicated to a central office. Recently, some instrumentation has been designed for installation in access holes (junction chambers) or for direct insertion into water distribution system pipes.

Vulnerability assessments performed by utilities and various research studies have identified that water distribution systems are vulnerable to intentional or accidental contamination. In addition to hardening systems to make it more difficult to contaminate a system, monitoring as part of a CWS has emerged as a logical approach to cope with potential contamination events. Monitors can also be used in a distribution system to provide real-time or near real-time information on water quality. The data can then be used to control treatment processes at a treatment plant or in the water distribution system. However, this type of program may not be practical for small systems.

SCADA is widely used in industrial environments and by larger water utilities to control and monitor their individual facility operations. However, water utilities typically do not use available SCADA systems for conventional water quality monitoring. Water utilities typically monitor water quality parameters by performing grab sampling on a scheduled or random basis that provides a periodic snapshot of the overall system. Current drinking water regulations require all public water systems to implement water quality monitoring for total coliform to ensure that good quality water is provided to consumers (USEPA 1996). Since the regulations do not clearly specify that real-time monitoring of water quality is required, utilities have been reluctant, for good reasons, to install and operate such devices.

After the events of 9/11, utilities have become more interested in the potential for continuous water quality monitoring. SCADA systems can assist in this function by constantly monitoring water quality within drinking water distribution systems. These systems can potentially reduce the risk of security-related threats or even nonsecurity-related threats, and detect undesirable water quality changes within a system (Meckes et al. 1998).

Users should evaluate monitoring data appropriately for errors and inconsistencies before commencing actions based on acquired data. Each component in a monitoring system is a potential source of error. For example, a remote monitoring system could have data errors for one or more of the following reasons: instrument errors and spikes, SCADA data errors related to system failure, backfilling caused by communication failure, timing errors, or missing data. It is important to validate data and understand routine changes in water quality related to system-specific operations.

Monitoring equipment should be chosen appropriately after establishing the monitoring requirements. The individual monitor characteristics, costs, and amenability to SCADA integration are vital to effective implementation. Each system should be individually examined and engineered for proper implementation.

The monitoring case studies presented in this chapter demonstrate the manner in which effective monitoring systems can be implemented in small, medium, and large distribution systems. However, if monitoring data are to be effectively used for responding to a contamination threat, it is important to understand the movement of water in the system. Based on the case studies cited in this chapter and information presented in this book, it is clear that linking monitoring and modeling data can assist in providing effective control and response strategies.

References

American Society of Civil Engineers (ASCE). 2004. *Interim Voluntary Guidelines for Designing an Online Contaminant Monitoring System.* Reston, Va.: ASCE.

Bahadur, R., W. Samuels, W. M. Grayman, D. Amstutz, and J. Pickus. 2003. PipelineNet: A Model for Monitoring Introduced Contaminants in a Distribution System. In *Proc. ASCE-EWRI World Water & Environmental Resources Congress, Philadelphia, Pa.* Reston, Va.: ASCE.

Berry, J., W. Hart, C. Phillips, and J. Uber. 2004. A General Integer-Programming-Based Framework for Sensor Placement in Municipal Water Networks. In *Proc. ASCE-EWRI World Water & Environmental Resources Congress, Salt Lake City, Utah.* Reston, Va.: ASCE.

Blackburn, B.G., G.F. Craun, J.S. Yoder, V. Hill, R.L. Calderon, N. Chen, S.H. Lee, D.A. Levy, and M.J. Beach. 2004. Surveillance for Waterborne-Disease Outbreaks associated With Drinking Water—United States, 2001–2002. *Morbidity and Mortality Weekly Report,* 53(SS-8):23–45.

Camper, A.K., K. Brastrup, A. Sandvig, J. Clement, C.Spencer, and A.J. Capuzzi. 2003. Impact of Distribution System Materials on Bacterial Regrowth. *Journal AWWA,* 95(7):107–121.

Clark, R.M., J.A. Goodrich, and L.J. Wymer. 1993a. Effect of the Distribution System on Drinking Water Quality. *Journal of Water Supply: Research and Technology—AQUA,* 42(1):30–38

Clark, R.M., L.A. Rossman, and L.J. Wymer. 1993b. Modeling Contaminant Propagation in Drinking Water Distribution Systems: Regulatory Implications. In *Preconference Seminar Proc. "Configuring and Operating Distribution Systems to Maintain Water Quality," AWWA Annual Conf. and Exposition, San Antonio, Texas, June 6–10.* Denver, Colo.: AWWA.

Clark, R.M., G. Smalley, J.A. Goodrich, R. Tull, L.A. Rossman, J.T. Vasconcelos, and P.F. Boulos. 1994. Managing Water Quality in Distribution Systems: Simulating TTHM and Chlorine Residual Propagation. *Jour. Water Supply: Research and Technology—AQUA,* 43(4):182–191.

Clark, R. M., and W. M. Grayman. 1998. *Modeling Water Quality in Drinking Water Distribution Systems.* Denver, Colo.: AWWA.

Clark, R.M., G.S. Rizzo, J.A. Belknap, and C. Cochrane. 1999. Water Quality and the Replacement and Repair of Drinking Water Infrastructure: The Washington, DC, Case Study. *Journal of Water Supply: Research and Technology-AQUA,* 48(3):106–114.

Clark R.M., W.M. Grayman, S.G. Buchberger, Y. Lee, and D.J. Hartman. 2004a. Drinking Water Distribution Systems: An Overview. In *Water Supply Systems Security,* ed. L.W. Mays. New York: McGraw-Hill, 4.1–4.49.

Clark, R.M., S. Panguluri, and R.C. Haught. 2004b. Remote Monitoring and Network Models: Their Potential for Protecting U.S. Water Supplies. In *Water Supply Systems Security,* ed. L.W. Mays. New York: McGraw-Hill, 14.1–14.22.

Connell, G.F. 1998. European Water Disinfection Practice Parallel U.S. Treatment Methods. *Drinking Water and Health Quarterly,* 4(3). http://www.waterandhealth.org/newsletter/new/summer-1998/disinfection.html.

Craun, G. F., and R. Calderon. 2001. Waterborne Disease Outbreaks Caused by Distribution System Deficiencies. *Journal AWWA,* 93:9:64–75.

Gullick, R.W., W.M. Grayman, R.A. Deininger, and R.M. Males. 2003. Design of Early Warning Monitoring Systems for Source Waters. *Journal AWWA,* 95(11):58–72.

International Life Sciences Institute (ILSI). 1999. *Early Warning Monitoring to Detect Hazardous Events in Water Supplies.* Washington, D.C.: ILSI Press.

Johansson, P. 1991. *An Introduction to Modern Welfare Economics.* Cambridge, UK: Cambridge University Press.

Kessler, A., A. Ostfeld, and G. Sinai. 1998. Detecting Accidental Contaminations in Municipal Water Networks. *Journal of Water Resources Planning and Management,* 124(4):192–198.

Liang, L., and P.C. Singer. 2003. Factors Influencing the Formation and Relative Distribution of Haloacetic Acids and Trihalomethanes in Drinking Water. *Environ. Sci. Technol.*, 37(13):2920–2928.

Meckes, M. C., J. S. Mattingly, G. J. Papadopoulos, M. Dosani, and S. Panguluri. 1998. Real Time Water Quality Monitoring of a Water Distribution System Using Remote Telemetry. In *Proc. AWWA Distribution System Symposium, Austin, Texas. September 20–22.* Denver, Colo.: AWWA.

National Research Council (NRC). 2006. *Drinking Water Distribution Systems: Assessing and Reducing Risks.* Washington, D.C.: National Academies Press.

Ostfeld, A. 2004. Optimal Monitoring Stations Allocations for Water Distribution System Security. In *Water Supply Systems Security*, ed. L.W. Mays. New York: McGraw-Hill, 16.1–16.15.

Ostfeld, A., and E. Salomons. 2004. Optimal Layout of Early Warning Detection Stations for Water Distribution Systems Security. *Journal of Water Planning and Resources Management,* 130(5):377–385.

Owens, J. 2001. A Review of Federal Drinking Water Regulations in the US. In *Controlling Disinfection By-Products and Microbial Contaminants in Drinking Water*, ed. R.M. Clark and B. Boutin, EPA/600/R-01/110. Cincinnati, Ohio: USEPA Office of Research and Development, 2-1–2-14.

Panguluri, S., R.C. Haught, M.C. Meckes, and M. Dosani. 1999. Remote Water Quality Monitoring of Drinking Water Treatment Systems. In *Proc. AWWA Water Quality Technology Conference, Denver, Colo., November 1999.* Denver, Colo.: AWWA.

Panguluri, S., W.M. Grayman, and R.M. Clark. 2005. *Distribution System Water Quality Report: A Guide to the Assessment and Management of Drinking Water Quality in Distribution Systems.* Cincinnati, Ohio: USEPA Office of Research and Development.

Premazzi, G., and E. Hargesheimer. 2002. Introduction. In *Online Monitoring for Drinking Water Utilities: Comparative Research Report,* ed. E. Hargestheimer, O. Conio, and J. Popovicova. Denver, Colo.: AwwaRF, 1–6.

Safe Drinking Water Act (SDWA). 1974. Public Law 93-523.

Uber, J., F. Shang, M. Ploycarpou, and Z. Wang. 2003. *Feedback Control of Booster Chlorination Systems.* Denver, Colo.: AwwaRF.

US Environmental Protection Agency (USEPA). 1989a. National Primary Drinking Water Regulations: Total Coliforms (Including Fecal Coliforms and *E. coli*); Final Rule. *Federal Register*, 54, 27544.

USEPA. 1989b. National Primary Drinking Water Regulations; Filtration, Disinfection; Turbidity, *Giardia lamblia*, viruses, *Legionella*, and Heterotrophic Bacteria; Final Rule. *Federal Register,* 54 (No. 124), 27486.

USEPA. 1996. *Drinking Water Regulations and Health Advisories.* EPA 822-B-B-96-002. Washington, D.C.: USEPA, Office of Water.

USEPA. 2003. *Draft Stage 2 Disinfectants and Disinfection Byproducts Initial Distribution System Evaluation Guidance Manual.* Washington, D.C.: USEPA.

USEPA. 2003/2004. Overview and Modules 1 through 6. In *Response Protocol Toolbox*. http:// cfpub.epa.gov/ safewater/watersecurity/ home.cfm?program_id=8#response_toolbox.

Watson, J., H.J. Greenberg, and W.E. Hart. 2004. A Multiple-Objective Analysis of Sensor Placement Optimization in Water Networks. In *Proc. ASCE-EWRI World Water & Environmental Resources Congress, Salt Lake City, Utah.* Reston, Va.: ASCE.

Zhang, W., and F.A. DiGiano. 2002. Comparison of Bacterial Regrowth in Distribution Systems Using Free Chlorine and Chloramine: A Statistical Study of Causative Factors. *Water Research*, 36(6):1469–1482.

Chapter 5

Hydraulic Modeling Principles

Introduction

Throughout history, people have been concerned with controlling floods, irrigation, and water supply. In the Middle Ages, pipelines were built sporadically in the Middle East to supply water from springs to nearby forts and palaces by gravity flow. More than 150 years ago, developments in iron and steel manufacturing and the introduction of efficient pumps allowed the building of pipelines to become common.

Despite its history, hydraulics is still somewhat empirical, with many indeterminate factors. However, enormous progress has been made during the last century in understanding the fundamental laws of fluid mechanics. Powerful mathematical techniques are available for solving these problems, but many practical hydraulics problems still do not have theoretical solutions. The amount of energy required to move water from one point to another and to provide the needed amount of water has become a major issue for designers. Because the amount of water moved is easily measurable, a great deal of experimental data have been collected on pipe flow. A consequence is that engineers and scientists have put these data into the form of graphs and empirical design formulas. Many of these equations are used in this chapter.

Hydraulic Principles

The theory and application of hydraulic models are thoroughly explained in many widely available references (Walski et al. 2003, AWWA 2004, Larock et al. 2000). Essentially, three basic relations are used to calculate fluid flow in a pipe network. These relationships are:

Conservation of Mass: This principle requires that the sum of the mass flows in all pipes entering a junction must equal the sum of all mass flows leaving the junction. Because water is essentially an incompressible fluid, conservation of mass is equivalent to conservation of volume.

Conservation of Energy: Three types of energy occur in a hydraulic system: kinetic energy associated with the movement of the fluid, potential energy associated with the elevation, and pressure energy. In water distribution networks, energy is referred to as "head," and energy losses (or head losses) within a network are associated primarily with friction along pipe walls and turbulence.

Pipe Friction Head Loss: A key factor in evaluating the flow through pipe networks is the ability to calculate friction head loss (Jeppson 1976). Three empirical equations commonly used are the Darcy–Weisbach, Hazen–Williams, and Manning equations. All three equations relate head or friction loss in pipes to the velocity, length of pipe, pipe diameter, and pipe roughness. A fundamental relationship that is important for hydraulic analysis is the Reynolds number (Re), which is a function of the kinematic viscosity of water (resistance to flow), velocity, and pipe diameter. The most widely used head loss equation in the United States is the Hazen–Williams equation, though the Darcy–Weisbach equation is generally considered to be theoretically more rigorous. The differences between the use of these two equations is typically insignificant under most circumstances.

A distribution system is represented in a hydraulic model as a series of links and nodes. Links represent pipes whereas nodes represent junctions, sources, tanks, and reservoirs. Valves and pumps are represented as either nodes or links depending on the specific software package. Figure 5-1 illustrates a simple link–node representation of a water distribution system.

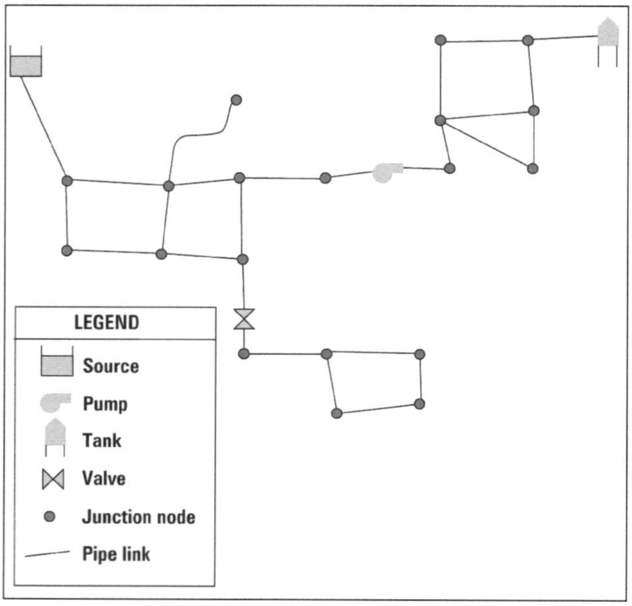

Figure 5-1. Simple link–node representation of a water distribution system (Panguluri et al. 2005)

As mentioned previously, two types of analyses may be conducted on drinking water distribution systems: steady-state and EPS. In a steady-state analysis, all demands and operations are treated as constant over time and a single solution is generated. In the EPS mode, variations in demand, tank water levels, and other operational conditions are simulated by a series of steady-state analyses that are linked together. Each steady-state solution in the EPS mode involves a separate solution of the set of nonlinear equations. EPS is used as the basis for water quality modeling. Though the EPS solution does introduce some approximations and ignores the transient phenomena resulting from sudden changes (e.g., a pump being turned on), these more refined assumptions are generally not considered significant for most distribution system studies.

Friction Head Loss Formulas

As mentioned, a key factor in evaluating the flow through pipe networks is the ability to calculate friction head loss (Jeppson 1976). The three equations discussed in this chapter are the Darcy–Weisbach, the Hazen–Williams, and the Manning equations.

The Darcy–Weisbach equation is:

$$h_f = f(L/D)(V^2/2g) \qquad \text{(Eq 5-1)}$$

Where:

h_f = head loss, in ft (m)
f = dimensionless friction factor
D = pipe diameter, in ft (m)
L = length of pipe, in ft (m)
V = average velocity of flow, in ft/sec (m/sec)
g = the acceleration of gravity (32.2 ft/s^2 for English units and for SI units is 9.81 m/s^2)

A fundamental relationship that is important for hydraulic analysis is the Reynolds number, as follows:

$$\text{Re} = VD/v \quad \text{(Eq 5-2)}$$

Where:

Re = Reynolds number (dimensionless)
v = kinematic viscosity, in ft^2/sec (m^2/sec)
V and D are as defined earlier.

Table 5-1 contains various equations that can be used for calculating f in the Darcy–Weisbach equation. Although the Darcy–Weisbach equation is fundamentally sound, the most widely used equation is the Hazen–Williams equation:

$$Q = 1.318CAR^{0.63}S^{0.54} \quad \text{(Eq 5-3)}$$

Where:

Q = flow, in ft^3/sec (m^3/sec)
C = Hazen–Williams roughness coefficient (dimensionless)
A = cross-sectional area, in ft^2 (m^2)
R = hydraulic radius (D/4)(m)
S = slope of the energy grade line (h_f/L[ft/ft])(m/m)

If the head loss is desired and Q is known, the Hazen–Williams equation for a pipe can be written as

$$h_f = (4.73L/C^{1.852}D^{4.87})Q^{1.8152} \quad \text{(Eq 5-4)}$$

Where the variables in Eq 5-4 are as defined previously.

Another empirical equation is the Manning equation for pipes, which has been solved for h_f as follows:

$$h_f = (4.637n^2L/D^{5.333})Q^2 \quad \text{(Eq 5-5)}$$

Where:

n = empirical constant
and the other variables are as defined earlier.

Table 5-1. Summary of friction factor equations for Darcy–Weisbach equation $h_f = f \frac{L}{D}\frac{V^2}{2g}$

Type of Flow	Equation Giving *f*	Range of Application
Laminar	$f = 64/\text{Re}$	Re < 2,100
Hydraulically smooth or turbulent smooth	$f = 0.316/\text{Re}^{0.25}$	4,000 < Re < 10^5
	$\frac{1}{\sqrt{f}} = 2\log_{10}(\text{Re}\sqrt{f}) - 0.8$	Re > 4,000
Transition between hydraulically smooth and wholly rough	$\frac{1}{\sqrt{f}} = 2\log_{10}\left(\frac{e/D}{3.7} + \frac{2.52}{\text{Re}\sqrt{f}}\right)$	Re > 4,000
	$= 1.14 - 2\log_{10}\left(\frac{e}{D} + \frac{9.35}{\text{Re}\sqrt{f}}\right)$	
Hydraulically rough or turbulent rough	$\frac{1}{\sqrt{f}} = 1.14 - 2\log_{10}(e/D)$	Re > 4,000
	$= 1.14 + 2\log_{10}(D/e)$	

Where:

h_f = head loss, in ft (m)
f = dimensionless friction factor
L = length of pipe, in ft (m)
D = diameter of pipe, in ft (m)
V = average velocity of flow, in ft/sec (m/sec)
g = the acceleration of gravity, in ft/sec^2 (m/sec^2)
Re = Reynolds number (dimensionless)

Table 5-2 contains values for the Hazen–Williams coefficient and the *n* values for various materials.

Table 5-2. Values of the Hazen–Williams coefficient C_{HW} and Manning's *n* for common pipe materials

Type of Pipe	C_{HW}	*n*
Polyvinyl chloride (PVC) pipe	150	0.008
Very smooth pipe	140	0.011
New cast iron or welded steel	130	0.014
Wood or concrete	120	0.016
Clay or new riveted steel	120	0.017
Old cast iron or brick	100	0.020
Badly corroded cast iron or steel	80	0.035

Source: Jeppson 1976.

Analysis Methods

Analyzing for the flow in pipe networks, particularly if a large number of pipes are involved, is a complex process. Deciding which pipes should be included in the analysis can be a matter of judgment. It may not be possible or practical to include all the pipes that deliver water to the consumer. Analysis frequently is conducted only on the major transmission lines in the network or on the pipes that carry water between separate sections of the network. This process is called *skeletonization*.

As mentioned previously, two types of analyses are usually conducted on drinking water distribution systems—steady state and dynamic—which were introduced previously. The dynamic analysis or extended period simulation (EPS) deals with unsteady flows or transient problems. Steady-state analysis is considered solved when the flow rate in each pipe is calculated based on a specific usage or consumption. The supply from reservoirs, storage tanks, and/or pumps is generally the inflow or outflow from some point in the network. If flow rates are known, then pressures or head losses can be computed throughout the system. If the heads or pressures are known at each pipe junction or network node, the flow rates can be computed in each pipe.

A steady-state analysis is needed for each demand or consumption pattern. The addition of new service areas, pumps, or storage tanks changes the system and requires a new steady-state analysis. Dynamic analysis or EPS is often simply a series of steady-state analyses linked by specified conditions. The oldest method of solving steady-state flow in pipes is the Hardy Cross method (Cross 1936). It was originally developed for solution by hand but has been programmed for solution using computers. However, when applied to large networks or for certain conditions, it might be slow or even fail to converge. More recently, the Newton–Raphson method and the "linear theory method," both described later in this chapter, have been applied to network solutions.

Other issues to be considered include the calibration and verification of the hydraulic assumptions. The issues of steady state, EPS, analysis, skeletonization, and calibration are discussed in this chapter.

Reducing Network Complexity

Pipe networks may include pipes in series, parallel, or branches (like branches of a tree). The network may also use elbows, valves, and other devices that cause local disturbances and head losses. These factors can frequently be combined with or converted into equivalent pipes, which is very useful in simplifying networks. The major methods of simplification follow.

Pipes in Series

For pipes in series, the same flow must pass through both pipes, therefore an equivalent head loss is the sum of the head losses for all the pipes being considered as part of the equivalent pipes. For example, Figure 5-2 depicts the head loss relationships for the pipes in series at different diameters.

The head loss for the equivalent pipe is

$$h_t = \Sigma h_i \qquad \text{(Eq 5-6)}$$

Where:

h_t = total head loss through both pipes
h_i = head loss through individual pipes

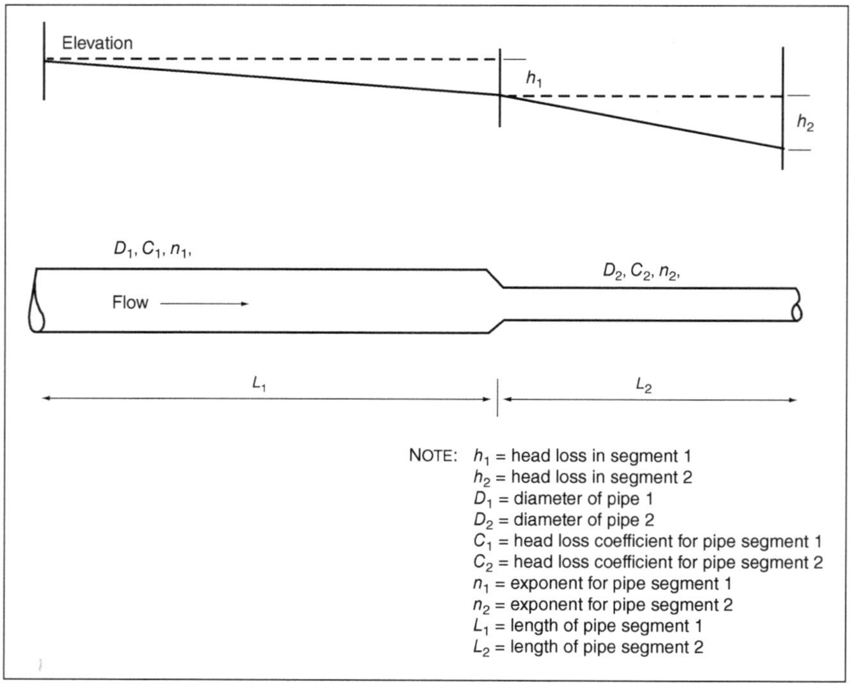

Figure 5-2. Pipes of different diameters in series

If we assume an exponential head loss relationship, such as in the Hazen–Williams formula, then

$$C_tQ^n = C_1Q^{n1} + C_2Q^{n2} + \ldots \quad \text{(Eq 5-7)}$$

Where:

C_t = empirical constant for head loss for entire length of pipe
Q = flow, in ft^3/sec (m^3/sec)
n = empirical constant
C_i = empirical constant for head loss for each length of pipe

If all of the *n*s are equal, then C_t equals the sum of the C_i for the individual pipes in series or

$$C_t = C_1 + C_2 + \ldots \quad \text{(Eq 5-8)}$$

Pipes in Parallel

Two or more pipes in parallel can also be replaced by an equivalent pipe. In this case, the head loss between junctions where the pipes part and then join again must be equal. Although using equivalent pipes may be effective from a hydraulic viewpoint, this practice can be misleading when modeling water quality. Water quality modeling accuracy depends on accurate measurement of velocity.

Figure 5-3 illustrates the equivalent pipe approach.

$$h_1 = h_2 = h_3 = \ldots = h_i \quad \text{(Eq 5-9)}$$

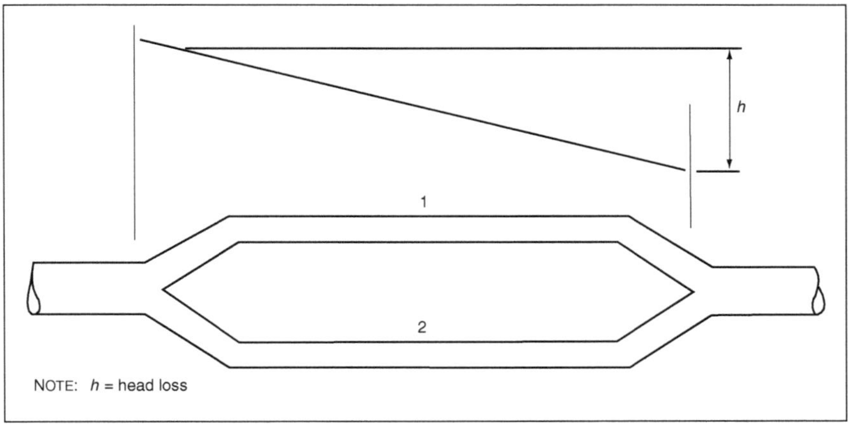

Figure 5-3. Head loss over a set of parallel pipes

The flow must equal the sum of the individual flows or

$$Q_t = Q_1 + Q_2 + \ldots + Q_i \quad \text{(Eq 5-10)}$$

or

$$Q_t = \Sigma Q_i \quad \text{(Eq 5-11)}$$

Where:

Q_t = total flow, in ft³/sec
Q_i = flow, in ft³/sec in each parallel pipe

Using the exponential form for head loss yields

$$C_t Q_t^{\,nt} = C_1 Q_1^{\,n1} + C_1 Q_2^{\,n2} + \ldots + C_i Q_i^{\,ni} \quad \text{(Eq 5-12)}$$

$$C_t Q_t^{\,nt} = \Sigma C_i Q_i^{\,ni} \quad \text{(Eq 5-13)}$$

Where:

$C_t Q_t^{\,n}$ = sum of head loss in parallel pipes
$C_i Q_i^{\,n}$ = head loss in each parallel pipe

If we assume the exponential form, then

$$\left(\frac{h_t}{C_t}\right)^{\frac{1}{n_t}} = \left(\frac{h_1}{C_1}\right)^{\frac{1}{n_1}} + \left(\frac{h_2}{C_2}\right)^{\frac{1}{n_2}} + \ldots \quad \text{(Eq 5-14)}$$

Since Eq 5-9 holds and assuming the exponents are equal, then

$$\left(\frac{1}{C_t}\right)^{\frac{1}{n}} = \left(\frac{1}{C_1}\right)^{\frac{1}{n}} + \left(\frac{1}{C_2}\right)^{\frac{1}{n}} + \ldots \qquad \text{(Eq 5-15)}$$

$$\left(\frac{1}{C_t}\right)^{\frac{1}{n}} = \Sigma_i \left(\frac{1}{C_i}\right)^{\frac{1}{n}} \qquad \text{(Eq 5-16)}$$

If the Darcy–Weisbach equation is used, it is common to assume n is equal for all pipes and Eq 5-16 can be used to solve for the C_t in the equivalent pipe.

Branching System

In a branching system, a number of pipes are connected to a larger pipe in the form of a tree (Figure 5-4). If the flow is from the larger pipe to the smaller laterals, the flow rate can be calculated in any pipe as the sum of the downstream consumption or demand. If the laterals supply water to the main, then the same approach can be taken. When a system is analyzed, frequently only the larger pipes are used in the network analysis.

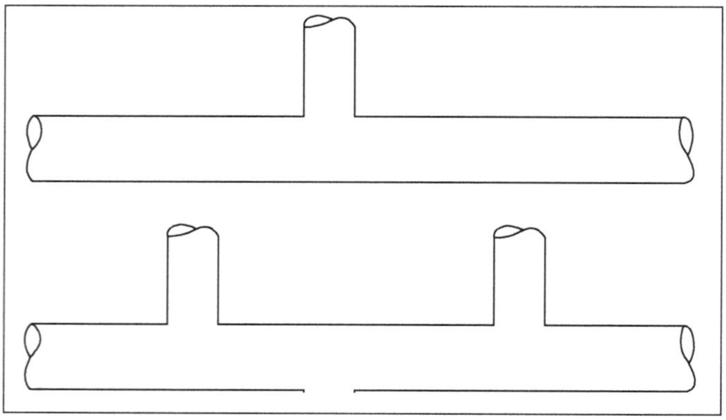

Figure 5-4. Typical branching pipes

Minor Losses

An equivalent pipe can be used in a network to approximate the minor losses associated with valves, meters, elbows, or other devices. Equivalent pipes are formed by adding a length to the actual pipe length that will result in the same head loss as in the component. Frequently, pipe roughness is adjusted to account for minor losses when modeling. The Darcy–Weisbach or Hazen–Williams equations can be used to compute the head loss.

Equations Describing Flow

Flow analysis in pipe networks is based on basic continuity and energy laws. The mass weight or volumetric flow rate into a junction must equal the mass weight or volumetric flow rate out of a junction, including demand. This relationship can be expressed as follows:

$$\Sigma Q_{out} - \Sigma Q_{in} = D \qquad \text{(Eq 5-17)}$$

Where:

Q_{out} = flow away from junction
Q_{in} = flow into junction
D = demand at the junction

In addition to continuity equations that must be satisfied, the head loss or energy equations must be satisfied. If one sums the head loss around a loop, the net head loss must equal zero. This relationship gives N equations of the form

$$\begin{aligned}
\sum_i h_i &= 0 \\
\sum_i^2 h_i &= 0 \\
&\vdots \\
\sum^N h_i &= 0
\end{aligned} \qquad \text{(Eq 5-18)}$$

Where:

N = the number of nonoverlapping (or independent) loops in the network, and the summation on i is over the pipes in the loops 1, 2...N

Using the exponential formula $h_i = C_i Q_i^{ni}$ can be written in terms of the flow rate, or

$$\begin{aligned}
\sum_i C_i Q_i^{ni} &= 0 \\
\sum_i^2 C_i Q_i^{ni} &= 0 \\
&\vdots \\
\sum^N C_i Q_i^{ni} &= 0
\end{aligned} \qquad \text{(Eq 5-19)}$$

The following equation describes a pipe network with junctions J, nonoverlapping loops N, and pipes P:

$$P = (J - 1) + N \qquad \text{(Eq 5-20)}$$

The flow rate in each pipe can be considered unknown, resulting in P unknowns. The number of independent equations that can be obtained for a network described above is $(J - 1) + N$. Therefore, the number of independent equations is equal in number to the unknown flow rates in the P pipes. The $(J - 1)$ continuity equations are linear and N energy (or head losses) equations are nonlinear. Many networks consist of hundreds of pipes, therefore systematic networks that use computers are needed for solving these systems of simultaneous quasi-linear equations.

Figure 5-5 is a simple network with one demand and one source. The equations that describe energy loss are shown below:

$$\begin{aligned}
h_{12} &= C_{12}Q_{12}{}^{n}\\
h_{23} &= C_{23}Q_{23}{}^{n}\\
h_{45} &= C_{45}Q_{45}{}^{n}\\
h_{56} &= C_{56}Q_{56}{}^{n}\\
h_{14} &= C_{14}Q_{14}{}^{n}\\
h_{25} &= C_{25}Q_{25}{}^{n}\\
h_{36} &= C_{36}Q_{36}{}^{n}
\end{aligned} \qquad \text{(Eq 5-21)}$$

Where:

h_{ij} = the head loss in link ij in the Hazen–Williams equation
C_{ij} = the coefficient for the pipe ij
Q_{ij} = the flow in pipe link ij
n = the exponent in the Hazen–Williams equation

The conservation of mass equations are

$$\begin{aligned}
Q &= Q_{12} + Q_{14}\\
Q_{12} &= Q_{25} + Q_{23}\\
Q_{23} &= Q_{36}\\
Q_{14} &= Q_{45}\\
Q_{56} &= Q_{45} + Q_{25}\\
Q &= Q_{56} + Q_{36}
\end{aligned} \qquad \text{(Eq 5-22)}$$

Between Eq 5-21 and Eq 5-22, there are 13 equations and 13 unknowns (6 unknown heads and 7 unknown flow rates).

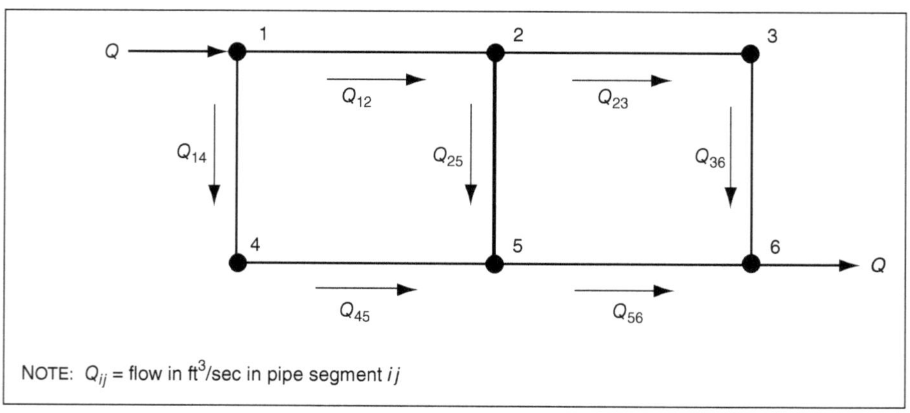

Figure 5-5. Steady-state network hydraulics

Heads at Junctions as Unknowns

If the head (either the total head or the piezometric head, since the velocity head is generally ignored in determining heads or pressure in pipe networks) at each junction is initially considered unknown instead of the flow rate in each pipe, the number of simultaneous equations that must be solved can be reduced in number (Jeppson 1976). The reductions in number of equations, however, are at the expense of not having some linear equations in the system.

To obtain the system of equations that contains the heads at the junctions of the network as unknowns, the $J - 1$ independent continuity equations are written as before. Thereafter, the relationship between the flow rate and head loss is substituted into the continuity equations. In writing these equations, it is convenient to use a double subscript for the flow rates. These subscripts correspond to the junctions at the ends of the pipe. The first subscript is the junction number from which the flow comes. The second number is the junction number to which the flow goes. Thus Q_{12} represents the flow in the pipe connecting junctions 1 and 2, assuming the flow is from junction 1 to junction 2. If the flow is actually in this direction, Q_{12} is positive and Q_{21} equals minus Q_{12}. Solving for Q from the exponential formula (using the double subscript notation) gives

$$Q_{ij} = \left(\frac{h_{ij}}{C_{ij}}\right)^{\frac{1}{n}} \qquad \text{(Eq 5-23)}$$

$$Q_{ij} = \left(\frac{(h_i - h_j)}{C_{ij}}\right)^{\frac{1}{n}}$$

Substituting Eq 5-23 into Eq 5-17 yields

$$C = \left[\Sigma\left(\frac{(h_i - h_j)}{C_{ij}}\right)^{\frac{1}{n}}\right]_{out} - \left[\Sigma\left(\frac{(h_i - h_j)}{C_{ij}}\right)^{\frac{1}{n}}\right]_{in} \qquad \text{(Eq 5-24)}$$

Equations of the form of Eq 5-24 are written at $J - 1$ junctions, producing a system of $J - 1$ nonlinear equations. Upon solving this nonlinear system of equations, the pressure at any junction j can be computed by subtracting the junction elevation from the head h_i and then multiplying this difference by the specific weight of water. To determine the flow rates in the pipes of the network, known heads are substituted into Eq 5-23.

Corrective Flow Rates Around Loops of Network Considered Unknowns

Since the number of junctions minus 1 (i.e., $J - 1$) will be less than the number of pipes in a network by the number of loops N in the network, the last set of head loss equations will generally be less in number than the system of mass balance equations (Jeppson 1976). The head loss equations will be nonlinear, but the flow equations will be linear with the exception of the energy equation, which will be nonlinear. Another approach is to write a series of equations that include a corrective flow rate in each loop as an unknown. This set of equations will be referred to as the ΔQ-equations. Since N basic loops occur in a network, the ΔQ-equations consist of N equations, all of which are nonlinear.

It is not difficult to establish an initial flow in each pipe that satisfies the continuity equations at each junction (which must also satisfy the jth junction continuity equation). These initial flow estimates generally will not simultaneously satisfy the N head loss equations and must, therefore, be corrected before they equal the true flow rates in the pipes. A flow rate adjustment or ΔQ can be added (accounting for sign) to the initially assumed flow in each pipe forming a loop of the network without violating continuity conditions at the junctions. This fact suggests establishing N energy (or head loss) equations around the N loops of the network in which the initial flow plus the correction rate ΔQ is used as the true flow rate in the head loss equations. Once these head loss equations are solved, the $J - 1$ continuity equations would remain satisfied as they initially were. The corrective loop flow rates ΔQ_i may be taken as positive or negative around the basic loops, but the sign convention must be consistent around any particular loop, and preferably in the same direction of all loops of the network. For purposes of this analysis, clockwise directions will be considered positive.

Let's assume an initial flow in a pipe of Q_{ij} (assume P pipes) and that the corrective loop flow rates will be denoted by ΔQ_{ij}. The energy equation for Figure 5-5 would be written as follows:

$$C_{12}(Q_{12} + \Delta Q_{12})^{n}{}_{12} + C_{25}(Q_{25} + \Delta Q_{25})^{n}{}_{25} + \quad \text{(Eq 5-25)}$$

$$C_{54}(Q_{54} + \Delta Q_{54})^{n}{}_{54} + C_{41}(Q_{41} + \Delta Q_{41})^{n}{}_{41} = 0$$

$$C_{23}(Q_{23} + \Delta Q_{23})^{n}{}_{23} + C_{36}(Q_{36} + \Delta Q_{36})^{n}{}_{36} +$$

$$C_{65}(Q_{65} + \Delta Q_{65})^{n}{}_{65} + C_{52}(Q_{52} + \Delta Q_{52})^{n}{}_{52} = 0$$

The first step in solving these equations is to provide initial estimates of the flow rate in each pipe that satisfy the continuity equations at the junctions and then solving for ΔQ_{ij}. Once the solution to ΔQ_{ij} is found, the flow rate for each pipe can be quickly determined. Several common techniques are used for solving the sets of equations discussed in this section. These include the linear method and the Newton–Raphson method, which will be discussed in the following sections.

Solution Techniques

Linear Theory

Linear theory transforms the N nonlinear loop equations into linear equations by approximating the head in each pipe by

$$h_{ij} = \{[C_{ij}Q_{ij}(\mathrm{e})]^{n-1}\}Q_{ij} \quad \text{(Eq 5-26)}$$

Where:

h_{ij} = the head loss in link ij in the Hazen–Williams equation
C_{ij} = the coefficient for the pipe ij
Q_{ij} = the flow in pipe link ij
n = the exponent in the Hazen–Williams equation
$Q_{ij}(\mathrm{e})^{n-1}$ = the initial estimate of flow in pipe ij

A coefficient is defined for each pipe as the product of C_{ij} multiplied by $[Q_{ij}(e)]^{n-1}$, an initial estimate of the flow rate in that pipe (Jeppson 1976). Combining the equation set represented by Eq 5-26 with the $J-1$ junction continuity equations provides a system of L linear equations that can be solved by linear algebra. After a series of iterations, the $Q_{ij}(e_k)s$ where $k = 1....K$ will converge to equal the $Q_{ij}s$. Pumps, flow control valves, pressure-sustaining valves, and pressure-reducing valves can be incorporated with the linear theory method.

Newton–Raphson Method

The Newton–Raphson technique is widely applied and can be used to solve any of the following types of equations describing flow in pipes: the flow rate in each pipe, the head at each junction, and the correct flow rate around each loop. In applying the Newton–Raphson method, a solution to the equation $F(x) = 0$ is obtained by applying the iterative formula $x^{m+1} = x^m - F(x^m)/F^1(x^m)$ (Jeppson 1976).

Network Characterization

Engineering analysis of water distribution systems is frequently limited to the solution of the hydraulic network problem, i.e., given the physical characteristics of a distribution system modeled as a node–link network and the demands at nodes (junctions where network components connect to one another), the flows in links (a connection of any two nodes), and head at all nodes of the network are determined. This problem is formulated as a set of simultaneous nonlinear equations, and a number of well-known solution methods exist, many of which have been coded as computer programs known as *hydraulic network models* (Cesario 1995).

Mathematical methods for analyzing the flow and pressure in networks have been in use for more than 50 years. They are generally based on well-accepted hydrodynamic equations. Computer-based models for performing this type of analysis were first developed in the 1950s and 1960s and greatly expanded and made more available in the 1970s and 1980s. Currently, dozens of such models are readily available on systems ranging from personal computers to supercomputers.

Hydraulic models were developed to simulate flow and pressures in a distribution system either under steady-state conditions or under time-varying demand (extended-period simulation) and operational conditions. Hydraulic models may also incorporate optimization components that aid the user in selecting system parameters that result in the best match between observed system performance and model results.

Hydraulic Transients

Many recently developed water quality models have incorporated surge analysis, which is a frequently overlooked problem in drinking water distribution systems. The intent is to protect the system from excessive transient or water hammer conditions. When velocities in a pipe system change so rapidly that the elastic properties of the pipe and liquid must be considered, a phenomenon known as water hammer occurs. In order to understand this phenomenon, we can consider a simple example.

Assume that steady flow is taking place in the pipe at velocity V. The head everywhere in the pipe is H in the absence of friction. Assume we can completely close the valve instantaneously, and at the valve the water velocity will suddenly be forced to zero. As we can show by hydraulic principles, the head at the valve will abruptly increase by the following amount:

$$H = aV/g \tag{Eq 5-27}$$

where a is the pressure wave propagation speed and g is the gravitational constant.

In equation 5-27

$$a = \frac{V_s}{\sqrt{1+\dfrac{kd}{Ee}}} \tag{Eq 5-28}$$

in which k is the modulus of compression of water, d is the internal pipe diameter, E is the modulus of elasticity of the pipe, e is the pipe wall thickness, and V_s is the velocity of sound in water (Larock et al. 2000).

The increased head creates two other changes at the valve. The pressure increase slightly enlarges the pipe and also increases the density of the fluid. How much the pipe expands depends on the diameter and thickness of the pipe and on the compressibility of the pipe material and the liquid. Normally this change is less than one-half percent.

The rise in pressure head causes a sharp-fronted pressure wave to propagate upstream at speed a toward the reservoir. The wave front reaches the reservoir V_a seconds after valve closure. At that instant, the velocity is zero throughout the pipe, the pressure head is everywhere $H + \Delta H$, the pipe is enlarged, and the fluid is compressed. The fluid in the pipe near the reservoir connection is not in equilibrium since the reservoir pressure head is only H. Hence, fluid begins to flow toward the region of lower head (the reservoir) as the distended pipeline forces flow in that direction. In the absence of friction, this leftward velocity is equal in magnitude to the original steady velocity as it is driven by the same head increment ΔH, and the source of the liquid for this flow is the compressed liquid that is stored in the enlarged pipe cross section under the increased pressure head.

The process continues to evolve with time. At time $2L/a$ after the beginning of the surge event, the pressure throughout the pipe has returned to its original value, but with the velocity reversed from its original direction. At this instant, the store of compressed liquid is exhausted, and the pressure wave appears to undergo a reflection. That is, the pressure head drops an amount ΔH below the original constant head, and this pressure drop and the closed valve cause the velocity behind the wave front to return to zero. Behind this negative wave, the pipe cross section shrinks and the liquid expands. By time $3L/a$, this negative wave has reached the reservoir, and the velocity is zero everywhere. However, the pressure head at the reservoir is again not in equilibrium with the reservoir head, so fluid is drawn from the reservoir into the pipe at velocity V. Behind the new advancing wave, the head is in equilibrium with the reservoir head. At time $4L/a$, the wave has reached the valve; at this instant, all variables have returned to the original steady state that existed before the valve was closed. This time interval is one full cycle in a hydraulic transient. In the absence of friction, these waves continue without abating.

The disturbance is not initially noticed everywhere; instead the fact that the valve is suddenly closed travels at the finite wave propagation speed a to the other locations and arrives at the midpoint in the pipe in half the time that is needed to traverse the entire pipe. The sudden increase in head at the valve, remains in place at the valve until one round-trip wave travel time has elapsed, and only then is the returning wave from the reservoir able to reduce the head there. In fact, the head increment ΔH need not be created instantaneously for the full incremental head to be present at the valve; it is only necessary for a set of incremental increases in head that sum to ΔH to be developed at the valve in a time interval that is less than this round-trip travel time $2L/a$ for the full increment in head ΔH to be present at the valve for a while. It is clear that, owing to the manner in which a valve decreases the discharge in a pipeline by creating large head losses, it may be necessary to close a valve in a time that is much greater that $2L/a$ if we are to avoid the creation of large transient pressures.

Surge or transient analysis is essential to estimate the worst-case scenarios in the distribution system. Transients occur whenever flow conditions are altered but are most severe when rapid changes occur, such as those resulting from power failure, emergency valve operations, or firefighting. These changes are generally characterized by fluctuating pressures and velocities and are critical precisely because pressure variations can be of high magnitude, possibly large enough to break or damage pipes or other equipment or to greatly disrupt delivery conditions. Regions that are particularly susceptible to transients are high-elevation areas, locations with either low or high static pressures, and regions far removed from overhead storage (Boulos et al. 2006, Wood et al. 2005).

Application of Models

The following steps should be followed in applying network models (Clark et al. 1988):

1. *Model selection.* Define model requirements and select a model (hydraulic and/or water quality) that fits your requirements, style, budget, and so on.
2. *Network representation.* Accurately represent the distribution system components in the model.
3. *Calibration.* Adjust model parameters so that predicted results adequately reflect observed field data.
4. *Verification.* Independently compare model and field results to verify the adequacy of the model representation.
5. *Problem definition.* Define the specific design or operational problem to be studied and incorporated (i.e., demands and system operation) into the model.
6. *Model application.* Use the model to study the specific problem or situation.
7. *Display and analysis of results.* Following the model application, display and analyze the results to determine how reasonable the results are. Finally, translate the results into a solution to the problem.

As mentioned in previous chapters, all water distribution system models represent the network as a series of connected links and nodes. Links are defined by the two end nodes and generally serve as conveyance devices (e.g., pipes), while nodes represent point components (e.g., junctions, tanks, wells, and treatment plants). Pumps and valves may be represented by links or nodes. Water demands are aggregated and assigned to nodes; an obvious simplification of real-world situations in which individual house taps are distributed along a pipe rather than at junction nodes.

A minimal level of skeletonization should include all pipes and features of major concern. Nodes are usually placed at pipe junctions, where pipe characteristics may change in diameter, C value (roughness), or material of construction. Nodes may also be placed at locations of known pressure or where pressure valves are desired.

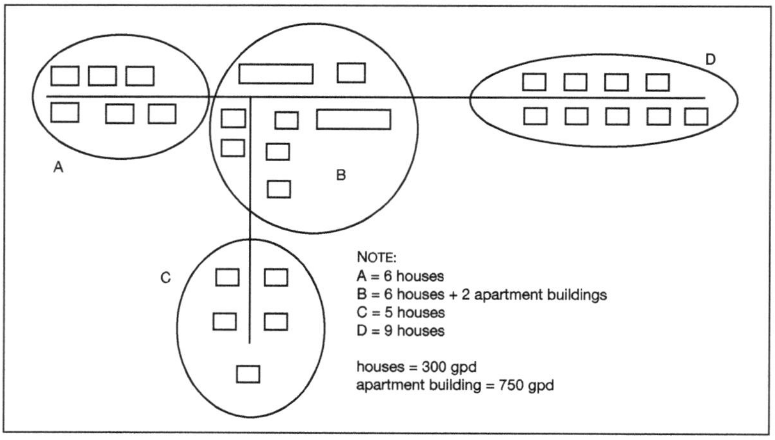

Figure 5-6. Usage by counting buildings

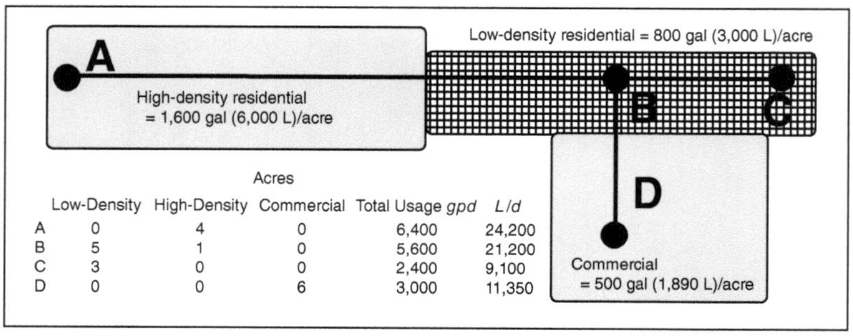

Figure 5-7. Water usage by land use

Network components include pipes, reservoirs, pumps, pressure-relief valves, control valves, altitude valves, check valves, and pressure-reducing valves. For steady-state modeling, a key input is information on water consumption. A universal adjustment factor should be used so that total usage in the model corresponds to total production. Figures 5-6 and 5-7 illustrate two methods of estimating water use.

Basic Hydraulic Model Input Characterization

Building a network model, particularly if a large number of pipes is involved, is a complex process. The following categories of information are needed to construct a hydraulic model:

- Characteristics of the pipe network components (pipes, pumps, tanks, valves).
- Water use (demands) assigned to nodes (temporal variations required in EPS).
- Topographic information (elevations assigned to nodes).
- Control information that describes how the system is operated (e.g., mode of pump operation).
- Solution parameters (e.g., time steps, tolerances as required by the solution techniques).

Commonly used methods for these inputs are briefly described in the following subsections.

Pipe Network Inputs

Construction of the pipe network and its characteristics may be done manually or through use of existing spatial databases stored in GIS or CAD packages. Most commonly, GIS or CAD packages are used in this process and are described in more detail later in this book. The initial step in constructing a network model is to identify pipes to be included in the model. Nodes are usually placed at pipe junctions, or at major facilities (tanks, pumps, control valves), or where pipe characteristics change in diameter, *C* value (roughness), or material of construction. Nodes may also be placed at locations of known pressure, at sampling locations, or at locations where water is used (demand nodes). The required pipe network component information includes the following:

- Pipes (length, diameter, roughness factor).
- Pumps (pump characteristic curve).
- Valves (setting and diameter).
- Tanks (cross section information, initial, minimum and maximum water levels).

Water Demand Inputs

Water consumption or water demand is the driving force behind the operation of a water distribution system. Any location at which water leaves the system can be characterized as a demand on the system. The water demands are aggregated and assigned to nodes, which represent an obvious simplification of real-world situations in which individual house taps are distributed along a pipe rather than at junction nodes. It is important to be able to determine the amount of water being used, where it is being used, and how this usage varies with time (Walski et al. 2003). Demand may be estimated by a count of structures of different types using a representative consumption per structure, by meter readings and the assignment of each meter to a node, and by general land use. A universal adjustment factor should be used to account for losses and other unaccounted-for water usage so that total usage in the model corresponds to total production.

In order to use a model in the EPS mode, information on temporal variations in water usage over the period being modeled is required. Spatially different temporal patterns can be applied to the individual network nodes. The best available information should be used for developing temporal patterns in order to make EPS most effective. For example, some users may have continuous water metering data, while others may use literature values as a first approximation for estimating residential temporal patterns. Temporal patterns also vary with climate. For example, lawn watering in summer months will cause a spike in usage of water during that time period. SCADA systems can be used to estimate active tank volumes and pump station changes and therefore can be used to estimate systemwide temporal patterns.

A typical hierarchy for assigning demands includes the following:

- *Baseline demands:* Baseline demands usually correspond to consumer demands and unaccounted-for water associated with average day conditions. This information is often acquired from a water utility's existing records, such as customer meter and billing records. Although the spatial assignment of these demands is extremely important and should include the assignment of customer classes such as industrial, residential, and commercial use, actual metering data should be used when available.
- *Seasonal variation:* Water use typically varies over the course of the year with higher demands occurring in warmer months. When developing a steady-state model, the baseline (average day) demand can be modified by multipliers in order to reflect

other conditions such as maximum day demand, peak-hour demand, and minimum day demand.

- *Fire demands:* Water provided for fire services can be the most important consideration in developing design standards for water systems. Typically, a system is modeled corresponding to maximum-use conditions, with needed fire flow added to a single node at a time. It is not uncommon for a requirement that multiple hydrants be flowing simultaneously.
- *Diurnal variation:* All water systems are unsteady because of continuously varying demands. It is important to account for these variations in order to achieve an adequate hydraulic model. Diurnal varying demand curves should be developed for each major consumer class or geographic zones within a service area. For example, diurnal demand curves might be developed for industrial establishments, commercial establishments, and residences. Large users such as manufacturing facilities may have unique usage patterns.

Topographical Inputs

Hydraulic models use elevation data to convert heads to pressure. Actual pipe elevations should be used to establish the correct hydraulic grade line. Elevations are assigned to each node in a network where pressure information is required. Various techniques are used to determine elevation information including the following:

- *Topographical maps:* Paper topographical maps produced by the United States Geological Survey (USGS) or other local agencies may be used to manually interpolate elevations for nodes. The relative accuracy depends upon the degree of topography in the area, the contour elevations on the map, and the manual takeoff methods used.
- *Digital elevation models (DEMs):* USGS and other agencies produce digital files containing topographical information. When used with various software tools, elevation information can be directly interpolated and assigned to nodes based on the coordinates of the nodes. The accuracy of this process depends upon the degree of detail in the DEM.
- *Global positioning systems (GPSs) or other field survey methods:* Standard field surveying techniques or modern surveying methods employing a GPS satellite can be used to measure elevations at nodes. The modern GPS units can calculate elevation by using four or more satellites. However, elevation is the most difficult calculation for a GPS unit, and depending on the location surveyed, it may be prone to significant error.

Model Control Inputs

In order to apply an EPS model, it is necessary to define a set of rules that tells the model how the water system operates. This may be as simple as specifying that a particular pump operates from 7:00 a.m. to 10:00 a.m. each day. Alternatively, it may be a set of complex "logical controls" in which operations such as pump off/on, pump speed, or valve status are controlled using Boolean operators (including if-then-else logic) for factors such as tank water levels, node pressures, system demand, and time of day (Grayman and Rossman 1994). For water systems that operate automatically based on a set of rules, determination of these rules is quite straightforward. For manual systems, the rules must be determined by interviews with system operators.

Extended-Period Simulation (EPS) Solution Parameters

Solution techniques used to iteratively solve the set of nonlinear equations typically have various global parameters that control the solution technique. These parameters may be time-step lengths for EPS runs or tolerance factors that tell the model when a solution is considered to have converged. The user must specify the values for the solution parameters or (as is frequently done) accept the default values that are built into the software products. The specific solution parameters vary among solution techniques and specific software products.

General Criteria for Model Selection and Application

The initial step in modeling is to define the basic scope and needs of the modeling process and to select an appropriate software package that will satisfy both the specific needs of the current project and likely future needs. Factors that may enter into the selection of a software package include:

- Technical features.
- Training/support and manuals.
- User interface.
- Integration with other software (such as GIS, CAD).
- Compatibility with EPANET.
- Cost.
- Response from existing users.

A summary of major available hydraulic–water quality modeling software is provided in a following section. Once a suitable model has been selected, the following steps should be followed in applying network models (Clark and Grayman 1998):

- Develop the basic network model.
- Calibrate and validate the model.
- Establish clear objectives and apply the model in a manner to meet the objectives.
- Analyze and display the results.

Developing a Basic Network Model

The basic network model inputs should be first characterized using the techniques described previously. The model should be developed based on accurate, up-to-date information. Information should be entered carefully and checked frequently. Following the entry of the data, an initial run of the model should be made to check for reasonableness.

Model Calibration and Validation

Calibration is an integral aspect of the art of modeling water distribution systems. Model calibration is the process of adjusting model input data (or, in some cases, model structure) so that the simulated hydraulic and water quality output sufficiently mirrors observed field data. Depending on the degree of accuracy desired, calibration can be difficult, costly, and time-consuming. The extent and difficulty of calibration are minimized by developing an accurate set of basic inputs that provide a good representation of the real network and its components.

A traditional technique for calibration is the use of fire-flow tests. In a fire-flow test, the system is stressed by opening hydrants to increase flows in small parts of the system. This results in increased head loss in pipes in the vicinity of the test. Pressures and flow are then measured in the field. Model parameters, such as roughness factors (C), demands, and valve positions, are adjusted so that the model adequately reflects the field data. Another common calibration technique is to measure predicted tank/reservoir levels derived from computer simulations against actual tank levels during a given period of record. For example, using water level, pressure, or flow data from SCADA systems or from on-line pressure and tank-level recorders, model parameters (such as roughness, water demands, and pump controls) can be adjusted in the simulation model until the model results match the actual tank level and other continuous information for the defined criteria. The resulting optimal parameter values should be checked to ensure that the values are realistic. Sophisticated commercial hydraulic models may also incorporate optimization components that aid the user in selecting system parameters resulting in the best match between observed system performance and model results (Walski et al. 2003).

Model validation is the step that follows calibration and uses an independent field data set to verify that the model is well calibrated. In the validation step, the calibrated model is run under conditions differing from those used for calibration and the results compared to field data. If the model results closely approximate the field results (visually) for an appropriate time period, the calibrated model is considered to be validated. Significant deviations indicate that further calibration is required. Automatic calibration procedures should be used carefully. For example it may seem that the *C* factor needs to be adjusted when in actuality a valve closed. A variety of calibration and validation techniques suitable to both large and small water utilities are discussed later.

Another rigorous methodology for calibration and validation is the use of tracers. Concentrations of naturally occurring materials or added chemical tracers may be measured in the field and the results used to calibrate hydraulic and water quality models.

Establishing Objectives and Model Application

Prior to applying the model, the specific modeling objectives should be clearly established. The objectives may include specification of particular water demand and operational modes. Based on these specifications, a series of scenarios can be defined and the model applied appropriately. Some software products contain a scenario manager that helps the user to define and manage a large number of specific model runs. Additional scenarios can be developed in order to test the sensitivity of the system to variations in model parameters that are not known with certainty.

Analysis and Display of Results

Water distribution system models generate a large amount of output. The amount of calculated information increases with increasing model size and, for EPS, the duration of the model run. Modern water distribution system analysis software typically provides a range of graphical and tabular displays that help the user wade through the large amount of output data so that it may be efficiently analyzed. Figures 3-2, 3-3, and 3-4 contain examples of various graphical and tabular outputs generated by the EPANET software. These outputs represent a small subset of types of graphics generated by most modeling software. The output should be analyzed to ensure that the model is operating properly and to extract the information required in order to analyze the specific problem being studied.

Modeling Temporal Variations

We know that most phenomena and behavior vary over time in the real world. Models that assume no variation over time are referred to as steady-state models, which provide a snapshot at a single point in time. In distribution system models, steady-state models assume that demands are constant and operations are constant (e.g., tank water-level elevations remain constant and pumps are either on or off at a constant speed). Though such assumptions may not be valid over long periods of time, steady-state models can provide some useful information concerning the behavior of a network under various representative conditions, including fire demand, nighttime low demands, and so on.

Models that allow for variations over time are referred to as temporally dynamic models. Most network distribution hydraulic models incorporate temporal variation by stringing together a series of steady-state solutions and refer to this method as extended-period simulation (EPS). For example, in the EPS mode, demand at a node may be assumed to be 100 gpm (378 L/min) from 2:00 p.m. to 3:00 p.m. and then 150 gpm (567 L/min) from 3:00 p.m. to 4:00 p.m. Thus, one steady-state solution may be assumed from 2:00 p.m. until 3:00 p.m., and a different solution may be calculated from 3:00 p.m. to 4:00 p.m. Further temporal dynamics may be incorporated by checking the water level in a tank and, if the water level reaches a maximum allowable level at 3:20 p.m., then another steady-state solution is started at 3:20 p.m. with the tank discharging instead of filling. Though the EPS solution does introduce some approximations and totally ignores the transient phenomena resulting from sudden changes, such as a pump being turned on, these more refined assumptions are generally not considered significant for most distribution system studies. Figure 5-8 illustrates systemwide demand variation.

Some of the characteristics of the EPS mode follow:

- Requires information on temporal variations in water usage over the period being modeled.
- Permits temporal patterns to be defined for groups of nodes (spatially different patterns can be applied to a given node).
- Uses the best available information on temporal patterns that can be estimated (for example, some users have continuous meters).
- Can sometimes use literature values for a first guess at residential patterns (which may vary with climate).
- Can use analysis of information from supervisory control and data acquisition (SCADA) to estimate systemwide temporal pattern.

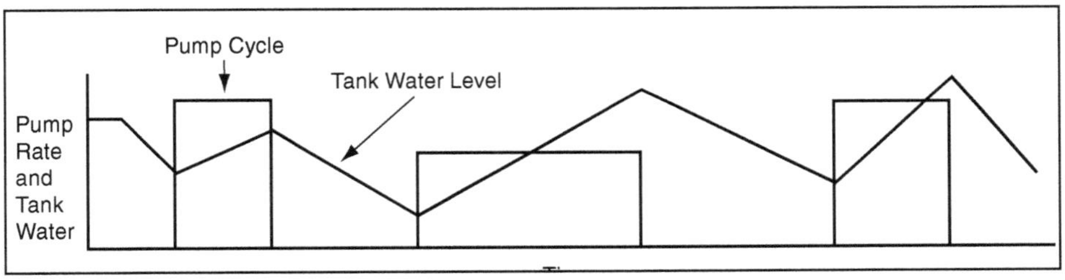

Figure 5-8. Systemwide temporal analysis

Fire-Flow Provisions

As has been mentioned, a major function of most distribution systems is to provide adequate standby fire flow, the standards for which are governed by the National Fire Protection Association (Cote et al. 2008). Fire-flow requirements for a single family house may vary from 750 to 1,500 gpm; for multifamily structures, the values range from 2,000 to 5,000 gpm; for commercial structures, the values range from 2,000 to 10,000 gpm; and for industrial structures, the values range from 3,000 to over 10,000 gpm (AWWA 2008). The duration for which these fire flows must be sustained normally range from 3 to 8 hr. The minimum allowable residual pressure during a fire is normally 20 psi, or 14 m. In order to satisfy this need for adequate standby capacity and pressure, most distribution systems use standpipes, elevated tanks, and large storage reservoirs.

The sizing of water mains is based on a variety of water quantity requirements such as projected demand, pressure requirements, velocity requirements, and pipe carrying capacity, in addition to fire protection requirements set by the Insurance Services Office, which represents the fire insurance underwriters in the United States (AWWA 2010, Von Huben 2005). The minimum flow that the water system can sustain for a specific period of time governs its fire protection rating, which then is used to set the fire insurance rates for the communities that are served by the system. As a consequence, fire flow governs much of the design of a distribution system, especially for smaller systems. A study conducted by the American Water Works Association Research Foundation (AwwaRF) confirmed the impact of fire-flow capacity on the operation of, and the water quality in, drinking water networks (Snyder et al. 2002). It found that although the amount of water used for firefighting is generally a small percentage of the annual water consumed, the required rates of water delivery for firefighting have a significant and quantifiable impact on the size of water mains, tank storage volumes, water age and capital, and operating and maintenance costs. Generally, nearly 75 percent of the capacity of a typical drinking water distribution system is devoted to firefighting (Walski et al. 2001). Some of the proprietary models available on the market have the capacity to incorporate flow requirements in network designs.

Snyder et al. (2002) recommended that water systems evaluate existing storage tanks to determine if modification or elimination of the tanks were feasible. Water-efficient fire-suppression technologies exist that use less water than conventional standards. In particular, the universal application of automatic sprinkler systems provides the most proven method for reducing loss of life and property caused by fire, while at the same time providing faster response to the fire and requiring significantly less water than conventional firefighting techniques. Snyder et al. (2002) also recommended that the universal application of automatic fire sprinklers be adopted by local jurisdictions for homes as well as in other buildings.

Flushing Water Mains

Flushing is important for both maintaining and recovering the integrity of distribution systems because it is the primary means by which to remove contaminated water from the water system. It is used in association with the cleaning and disinfection of water mains following pipe installation, repair, and replacement. Flushing is common in areas with repeat customer complaints about color, taste, or odor; in dead ends mains; and in storage facilities. Its importance with respect to maintaining hydraulic integrity is that flushing removes accumulated sediment and corrosion products that reduce the water carrying capacity of the pipe.

Deposits form over time in water distribution systems via precipitation, microbiological activity, and corrosion, which will increase the hydraulic roughness and reduce the cross-sectional area and hydraulic carrying capacity of the mains. A loss in carrying capacity can lead to low pressures as well as to the inability of the system to deliver fire flows, especially during high-demand periods. Flushing the water mains is one tool utilities can use to maintain water quality and water capacity because it improves the flow of water through the distribution system, allowing it to work at capacity.

Flushing is performed by isolating sections of the water distribution system and opening fire hydrants (or flushing valves) to cause a large volume of flow (large velocity) to pass through the isolated pipelines so that a scouring action is created. Water is then discharged through a hydrant, which in turn removes any material buildup from the pipe. When flushing pipes, it is important to ensure that the flushing velocity is sufficient to suspend loose sediments. Flushing should continue until the water has cleared and disinfectant residual has reached normal expected levels. To minimize any negative environmental impacts (as flushed water may be high in suspended solids and other contaminants that can harm water bodies), flushed water is normally discharged into sanitary sewers or stormwater management facilities such as detention ponds. It is important to optimize a flushing program in a water system, as excessive flushing can waste significant volumes of water (NRC 2006). As with fire-flow analysis, many proprietary water quality models incorporate flushing as a standard routine.

References

American Water Works Association (AWWA). 2004. *AWWA Manual M32, Computer Modeling of Water Distribution Systems.* 2nd ed. Denver, Colo.: AWWA.

AWWA. 2008. *AWWA Manual M31: Distribution System Requirements for Fire Protection.* 4th ed. Denver, Colo.: AWWA.

AWWA. 2010. *Water Transmission and Distribution.* 4th ed. Denver, Colo.: AWWA.

Boulos, P.F., K.E. Lansey, and B.W. Karney. 2006. *Comprehensive Water Distribution Systems Analysis Handbook for Engineers and Planners.* 2nd ed. Broomfield, Colo.: MWH Soft, 660 pp.

Brandt, M., J. Clement, J. Powell, R. Casey, D. Holt, N. Harris, and C.T. Ta. 2004. *Managing Distribution Retention Time to Improve Water Quality: Phase I.* Denver, Colo.: AwwaRF.

Cesario, L. 1995. *Modeling Analysis and Design of Water Distribution Systems.* Denver, Colo.: AWWA.

Clark, R.M., W.M. Grayman, R.M. Males, and J. Coyle. 1988. Modeling Contaminant Propagation in Drinking Water Distribution Systems. *Journal of Water Supply: Research and Technology–AQUA*, 37(3):137–151.

Clark, R.M., and W.M. Grayman. 1998. *Modeling Water Quality in Drinking Water Systems.* Denver, Colo.: AWWA.

Cote, A.E., J.R. Hall, P. Powell, C.C. Grant, and R.E. Solomon. 2008. *Fire Protection Handbook 2008.* Quincy, Mass: National Fire Protection Association.

Cross, H. 1936. *Analysis of Flow in Networks of Conduits or Conductors.* Univ. of Ill. Eng. Experiment Station Bulletin 286. Urbana: University of Illinois.

Grayman, W.M., and L.A. Rossman. 1994. Integrating a Network Model With an Expert System. In *Proc. AWWA Computer Conference.* Denver, Colo.: AWWA.

Jeppson, R.W. 1976. *Analysis of Flow in Pipe Networks.* Ann Arbor, Mich.: Ann Arbor Science.

Larock, B.E., R.W. Jeppson, and G.Z. Watters. 2000. *Hydraulics of Pipeline Systems.* Boca Raton, Fla.: CRC Press, 12–26.

National Research Council (NRC). 2006. *Drinking Water Distribution Systems: Assessing and* Reducing *Risks.* Washington, D.C.: National Academies Press, 15–46.

Snyder, J.K., A.K. Deb, F.M. Grablutz, S.B. McCammon, W.M. Grayman, R.M. Clark, D.A. Okun, S.M. Tyler, and D. Savic. 2002. *Impacts of Fire Flow on Distribution System Water Quality, Design, and Operation.* Denver, Colo.: AwwaRF.

Von Huben, H., ed. 2005. *Water Distribution Operator Training Handbook.* 3rd ed. Denver, Colo.: AWWA.

Walski, T.M., D.V. Chase, and D.A. Savic. 2001. *Water Distribution Modeling.* Waterbury, Conn.: Haestad Press.

Walski, T.M., D.V. Chase, D.A. Savic, W. Grayman, S. Beckwith, and E. Koelle. 2003. *Advanced Water Distribution Modeling and Management.* Waterbury, Conn.: Haestad Press.

Wood, D.J., S. Lingireddy, and P.F. Boulos. 2005. *Pressure Wave Analysis of Transient Flow in Pipe Distribution Systems.* Broomfield, Colo.: MWH Soft, 213 pp.

Chapter 6

Water Quality Modeling Principles

Introduction

Water quality models traditionally use the output of hydraulic models in conjunction with additional inputs (described later in this section) to predict the temporal and spatial distribution of a variety of constituents within a distribution system. These constituents include:

- The fraction of water originating from a particular source.
- The age of water (e.g., duration since leaving the source).
- The concentration of a nonreactive constituent or tracer compound either added to or removed from the system (e.g., chloride or fluoride).
- The concentration of a reactive compound including the concentration of a secondary disinfectant with additional input of its loss rate (e.g., chlorine or chloramines) and the concentration of disinfection by-products with their growth rate (e.g., THMs).

Various water quality processes are occurring in water distribution systems that can lead to introduction of contaminants and water quality transformations as water moves through the distribution system. Cross connections, failures at the treatment barrier, and transformations in the bulk phase can all degrade water quality. Corrosion, leaching of pipe material, biofilm formation, and scour can occur at the pipe wall to degrade water quality. Bacteriological quality changes may cause aesthetic problems involving taste and odor development, discolored water, and other adverse impacts (NRC 2006).

In addition to the basic hydraulic modeling equations discussed previously, water quality models employ various mathematical equations that are based on conservation of constituent mass. These models represent the following phenomena occurring in a distribution system (Rossman 2000):

- *Advective transport of mass within pipes:* A dissolved substance will travel down the length of a pipe with the same average velocity as the carrier fluid while at the same time reacting (either growing or decaying) at some given rate. Longitudinal dispersion is not an important transport mechanism in turbulent flow, which is normal inside transmission mains under most operating conditions. It may, however, be an important factor in dead-end pipes or in low and intermittent flow scenarios.
- *Mixing of mass at pipe junctions:* All water quality models assume that, at junctions receiving inflow from two or more pipes, the mixing of fluid is complete and instantaneous. Thus, the concentration of a substance in water leaving the junction is simply the flow-weighted sum of the concentrations in the inflowing pipes.
- *Mixing of mass within storage tanks:* Most water quality models assume that the contents of storage tanks are completely mixed.

- *Reactions within pipes and storage tanks:* While a substance moves down a pipe or resides in storage, it can undergo reaction. The rate of reaction, measured in mass reacted per volume of water per unit of time, will depend on the type of water quality constituent being modeled. Some constituents, such as fluoride, do not react and are termed "conservative." Other constituents, such as chlorine residual, decay with time; while the generation of DBPs, such as THMs, may increase over time. Some constituents, such as chlorine, will react with materials both in the bulk liquid phase and at the liquid–pipe wall boundary.

Water quality models represent these phenomena (transport within pipes, mixing at junctions and storage tanks, and reaction kinetics in the bulk liquid phase and at the liquid–pipe wall boundary) with a set of mathematical equations. These equations are then solved under an appropriate set of boundary and initial conditions to predict the variation of water quality throughout the distribution system.

Several solution methods are available for dynamic water quality models (Rossman and Boulos 1996). All of these methods require that a hydraulic analysis be run first to determine how flow quantities and directions change from one time period to another throughout the pipe network. The water quality constituent is subsequently routed through each pipe link and then mixed at downstream nodes with other inflows into the node. For nonconservative substances, concentrations are continuously adjusted to accommodate the decay or growth of the constituent with time. This concentration is then released from the node into its outflowing pipes. This process continues for all pipes and for the duration of the simulation.

The methods described above are also applied when modeling water age and source-tracing in water quality models. Water age is equivalent to modeling a reactive constituent that ages and combines linearly. For example, for every hour that a "packet" of water spends in a tank, its age will increase by one hour. Additionally, combining a volume of water that is four days old with a similar volume of water that is eight days old will result in an average age of six days. When modeling the fraction of water coming from a designated source (source tracing), this parameter is modeled as a conservative substance and is linearly combined. For example, combining a volume of water that is entirely from the designated source with a similar volume of water from a different source will provide a mixed volume calculated as 50 percent from the designated source.

Water Quality Models Inputs and Application

In addition to the basic hydraulic model inputs described in chapter 5, the water quality models require the following data elements to simulate the behavior in a distribution system:

- *Water quality boundary conditions:* A water quality model requires the quality of all external inflows to the network and the water quality throughout the network be specified at the start of the simulation. Data on external inflows can be obtained from existing source monitoring records when simulating existing operations or could be set to desired values to investigate operational changes. Initial water quality values can be estimated based on field data. Alternatively, best estimates can be made for initial conditions. Then the model is run for a sufficiently long period of time under a repeating pattern of source and demand inputs so that the initial conditions, especially in storage tanks, do not influence the water quality predictions in the distribution system. The water age and source tracing options only require input from the hydraulic model.
- *Reaction rate data:* For nonconservative substances, information is needed on how the constituents decay or grow over time. Modeling the fate of a residual disinfectant is

one of the most common applications of network water quality models. The two most frequently used disinfectants in distribution systems are chlorine and chloramines (a reactant of chlorine and ammonia). Free chlorine is more reactive than chloramine, and its reaction kinetics have been studied more extensively. Studies have shown that two separate reaction mechanisms occur for chlorine decay, one involving reactions within the bulk fluid and another involving reactions with material on or released from the pipe wall (Vasconcelos et al. 1997). Bulk decay is typically represented as a first-order exponential decay function with a single decay coefficient specified to represent the decay over time. In some circumstances, this function does not adequately represent the observed decay characteristics, and more complex formulations may be used to describe the decay. Wall reaction represents the disinfectant decay caused by contact with oxidizable substances at the pipe wall, such as corrosion products or biofilm. The most widely used approach for representing wall demand considers two interacting processes: transport of the disinfectant from the bulk flow to the wall and interaction with the wall (Rossman et al. 1994). Recent studies have suggested that this formulation may not adequately represent the actual wall demand processes and that further research is needed (Clark et al. 2005, Grayman et al. 2002, DiGiano and Zhang 2004). Little study has been conducted on the nature of the wall reaction in chloraminated systems. A limited amount of modeling of the growth of DBPs (most notably THMs) has been performed assuming an exponential growth approaching a maximum value corresponding to the THM formation potential. Both the formation potential and the growth rate constant must be specified in this type of model (Clark et al. 1996). Extensive research has been done on biofilm formation in distribution systems, and this has led to the development of several theoretical models of this phenomenon (Powell et al. 2004). However, these models are generally quite complex involving many parameters that are difficult to determine, and thus are not ready for inclusion in a general water distribution system model.

Water Quality Fundamentals

In order to model water quality within distribution systems, the concentration of a particular substance must be calculated as it moves through the system from various points of entry (e.g., treatment plants) and on to water users. This movement is based on three principles:

1. Conservation of mass within differential lengths of pipe.
2. Complete and instantaneous mixing of the water entering pipe junctions.
3. Appropriate kinetic expressions for the growth or decay of the substance as it flows through pipes and storage facilities.

This change in concentration of a substance can be expressed by the following differential equation:

$$\frac{\partial C_{ij}}{\partial_t} = -v_{ij}\frac{\partial C_{ij}}{\partial_x} + k_{ij}C_{ij} \quad \text{(Eq 6-1)}$$

Where:

C_{ij} = substance concentration (g/m^3) at position x and time t in the link between nodes i and j

v_{ij} = flow velocity in the link (equal to the link's flow divided by its cross-sectional area) (m/sec)

k_{ij} = rate at which the substance reacts within the link (sec^{-1})

This equation shows that the rate at which the mass of material changes within a small section of pipe equals the difference in mass flow into and out of the section plus the rate of reaction within the section. It is assumed that the velocities in the links are known beforehand from the solution to a hydraulic model of the network. In order to solve Eq 6-1, we need to know C_{ij} at $x = 0$ for all times (a boundary condition) and a value for k_{ij}.

Eq 6-2 represents the concentration of material leaving the junction and entering a pipe.

$$C_{ij}|_{x=0} = \frac{\sum_k Q_{ki} C_{kj} \mid_{x=L}}{\sum_k Q_{kj}} \qquad \text{(Eq 6-2)}$$

Where:

$C_{ij}|_{x=0}$ = the concentration at the start of the link connecting node i to node j in mg/L (i.e., where $x = 0$)

$C_{kj}|_{x=L}$ = the concentration at the end of a link, in mg/L

Q_{ki} = flow from k to i

Eq 6-2 implies that the concentration leaving a junction equals the total mass of a substance mass flowing into the junction divided by the total flow into the junction.

Storage tanks can be modeled as completely mixed, variable volume reactors in which the change in volume and concentration over time are as follows:

$$\frac{dV_s}{dt} = \sum_k Q_{ks} - \sum_i Q_{sj} \qquad \text{(Eq 6-3)}$$

$$\frac{d(V_s C_s)}{dt} = \sum_k Q_{ks} C_{ks} \mid_{x=L} - \sum_i Q_{sj} C_s + k_{ij}(C_s) \qquad \text{(Eq 6-4)}$$

Where:

C_s = the concentration for tank s, in mg/L

dt = change in time, in sec

Q_{ks} = flow from node k to s, in ft^3/sec (m^3/sec)

Q_{sj} = flow from node s to j, in ft^3/sec (m^3/sec)

dV_s = change in volume of tank at node s, in ft^3 (m^3)

V_s = volume of tank at node s, in ft^3 (m^3)

C_{ks} = concentration of contaminant at end of links, in mg/ft^3 (mg/m^3)

k_{ij} = decay coefficient between node i and j, in sec^{-1}

Previous chapters have described the historical development and application of water quality contaminant propagation models. This chapter focuses on some of the models and approaches used in modeling contaminant propagation in drinking water distribution systems. Most of the discussion will be focused on a US Environmental Protection Agency (USEPA) developed hydraulic/contaminant propagation model called EPANET (Rossman 1994), which is based on mass transfer concepts (transfer of a substance through another on a molecular scale). Another approach to water quality contaminant propagation developed by Biswas at al. (1993) uses a steady-state transport equation that takes into account the simultaneous corrective transport of chlorine in the axial direction, diffusion in the radial direction, and consumption by first-order reaction in the bulk liquid phase. Islam (1995) developed a model called QUALNET, which predicts the temporal and spatial distribution of chlorine in a pipe network under slowly varying unsteady flow conditions.

EPANET

As mentioned previously, the EPANET hydraulic model has been a key component in providing the basis for water quality modeling. Many hydraulic models are commercially available that also incorporate water quality models. EPANET is a computer program based on the EPS approach to solving hydraulic behavior of a network. In addition, it is designed to be a research tool for modeling the movement and fate of drinking water constituents within distribution systems. EPANET calculates all flows in cubic feet per second (ft^3/s) and has an option for accepting flow units in gallons per minute (gpm), million gallons per day (mgd), or liters per second (L/sec). Instructions on accessing EPANET through the Internet are discussed later in this book.

EPANET uses the Hazen–Williams formula, the Darcy–Weisbach formula, or the Chezy–Manning formula for calculating the head loss in pipes. Pumps, valves, and minor loss calculations in EPANET are also consistent with the convention established in the previous chapters. All nodes have their elevations above sea level specified, and tanks and reservoirs are assumed to have a free water surface. The hydraulic head is simply the elevation of the surface above sea level. The surface of tanks is assumed to change in accordance with the following equation:

$$\Delta y = (q/A)\,\Delta t \qquad \text{(Eq 6-5)}$$

Where:

Δy = change in water level, in ft (m)
q = flow rate into (+) or out of (–) tank, in ft^3/s (m^3/s)
A = cross-sectional area of the tank, in ft (m)
Δt = time interval, in sec

It is assumed that water usage rates, external water supply rates, and source concentrations at nodes remain constant over a fixed period of time, although these quantities can change from one period to another. Nodes are junctions where network components connect to one another (links), as well as to tanks and reservoirs. The default period interval is one hour but can be set to any desired value. Various consumption or water usage patterns can be assigned to individual nodes or groups of nodes.

EPANET solves a series of equations for each link using the gradient algorithm. Gradient algorithms provide an interactive mechanism for approaching an “optimal solution” by calculating a series of slopes that lead to improved solutions. Flow continuity is maintained at all nodes after the first iteration. The method easily handles pumps and valves.

The set of equations solved for each link (between nodes *i* and *j*) and each node *k* is as follows:

$$h_i - h_j = f(q_{ij}) \tag{Eq 6-6}$$

$$\textstyle\sum_i q_{ik} - \sum_j q_{kj} - Q_k = 0 \tag{Eq 6-7}$$

Where:

q_{ij} = flow in link connecting nodes *i* and *j*, in ft³/s (m³/s)
h_i = hydraulic grade line elevation at node *i* (equal to elevation head plus pressure head), in ft (m)
Q_k = flow consumed (+) or supplied (–) at node *k*, in ft³/s (m³/s)
$f(\cdot)$ = functional relation between head loss and flow in a link

The set of equations for each storage node (tank or reservoir) in the system is as follows:

$$\frac{\partial y_s}{\partial_t} = \frac{q_s}{A_s} \tag{Eq 6-8}$$

$$q_s = \textstyle\sum_i q_{is} - \sum_j q_{sj} \tag{Eq 6-9}$$

$$h_s = E_s + y_s \tag{Eq 6-10}$$

Where:

y_s = height of water stored at node s, in ft (m)
A_s = cross-sectional area of storage node s (infinite for reservoirs), in ft² (m²)
E_s = elevation of node s, in ft (m)
q_s = flow into storage node s, in ft (m)
t = time, in sec
h_s = height of water in storage tank, in ft (m)

Eq 6-8 expresses conservation of water volume at a storage node. Equations 6-9 and 6-10 express the same relationship for pipe junctions, and Eq 6-6 represents energy loss or gain caused by flow within a link. For known initial storage node levels y_s, at time zero, Eq 6-6 and Eq 6-7 are solved for all flows q_{ij} and heads h_i using Eq 6-10 as a boundary condition. This is called *hydraulically balancing* the network and is accomplished by using an iterative technique to solve the resulting nonlinear equations.

After a network hydraulic solution is obtained, flow into (or out of) each storage node q_s is found using Eq 6-9 and used in Eq 6-8 to find new storage elevations after a time step *dt*. This process is then repeated for all following time steps for the remainder of the simulation period.

Water Quality Simulation

EPANET uses the flows from the hydraulic simulation to track the propagation of contaminants through a distribution system. A conservation of mass equation is solved for the substance within each link between nodes *i* and *j* as follows:

$$\frac{\partial c_{ij}}{\partial_t} = \frac{q_{ij}}{A_{ij}}\left(\frac{\partial c_{ij}}{\partial x_{ij}}\right) + \Theta(c_{ij}) \quad \text{(Eq 6-11)}$$

Where:

c_{ij} = concentration of substance in link i, j as a function of distance and time (i.e., $C_{ij} = C_{ij}[X_{ij}, t]$), in mass/ft^3 (mass/m^3)
x_{ij} = distance along link i, j, in ft (m)
q_{ij} = flow rate in link i, j at time t, in ft^3/s (m^3/s)
a_{ij} = cross-sectional area of link i, j, in ft^2 (m^2)
$\Theta(c_{ij})$ = rate of reaction of constituent within link i, j, in mass/ft^3/d (mass/m^3/d)

Eq 6-11 must be solved with known initial conditions at time zero. The following boundary condition at the beginning of the link, i.e., at node i where $x_{ij} = 0$ must hold:

$$c_{ij}(0, t) = \frac{\Sigma_k q_{ki} c_{ki}(L_{ki}, t) + M_i}{\Sigma_k q_{ki} + Q_{si}} \quad \text{(Eq 6-12)}$$

The summations are made over all links k, i that flow into the head node i of the link i, j.

Where:

L_{ki} = the length of link k, i
M_i = the substance mass introduced by any external source at node i
Q_{si} = the source's flow rate

The boundary condition for link k, i depends on the concentrations at the head of the nodes of all links k, i that flow into link i, j. Equations 6-11 and 6-12 form a coupled set of differential/algebraic equations over all links in the network.

These equations are solved within EPANET by using the discrete volume method (DVM), which is described later in this chapter.

Water quality time steps are chosen to be as large as possible without causing the flow volume of any pipe to exceed its physical volume. Therefore the water quality time step dt_{wq} source cannot be larger than the shortest time of travel through any pipe in the network, i.e.,

$$dt_{wq} = Min\left(\frac{V_{ij}}{q_{ij}}\right) \text{for all pipes } i, j \quad \text{(Eq 6-13)}$$

Where:

V_{ij} = the volume of pipe i, j
q_{ij} = the flow rate of pipe i, j

Pumps and valves are not part of this determination because transport through them is assumed to occur instantaneously. Based on this water quality time step, the number of volume segments in each pipe (n_{ij}) is

$$n_{ij} = INT\left(\frac{V_{ij}}{q_{ij}dt_{wq}}\right) \quad \text{(Eq 6-14)}$$

Where:

$INT(x)$ = the largest integer less than or equal to x.

There is both a default limit of 100 for pipe segments or the user can set dt_{wq} to be no smaller than a user-adjustable time tolerance.

Reaction Rate Model

Eq 6-1 provides a mechanism for evaluating the reaction of a substance as it travels through a distribution system (Rossman et al. 1994). Reaction can occur in the bulk phase or with the pipe wall. EPANET models both types of reactions using first-order kinetics. In general, within any given pipe, material in the bulk phase and at the pipe wall will decrease according to the following equation:

$$\Theta(c) = -k_b c - \left(\frac{k_f}{R_H}\right)(c - c_w) \quad \text{(Eq 6-15)}$$

Where:

$\Theta(c)$ = total reaction rate
k_b = first-order bulk reaction rate, in constant/sec
c = substance concentration in bulk flow, in mass/ft^3 (mass/m^3)
k_f = mass transfer coefficient between bulk flow and pipe wall, in ft/sec (m/sec)
R_H = hydraulic radius of pipe (pipe radius/2), in ft (m)
c_w = substance concentration at the wall, in mass/ft^3 (mass/m^3)

Assuming a mass balance for the substance at the pipe wall yields:

$$k_f(c - c_w) = k_w - c_w \quad \text{(Eq 6-16)}$$

Where:

k_w = rate of chlorine demand at wall (wall demand), in ft/sec (m/sec)
i = an overall first-order reaction rate is equal to:

$$K = k_b + (k_w k_f)/R_H(k_w + k_f) \quad \text{(Eq 6-17)}$$

Where:

K = the overall reaction rate coefficient

Eq 6-16 pertains to the growth or decay of a substance, with mass transfer to or from the pipe wall depending on whether the sign of the equation is positive or negative. A negative sign means decay and a positive sign means growth.

EPANET uses three coefficients to describe reactions within a pipe. These are the bulk rate constant k_b and its wall rate constant k_w, which must be determined empirically and supplied as input to the model. The mass transfer coefficient is calculated internally by EPANET.

Other Features

EPANET can also model the changes in age of water and travel time to a node. The percentage of water reaching any node from any other node can also be calculated. Source tracing is a useful tool for showing the degree to which water from a given source blends with that from other sources, and how this pattern changes over time.

Inputs and Outputs

The following input steps should be taken before using EPANET:

1. Identify all network components and their connections. These components include pipes, pumps, valves, storage tanks, and reservoirs.
2. Assign unique identification (ID) numbers to all nodes. These numbers must be between 1 and 2,147,483,647 but do not have to be in any specific order.
3. Assign ID numbers to each link (pipe, pump, or valve). Both a link and a node can have the same ID number.
4. Collect information on the following system parameters:
   a. Diameter, length, roughness, and minor loss coefficient for each pipe.
   b. Characteristic operating curve for each pump.
   c. Diameter, minor loss coefficient, and pressure or flow setting for each control valve.
   d. Diameter and lower and upper water levels for each tank.
   e. Control rules that determine how pump, valve, and pipe settings change with time, tank water levels, or nodal pressures.
   f. Changes in water demands for each node over the time period being simulated.
   g. Initial water quality at all nodes and changes in water quality over time at source nodes.

Output from EPANET includes

- Color-coded network maps.
- Time series plots.
- Tabular reports.

Table 6-1 contains a summary of input parameter units for EPANET.

Convective Transport Model

Biswas et al. (1993) developed a steady-state transport equation that takes into account the simultaneous convective transport of chlorine in the axial direction, diffusion in the radial direction, and consumption by first order in the bulk liquid phase. Different wall conditions are considered in the model, including a perfect sink, no wall consumption, and partial wall consumption. The transport equations are developed in terms of the dissolved chlorine concentrations $C^*(HOCl^+ + OCl^-)$ in mg/L^{-1}. Figure 6-1 is a schematic of the differential volume (DV) in a cylindrical pipe. Assuming that flow is in the axial direction alone, a mass balance for this control volume (CV) gives:

$$\begin{aligned}\partial(2\pi r^*\Delta r^*\Delta X^* C^*)/\partial t^* = {} & (U^* f(r^*) C^* 2\pi\Delta r^*)_{X^*} \\ & - (U^* f(r^*) C^* 2\pi r^*\Delta r^*)_{X^*+\Delta X^*} + (-D(\partial C^*/\partial X^*)2\pi r^*\Delta r^*)_{X^*} \\ & - (-D(\partial C^*/\partial X^*)2\pi r^*\Delta r^*)_{X^*+\Delta X^*} - (-D(\partial C^*/\partial X^*)2\pi r^*\Delta r^*)_{X^*+\Delta X^*} \\ & - (-D(\partial C^*/\partial r^*)2\pi r^*\Delta X^*)_{r^*} - (D(\partial C^*/\partial r^*)2\pi r^*\Delta X^*)_{r^*+\Delta r^*} \\ & - C^* 2\pi r^*\Delta r^*\Delta X^*\end{aligned} \qquad \text{(Eq 6-18)}$$

Table 6-1. Summary of input parameter units

Parameter	US Customary Units	SI (Metric) Units
Junction elevation	feet	meters
Junction demand	flow units	flow units
Tank bottom elevation	feet	meters
Tank levels	feet above bottom	meters above bottom
Tank diameter	feet	meters
Tank minimum volume	cubic feet	cubic meters
Junction/tank quality		
• Chemical	milligrams/liter (or user supplied)	same as US customary
• Age	hours	same as US customary
• Source trace	percent	same as US customary
Pipe length	feet	meters
Pipe diameter	inches	millimeters
Pipe roughness		
• Hazen–Williams	none	none
• Darcy–Weisbach	millifeet	millimeters
• Chezy–Manning	none	none
Minor loss coefficient	none	none
Pump power rating	horsepower	kilowatts
Pump head	feet	meters
Pump flow	flow units	flow units
Pump speed setting	none	none
Valve diameter	inches	millimeters
Valve pressure setting	pounds per square inch	meters
Valve flow setting	flow units	flow units
Bulk reaction coefficient	$days^{-1}$	$days^{-1}$
Wall reaction coefficient	feet/day	meters/day
Specific gravity	none	none
Viscosity	square feet per second	square meters per second
Diffusivity	square feet per second	square meters per second

Where:

U^* = the average flow velocity throughout the distribution system, in cm/s
$f(r^*)$ = the flow parameter term depending on the flow regime. For laminar flow: $f(r^*) = 2[1 - (r^*/r_0^*)^2]$; for plug flow: $f(r^*) = 1$
r_0^* = the pipe radius, in cm
D = the effective diffusivity of chlorine species in water, in cm^2/s
k = the first-order decay rate constant in the bulk water (sec^{-1})
r^* = the radial distance from the center of the pipe, in cm
X^* = the axial distance from the inlet along the pipe, in cm
Δr^* and ΔX^* = incremental distances at r^* and X^*, respectively
C^* = the concentration of chlorine

The term on the left-hand side denotes a net accumulation of chlorine in the DV with time, the first two terms on the right-hand side account for increases in concentration due to the inflow of water into the DV, the next two terms account for increases due to diffusion of chlorine into the DV in the axial direction, the fifth and sixth terms account for increases due to diffusion of chlorine into the DV in the radial direction, and the last term on the right side accounts for the reduction of chlorine in the DV due to consumption by both chemical and microbiological reactions in the bulk water. Dividing the left side and right side by $2\pi r^*\Delta r^*\Delta X^*$ and taking limits as Δr^* and ΔX^* go to zero, Eq 6-18 becomes

$$\partial C^*/\partial t^* = - \partial(U^*f(r^*)C^*)/\partial X^* + \partial(D\partial C^*/\partial X^*)/\partial X^* + (1/r^*)(\partial(r^*D\partial C^*/\partial r^*)/r^*) - kC^* \quad \text{(Eq 6-19)}$$

where the variables are as previously defined.

Assuming quasi-steady conditions ($\partial C^*/\partial t^* = 0$) and rearranging Eq 6-19, we have

$$\partial(U^*f(r^*)C^*)/\partial X^* = \partial(D\partial C^*/\partial X^*)/\partial X^* + (1/r^*)\,(\partial(r^*D\partial C^*/\partial r^*)/\partial r^*) - kC^* \quad \text{(Eq 6-20)}$$

The boundary conditions are as follows

$$X^* = 0,\ C^* = C_0^* \text{ at } 0 \le r^* \le r^* \text{ entry conditions} \quad \text{(Eq 6-21)}$$
$$r^* = 0,\ \partial C^*/\partial r^* = 0 \text{ at } 0 \le X^* \le L^* \text{ symmetry condition}$$
$$r^* = r_0^*,\ D\partial C^*/\partial r^* = -V_d^*C^* \text{ at } 0 \le X^* \le L^* \text{ wall condition}$$

Where:

C_0^* = the inlet chlorine concentration, in mg/L
L^* = the pipe length, in cm
V_d^* = an empirical parameter proportional to the degree of absorption of the pipe surface, the "consumption rate" of chlorine at the pipe wall, in cm/s

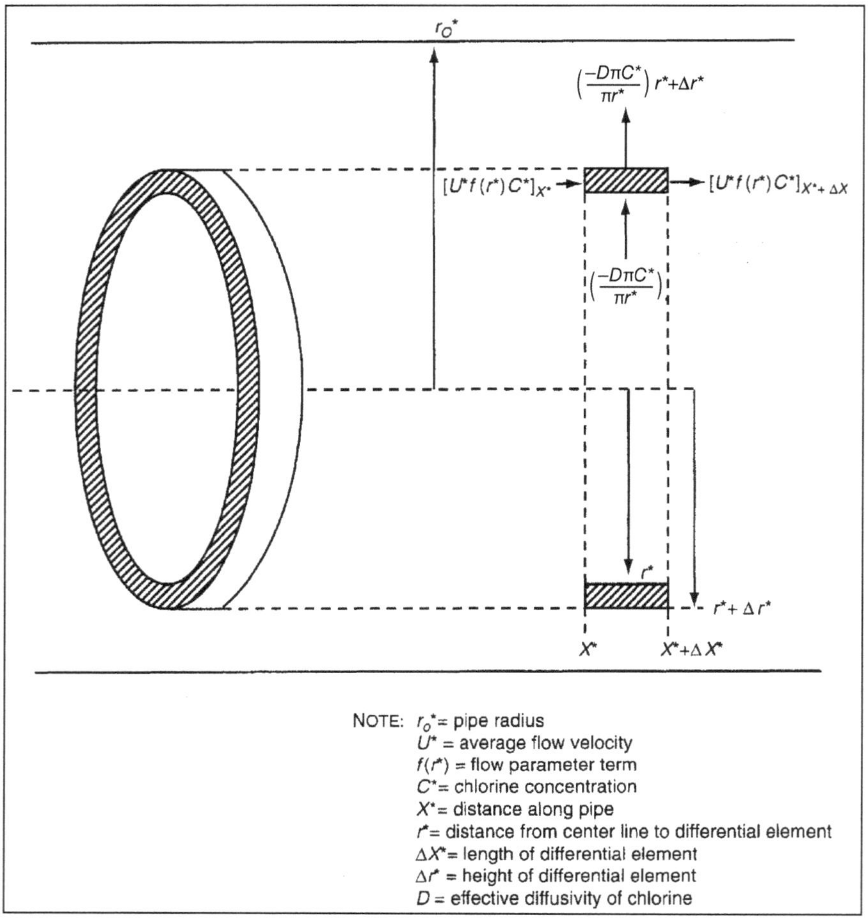

Figure 6-1. Schematic of the differential volume (DV) in a cylindrical pipe

The consumption process at the pipe wall can be compared to the transfer of mass or heat from the bulk liquid phase to the pipe surface. The following three wall conditions may be used, depending on the wall characteristics:

1. Perfect sink, $V_d^* \to \infty$, i.e., $C^* = 0$
2. No consumption at pipe wall, $V_d^* = 0$, i.e., $\partial C^*/\partial r^* = 0$
3. Partial consumption, V_d^* is some finite (nonzero) value

Assuming that the time period over which the experiments are done, U^* and D remain constant (quasi-steady state), and Eq 6-20 can be rewritten as follows:

$$\frac{f(r)\partial C}{\partial X} = \left(\frac{1}{\mathrm{Pe_a}}\right)\left(\frac{\partial^2 C}{\partial X^2}\right) + \frac{A_0}{r}\frac{\partial}{\partial r}\left(r\frac{\partial c}{\partial r}\right) = A_1 C \qquad \text{(Eq 6-22)}$$

Where:

$$C = \frac{C^*}{C_0^*}$$

$$X = \frac{X^*}{L^*}$$

$$r = \frac{r^*}{r_0^*}$$

$$\text{Pe}_\text{a} = \frac{L^* U^*}{D}$$

$$A_0 = \frac{L^* D}{r^{*2} U^*} = \frac{\Pi L^* D}{Q} \quad (Q = \text{flow rate throughout the system})$$

$$A_1 = \frac{kL^*}{U^*}$$

For distribution systems, the axial Peclet number Pe_a is typically large (>10^6), thus the axial diffusion term can be neglected with respect to the axial convective term. Turbulent flow conditions are often encountered, and then $f(r)$ is equal to 1, or the axial velocity can be assumed to be independent of the radial location for most parts of the cross section. Representing the axial velocity by a constant [$f(r) = 1$] average value (U^*), the governing equation for chlorine transport in pipes under these conditions is

$$\frac{\partial C}{\partial x} = \left(\frac{A_0}{r}\right)\left(\partial \frac{r\partial C}{\partial r}\right) - A_1 C \qquad \text{(Eq 6-23)}$$

with the boundary conditions

$$X = 0, C = 1, \text{ at } 0 \le r \le 1 \quad \text{(Eq 6-24)}$$

$$r = 0, \frac{\partial C}{\partial r} = 0, \text{ at } 0 \le X \le 1$$

$$r = 1 \frac{\partial C}{\partial r} = -\left(\frac{V_d^* r_o^*}{D}\right) C$$

$$= -A_2 C \text{ at } 0 \le X \le 1$$

Three dimensionless parameters A_0, A_1, and A_2 govern the chlorine decay in the distribution system. The parameter A_0 accounts for the radial diffusion and depends on the pipe length, on the effective diffusivity of chlorine, and on the flow rate throughout the system. The variable A_1 depends on the reactivity of chlorine with species such as viable cells or chemical compounds in the bulk liquid phase and on the residence time in the system. The variable A_2 is a wall consumption parameter depending on the wall consumption rate, the pipe radius, and the effective diffusivity of chlorine.

Since the axial velocity is represented by an average quantity, Eq 6-23 can be solved analytically. Using separation of variables, a solution to Eq 6-23 is as follows:

$$C(X, r) = 2\Sigma_{n=1}^{\infty} \frac{\lambda_n J_o(\lambda_n r) J_1(\lambda_n)}{(\lambda_n^2 + A_2^2) J_o^2(\lambda_n)} \times \exp\left[-A_1 + \lambda n^2 A_0)X\right] \qquad \text{(Eq 6-25)}$$

Where:

J_0 and J_1 = the Bessel functions of the first kind of order zero and 1, respectively
λ_ns = the roots of $\lambda_n J_1(\lambda_n) = A_2 J_0(\lambda_n)$

Through a series of simplifying assumptions to Eq 6-25

$$C_{av} = \frac{\exp(-A_1 X)}{(1+\varepsilon)} \qquad \text{(Eq 6-26)}$$

In Eq 6-22, an expression for ε for A_0 in the range of 10^{-4} to 100 is given by

$$\varepsilon = 2.4416\, A_0 A_2 - 0.1559\, A_0 A_2^{\,2} \qquad 0.01 \le A_2 < 10 \qquad \text{(Eq 6-27)}$$

$$\varepsilon = 10.105\, A_0 + 0.0014\, A_2 + 0.31\, A_0^{\,2} A_2 \qquad 0.01 \le A_2 < 10$$

The average concentration at any location can be determined using Eq 6-26 and Eq 6-27 provided the three nondimensional parameters A_0, A_1, and A_2 are known.

Once the flow rates throughout the system are known (as can be computed provided there is an effective diffusivity of chlorine in water), D is also known. The value of D depends on ionic and molecular transport and on the flow regime. In actual practice, ionic transport has to be accounted for to determine the diffusivity (Lu 1991). If turbulent flow conditions prevail, the eddy diffusivity has to be used as it typically is greater than the molecular diffusivity ($D = 1.25 \times 10^{-5}$ $cm^2 s^{-1}$).

Field data can be used to compute the three nondimensional parameters A_0, A_1, and A_2. Concentrations would need to be measured in at least three locations (in addition to the inlet), and Eq 6-26 and Eq 6-27 can then be solved to determine A_0, A_1, and A_2. If data are available at more than three locations, a least-squares procedure can be used to determine the best-fit values for A_0, A_1, and A_2.

Determination of Parameters From Field Data

As described earlier, researchers have used a simple decay rate expression (as in Eq 6-17) to fit field data and report values of k. The value for k is a composite rate consumption that combines the effect of bulk liquid and wall consumption. As wall consumption and diffusion are not accounted for, different values of k are reported for different conditions. The model developed is also demonstrated by solving the equations for the conditions of a field sampling study conducted in the South Central Connecticut Regional Water Authority (SCCRWA). A schematic of the distribution system composed of 8-in. (200-mm) and 12-in. (300-mm) mains is shown in Figure 6-2.

The Brushy Plains/Cherry Hill service area operates under two basic scenarios: pumps on, when the tank is filling; and pumps off, when the tank is emptying. These scenarios hydraulically define links of pipe over which chlorine consumption can be measured. Chlorine concentrations at the inlet and outlet points of these pipe segments are shown in milligrams per liter in Table 6-2.

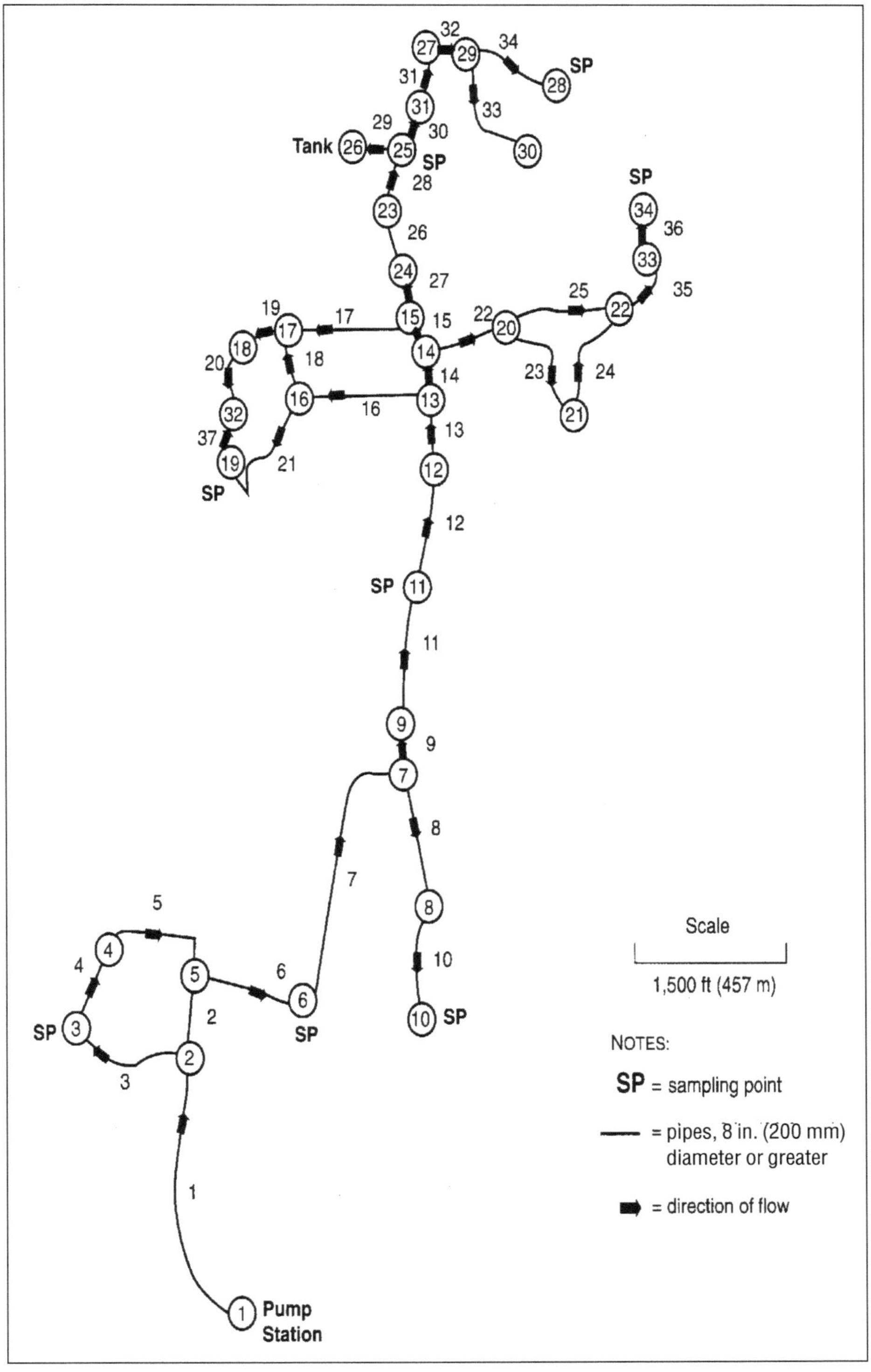

Figure 6-2. Link–node representation of the Cherry Hill/Brushy Plains network with "pumps-on" scenario

Table 6-2. Chlorine concentrations at inlet and outlet of unidirectional flow segments

Pipes in Segment	Chlorine Concentration at Segment, in *mg/L* Inlet	Outlet
1 and 3	1.08	1.00
7, 8, and 10	1.00	0.32
7, 9, and 11	1.00	0.98
12, 13, 16, and 21	0.98	0.16
12, 13, 14, 15, 26, 27, and 28	0.98	0.94

Five unidirectional flow segments were identified (shown in Table 6-2) wherein inlet and outlet chlorine concentrations were measured. The geometrical and flow parameters are listed in Table 6-3. As indicated in Table 6-3, the flow conditions in the pipe are turbulent, and hence the eddy diffusivity was used to estimate the diffusion coefficient, $D_{eddy} = 1.233 \times 10^{-2}\, U^{*}r_o^{*}$ at 78°F (25°C). Using these parameters, A_0 is computed for the different pipes and the values listed in Table 6-3, and k was estimated to be $6.4 \times 10^{-6}\mathrm{s}^{-1}$ from bench kinetic tests performed with water samples taken at the inlet to the network. Assuming that the k value remains the same for the pipes, A_1 was computed for the different pipes and the values are listed in Table 6-3 (Biswas et al. 1993). This assumption is expected to be valid for the main network branch; however, it was not validated by performing a kinetic test with a water sample taken from a downstream sampling point. With A_1 and A_2 known, an iterative procedure using Eq 6-26 and Eq 6-27 was used to compute V_d^{*} and ε. The calculation is demonstrated for the segment with pipes 7, 9, and 11. The superscripts on the variables indicate the pipe number. At the outlet of the pipe, $X-1$, and Eq 6-26 gives

$$C_{av}^{7} = \frac{\exp(-A_1^{7})}{(1+\varepsilon^{7})} \qquad \text{(Eq 6-28)}$$

$$C_{av}^{9} = \frac{\exp(-A_1^{9})}{(1+\varepsilon^{9})}$$

$$C_{av}^{11} = \frac{\exp(-A_1^{11})}{(1+\varepsilon^{11})}$$

The measured concentration at the outlet to pipe 11 divided by the concentration at the inlet to pipe 7 is

$$C_{av}^{7} C_{av}^{9} C_{av}^{11} = \frac{0.98}{1.00} \qquad \text{(Eq 6-29)}$$

$$= \frac{\exp(-A_1^{7})\exp(-A_1^{9})\exp(-A_1^{11})}{(1+\varepsilon^{7})(1+\varepsilon^{9})(1+\varepsilon^{11})}$$

The unknowns, ε^7, ε^9 and ε^{11} in Eq 6-29 are expressed as a function of A_0 and A_2 using Eq 6-27. The A_0s are known (Table 6-3) and A_2 can be expressed as a function of the unknown chlorine consumption rate at the pipe wall, V_d^*, and other known parameters. Similar expressions are developed for the other segments listed in Table 6-3. The value V_d^* is expected to be the same for pipes made of the same material in the main branch of the network. A spreadsheet program used wherein values V_d^* could be entered and then A_2 and ε computed Eq 6-27. The outlet concentrations were then estimated using expressions similar to Eq 6-28, and then compared to the measured concentration ratios (Table 6-3). A value of 1.74e^{-7} m/s, for the main branch pipes led to good agreement with the measurements. However, different values of V_d^* were obtained for the pipes on dead ends or off the main network branch (pipes 3, 8, 10, and 21). Pipes 3, 8, 10, and 21 have high values indicating that wall consumption is significant. These are off-branch pipes (Figure 6-2) and qualitative observations indicate that significant biofilm growth occurs on such pipes.

QUALNET

Islam (1995) developed a model called QUALNET, which predicts the spatial and temporal distribution of chlorine residuals in pipe networks under slowly varying unsteady flow conditions. Unlike other available models, which use steady-state or extended-period simulation (EPS) of steady flow conditions, QUALNET uses a lumped-system approach to compute unsteady flow conditions and includes dispersion and decay of chlorine during travel in a pipe. The pipe network is first analyzed to determine the initial steady-state conditions. The slowly varying conditions are then computed by numerically integrating the governing equations by an implicit-finite-difference, a scheme subject to the appropriate boundary conditions. The one-dimensional dispersion equation is used to calculate the concentration of chlorine over time during travel in a pipe, assuming a first-order decay rate.

Numerical techniques are used to solve the dispersion, diffusion, and decay equations. Complete mixing is assumed at the pipe junctions. The model has been verified by comparing the results with those of EPANET for two typical networks. The results are in good agreement at the beginning of the simulation model for unsteady flow; however, chlorine concentrations at different nodes vary when the flow becomes unsteady and when reverse flows occur. The model may be used to analyze the propagation and decay of any other substance for which a first-order reaction rate is valid. The dynamic equation describing the slowly time-varying flow in a pipe may be written as follows (Chaudhry 1987):

$$\frac{L}{gA}\frac{dQ}{dt} = H_1 - H_2 - \mathrm{k}Q|Q^{\mathrm{m}} \qquad \text{(Eq 6-30)}$$

Table 6-3. Model parameters for different pipes in the network

Pipe	Length, m	Radius, m	Flow Velocity, m/s	Diffusion Coefficient, m^2/sec	Reynolds Number	V_d*, m/sec	A_0	A_1	A_2	E
1	731.5	0.152	0.546	1.03e−3	1.75e+5	1.00e−7	5.92e+1	8.58e−3	1.49e−5	2.2e−3
3	396.2	0.102	0.195	2.46e−4	4.19e+4	1.15e−6	4.79e+1	1.30e−2	4.78e−4	5.6e−2
7	822.9	0.152	0.512	9.62e−4	1.64e+5	1.00e−7	6.66e+1	1.03e−2	1.58e−5	2.6e−3
8	365.8	0.152	0.014	2.58e−5	4.40e+3	1.00e−7	2.96e+1	1.71e−1	5.91e−4	4.3e−2
9	121.9	0.152	0.494	9.28e−4	1.58e+5	1.00e−7	9.86e+0	1.58e−3	1.64e−5	4.0e−4
10	304.8	0.102	0.014	1.80e−5	3.08e+3	2.3e−6	3.68e+1	1.36e−1	1.30e−2	1.17
11	213.4	0.152	0.485	9.11e−4	1.55e+5	1.00e−7	1.73e+1	2.82e−3	1.67e−5	7.1e−4
12	579.1	0.152	0.457	8.59e−4	1.47e+5	1.00e−7	4.69e+1	8.11e−3	1.77e−5	2.0e−3
13	182.9	0.152	0.445	8.36e−4	1.43e+5	1.00e−7	1.48e+1	2.63e−3	1.82e−5	6.6e−4
14	121.9	0.152	0.372	6.99e−4	1.19e+5	1.00e−7	9.86e+0	2.10e−3	2.18e−5	5.3e−4
15	91.4	0.152	0.329	6.19e−4	1.06e+5	1.00e−7	7.40e+0	1.78e−3	2.46e−5	4.5e−4
16	457.2	0.102	0.168	2.11e−4	3.60e+4	1.00e−7	5.52e+1	1.75e−3	4.84e−5	6.5e−3
21	426.7	0.102	0.049	6.14e−5	1.05e+4	2.2e−5	5.15e+1	5.60e−2	3.66e−2	4.589
26	182.9	0.152	0.329	6.19e−4	1.06e+5	1.00e−7	1.48e+1	3.56e−3	2.46e−5	8.9e−4
27	76.2	0.152	0.338	6.36e−4	1.09e+5	1.00e−7	6.17e+0	1.44e−3	2.40e−5	3.6e−4
28	91.4	0.152	0.323	6.07e−4	1.04e+5	1.00e−7	7.40e+0	1.81e−3	2.51e−5	4.5e−4

Source: Biswas et al. 1993.

Where:

L = pipe length
g = acceleration due to gravity
A = pipe cross-sectional area, in ft^2 (m^2)
Q = rate of discharge, in ft/sec (m/sec)
t = time
H_1 = piezometric head on the upstream end of pipe, in ft (m)
H_2 = piezometric head on the downstream end of pipe, in ft (m)
k (L/sec) and m = constants in a head loss formula of an exponential type (i.e., Darcy–Weisbach or Hazen–Williams)

The absolute value allows for a proper sign for the head loss term in case the flow reverses during the simulation.

The continuity equation describing the mass conservation at a junction of pipes may be written as

$$\Sigma Q_j + \Sigma Q_0 = 0 \qquad \text{(Eq 6-31)}$$

in which the inflows are considered positive and the outflows are considered negative.

A computational technique developed by Holloway (1985) is used to solve the simulation of slowly varying flows in a pipe. For a first-order decay rate, the one-dimensional dispersion of a constituent in a pipe is modeled by the following equation:

$$\frac{\partial C}{\partial t} + V\frac{\partial C}{\partial x} = D\frac{\partial^2 C}{\partial x^2} - k_1 C \quad \text{(Eq 6-32)}$$

Where:

C = constituent concentration
V = advective velocity, in ft/sec (m/sec)
D = dispersion coefficient, in ft^2/sec (m^2/sec)
x = spatial coordinate along the pipe axis, in ft (m)
k_1 = first-order reaction rate coefficient, in sec^{-1}

In order to avoid problems with numerical dispersion in solving Eq 6-32, it is split into the following two separate equations:

$$\frac{\partial C}{\partial t} + V\frac{\partial C}{\partial z} = 0 \quad \text{(Eq 6-33)}$$

$$\frac{\partial C}{\partial t} - D\frac{\partial^2 C}{\partial x^2} - k_1 C = 0 \quad \text{(Eq 6-34)}$$

Eq 6-9 can be discretized first to convert the mass of constituents a certain distance downstream for each time step so that no numerical diffusion occurs. Then the advected concentrations are diffused longitudinally by solving Eq 6-34 at each spatial grid point, including their decay. The final concentration field is thus obtained from the entire time increment. The diffusion coefficient was estimated by the following equation:

$$D = 10.1\, \tau\sqrt{\frac{\tau_0}{\rho}}\left(\frac{\text{cm}^2}{\text{sec}}\right) \quad \text{(Eq 6-35)}$$

Where:

τ = pipe radius
τ_0 = shear stress at the wall
ρ = density of the fluid

The Warming–Kutler–Lomax scheme was used for the solution. QUALNET was calibrated on sample networks against EPANET. The results compared favorably.

Solution Methods

The water quality simulation process used in the previous models is based on a one-dimensional transport model, in conjunction with the assumption that complete mixing of material occurs at the junction of pipes. These models consist of moving the substance concentrations forward in time at the mean flow velocity while undergoing a concentration change based on kinetic assumptions. The simulation proceeds by considering all the changes to the state of the system as the changes occur in chronological order. Based on this approach, the advective movement of substance defines the dynamic simulation model. Most water quality simulation models are interval oriented, which in some cases can lead to solutions that are either prohibitively expensive or contain excessive errors.

Several different numerical methods can be used to solve contaminant propagation equations (Rossman and Boulos 1996, Lansey and Boulos 2005). Four commonly used techniques are Eulerian finite-difference method (FDM), Eulerian discrete volume method (DVM), Lagrangian time-driven method (TDM), and Lagrangian event-driven method (EDM).

The FDM approximates derivatives with finite-difference equivalents along a fixed grid of points in time and space. Islam (1995) used this technique to model chlorine decay in distribution systems as part of the QUALNET model discussed earlier.

The DVM divides each pipe into a series of equally sized, completely mixed volume segments. At the end of each successive water quality time step, the concentration within each volume segment is first reacted and then transferred to the adjacent downstream segment. This approach was used in the models that were the basis for the early USEPA studies.

The TDM tracks the concentration and size of a nonoverlapping segment of water that fills each link of a network. As time progresses, the size of the most upstream segment in a link increases as water enters the link. An equal loss in size of the most downstream segment occurs as water leaves the link. The size of these segments remains unchanged.

The EDM is similar to TDM, except that rather than update an entire network at fixed time steps, individual link–node conditions are updated only at times when the leading segment in a link completely disappears through this downstream node.

Event-Driven Method

Boulos et al. (1995) first proposed the event-driven method (EDM). This is extremely simple in concept and is based on a next-event scheduling approach. In this method, the simulation clock time is advanced to the time of the next event to take place in the system. The simulation scheduled is executed by carrying out all the changes to a system associated with an event, as events occur in chronological order. Since the only factors affecting the concentration at any node are the concentrations and flows at the pipes immediately upstream of the given node, the only information that must be available during the simulation is the different segment concentrations. The technique makes the water quality simulation process very fast and efficient.

The advective transport process is dictated by the distribution system demand. The model follows a front tracking approach and explicitly determines the optimal pipe segmentation scheme with the smallest number of segments necessary to carry out the simulation process. To each pipe, pointers (concentration fronts), whose function is to delineate volumes of water with different concentrations, are dynamically assigned. Particles representing substance injections are processed in chronological order as they encounter the nodes. All concentration fronts are advanced within their respective pipes based on their velocities. As the injected constituent moves through the system, the position of the concentration fronts defines the spatial location behind which constituent concentrations exist at any given time. The concentration at each affected node is then given in the form of a time–concentration histogram.

The primary advantage of this model is that it allows for dynamic water quality modeling that is less sensitive to the structure of the network and to the length of the simulation process.

In addition, numerical dispersion of the concentration front profile resolution is nearly eliminated. The method can be readily applied to all types of network configurations and dynamic hydraulic conditions and has been shown to exhibit excellent convergence characteristics.

Commercial Hydraulic–Water Quality Modeling Software

In addition to EPANET, several commercial software packages are widely used in the United States and internationally. Most of these packages are based on the EPANET formulation and include value-added components as parts of a graphical user interface that increases the capability of the software. Examples of such value-added components that are part of one or more of the commercially available software packages include:

- *Scenario manager:* Manage inputs and outputs of a group of model runs.
- *Calibration optimization:* Utilize genetic algorithm optimization technique to determine model parameters that best fit a set of field data.
- *Design optimization:* Utilize genetic algorithm optimization techniques to select pipe sizes that minimize costs or other selected objectives.
- *Integration with GIS or CAD:* Water distribution model directly integrates with GIS or CAD to assist in constructing or updating models.

Integration with GIS or CAD allows the user to:

- Display results.
- Make use of flexible output graphics and provides convenient ways to modify parameters for graphical displays of output data.
- Consider energy management by calculating energy use for selected alternatives.
- Conduct automated fire-flow analysis by assessing the availability of fire flow at a range of nodes and determining whether a system meets fire-flow requirements.
- Consider water security and vulnerability assessment methods, incorporate skeletonization, and demand allocation tools.

Table 11-1 (see p. 231) provides a summary listing of major commercial software and a Web link where additional details may be obtained on specific features and current version availability/pricing.

References

Biswas, P., C. Lu, and R.M. Clark. 1993. A Model for Chlorine Concentration Decay in Drinking Water Distribution Pipes. *Water Research*, 27(12):1715–1724.

Boulos, P.F., T. Altman, P.A. Jarrige, and F. Collevati. 1995. Discrete Simulation Approach for Network–Water-Quality Models. *Jour. Water Resources Planning and Management*, 121(1):49–60.

Chaudhry, M.H. 1987. *Applied Hydraulic Transients*, 2nd ed. New York: Van Nostrand Reinhold.

Clark, R.M., H. Pourmaghaddas, L.G. Wymer, and R.C. Dressman. 1996. Modeling the Kinetics of Chlorination By-Product Formation: The Effects of Bromide. *Journal of Water Supply: Research and Technology–AQUA*, 45(1):1–8.

Clark R.M., R.C. Haught, and W.M. Grayman. 2005. Predicting Chlorine Residuals in Unlined Metallic Pipes. In *Proc. ASCE-EWRI World Water & Environmental Resources Congress, Anchorage, AK.* Reston, Va.: ASCE.

DiGiano, F.A., and W. Zhang. 2004. Uncertainty Analysis in a Mechanistic Model of Bacterial Regrowth in Distribution System. *Journal of Environmental Science and Technology*, 38(22):5925–5931.

Grayman, W.M., L.A. Rossman, M.A. Gill, Y. Li, and D.E. Guastella. 2002. Measuring and Modeling Disinfectant Wall Demand in Metallic Pipes. In *Proc. EWRI Conference on Water Resources Planning and Management.* Reston, Va.: ASCE.

Holloway, M.B. 1985. *Dynamic Pipe Network Computer Model.* PhD thesis. Pullman, Wash: Washington State University.

Islam, M.R. 1995. *Modeling of Chlorine Concentration in Unsteady Flows in Pipe Networks*. PhD thesis. Pullman: Washington State University.

Lansey, K.E., and P.F. Boulos. 2005. *Comprehensive Handbook on Water Quality Analysis in Distribution Systems*. Broomfield, Colo.: MWH Soft, 448 pp.

Lu, C. 1991. *Theoretical Study at Particle, Chemical, and Microbial Transport in Drinking Water Distribution Systems*. PhD thesis. Cincinnati, Ohio: University of Cincinnati.

National Research Council (NRC). 2006. *Drinking Water Distribution Systems: Assessing and Reducing Risks*. Washington, D.C.: National Academies Press, 404 pp.

Powell, J., J. Clement, M. Brandt, R. Casey, D. Holt, W. Grayman, and M. LeChevallier. 2004. *Predictive Models for Water Quality in Distribution Systems*. Denver, Colo.: AwwaRF.

Rossman, L.A. 1994. *EPANET User's Manual*. Cincinnati, Ohio: USEPA.

Rossman, L.A., R.M. Clark, and W.M. Grayman. 1994. Modeling Chlorine Residuals in Drinking Water Distribution Systems. *Jour. Environ. Eng.*, 120(4):803–820.

Rossman, L.A., and P.F. Boulos. 1996. Numerical Methods for Modeling Water Quality in Distribution Systems: A Comparison. *Journal Water Resources Planning and Management*, 122(2):137–146.

Rossman, L.A. 2000. *EPANET Version 2 User's Manual*. Cincinnati, Ohio: USEPA Water Supply and Water Resources Division.

Vasconcelos, J.J., L.A. Rossman, W.M. Grayman, P.F. Boulos, and R.M. Clark. 1997. Kinetics of Chlorine Decay. *Journal AWWA*, 89(7):54–65.

Chapter 7

Modeling Individual Constituents in Drinking Water: Total Trihalomethanes and Disinfectant Decay

Introduction

An important aspect of modeling water quality in drinking water distribution systems is the ability to predict the fate and transport of disinfection residuals, disinfectant decay, and the formation of disinfection by-products. Before leaving the treatment plant, drinking water is generally disinfected as a final step before being transported through the distribution system to the consumer. It is presumed that a detectable disinfectant residual will minimize the potential for waterborne disease and biofilm growth in the system. Maintenance of disinfectant residuals is generally considered to be a water quality goal in the United States, and most US water systems attempt to maintain a detectable residual throughout the distribution system. Chlorine has been the disinfectant of choice for drinking water utilities in the United States; however, interest is growing in the use of chloramines. The chemistry of both chlorine and chloramines in drinking water will be discussed in this chapter as well as the tendency of both disinfectants to form disinfection by-products.

As dissolved chlorine travels through the distribution system, it reacts with natural organic matter (NOM) in the bulk water, with biofilm and tubercles on the pipe walls, or with the pipe wall material itself. This reaction results in a decrease in chlorine residual and a corresponding increase in disinfection by-products (DBPs), with the amounts depending on the residence time in the network pipes and holding time in storage facilities. Understanding these reactions assists water utility managers in delivering high-quality drinking water and in meeting regulatory requirements under the Safe Drinking Water Act (SDWA) and its amendments.

Drinking water chlorination, however, poses a dilemma. Chemical disinfection reduces risk of infectious disease, but the interaction between chemical disinfectants and precursor materials in source water results in the formation of potentially harmful chemical by-products. Although disinfection of public drinking water has dramatically reduced disease outbreaks attributable to waterborne pathogens, the identification of chloroform, a DBP, in drinking water (Rook 1974, Bellar and Lichtenberg 1974) raised questions about possible health risks posed by DBPs. Since 1974, additional DBPs have been identified and concerns have intensified about health risks resulting from exposures to them. Although a causal relationship between DBP exposures and these health risks has not been conclusively established, risk managers have responded, in the interest of protecting public health, by developing alternative treatment systems and issuing rules and regulations designed to maintain protective levels of disinfection while reducing potentially harmful levels of DBPs.

Depending on the level and type of treatment applied, drinking water will exert a chlorine demand because of reactions with natural organic matter (NOM) and other constituents. Therefore, the applied disinfectant dose must be sufficient to meet the inherent demand in the treated water, to provide sufficient protection against microbial infection, and at the same time to minimize the production of DBPs. Consequently, much research has been invested in attempting to characterize the nature of this chlorine demand and the conditions that govern the formation of DBPs in drinking water. One aspect of this research is the development of mathematical models for predicting chlorine decay, and for predicting the formation of disinfection by-products, both within the treatment plant and within the piping network that delivers treated water to consumers. These models allow one to evaluate alternative treatment and operational strategies to best meet the dual goals of providing both microbiologically and chemically safe water.

Although chloramines are less reactive than free chlorine, they also react with NOM and reduced metals in the aqueous phase, as well as with Fe(II) in pipe deposits, leading to the formation of ammonia. Chloramine losses should decrease with increasing pH because both autodecomposition and its reaction with NOM become slower (NRC 2006).

This chapter will review current and historical research efforts related to the development of models for predicting chlorine and chloramine decay and the formation of DBPs. It covers both the chemical kinetics of chlorine and chloramine decay and DBP formation in bulk solution as well as the dynamics of these processes as they occur in a distribution system. In the latter case, more than just bulk water reactions are at work. Reactions at the pipe wall, hydraulic behavior, and utilization of storage reservoirs will also have significant influences on both the spatial and temporal distribution of chlorine and chloramine residuals and DBP concentrations.

Water quality modeling has the potential for providing insight into the factors that influence the variables affecting changes in distribution systems' water quality. Understanding the factors that influence the formation of total trihalomethanes (TTHMs) and maintenance of chlorine residuals is of particular interest. The modeling program EPANET has proven to be especially useful for modeling both TTHM formation and the propagation and maintenance of disinfectant residuals.

One of the first studies to address these issues was conducted by the US Environmental Protection Agency (USEPA) in collaboration with the North Marin Water District (NMWD) in California (Clark et al. 1994). Another study conducted jointly by USEPA and the Awwa Research Foundation (AwwaRF, now called the Water Research Foundation [WaterRF]) examined these same issues (Vasconcelos et al. 1996). In the AwwaRF/USEPA study, various types of models were evaluated to describe both of these phenomena.

Chemistry of Chlorine in Water

In order to understand the nature of the risk tradeoffs associated with the loss of chlorine residuals and the formation of disinfection by-products, it is necessary to review some of the chemistry involved in this process.

Chlorine gas dissolved in water forms hypochlorous acid (HOCl) according to the reaction:

$$Cl_2 + H_2O \rightarrow HOCl + H^+ + Cl^- \qquad \text{(Eq 7-1)}$$

HOCl is a weak acid that dissociates to hypochlorite ion OCl^- depending on pH (Shukairy 1994). At pH 6, a chlorine solution is about 95 percent HOCl, while at pH 9 it contains about 95 percent OCl^-. HOCl has stronger disinfecting power than OCl^-. Chlorine is a strong oxidizing agent that readily reacts with a number of inorganic and organic constituents in water. These reactants make up the chlorine demand of a water. Because these reactants are present in different concentrations and have different degrees of reactivity, the loss of chlorine over time is a gradual process. Half-lives of chlorine in treated water (i.e., the time it takes for 50 percent of the initial chlorine to disappear) can vary from several hours to several days.

Chlorine and the hypobromous acid formed when bromide reacts with HOCl can react with NOM to produce literally hundreds of halogenated by-products, most of which are present only in trace amounts (Shukairy et al. 1994a). The by-products produced in greatest quantities are the trihalomethanes (THMs). These are methane (CH_4) molecules in which three of the hydrogen atoms have been replaced by some combination of chlorine and bromine atoms. Total THMs (TTHMs) are regulated by USEPA to be below 80 μg/L in all community water supplies. THMs constitute only 20 to 40 percent of the total organic halides produced by chlorination. Other classes of by-products of concern include haloacetic acids, haloacetonitriles, cyanogen halides, halopicrins, and chloral hydrate.

Chorine demand and the formation of DBPs are influenced by both the condition of treatment as well as the constituents in the raw and treated water. Treatment processes change the concentration of drinking water constituents and are likely to change the chlorine demand and DBP formation potential of a given water. Therefore, when modeling chlorine decay and the formation of DBPs, it is important to find modeling input parameters that can reflect these changes in water quality characteristics. Some common surrogate parameters for NOM are total organic carbon (TOC) and spectral absorbance. TOC concentration is indicative of the mass of organic material, whereas spectral absorbance relates more to specific structure and functional groups. For NOM, the most commonly used spectral absorbance is the ultraviolet absorbance (UVA) at a wave length of 254 nm, which measures conjugated double bonds. The specific UVA (SUVA) is the ratio of UVA to the dissolved fraction of the TOC (DOC). Its values give an indication of the NOM's nature, with higher values indicating a more aromatic character (Amy 1993).

For most waters, the reactions of chlorine with NOM comprise the majority of the chlorine demand. Three categories of reactions that occur between organic compounds and chlorine species are as follows (Brezonik 1994):

- Oxidation of organic functional groups
- Addition to double bonds
- Electrophelic substitution

The time scale of the chlorine oxidation reactions varies widely depending on the nature of the organic compound.

Chlorinated organic compounds are not formed when chlorine fully oxidizes organic matter to produce CO_2. Rather, reactions in which chlorine is added or substituted into the structure of another molecule, as in the formation of chlorinated organics, yield many of the DBPs of concern. Precursor compounds that can react with chlorine to form DBPs are found in all waters and are part of the NOM. The amount of precursor material depends on the source water and the water quality characteristics. A major precursor group is humic material, which is predominantly the end product of biological decay and tends to be fairly stable.

Other components of NOM include hydrophilic acids, carbohydrates, carboxcylic acids, and amino acids. However, humic material represents the dominant fraction, often comprising 50 to 70 percent of the dissolved organic carbon (DOC). Humic materials have shown significant potential for the formation of THMs and other chlorinated by-products.

Chlorine also reacts with various inorganic compounds. For example, chlorine reacts with ammonia to form different species of chloramines. Aqueous chlorine converts Br^- into hypobromous acid ($HOBr^-$), which then attacks organic compounds to form brominated DBPs.

Chlorine will oxidize soluble iron and manganese to the insoluble ferric and manganic forms, respectively. Hydrogen sulfide reacts with chlorine, and successful removal of H_2S by conventional treatment with prechlorination has been reported.

Some of the factors that influence both the formation of disinfection by-products and the demand for chlorine are as follows:

- *Disinfectant dose:* Several studies have shown that the formation of disinfection by-products increases with chlorine concentration.
- *Reaction time:* A longer reaction time generally leads to both higher disinfectant demand and greater DBP formation.
- *pH:* It has been observed that, for chlorine reactions, a shift in pH has little effect on chlorine demand. However, increase in TTHM formation has been observed with increases in pH. The sum of HAA6 (the sum of six of the nine HAA species) has been found to decrease when pH increases.
- *Temperature:* An increase in temperature has been shown to cause an increase in both DBP formation and chlorine demand.

As mentioned previously, chlorine residuals are important to ensure the microbial safety of distributed drinking water. Requirements for disinfection of drinking water are defined in the Surface Water Treatment Rule (SWTR). According to the SWTR, treatment including disinfection must reliably achieve at least a 3-log (99.9 percent) removal and/or inactivation of *Giardia lamblia* cysts, and a 4-log (99.99 percent) removal and/or inactivation of viruses prior to the delivery of water to the first consumer (Clark and Feige 1993). A control parameter frequently considered and specified in the SWTR is the *CT* concept. *CT* values for purposes of the SWTR are defined as the product of the disinfectant residual concentration (mg/L) and contact time (min) measured at peak hourly flow. Contact time is measured from the point of disinfectant application to the first customer (Clark and Feige 1993). Different disinfectants require different *CT* values because of their variability of action against different types of organisms. For example, chlorine is relatively ineffective against some protozoans such as *Cryptosporidium* but is generally very effective against most bacteria and viruses.

In addition to pathogen inactivation at the treatment plant, maintenance of chlorine residuals throughout a distribution system is important to minimize microbial regrowth and to provide protection against pathogen intrusion. The SWTR requires that systems maintain a residual of at least 0.2 mg/L throughout. Yet the recently promulgated maximum concentration limit on free chlorine has been set at 4 mg/L. These requirements can place severe demands on utilities that experience high chlorine demands in their distribution systems, forcing them to use booster chlorination out in the system or switch to a more stable secondary disinfectant such as chloramine.

Chemistry of Chloramines in Water

Because of increasingly stringent drinking water regulations, interest is growing in the use of chloramines as a primary disinfectant. Chloramines are generally less reactive than free chlorine, but they tend to be more unstable because they undergo autodecomposition (NRC 2006). Chloramine also reacts with NOM and reduced metals in the aqueous phase, as well as with Fe(II) in pipe deposits, leading to the formation of ammonia. Trace levels of free chlorine tend to come to equilibrium with monochloramine, which results in a slow monochloramine loss caused by reaction with NOM. Chloramine losses should decrease with increasing pH because both autodecomposition and its reaction with NOM become slower. Some of the key reactions governing chloramine in water are shown below:

- Chloramine autodecomposition

$$3\,NH_2Cl \rightarrow N_2 + NH_4^+ + 3\,Cl^- + H^+ \qquad \text{(Eq 7-2)}$$

- Oxidation of organic matter by chloramines

$$0.1\,C_5H_7O_2N + NH_2Cl + 0.9\,H_2O \rightarrow 0.4\,CO_2 + 0.1\,HCO_3^- + 1.1\,NH_4^+ + Cl^- \qquad \text{(Eq 7-3)}$$

- Reaction of chloramine with corrosion products at pipe wall

$$0.5\,NH_2Cl + H^+ + Fe^{2+} \rightarrow Fe^{3+} + 0.5\,NH^{4+} + 0.5\,Cl^- \qquad \text{(Eq 7-4)}$$

- Oxidation of nitrite by chloramines

$$NH_2Cl + NO_2^- + H_2O \rightarrow NH_3 + NO_3^- + HCl \qquad \text{(Eq 7-5)}$$

Nitrification

An important process associated with chloramination is biological nitrification, which is a process in which bacteria oxidize reduced nitrogen compounds (e.g., ammonia) to nitrite and then nitrate. It is an important process associated with nitrifying bacteria in distribution systems and long retention times in water supply systems practicing chloramination. An important problem associated with nitrification is loss of the chloramine disinfectant residual. A reduction in ammonia results in an increased ratio of chlorine to ammonia nitrogen, which controls the stability of monochloramine. If the ratio approaches 1.5 on a molar basis, then a rapid loss of monochloramine occurs, which is related to the oxidation of N(III) to primarily nitrogen gas and the release of more ammonia. The released ammonia can be further oxidized by the nitrifying organisms. A positive feedback loop is established, and the loss of disinfectant residual allows for the increased activity of nitrifiers and may lead to the increased occurrence of microorganisms such as coliforms and heterotrophic bacteria. The loss of chloramine residual is probably the most significant health threat that could result from nitrification.

Modeling the Decay of Chlorine Residuals

Because of the importance of disinfection, a number of investigators have conducted research into the development of models to predict chlorine decay in drinking water. In one of earliest attempts to model chlorine decay, Feben and Taras (1951) developed the following model:

$$D_t = D_1 t^n \tag{Eq 7-6}$$

In Eq 7-6, D_t = the chlorine consumed at time t (hr), D_1 = the chlorine consumed after 1 hr, and n = a constant characteristic of a given water. The 1-hr chlorine demand and n must be determined experimentally for a given water.

The long-term decay of chlorine, in both distributed drinking water and in natural waters receiving chlorinated discharges, has been typically modeled using first-order kinetics (Johnson 1978):

$$C_t = C_0 \exp(-kt) \tag{Eq 7-7}$$

where C_t = chlorine concentration at time t, C_0 = initial concentration, t = time (hr), and k = a first-order reaction rate coefficient (hr^{-1}). The rate coefficient k is highly dependent on the source and degree of treatment received by the water. Although simple to apply, the first-order model is not capable of reproducing the higher decay rate observed in the initial stages of chlorination or the slow tailing off of the rate at very long reaction times.

Haas and Karra (1984) investigated several models to describe chlorine decay including:

- First-order decay
- Power-law decay (nth order)
- First-order decay with stable components
- Power-law decay with stable components (nth order)
- Parallel first-order decay

They found that the parallel first-order decay model yielded the best results. This model assumes that two constituents occur in water that react with chlorine: (1) fast-reacting components, which exert some initial decay, and (2) slow-reacting components, which are responsible for long-term chlorine demand. This model was applied by Vasconcelos et al. (1996) in a project conducted jointly by the US Environmental Protection Agency and the American Water Works Association, which will be discussed later.

A model by Qualls and Johnson (1983) described the short-term chlorine consumption by fulvic acids during the first 5 min of a reaction. This model (Eq 7-7) was originally developed for cooling water systems, but then applied to disinfection of natural waters:

$$d\text{Cl}/dt = \text{k}_1[\text{Cl}][\text{F}_1] + \text{k}_2[\text{Cl}][\text{F}_2] \tag{Eq 7-7}$$

The chlorine decay is described by the sum of two first-order equations, in which the first part describes a rapid decay within the first 30 sec, and the second simulates a slower decay from 30 sec to 5 min. In Eq 7-7, [Cl] is the free residual chlorine, k_1 and k_2 are rate constants for the fast and slow reactions, respectively, and $[F_1]$ and $[F_2]$ are the concentrations of reactive sites on the fulvic acids for the fast and slow reaction, respectively.

Ventresque et al. (1990) conducted a study at the Choisy-le-Roi water treatment plant near Paris to identify the organic components that react with chlorine. The plant consists of preozonation, coagulation, sedimentation, sand filtration, ozonation, granular activated carbon (GAC) bioadsorption, and postchlorination. The water is dechlorinated before being distributed to the consumer. The authors applied a second-order kinetic model to the long-term chlorine demand data. They analyzed the kinetics at every stage of the plant and found that the initial chlorine demand and the kinetic constants of ozone/GAC–treated water were always lower then those obtained from sand-filtered water. These results demonstrated the effect of activated carbon on the removal of organic matter.

Jadas-Hécart et al. (1992) attempted to identify the organic compounds that react with chlorine from the Seine River entering the final treatment stage at the Choisy-le-Roi plant. They studied the chlorine consumption kinetics of samples of water taken after overall treatment. They divided the chlorine decay into two phases. An initial phase of immediate consumption during the first 4 hr was called the initial chlorine demand. The second slower consumption phase after the first 4 hr was defined as the long-term demand. The long-term demand (LTD) was interpreted with the following kinetic equation:

$$\frac{dx}{dt} = \mathrm{k}(a - x)^{\alpha}\,(b - \frac{x}{n}) \qquad \text{(Eq 7-8)}$$

where x = chlorine consumed after $(t - 4)$ hr, k = the rate constant, a = the total residual chlorine at 4 hr, b = maximum potential chlorine demand, n = stoichiometry coefficient, and α, β = partial orders of reaction. This model cannot be applied from time zero of chlorination because the long-term demand obeys a different reaction order then the initial chlorine demand.

The water treatment plant (USEPA 1992) model described chlorine decay by dividing the decay curve into three components. These included an initial (t < 5 min) reaction, a second-order reaction (5 min < t < 5 hr), and a first-order reaction (t > 5 hr). These three phases are described by the following equations:

$$t < 5\ \text{min}\quad \ln(C_0 - C_1 - 7.6\text{NH}_3) = -\,0.62 + 0.522\ln(C_0/\text{TOC}) + 0.302\ln(\text{UV}_{254}) + 0.842\ln(\text{TOC}) \qquad \text{(Eq 7-9)}$$

$$5\ \text{min} < t < 5\ \text{hr} \qquad (1/C_t) = (1/C_1) + k_1 t \qquad \text{(Eq 7-10)}$$

$$t > 5\ \text{hr} \qquad C_t = Ae^{-k2t} \qquad \text{(Eq 7-11)}$$

C_t is the chlorine residual at 5 min,and k_1 and k are second-order and first-order rate parameters. A is obtained by setting Eq 7-10 and Eq 7-11 equal to each other at t = 5 hr. These equations apply only when the initial chlorine/TOC ratio is ≥ 1:1.

Zhang et al. (1992) conducted a study of chlorine modeling in sand-filtered water (before postchlorination) at the Macao treatment plant, which draws water from the estuary of the West River (the main stream of the Pearl River in South China). The study indicated that the chlorine consumption in sand-filtered water can be divided into two phases: an initial chlorine consumption during the first hour, which corresponds to the contact time in the reservoir of the treatment plant, and a long-term chlorine consumption after 1 hr in the network. This second component is interpreted in terms of an apparent first-order equation. According to the experiments performed on steel and asbestos–cement pipes (diameter > 250 mm), the chlorine consumption by pipes is negligible. The authors suggested more experiments to be performed to verify the eventual influence of water velocity and diameter of pipe on the chlorine consumption by pipe itself. The chlorine disappearance in the network of Macao can be modeled as a first-order reaction.

To describe the entire disinfectant reaction for one groundwater treated in a particular plant, Lyn and Taylor (1993) calculated the chlorine residual (CLR) as a function of chlorine dose, DOC, temperature, and time, using an empirical constant applicable only for that particular water. No general equation for estimating the constants for other waters were given.

$$\begin{aligned} CLR = 0.285(Cl_2 - \text{dose})^{1.631}\,(DOC)^{-0.313}\,\text{Temp})^{-0.176} \\ (DOC)^{-0.241}\,(\text{Temp})^{0.101}\,(\text{Time})^{0.265} \end{aligned} \qquad \text{(Eq 7-12)}$$

Biswas et al. (1993) proposed (as discussed earlier) an equation that represents the average loss of chlorine through a pipe as follows:

$$C(X, r) = 2\Sigma_{n=1}^{\infty} \frac{\lambda_n J_o(\lambda_n r) J_1(\lambda_n)}{(\lambda_n^2 + A_2^2) J_o^2(\lambda_n)} \times \exp\,[-A_1 + \lambda n^2 A_0)X] \qquad \text{(Eq 7-13)}$$

Where:

J_0 and J_1 = the Bessel functions of the first kind of order zero and 1, respectively

λ_ns = the roots of $\lambda_n J_1(\lambda_n) = A_2 J_0(\lambda_n)$

A_1, A_2, and A_3 = nondimensional parameters to be determined empirically

Rossman et al. (1994) proposed within EPANET (as discussed earlier) a mechanism for considering the loss (or growth) of a substance by reaction as it travels through the distribution system. Reaction is assumed to occur both within the bulk flow and within the pipe-wall material, based on using first-order kinetics. The general expression for chlorine decay in the bulk flow and at the pipe wall is

$$\frac{dc}{dt} = k_b c - \frac{k_f}{r_h}(c - c_w) \qquad \text{(Eq 7-14)}$$

Where:

dc/dt = rate of chlorine decay, in mg/L/d

c = chlorine concentration in the bulk flow, in mg/L

t = time, in days

k_f = mass transfer coefficient, in ft/d

r_h = hydraulic radius

c_w = chlorine concentration at the pipe wall, in mg/L

k_b = bulk phase chlorine decay coefficient, in day^{-1}

The first term in Eq 7-14 is the bulk flow reaction, while the second term includes c_w, which represents the rate at which material is transported between the bulk flow and reaction sites at the pipe wall. If it is assumed that the rate at the wall is first-order with respect to c_w and that it proceeds at the same rate as material is transported to the wall so that no accumulation occurs, the following mass balance at the wall holds:

$$k_f(c - c_w) = k_w c_w \qquad \text{(Eq 7-15)}$$

Where:

k_w = wall reaction constant, in ft/d (m/d)
and other variables as defined previously.

Solving for c_w and substituting into Eq 7-14 yields the following reaction rate expression:

$$\frac{dc}{dt} = -k_b c - \frac{k_w k_f c}{r_h(k_w + k_f)} \quad \text{(Eq 7-16)}$$

$$k_t = k_b + \frac{k_w k_f}{r_h(k_w + k_f)} \quad \text{(Eq 7-17)}$$

Substituting Eq 7-17 into Eq 7-14 yields

$$\frac{dc}{dt} = -k_t c \quad \text{(Eq 7-18)}$$

For steady-state flow conditions in a single pipe, this rate expression yields a first-order decay model for chlorine, i.e.:

$$C_t = C_0 e^{-k_t t} \quad \text{(Eq 7-19)}$$

Where:

C_t = concentration of chlorine at any time t, in mg/L
C_0 = initial concentration of chlorine, in mg/L

However, the decay constant in this model is now a function of a bulk decay rate constant, a wall decay rate constant, the molecular diffusivity of chlorine, the water's kinematic viscosity, velocity, and the pipe radius (Clark et al. 1995).

Dugan et al. (1995) proposed a saturation model based on the Michaelis–Menton equation to predict the entire chlorine decay curve with one equation. TOC was chosen as the predictive water quality parameter for the saturation model because it represents the compounds exhibiting chlorine demand. The following model was proposed:

$$\frac{dC}{dt} = -(\mathrm{k} * \mathrm{TOC} * C_t)/(\mathrm{K} * \mathrm{TOC} + C_t) \quad \text{(Eq 7-20)}$$

$$C_t = \mathrm{K} * \mathrm{TOC} * \ln(C_0 / C_t) - \mathrm{k} * \mathrm{TOC} * t + C_0 \quad \text{(Eq 7-21)}$$

$$\mathrm{K} = a + b * (C_0 / \mathrm{TOC}) \quad \text{(Eq 7-22)}$$

$$\mathrm{k} = c + d * (C_0 / \mathrm{TOC}) \quad \text{(Eq 7-23)}$$

Eq 7-21 is used to predict the entire chlorine decay curve, and Eq 7-22 and Eq 7- 23 are used to estimate the rate constants k and K.

Koechling (1998) developed a modified version of the saturation model that made the rate coefficients a function of the water's DOC and UVA:

$$\frac{dC_t}{dt} = \frac{-k_2 \mathrm{SUVA}_0 C_t}{k_1 + C_t} \qquad \text{(Eq 7-24)}$$

Where

k_1 = 4.98 UVA_0 + 1.91 DOC
k_2 = −0.32 $\mathrm{UVA}_0^{1.365}$
SUVA_0 = UVA_0 / DOC
and C_t = chlorine concentration at time t, DOC = dissolved organic carbon (mg/L), UVA_0 = UVA at time 0 (cm^{-1}).

This model gave excellent results for a large range of waters tested.

Chambers et al. (1995) conducted a study to test the validity of the exponential decay expression for free and total chlorine modeling using two sample networks and using proprietary models. A sampling program was devised to collect information to calibrate the models: Rate constants for free and total chlorine were calculated. The results showed that the exponential decay model is appropriate for modeling chlorine in distribution systems and that it is possible to produce useful water quality models. They also showed that the rate constants for the water in the network were different than the rate constants collected in bench-top experiments.

Clark (1998) developed an equation for chlorine decay based on the concept of competing reacting substances and on the assumption that the balanced reaction equation be represented by:

$$aA + bB \rightarrow pP \qquad \text{(Eq 7-25)}$$

where A represents chlorine, B the collection of all reacting substances, P the collection of all reaction products, and a, b, and p are stoichiometric coefficients. In Eq 7-25, if A and B are the reacting substances and P is the product of the reaction then the rate of reaction is given by:

$$dC_A/dt = -k_A C_A C_B$$

or

$$dC_B/dt = -k_B C_A C_B \qquad \text{(Eq 7-26)}$$

or

$$dC_p/dt = k_p C_A C_B$$

where k_A, k_B, and k_P are second-order reaction rate coefficients. By equating the amount of A that reacts in a given time to the equivalent amount of B and integrating the result, the following equation for chlorine at any time t is obtained:

$$C_t = K / \{1 - \mathrm{R}^{-ut}\} \qquad \text{(Eq 7-27)}$$

where R (dimensionless), K (mg/L), and u (min^{-1}) are parameters to be estimated, and t = the time of reaction in min. In Eq 7-27, the value for the rate constants can be rewritten as follows:

$$u = M(1-K) \quad \text{(Eq 7-28)}$$

Where:

$$M = \frac{k_A b C_0}{a} \quad \text{(Eq 7-29)}$$

and C_0 is the chlorine concentration at time 0

Clark and Sivaganesan (1998) used Eq 7-27 to predict chlorine decay and TTHM formation in a number of field and laboratory data sets. A two-stage least-squares estimation procedure using the "Marquardt–Levenberg" method was used to estimate the model's coefficients. Predictive equations were developed for the parameters K, M, and R based on initial chlorine concentration (C_0), pH, TOC, and temperature (Temp in °C). The estimated parameters for Eq 7-27 are as follows:

$$K = e^{0.32}\,(C_0)^{-0.44}\,(\text{TOC})^{0.63}\,(\text{pH})^{-0.29}\,(\text{Temp})^{0.14}\ (R^2 = 0.71) \quad \text{(Eq 7-30)}$$

$$R = e^{1.49}\,(C_0)^{-0.48}\,(\text{TOC})^{0.18}\,(\text{pH})^{0.96}\,(\text{Temp})^{0.28}\ (R^2 = 0.78) \quad \text{(Eq 7-31)}$$

and

$$\text{Log}_e(\text{M}) = -2.46 - (0.19\ \text{TOC}) - (0.14\ \text{pH}) - (0.07\ \text{Temp}) + (0.01\ \text{Temp*pH})\ (R^2 = 0.42) \quad \text{(Eq 7-32)}$$

They found that a two-stage second-order model provided an excellent representation of bulk phase chlorine decay.

Powell et al. (2000a) observed that most modeling packages fail to account for temporal variability in chlorine decay characteristics. Frequently, chlorine decay is represented by the following first-order relationship:

$$C = C_0 \exp(-kt) \quad \text{(Eq 7-33)}$$

where C = chlorine concentration at time t (mg/L); C_0 = initial chlorine concentration (mg/L); and k = first-order chlorine decay constant (L/h). It is widely accepted that chlorine in distribution systems will decay either because of reactions with compounds contained within the bulk water or reactions at the pipe wall. Therefore, a simple method to represent this relationship is for this to define the overall decay constant as the sum of a bulk and wall decay constant (k_b and k_w):

$$k = k_b + k_w \quad \text{(Eq 7-34)}$$

The authors conducted laboratory investigations to study the factors that affect bulk chlorine decay and to develop functional relationships that can be used to enhance the durability of network models. All the research related exclusively to the decay of free chlorine. All the water samples were taken from distribution networks or water treatment works operated by Severn Trent Water Ltd. The investigators observed that k_b was affected by temperature, organic content (TOC and UV), and the initial chlorine concentration (C_0). They found a twofold increase in decay rates between 10 and 20°C, a linear increase with TOC or UV, and an inverse relationship with C_0. No relationship was observed between pH and k_b, possibly because of the small pH range observed. In all observed cases, rechlorination of a sample reduced the value of k_b. Some of the observed behavior contradicts the assumption of first-order kinetics. Second-order kinetics can partially explain this behavior. The authors concluded that the most pragmatic method is to assume first-order decay but to define the decay constant as a function of temperature, TOC and UV, and the initial chlorine concentration (C_0).

According to Powell et al. (2000b), drinking water can contain traces of many different compounds that have the potential to react with chlorine. Each of these compounds would have a different reaction mechanism. Most network models assume that chlorine decays by first-order kinetics with respect to chlorine only, as defined by the following equation:

$$C = C_0 \exp(-kt) \qquad \text{(Eq 7-35)}$$

where C = chlorine concentration at time t (mg/L); C_0 = initial chlorine concentration (mg/L); and k = first-order chlorine decay constant (L/h). This has led some researchers to model wall and bulk decay separately. The simplest method is to define the overall decay constant as the sum of a bulk and wall decay constant (k_b and k_w):

$$k = k_b + k_w \qquad \text{(Eq 7-36)}$$

Where bulk decay is given by k_b and the wall decay constant is k_w. The investigators conducted a series of experiments to compare the predictive performance of the various decay kinetic models that have been proposed and select the most suitable decay kinetic model for network modeling purposes. The objectives of this study were to measure chlorine decay in a large number of samples, representative of a wide range of water characteristics and environmental conditions. The investigators examined the performance of a number of alternative decay kinetic models against 207 determinations of bulk chlorine decay against time. Inevitably, the more parameters a model had, the better it would fit the observed data. However, the performance benefit over the simple first-order model was marginal, with all models giving an average R^2 fit in excess of 0.95. Measurement of longitudinal decay profiles in two different systems found the first-order model to be adequate for modeling in situ (bulk and wall) decay. Considerable variability was observed in the value of the best-fit first-order decay constants. Some of this variability is likely a result of changes in the concentration and chemical nature of the compounds that the chlorine is reacting with. A second-order model would be theoretically more stable under such variations; however, difficulty in defining the value of the decay parameters currently restricts its use for network modeling purposes.

Hallam et al. (2002) developed a method for measuring the in-situ chlorine decay in distribution system pipe. It was found that the pipe material has a strong influence on wall decay. It was also found that reactive pipe included unlined iron pipe, and unreactive pipes included PVC and cement-lined iron pipe. Results suggested that the wall decay of unreactive pipes is limited by the reactivity of the pipe material, but the wall decay of reactive pipes is limited by the transfer of chlorine to the pipe wall. The wall decay rate is inversely related to initial chlorine concentration for low-reactivity pipes. In general, water velocity increases wall decay rates though the statistical confidence is low for low-reactivity pipes. A moderate biofilm coating did not influence the wall decay rate for low-reactivity pipes.

Courtis et al. (2009) refined earlier work conducted by Hallam et al. (2002) using new data from the same water treatment works and distribution system. The investigators analyze 189 chlorine decay tests collected over a 2.5-year period. The work develops empirical relationships between the decay of free chlorine in the bulk water (k_B) at various locations and the final water chlorine concentration, TOC, temperature, and a parameter reflecting the cumulative concentration of chlorine applied to the water. Previous work produced an empirical method for predicting the bulk decay of chlorine in distribution systems but did not reflect the water's chlorination concentration history. New data presented show that TOC and temperature are usually effectively constant through a given system for a given day. Therefore, the main effect controlling chlorine uptake in distribution is the change to the organic carbon matrix of the water through chlorine-driven oxidation processes. The complex chemistry of organics oxidation means that neither TOC nor the number of rechlorination stages appears to be an appropriate independent parameter. Consequently, it should not be surprising that new data show that the previous approach was found to lack precision and accuracy for the conditions observed. The revised approach to bulk decay determination has resulted in a practical, cost-effective methodology that accommodates the effects of seasonal variations in the water treatment works and the distribution system. Embedded within this approach is the inherent assumption that wall and mixing effects on k_B are implicitly incorporated in the equation coefficients.

Boccelli et al. (2003) extended the concept of chlorine as a first-order reaction in drinking water to include relatively simple dynamic models that include a second hypothetical reactive species. Their research uses this approach to account for discontinuities associated with rechlorination events, such as those caused by booster chlorination and by mixing at distribution system junction nodes. Their analysis showed that the reactive species model will always represent chlorine decay better than, or as well as, a first-order model, under single-dose or rechlorination conditions. These results are confirmed by experiments on five different natural waters and further demonstrate that the reactive species model can be significantly better under some rechlorination conditions. Trihalomethane (THM) formation was also monitored, and results show that a linear relationship between total THM (TTHM) formation and chlorine demand is appropriate under both single-dose and rechlorination conditions. This linear relationship was estimated using the modeled chlorine demand from a calibrated reactive species model and using the measured chlorine demand, both of which adequately represented the TTHM formation.

Huang and McBean (2006, 2008) built on the model based on the concept of competing reacting substances proposed by Clark (1998). A corrected mathematical formulation is developed and, because the recent findings suggested that not all natural organic matter (NOM) is involved in the chlorine decay process, an additional parameter was introduced. A parameter assignment method, employing Bayesian statistical analysis and incorporating Monte Carlo Markov Chain (MCMC) modeling with Gibbs sampling to make inferences, was used in the estimation of model parameters. Three parameters are estimated for the model, namely the ratio of chlorine to TOC, the chlorine reaction rate, and a fraction factor of TOC that represents the true amount of TOC involved in chlorine decay process. Water samples taken from Goderich, Ont., Canada, in the summer of 2005 are used for estimating the parameters. Bayesian parameter estimation methods were used in conjunction with a combination of domain knowledge and observed data to establish a probability distribution, known as the posterior, to describe the unknown parameters in the model developed by domain experts. This method is subjective since the domain knowledge might be subjective. However, it is also the means by which Bayesian statistics manages the uncertainty of the observed data and the uncertainty of the unknown parameters. The posterior distribution combines the available prior information (domain knowledge), expressed as a given distribution, and the information in the observed data, expressed as the likelihood (joint density), and the proportionality to the product of the two. Their findings suggest that

- A second-order chlorine decay model can be established that considers the reactant in the finished water.
- The key parameters, namely, the ratio of the quantity of the two reactants, the chlorine decay reaction rate, and the fraction factor (which shows how much reactant [TOC] is involved in the chlorine decay process), can be estimated using a Bayesian statistical approach.
- The uncertainty of both measurements and the various forms of TOC can be appropriately managed by implementing MCMC and Gibbs sampling.

Clark et al. (2010) conducted a study in parallel pipe loops located at the USEPA's T&E Facility in Cincinnati, Ohio (Rossman et al. 1999). Two loops were used in this study in a continuously recalculating "batch" mode (no outflow and no inflow). One was an unlined ductile-iron pipe and the other composed of PVC. Five experimental runs were conducted in each loop in an attempt to characterize the loss of chlorine in the loop as a function of velocity. Tests were performed at the recirculation flow rates of 0.011 m^3/min (3 gpm), 0.023 m^3/min (6 gpm), 0.038 m^3/min (10 gpm), and 0.379 m^3/min (100 gpm), and under no flow or stagnant conditions. The water temperature ranged between 12.439°C (55.6°F) and 26.14°C (74.7°F), and the pH of the water ranged between 8.19 and 9.06. The initial free chlorine concentrations were varied for several of the experiments, but Clark et al. (2010) focused on the studies in which free chlorine concentrations were held at approximately 2.0 mg/L.

In order to provide an analytical framework, a first-order chlorine residual decay over time was used, although evidence exists that bulk residual decay exhibits second-order kinetics (Clark 1998, Yang et al. 2007, Huang and McBean 2006 and 2008). For purposes of this analysis, the following is assumed:

$$C_t = C_0 e^{-(k_b + k_r)t} \qquad \text{(Eq 7-37)}$$

where C_t = the concentration of chlorine at any time t, C_0= the initial concentration of chlorine, k_b= the bulk decay constant, k_r = the wall decay constant, and t = time in seconds.

Taking the natural log of Eq 7-37 and rearranging yields:

$$\ln\left(\frac{C_t}{C_0}\right) = -(k_b + k_r)t \qquad \text{(Eq 7-38)}$$

The model developed for use in the study borrowed from both the Biswas et al. (1993) model and EPANET (Rossman et al. 1994). For purposes of this analysis, the following three modeling approaches will be examined:

- Wall reaction limited (EPANET).

$$C_t = C_0 e^{-(k_b + k_r)t} \qquad \text{(Eq 7-39)}$$

- Zero-order reaction limited.

$$C_t = C_0 e^{-k_b t} - \frac{2k_w}{ak_b}\left(1 - e^{-k_b t}\right) \qquad \text{(Eq 7-40)}$$

- Mass transfer limited.

$$C_t = C_0 e^{-\left(k_b + \frac{2k_{mt}}{a}\right) t} \quad \text{(Eq 7-41)}$$

In the above equations, C_t is the concentration of chlorine at any time t, k_b is the bulk demand, k_r is the wall demand, k_{mt} is the mass transfer coefficient, a is the pipe radius, and C_0 is the initial chlorine concentration. It was found that differences existed in the wall demand characteristics between unlined metallic pipe and PVC. For example, it was found that chlorine residual virtually disappears in the unlined metallic pipe in a short time but stayed at a relatively high level in the PVC pipe. THMs generated in both loops during the experiment were very similar, even though chlorine residual transport levels were quite different. For water utilities, these results imply that a change in pipe material can result in more effective maintenance of chlorine residual. The experimental results reported (Clark et al. 2010) demonstrate that surface or pipe wall demand varies significantly among different types of pipe material, and therefore the type of pipe material can influence the transport of free chlorine in drinking water distribution system networks. These characteristics can have regulatory implications. For example, chlorine residuals dissipate rapidly in unlined ductile-iron pipe but are maintained in PVC pipe. It is also clear from the experimental results that the various models proposed for predicting chlorine transport have limitations. For example, based on the results presented, wall demand, which is a fundamental parameter in EPANET, is a fitting parameter and not necessarily constant under all flow conditions. The zero-reaction model, which is frequently mentioned in the literature, has a similar problem. The zero-reaction coefficient also varies with flow but, as pointed out by DiGiano and Zhang (2005), varies in a more consistent and predictable manner. The mass transfer model has similar limitations.

Modeling the Decay of Chloramine Residuals

While autodecomposition occurs via a complex set of reactions, the net loss of monochloramine occurs according to the stoichiometry (NRC 2006):

$$3\,NH_2Cl \rightarrow N_2 + NH_3 + 3\,Cl^- + 3\,H^+ \quad \text{(Eq 7-42)}$$

This reaction can be approximated (in the absence of other reactions) by a simple second-order relationship,

$$1/[NH_2Cl] - 1/[NH_2Cl]_0 = kvsc\,t \quad \text{(Eq 7-43)}$$

Yang et al. (2008) developed a suspended growth nitrification model to describe nitrification dynamics in terms of chloramine, ammonia, nitrite, nitrate, and nitrifying bacteria concentrations in a pilot-scale water distribution system. The model was able to simulate the absence or occurrence of nitrification episodes observed in the pilot-scale system and to provide a mechanism for studying the regrowth and persistence of nitrifiers. In their restricted low-nutrient drinking water environment, growth kinetic parameters for nitrifiers were estimated to be significantly lower than ranges reported in the literature. Results suggested that heterotrophic growth can be a significant contributor to chloramine decay in some chloraminated distribution systems. From a water utility management perspective, the most important function of the model was to predict whether a nitrification episode would or would not occur at some point in the future. In addition to predicting nitrification absence and occurrence in the future, accurate projection of when and where nitrification will occur in full-scale systems would allow water utilities to implement control strategies such as flushing or boosting chlorine residuals before any measurable deterioration of water quality.

Westbrook and DiGiano (2009) conducted a study using bench-scale, pipe section reactors (PSRs) to determine chloramine surface reaction rate constants for cement-lined ductile-iron pipe and tuberculated, unlined cast-iron pipe. The reactor was constructed of a small section of pipe (2 ft) that is modified by inserting impeller and stirring motors in such a way that permitted measurement of the disinfectant decay rate as a function of water velocity. Small batches of water were used in which the chemical characteristics and temperature were easily controlled. The investigators used modified first-order decay models to characterize the loss of chloramine residual. Initial disinfectant concentration, pH, and temperature were measured. The experiments showed that in cast-iron (CI) pipe, chloramine decay was three orders of magnitude faster than bulk decay, and the reaction rate was first-order with respect to chloramine, unlike the zero-order reaction found for free chlorine in previous work (DiGiano and Zhang 2005). This difference was attributed to the slower reaction of chloramine than that of free chlorine, which therefore controlled the reaction rate instead of $Fe^{2}+$ in the corrosion scales. The rate of reaction at the CI pipe surface increased with decreasing pH. The investigators believe that one possible explanation is that lower pH causes an increase in $Fe^{2}+$ dissolution from corrosion scales with subsequent oxidation by chloramine. The dependence of the chloramine decay rate on water velocity suggests that both the rate of iron-corrosion product release and mass transfer are also important rate-limiting factors in CI pipe. The chloramine decay rate was much slower in the presence of ductile-iron (DI) pipe than it was in CI pipe; however, the rate was two orders of magnitude greater than the bulk decay rate. The investigators therefore concluded that interactions between chloramine and the cement lining are the predominant cause of chloramine decay. The rate of reaction at the DI pipe increased with decreasing pH. Because the surface reaction rate was much slower for DI pipe than for CI pipe, mass transfer was not rate controlling, as evidenced by no change in the decay rate over the range of water velocities investigated.

Zhang and Edwards (2009) conducted a study to test the hypothesis that chloramine may be less persistent than free chlorine in certain situations encountered in premise plumbing. The study was conducted using a large-scale pipe-rig setup designed to simulate operating conditions in a premise-plumbing system. The rig used a large reservoir to feed eight parallel pipe sections, each composed of a different type of material that might be used in home plumbing. Three rigs were used to explore extremes of disinfection encountered throughout water distribution systems. Water flowed through the pipe loops on a regular schedule of 2 min of hydraulic flushing (flow rate of 900 mL/min) after 8 hr of stagnation. The three pipe rigs were maintained at room temperature throughout the test. Initially, the eight pipe materials used in each rig were glass tubes covered with foil (control), cross-linked polyethylene (PEX), polyethylene (PE), chlorinated polyvinyl chloride (PVC), copper (Cu), copper with epoxy coating (CuE), galvanized iron, and stainless steel. Through modeling analysis, the investigators found that the relative efficacy of chlorine versus chloramine is a complex function of autodecomposition, reactions with corrosion by-products, nitrification rates, and the pipe materials themselves. They concluded the following:

- In the absence of nitrification, chloramine was always more persistent than chlorine, confirming expectations based on previous research.

- With rapid nitrification in relatively inert pipe materials (e.g., PVC, CuE, stainless steel), chloramine was less persistent than chlorine. In such situations, chloramine was less effective than free chlorine in reducing bulk water and biofilm HPCs.
- In reactive pipe materials (e.g., cast-iron, galvanized-iron, old lead), the relative disinfectant decay rate and the ability to control bacterial growth are strongly dependent on the nitrification. In this study, similar decay rates were observed for chlorine and chloramine in these materials, but in situations in which nitrification rates are high and corrosion rates are low, chlorine would tend to provide better control of microbial regrowth.

Modeling By-product Formation

Disinfection by-products (DBPs) in distribution systems may be formed because of the reactions of chemical disinfectants with NOM either in the bulk phase or possibly with pipe deposits (NRC 2006). Most studies have focused on formation of halogenated DBPs produced from reactions of NOM with free chlorine. USEPA currently regulates trihalomethanes (THMs) and haloacetic acids (HAAs), although over 600 DBPs have been identified (NRC 2006), including both chlorinated and brominated compounds. Brominated compounds result from the oxidation of bromide, which can be an important factor in determining DBP speciation even when found at the sub–milligram-per-liter level. Many of the DBPs formed in chlorinated systems are also the same as those observed in systems that chloraminate. However, the rates of formation of most DBPs are much slower in chloraminated systems, resulting in the reduced levels of many DBPs, especially THMs.

Given the relatively high reactivity of free chlorine with NOM, it is not surprising that a significant amount of DBPs are formed in the water treatment plant as the result of primary disinfection (i.e., disinfection at the treatment plant to meet *CT* requirements). DBP formation, however, continues in the distribution system.

To date, DBP modeling efforts are empirical and semimechanistic. Some have been developed that couple disinfectant loss to the formation of selected DBPs. It should be pointed out that while chlorine demand may be a useful measurement to correlate with DBP formation in the bulk water, this would not likely be the case if the demand were governed by inorganic constituents.

TTHM Formation Modeling

Trussell and Umphres (1978) reviewed the effect of preozonation, bromide, pH, and chlorine dose on the formation of TTHMs in natural waters and proposed a kinetic model that describes the formation of TTHMs. They concluded that some of the factors that might influence the rate of the TTHM reaction are pH, temperature, the level of precursor, the level of chlorine, and the level of bromide ion before chlorine addition. They proposed two equations: one describing the rate of chlorine (Cl) consumption, and one describing the rate of reduction of precursor or, inversely, the rate of TTHM production. Assuming that the reaction between chlorine residual and aquatic humus is related to the concentration of each of them, a simple relation is obtained for the rate of chlorine consumption.

$$\frac{d[\mathrm{Cl_2}]}{dt} = -k_1\,[\mathrm{Cl_2}][\mathrm{TOC}] \qquad \text{(Eq 7-44)}$$

If it is assumed that the action of the chlorine does not significantly reduce the total concentration of the humic precursor and it is assumed that the action of the chlorine does not reduce the precursor material present in the water, then the following equation can be used to estimate the evolution of TTHMs with time:

$$\frac{\mathrm{dTHM}}{dt} = -\frac{dC}{dt} = k_2(\mathrm{Cl}_2)(C)^m \qquad \text{(Eq 7-45)}$$

where m is the order of reaction with respect to the precursor concentration and C is the concentration of the organic precursor. Because C^m is assumed to be a constant, the growth of TTHM is seen to be first-order with respect to chlorine decay.

Kavanaugh et al. (1980) developed a two-parameter kinetic model for predicting THM formation in the distribution system following postchlorination. They hypothesized that THM formation can be described by the overall stoichiometric expression

$$3A + B \rightarrow C \qquad \text{(Eq 7-46)}$$

where A = HOCl, B = TOC, and C = TTHM. Based on this expression, 3 mol of hypochlorous acid react with 1 mol of carbon in the organic precursor material to form 1 mol of TTHM. The rate expression for the formation of TTHM is given by

$$\frac{dC}{dt} = k_n(B)(A)^m \qquad \text{(Eq 7-47)}$$

assuming that the rate of formation is first order with respect to TOC and m-th order with respect to HOCl. The chlorine concentration A can then be related stoichiometrically to C by defining a TTHM yield as the moles of TTHM formed per mole of Cl_2 consumed. The rate expression then becomes

$$\frac{dC}{dt} = k_n(B(A_0 - \frac{3C}{f})^m) \qquad \text{(Eq 7-48)}$$

When the free chlorine is exhausted, $f = \frac{3C}{A_0}$. The yield f can be determined by measuring the THM concentration when $A = 0$. The equation is therefore a three-parameter kinetic model with k_n as the rate constant, m the reaction order with respect to Cl_2, and f the TTHM yield that must be determined empirically for the particular system under investigation.

Amy et al. (1987) discussed the formulation and calibration of several models for predicting total trihalomahanes (TTHM) in untreated natural waters subjected to chlorination. Their general approach was to analyze specific portions of a large database derived from several natural water sources in order to isolate the effects of a given parameter on TTHM formation. Two general strategies were used in formulating the models. One employed multiple linear regression models using logarithmic transformations of both independent and dependent variables. The second developed multiple nonlinear regression models. Both models assume that a chlorine residual is maintained throughout a 168-hr reaction period and that TTHMs continuously increase with time. Quantities such as UVA, chlorine dose, temperature, and TOC were used as the independent variables.

Harrington et al. (1992) developed a computer program to simulate disinfection byproduct formation, removal of natural organic matter, inorganic water quality changes, and disinfectant decay in water treatment processes. Equations were developed that simulate the formation of total trihalomethanes (TTHMs), removal of total organic carbon (TOC), and ultraviolet absorbance by alum coagulation, as well as changes in alkalinity and pH. Model simulations were compared with limited sets of observed values. The central tendency of the model was to underpredict finished-water pH by 4 percent, finished-water TOC by 7 percent, and simulated distribution system TTHMs by 20–30 percent.

Cowman and Singer (1996) investigated the effect of bromide ion on the distribution of haloacetic acid (HAA) species resulting from the chlorination and chloramination of waters containing aquatic humic substances. Aquatic humic substances were extracted from a surface water and a groundwater and were chlorinated and chloraminated under standard conditions at pH 8 and pH 6 in the presence of bromide concentrations ranging from 0 to 25 μM (0–2 mg/L). The treated waters were analyzed for all nine of the HAA species containing bromine and chlorine. Bromochloro-, bromodichloro-, and dibromochloroacetic acid formed readily and constituted at least 10 percent of the total HAA concentration in waters containing as little as 1.2 μM (0.1 mg/L) bromide. The mixed bromochloro HAA species were major components of the total HAA concentration at bromide concentrations found in raw drinking waters. Distribution of the HAA species among the mono-, di-, and trihalogenated forms appeared to be independent of bromide concentration.

Clark et al. (1996) postulated that TTHM formation is governed by the following equation:

$$\mathrm{DBP} = \mathrm{DBP}_u - \mathrm{DBP}_u^t \qquad \text{(Eq 7-49)}$$

Where:

DBP_u = the ultimate formation potential of an individual DBP
DBP_u^t = the remaining formation potential at time t

They also made the first-order assumption

$$\frac{d\mathrm{DBP}_u^t}{dt} = -k\mathrm{DBP}_u^t \qquad \text{(Eq 7-50)}$$

Where:

DBP_u^t = DBP_u at $t = 0$

Integrating Eq 7-50 yields

$$\mathrm{DBP} = \mathrm{DBP}_u(1 - e^{-kt}) \qquad \text{(Eq 7-51)}$$

Eq 7-50 and Eq 7-51 illustrate that both the loss of chlorine residual and the formation of TTHMs are time-dependent and will vary throughout the system depending on propagation conditions. As mentioned earlier, a study to evaluate these changes was conducted by USEPA in the North Marin Water District (NMWD).

Clark et al. (1996) developed a first-order model to characterize the formation of brominated disinfection by-products. Using a data set generated by Pourmoghaddas et al. (1993), models were developed that describe the formation of THM and non-THM chlorination by-products and their speciation. The model, which considered pH, time, chlorine, and bromide concentration, demonstrates the effect of bromide concentration on $CHCl_3$ and $CHBr_3$ formation. The model shows that the concentration of $CHBr_2Cl$ and $CHBrCl_2$ increases to a maximum level for bromide concentrations of 2.5 mg/L and 0.5 mg/L, respectively, and then declines with increasing bromide levels. Concentrations of $CHCl_3$ consistently decline with increasing bromide concentration, and $CHBr_3$ consistently increases with increasing bromide concentration.

As a complement to his work on chlorine decay modeling, Clark (1998) proposed a TTHM formation model that said that TTHM formation was linearly proportional to chlorine consumption as follows:

$$\text{TTHM} = T\,(C_0 - (\{C_0(1-R)\}/\{1-\text{Re}^{-ut}\})) \qquad \text{(Eq 7-52)}$$

where TTHM = total trihalomethane at time t (μg/L), T = μg of TTHM produced per mg/L of chlorine consumed, C_0 = initial chlorine concentration (mg/L), and R, u = parameters from the chlorine decay equation (7-27). Clark and Sivaganesan (1998) verified this equation using field data from the study by Vasconcelos et al. (1996, 1997) and laboratory data collected by the Water Supply and Water Resources Division of USEPA.

Rossman et al. (1999) examined the factors that characterize the reaction conditions affecting treated water in a distribution system. Some of these factors are the pipe wall material, such as corrosion products and biofilm slime, which can exert a significant chlorine demand. If chlorine is being consumed within a pipe by non-DBP producing materials (such as ferrous corrosion products), then less is available to react with the water's natural organic matter. However, iron tubercles have also been shown to contain organic material that might include DBP precursors. Much of the study was devoted to studying the rate at which DBPs form in a pipe as compared to a glass bottle. One of the conclusions from their study was thatthe rate constants for chlorine decay in the pipe were an order of magnitude higher than in the bottle.

Much of the study was devoted to assessing the factors affecting the formation of disinfection by-products and will be discussed in more detail in the following section.

Carrico and Singer (2009) studied the effect of conventional and booster chlorination on chlorine residuals and trihalomethane (THM) formation in drinking water distribution systems using the EPANET hydraulic modeling software. Results from their research suggest that booster chlorination may allow utilities to meet disinfection goals better by carrying chlorine residuals to remote points in the distribution system. This would provide a more consistent chlorine residual and possibly reduce THM formation. A new version of EPANET, the EPANET Multi-Species eXtension model, or EPANET MSX, was used to compare chlorine decay caused by reactions in the bulk fluid and reactions occurring at the pipe wall. The results suggest that chlorine decay caused by wall reactions can be very significant at remote points in the distribution system. Additionally, if THMs are assumed to form primarily through reactions in the bulk fluid, use of the new EPANET Multi-Species eXtension model software allows for calculation of THM formation based solely on chlorine reactions in the bulk fluid rather than on overall chlorine decay. The investigators confirmed that THM formation can be modeled solely as a function of chlorine decay (Clark 1998, Clark and Sivaganesan 1998, Boccelli et al. 2003).

Exposure to DBPs in Distribution Systems

Traditionally, the US Environmental Protection Agency has established environmental regulatory programs for the protection of the outdoor environment from industrial and commercial sources of contamination. It has become increasingly clear, however, that indoor environments may be at risk from contaminants that are found outside the household. For example, drinking water can be a carrier of contaminants that volatilize when they enter a household, thereby subjecting the consumer to inhalation, ingestion, and dermal exposure (Clark and Goodrich 1992). A USEPA study found that personal or indoor exposures to many toxic or carcinogenic chemicals are greater than outdoor concentrations. It also found that for all chemicals, except for trihalomethanes, the air provided > 99 percent of the exposure. Water provided nearly all of the exposure to the trihalomethanes and more than half of most personal exposure to chloroform.

Clark et al. (1992) demonstrated that the dynamics of multiple source mixing and system operation in a water distribution system can affect lifetime exposures to volatile organic chemicals such as chloroform. The dose received will depend on one's location within the system and the timing of water use. Dynamic water quality models are valuable tools for conducting drinking water exposure assessments. These models help in interpreting the results of field monitoring sites so that a clearer picture of the distribution system emerges and provide time-varying estimates of drinking water quality at all locations within a service area so that more accurate exposure assessments can be made. It was observed that for adults, the inhalation pathway for chloroform contributes more to total exposure that does the ingestion pathway at most locations within the service area, although the opposite is true for infants. Showering is the greatest potential contributor to household inhalation exposure.

Schenck et al. (1998) conducted a study to assess the applicability of the model developed by Vartiainen and Liimatainen (1988) to source waters and water treatment practices in the United States. The model developed by Vartiainen and Liimatainen (1988) based on data collected in Finland relates mutagenicity, as determined in the Ames assay, to total organic carbon (TOC) concentration of the water, chlorine dose, and to a minor extent, the concentration of ammonia. It has been used as the basis for recent epidemiological studies conducted in Finland that have reported a positive correlation between the mutagenicity of chlorinated drinking waters and certain human cancers. In the work by Schenck et al. (1998), water samples were collected from three full-scale treatment plants and one pilot-scale plant in the United States. All the plants used chlorine exclusively for disinfection. One full-scale plant used groundwater, while the other plants used surface water. TOC and ammonia concentrations were determined analytically, and chlorine doses were obtained from the treatment plants. The water samples were concentrated by XAD resin adsorption for testing in the Ames assay. The observed levels of mutagenicity in the finished waters were 1.5 to 2-fold higher than those predicted using the Vartiainen and Liimatainen model. The authors concluded that further validation was needed before the Finnish model could be used to assess exposure to mutagenicity in chlorinated drinking waters in the United States.

According to Yang et al. (2007), the implementation of chloramine disinfection introduces free ammonia into drinking water distribution systems, which in turn serves as an energy source for ammonia-oxidizing bacteria that cause nitrification. Nitrification may be affected by many factors including temperature, pH, water age, chloramine residual, and chlorine-to-ammonia application ratio. The authors developed a simplified model that was able to predict the risk of nitrification occurrence in a pilot-scale system based on a combination of pH, temperature, and total chlorine residual. The objective of the model was to provide an easy-to-use tool for drinking water utilities to minimize the risk of nitrification occurrence based on ambient conditions (temperature and pH) and operating parameters (water age and chloramine residual).

References

Amy, G.L., P.A. Chadik, and Z.K. Chowdhury. 1987. Developing Models for Predicting Trihalomethane Formation Potential and Kinetics. *Journal AWWA*, 79(7):89–97.

Amy, G.L. 1993. Using NOM Characterization for the Evaluation of Treatment. In *Workshop Proc. Natural Organic Matter in Drinking Water: Origin, Characterization, and Removal, Sept. 19–22, Chamonix, France.* Denver, Colo.: AWWA, 19.

Bellar, T.A., and J.J. Lichtenberg. 1974. Determining Volatile Organics at Microgram-per-Liter Levels by Gas Chromatography. *Journal AWWA*, 66:739–744.

Biswas, P., C. Lu, and R.M. Clark. 1993. A Model for Chlorine Concentration Decay in Drinking Water Distribution Pipes. *Water Research*, 27(12):1715–1724.

Boccelli, D.L., Tryby, M.E., Uber, J.G., and Summers, R.S. 2003. A Reactive Species Model for Chlorine Decay and THM Formation Under Rechlorination Conditions. *Water Res.*, 37(11), 2654–2666.

Brezonik, P.L. 1994. *Chemical Kinetics and Process Dynamics in Aquatic Systems.* Chelsea, Mich.: Lewis Publishers.

Carrico, B., and P.C. Singer. 2009. Impact of Booster Chlorination on Chlorine Decay and THM Production: Simulated Analysis. *Journal of Environmental Engineering*, 135(10).

Chambers, V.K., J.D. Creasey, and J.S. Joy. 1995. Modeling Free and Total Chlorine Decay in Potable Water Distribution Systems. *Journal of Water Supply: Research and Technology–AQUA*, 44(2):60–69.

Clark, R.M., and J.A. Goodrich. 1992. Modeling Human Exposure to Contaminants from Drinking Water. *Journal of Water Supply: Research and Technology–AQUA*, 41(4):224–230.

Clark, R.M., L.A. Rossman, and J.A. Goodrich. 1992. Modeling the Variation in Human Exposure to Contaminants From Drinking Water. *Journal of Exposure Analysis and Environmental Epidemiology*, Suppl. 1, 159–175.

Clark, R.M, and W.A. Feige. 1993. Meeting the Requirements of the Safe Drinking Water Act. In *Strategies and Technologies for Meeting SDWA Requirements,* ed. R.M. Clark, R.S. Summers. Lancaster, Pa.: Technomic.

Clark, R.M., G. Smalley, J.A. Goodrich, R. Tull, L.A. Rossman, J. Vasconcelos, and P. Boulos. 1994. Managing Water Quality in Distribution Systems: Simulating TTHM and Chlorine Residual Propagation. *Journal of Water Supply: Research and Technology–AQUA*, 43(4):182–191.

Clark, R.M., L. Rossman, and L. Wymer. 1995. Modeling Distribution System Water Quality: Regulatory Implications. *Journal of Water Resources Planning and Management*, 121(6):423–428.

Clark, R.M., H. Pourmoghaddas, L.G. Wymer, and R.C. Dressman. 1996. Modeling the Kinetics of Chlorination By-Product Formation: The Effects of Bromide. *Journal of Water Supply: Research and Technology–AQUA*, 45(1):1–8.

Clark, Robert M. 1998. Chlorine Demand and TTHM Formation Kinetics: A Second-Order Model. *Journal of Environmental Engineering*, 124(1):16–24.

Clark, R.M., and M. Sivaganesan. 1998. Predicting Chlorine Residuals and the Formation of TTHMS in Drinking Water. *Journal of Environmental Engineering*, 124(12):1203–1210.

Clark, R.M., Y.J. Yang, C.A. Impellitteri, R.C. Haught, D.A. Schupp, S. Panguluri, and E.R. Krishnan. 2010. Chlorine Fate and Transport in Distribution Systems: Experimental and Modeling Studies. *Journal AWWA*, 102(5):144–155.

Courtis, B.J., J.R. West, and J. Bridgeman. 2009. Temporal and Spatial Variations in Bulk Decay Within a Water Supply System. *Journal of Environmental Engineering,* 135(3):147–152.

Cowman, G.A., and P.C. Singer. 1996. Effect of Bromide Ion on Haloacetic Acid Speciation Resulting from Chlorination and Chloramination of Aquatic Humic Substances. *Environ. Sci. & Techn.,* 30(1):16–24.

DiGiano, F.A., and Zhang, W. 2005. Pipe Section Reactor to Evaluate Chlorine–Wall Reaction. *Journal AWWA*, 97(1):74–85.

Dugan, N.R., R.S. Summers, R.J. Miltner, H.M. Shukairy. 1995. An Alternative Approach to Predicting Chlorine Residual Decay. In *Proc. AWWA–WQT Conference, Nov. 12–16, New Orleans, La.* Denver, Colo.: AWWA, 1317–1338.

Feben, D., and M.J. Taras. 1951. Studies on Chlorine Demand Constants. *Journal AWWA,* 43(11):922–932.

Haas, C.N., and S.B. Karra. 1984: Kinetics of Wastewater Chlorine Demand Exertion. *Journal WPCF,* 56(2):170–173.

Hallam, N.B., J.R. West, C.F. Forster, J.C. Powell, and I. Spencer. 2002. The Decay of Chlorine Associated With the Pipe Wall in Water Distribution Systems. *Water Research,* 36(13):3479–3488.

Harrington, G.W., Z.K. Chowdhury, and D.M. Owen. 1992. Developing a Computer Model to Simulate DBP Formation During Water Treatment. *Journal AWWA*, 84(11):78–87.

Huang, J.J., and E.A. McBean, 2006. Using Bayesian Statistics to Estimate the Coefficients of a Two-Component Second-Order Chlorine Bulk Decay Model for a Water Distribution System. *Water Res.*, 41(2):287–294.

Huang, J.J., and E.A. McBean. 2008. Using Bayesian Statistics to Estimate Chlorine Wall Decay Coefficients for Water Supply System. *Journal of Water Resources Planning and Management*, 134(2).

Jadas-Hécart, A., A. El Morer, M. Stitou, P. Bouillot, and B. Legube. 1992: Modelisation de la Démande en Chlore d'une Eau Traitée. *Water Research*, 26(8):1073.

Johnson, J.D. 1978. Measurement and Persistence of Chlorine Residuals in Natural Waters. In *Water Chlorination: Environmental Impact and Health Effects*, ed. R.L. Jolley. Ann Arbor, Mich.: Ann Arbor Science.

Kavanaugh, M.C., A.R. Trussell, J. Cromer, and R.R. Trussell. 1980. An Empirical Kinetic Model of Trihalomethane Formation: Applications to Meet the Proposed THM Standard. *Journal AWWA*, 72(10):578–582.

Koechling, M.T. 1998. *Assessment and Modeling of Chlorine Reactions With Natural Organic Matter: Impact of Source Water Quality and Reaction Conditions*. Ph.D. Thesis, Department of Civil and Environmental Engineering, University of Cincinnati, Cincinnati, Ohio.

Lyn, T.L., and J.S. Taylor. 1993. Modeling Compliance of Chlorine Residual and Disinfection By-Products. In *Proc. AWWA-WQT Conf., Nov. 7–11, Miami, Fla.*, 513–523. Denver, Colo.: AWWA.

National Research Council (NRC). 2006. *Drinking Water Distribution Systems: Assessing and Reducing Risks.* Washington, D.C.: National Academies Press, 15–46.

Pourmoghaddas, H., A.A. Stevens, R.N. Kinman, R.C. Dressman, L.A. Moore, and J.C. Ireland. 1993. Effect of Bromide Ion on the Formation of HAAs During Chlorination. *Journal AWWA*, 85(1):82–87.

Powell, J.C., N.B. Hallam, J.R. West, C.F. Forester, and J. Simms. 2000a. Factors Which Control Bulk Chlorine Decay Rates. *Water Research*, 34(1):117–126.

Powell, J.C., N.B. Hallam, J.R. West, C.F. Forester, and J. Simms. 2000b. Performance of Various Kinetic Models for Chlorine Decay. *Journal of Water Resources Planning and Management*, 126(1):13.

Qualls, R.G., and J.D. Johnson. 1983. Kinetics of the Short-Term Consumption of Chlorine by Fulvic Acid. *Environ. Sci. & Techn.*, 17(11):692–698.

Rook, J.J. 1974. Formation of Haloforms During Chlorination of Natural Waters. *Water Treatment Exam.*, 23:234.

Rossman, L.A., R.M. Clark. and W.M. Grayman. 1994. Modeling Chlorine Residuals in Drinking-Water Distribution Systems. *Journal of Environmental Engineering*, 120(4):803–820.

Rossman, L.A., R.A. Brown, P.C. Singer, and J.R. Nuckols. 1999. Kinetics of THM and HAA Production in a Simulated Distribution System. In *Proc. AWWA Water Quality Technology Conference, Oct. 31–Nov. 3, 1999, Tampa, Fla.* Denver, Colo.: AWWA.

Schenck, K.M, L.J. Wymer, B.W. Lykins Jr., and R.M. Clark. 1998. Application of a Finnish Mutagenicity Model to Drinking Waters in the US. *Chemosphere*, 37(3):451–464.

Shukairy, H.M., 1994: *The Control of Disinfection By-Product Formation by Ozonation and Biotreatment.* Ph.D. Dissertation, University of Cincinnati, Cincinnati, Ohio.

Shukairy, H.M., M.T. Koechling, and R.S. Summers. 1994a. Fractionated Organic Matter DBP Formation: Kinetics, Reactivity and Speciation. In *Proc. AWWA Annual Conf., June 19–23, New York.* Denver, Colo.: AWWA, 941–957.

Trussell, R.R. and M.D.Umphres. 1978. The Formation of Trihlomethanes. *Journal AWWA*, 70(11):604–612.

US Environmental Protection Agency (USEPA). 1992. *Water Treatment Plant Simulation Program User's Manual Version 1.21.* Washington, D.C.: USEPA Drinking Water Technology Branch, Drinking Water Standards Division, Office of Ground Water and Drinking Water.

Vasconcelos, J.J., P.F. Boulos, W.M. Grayman, L. Kiene, O. Wable, P. Biswas, A. Bhari, L.A. Rossman, R.M. Clark, and J.A. Goodrich. 1996. *Characterization and Modeling of Chlorine Decay in Distribution Systems.* Denver, Colo.: AwwaRF and AWWA.

Vasconcelos, J.J., L.A. Rossman, W.M. Grayman, P.F. Boulos, and R.M. Clark. 1997. Kinetics of Chlorine Decay. *Journal AWWA*, 89(7):54–65.

Vartiainen, T., and A. Liimatainen. 1988. Relations Between Drinking Water Mutagenicity and Water Quality Parameters. *Chemosphere*, 17(1):189–202.

Ventresque, C., G. Bablon, B. Legube, A. Jadas-Hécart, and M. Dore. 1990. Development of Chlorine Demand Kinetics in a Drinking Water Treatment Plant. In *Water Chlorination: Chemistry, Environmental Impact and Health Effects,* Vol. 6, eds. R.L. Jolley, L.W. Condie, J.D. Johnson, S. Katz, R.A. Minear, J.S. Mattice, and V.A. Jacobs. Chelsea, Mich.: Lewis Publishers, 715–728.

Westbrook, A., and F.A. DiGiano. 2009. Rate of Chloramine Decay at Pipe Surfaces. *Journal AWWA.* 101(7):59–70.

Yang, J., G.W. Harrington, D.R. Noguera, and K.K. Fleming. 2007. Risk Analysis of Nitrification Occurrence in Pilot-Scale Chloraminated Distribution Systems. *Journal of Water Supply: Research and Technology–AQUA,* 56(5):293–311.

Yang, J., G.W. Harrington, and D.R. Noquera. 2008. Nitrification Modeling in Pilot-Scale Chloraminated Drinking Water Distribution Systems. *Journal of Environmental Engineering,* 134(9).

Zhang, G.R., L. Keine, O. Wable, U.S. Chan, and J.P. Duget. 1992. Modeling of Chlorine Residual in the Water Distribution System Network of Macao. *Environmental Technology,* 13:937–946.

Zhang, Y., and M. Edwards. 2009. Accelerated Chloramine Decay and Microbial Growth by Nitrification in Household Plumbing. *Journal AWWA,* 101(11):51–62.

Chapter 8

Early Modeling Studies

Introduction

Several early water quality modeling studies were conducted, under the sponsorship of the USEPA, to validate and demonstrate the potential for modeling water quality in drinking water distribution systems. These five early studies played a major role in the development of water quality models for application to drinking water distribution systems and will be discussed in a historical context. The first of these studies was conducted by the USEPA in collaboration with New Vienna, Ohio, which is a village of 1,000 people. The distribution system was characterized in a link–node form (Clark and Males 1986, Clark and Males 1985). The model was used to evaluate the costs and the variations in water quality in the distribution system. The second study was a comprehensive water quality modeling study conducted under a cooperative agreement initiated between the North Penn Water Authority (NPWA), Lansdale, Pa., and the US Environmental Protection Agency (USEPA). It was a comprehensive project designed to investigate the feasibility of water quality modeling development and application to study the mixing of water from multiple sources. This project investigated the feasibility of the development and application of a steady-state water quality model. As the study progressed, it became obvious that the dynamic nature of both demand patterns and water quality variations required the development of a dynamic water quality model. In addition, techniques for semicontinuous monitoring of volatile organic contaminants (VOCs) were explored (Clark et al. 1988) in this study.

After the NPWA study concluded, it became apparent that there was need to develop more sophisticated approaches to water quality modeling including dynamic water quality modeling. Therefore another study was initiated by the USEPA and the University of Michigan in collaboration with the South Central Connecticut Regional Water Authority (SCCRWA) to further develop the concept of water quality modeling in drinking water distribution systems. A third study was conducted by the USEPA and the North Marin Water District, which at the time of the study served a suburban population of 53,000 in or near Novato, Calif. This study integrated the results from extensive field monitoring and modeling. It also provided a field-level demonstration of EPANET. A fifth study, which built on the NMWD study, was conducted by the USEPA in collaboration with the Awwa Research Foundation (now Water Research Foundation). The NPWA, the SCCRWA, the NMWD and the joint USEPA/AwwaRF (WaterRF) studies are discussed in this chapter.

North Penn Water Authority Study

At the time of the USEPA study, the NPWA served 14,500 customers in 10 municipalities. Average demand was 5 mgd (19 ML/d) (Clark et al. 1988). Water sources included a 1-mgd (3.7-ML/d) treated surface water source purchased from the Keystone Water Company and 4 mgd (15 ML/d) from 40 wells operated by NPWA. Figure 8-1 is a schematic representation of the 225 mi (362 km) of pipe in the NPWA distribution system. The figure shows the location of wells, the Keystone tie-in, and the three pressure zones: the Souderton zone, Lansdale low zone, and Hillcrest zone.

Surface water entered the NPWA system at the Keystone tie-in. The rate of flow into the system was determined by the elevation of the tank in the Keystone system and by a throttling valve at the tie-in. Flow was monitored continuously and was relatively constant. Water flowed into the Lansdale low-pressure zone, entered the Lawn Avenue tank, and was pumped into the Souderton zone. Additional water from the Hillcrest pressure zone entered the Lansdale system at the Office Hillcrest transfer point. This water was solely derived from wells in the Hillcrest zone. Except for unusual and extreme circumstances, such as fire or main breaks, water did not flow from the Souderton zone into the Lansdale low zone, nor from Lansdale low into Hillcrest.

Distinct chemical characteristics were associated with the Keystone water as compared with the well waters. Keystone water contained total trihalomethanes (TTHM) at

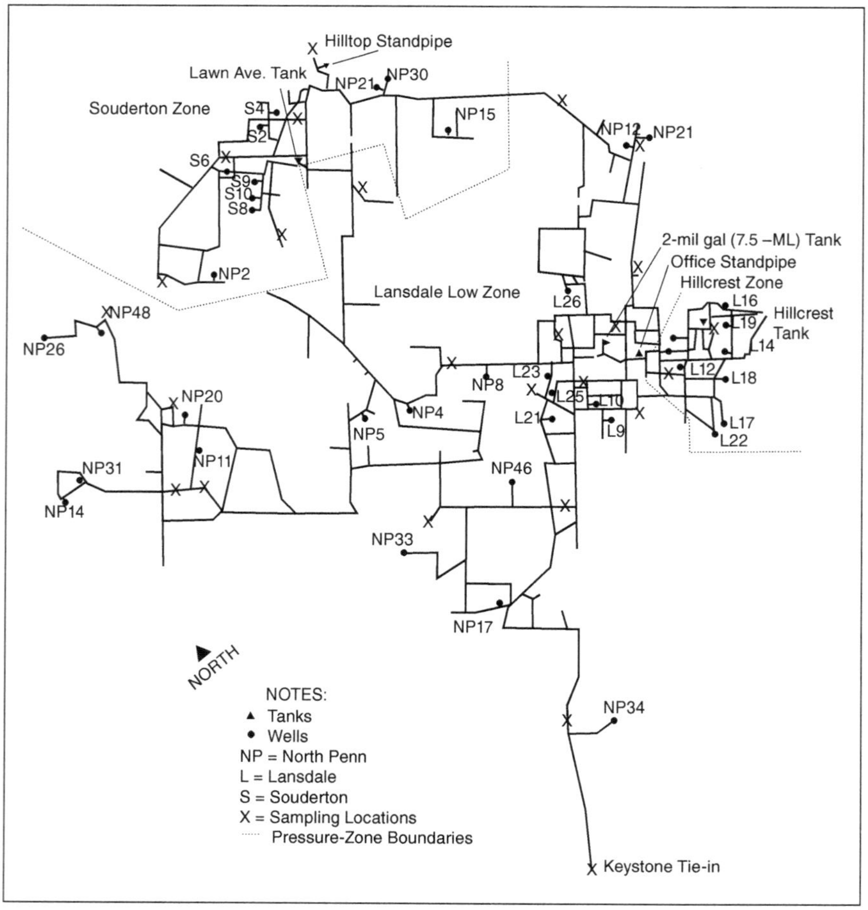

Figure 8-1. North Penn Water Authority distribution system

significantly higher levels than the well water. Certain wells showed the occurrence of trichloroethylene (TCE) and/or cis-1,2-dichloroethylene (cis-1,2-DCE). Inorganic chemicals also varied from well to well, and between the wells and Keystone.

Network Modeling

The NPWA distribution system was modeled in a network representation consisting of 528 links and 456 nodes. Water demands for modeling represented conditions from May to July 1984. The network hydraulic model used was the Wadiso model, which contained provisions for both steady-state and quasi-dynamic hydraulic modeling or EPS (Gessler and Walski 1985). The model was employed to study the overall sensitivity of the system to well pumping, demand, and other factors resulting in the development of a number of typical flow scenarios.

The modeling effort revealed that significant portions of the system were subject to flow reversals. This type of analysis was used to select sampling points for an in-depth sampling effort described later.

Dynamic Variations in Water Quality Data

To investigate the nature of water quality variability under dynamic conditions within the system, a sampling program was conducted at six sites. Site selection was based on spatial variations determined from historical data and modeling results. Figure 8-2 shows the various sampling points used in the field study. Figure 8-3 depicts the results of the intensive sampling program using total trihalomethanes (TTHMs) as a tracer. Laboratory and field evaluations demonstrated that the THMs in the Keystone water had reached their formation potential and were relatively stable. Any trihalomethanes (THMs) formed from well sources were relatively minor. Figure 8-3 also shows the variation in hardness of these same points (hardness was primarily associated with flow from the wells). At the Mainland sampling point, a flushing back and forth of water between the surface source and the well sources could be seen. The peaks of the TTHMs at Mainland were approximately 12 hr out of phase with the peaks from the wells. This indicated that water flow at this point was affected by the surface and groundwater sources. The results pointed out the problems in attempting to predict a dynamic situation using a steady-state approach and the dynamic nature of water movement in the distribution system.

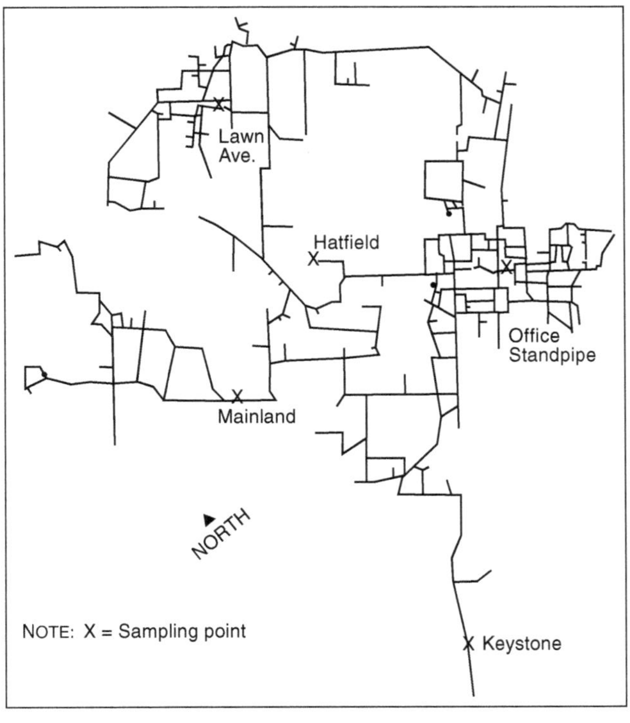

Figure 8-2. Sample points for field study

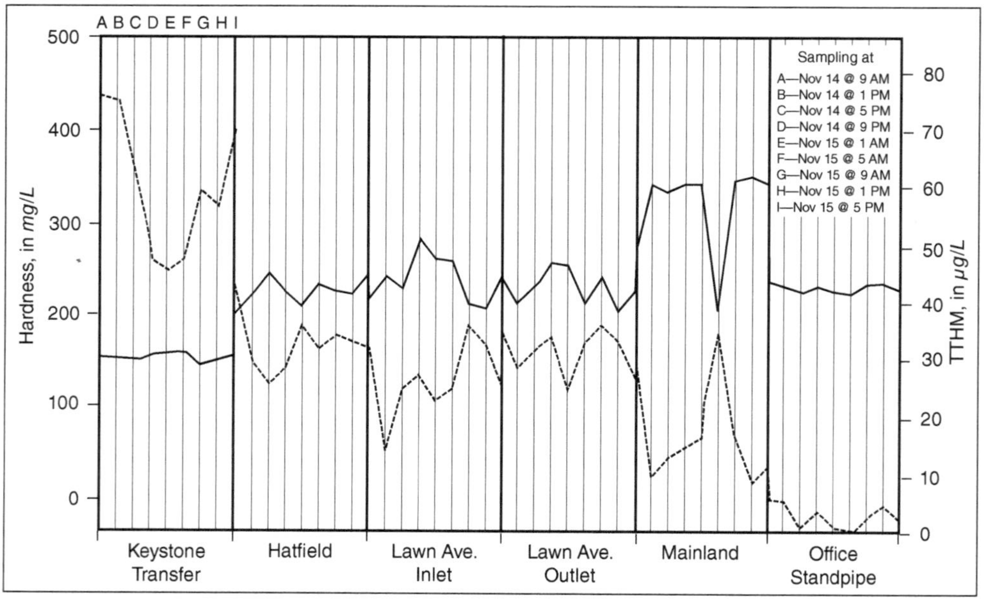

Figure 8-3. Results from six sampling stations in North Penn Water Authority

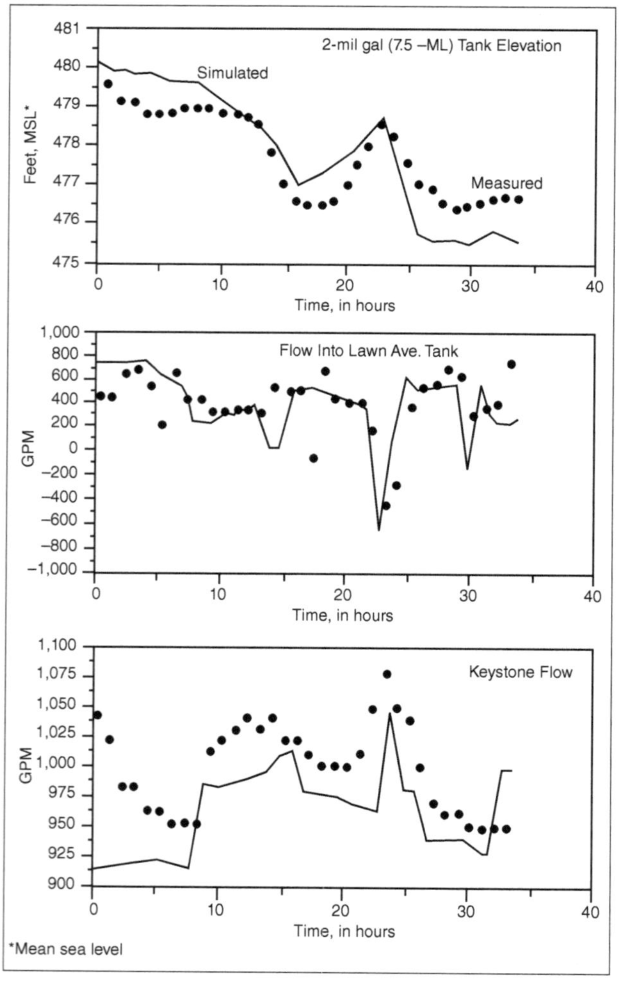

Figure 8-4. Comparison of measured and simulated hydraulic conditions

Dynamic Water Quality Algorithm Development

A dynamic water quality model was developed using a numerical routing solution to trace water quality through the network. Demands and inflows (both values and concentrations) were assumed constant over a user-defined period, and a quasi-dynamic externally generated hydraulic solution were used for each period. Thus, the flow and velocity for each link were known from the hydraulic solution for each time period. Each time period was evenly divided into an integer number of computational time steps. Each link was then divided into sublinks by a series of evenly spaced subnodes (though the distance between subnodes varied from link to link or for a link at different time periods) such that the travel time from a subnode (or node) to the adjacent subnode (or node) was approximately equal to Δt (increment of time).

Model Application

The dynamic water quality algorithm was implemented as a microcomputer-based dynamic water quality model written in FORTRAN, and the program, called the Dynamic Water Quality Model (DWQM), was applied in a full-scale demonstration on the NPWA distribution system.

A 34-hr period was simulated, corresponding to conditions present during the pilot-level sampling program conducted on Nov. 14–15, 1985. Parameters of the model were adjusted so that predicted tank levels and flows at selected sites represented those measured during the sampling period. A comparison of measured and modeled hydraulic conditions at three locations is shown in Figure 8-4. The resulting predicted time history of chloroform is plotted in Figure 8-5 along with field sampling results for three sampling stations.

Modeling results for TTHM are plotted in Figure 8-6 for three sampling stations, as are the sampling results. The results of the predicted hardness as compared to sampling measurements are presented in Figure 8-7.

The results of applying the dynamic water quality model to the North Penn system as reflected in Figures 8-5 and 8-6 indicate both close agreement between predicted and observed results and some anomalous behavior. For chloroform, TTHM, and hardness, the general levels of concentration compare very favorably to the observed values at the three selected sampling stations (Figures 8-5, 8-6, and 8-7). In each case, some differences occurred in the timing of peak or minimum values. When the spatial variations of predicted TTHM concentrations were compared to the historical average TTHM level, the same general patterns were apparent. Additionally, the predicted patterns bracketed the pattern corresponding to the long-term historical average; a result that would be expected because the two selected times correspond to the extreme spatial patterns during the sampling period.

South Central Connecticut Regional Water Authority Case Study

The North Penn case study, discussed previously, provided fertile ground for developing a dynamic water quality propagation model. In order to extend that application, USEPA initiated a cooperative agreement with the University of Michigan. Under the agreement, the university began a program with the South Central Connecticut Regional Water Authority (SCCRWA) to test the concepts developed previously in the North Marin case study including field studies to verify and calibrate the model (Clark and Goodrich 1993).

At the time of the study, the SCCRWA supplied water to approximately 95,000 customers (380,000 individuals) in 12 municipalities in the New Haven, Conn., area. The SCCRWA service area was divided into 16 separate pressure/distribution zones (see Figure 8-8).

Average production was 50 mgd (189 ML/d) with a safe yield of approximately 74 mgd (280 ML/d). Surface water sources included Lake Gaillard, Lake Saltonstall, Lake Whitney, and the West River system. Five well fields served as sources (North Cheshire, South Cheshire, Mt. Carmel, North Sleeping Giant, and South Sleeping Giant). Approximately 80 percent of the water in use in the system came from surface sources, with the remaining 20 percent from wells. All water was treated using chlorination, filtration, and a phosphate

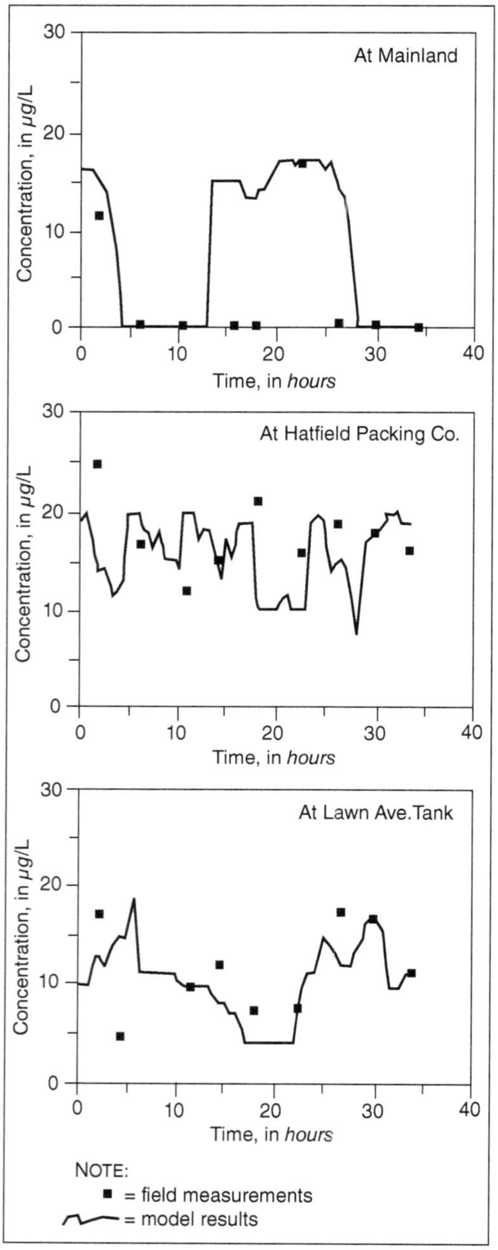

Figure 8-5. Comparison of modeled and measured chloroform at selected stations

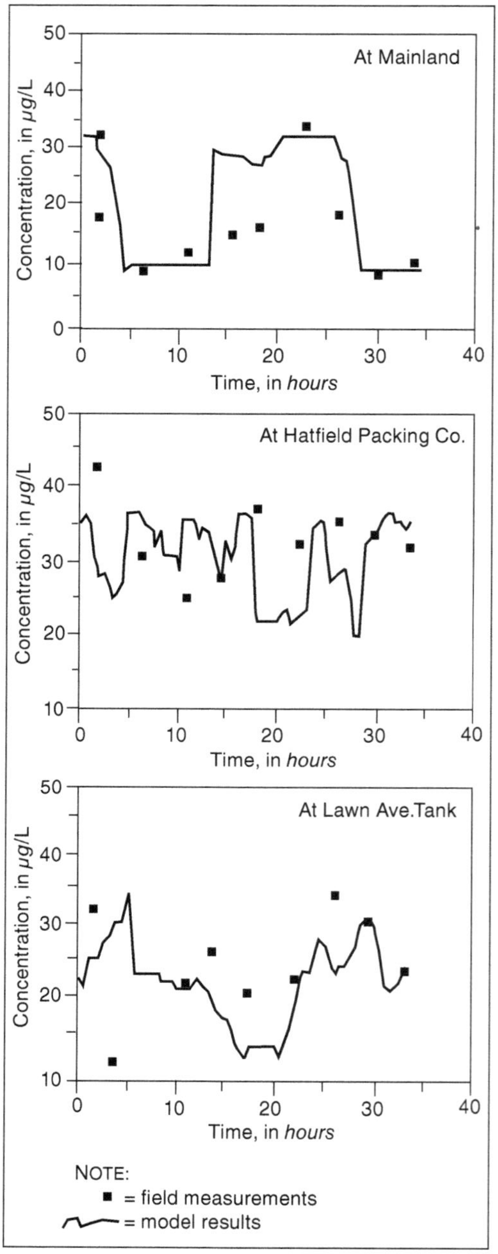

Figure 8-6. Comparison of modeled and measured TTHM at selected stations

corrosion inhibitor. The system included 22 pumping stations, 23 storage tanks, and approximately 1,300 mi (2,080 km) of water mains.

System Modeling

The preliminary efforts in developing and validating a model for the SCCRWA were concentrated on the Cheshire service area. This area is relatively isolated and provided a prototype

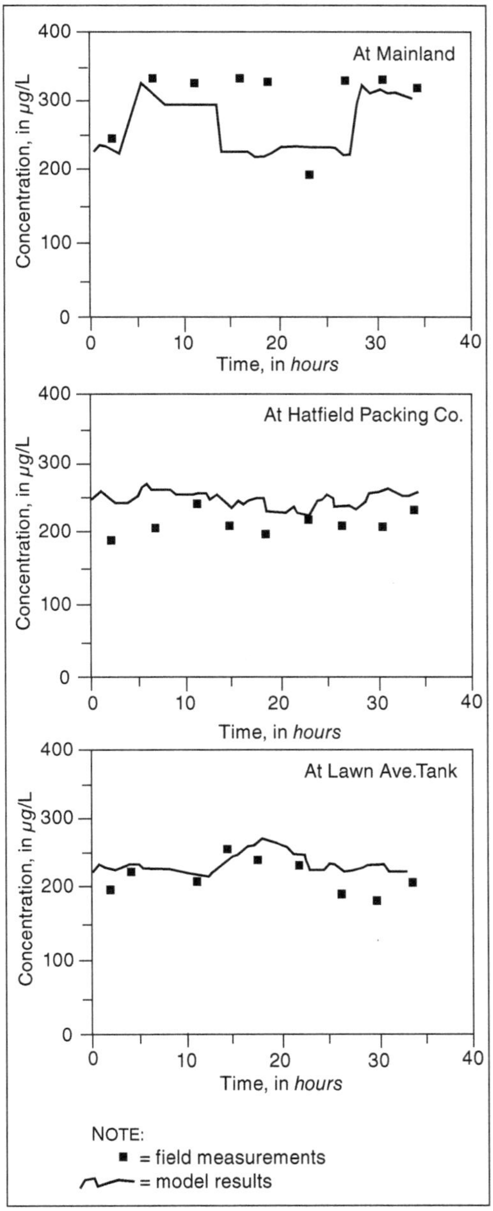

Figure 8-7. Comparison of modeled and measured hardness at selected stations

for modeling the remainder of the system. In order to validate the model, an extensive study was planned in which the fluoride feed at the North Well field was turned off and the propagation of the fluoride feedwater tracked through the system.

Hydraulic Modeling

Prior to the application of the water quality modeling effort and related field study, extensive hydraulic analyses were conducted on the system. For the preliminary modeling, the full SCCRWA system network was represented by approximately 520 nodes and 700 links. This representation was modified based on the most current maps supplied by the authority. For other service areas, distribution system maps were used to develop a network skeletonization at a level of detail consistent throughout the system. The network was drawn as a mylar overlay to the US Geological Survey (USGS) 1:24,000-scale topographic maps. In most cases, the system was represented by skeletonization (i.e., selectively choosing pipes based on their size and perceived impact as transmission mains). In a few cases, surrogate pipes were used to represent the effects of several pipes. For example, a single 14-in. (355-mm) pipe may be used to represent the effects of a pair of parallel 8-in. (200-mm) and 12-in. (305-mm) pipes. In these model applications, the hydraulic and water quality models were applied to the Cheshire system simulating the proposed fluoride study.

The first step in the modeling process after database development and hydraulic model implementation was to apply the steady-state and dynamic water quality models developed as a result of the North Penn study to the Cheshire system. This was designed to simulate the propagation of fluoride feedwater and thus select sampling locations when the fluoride feed was actually turned off.

Design of the Field Program

Based on the simulation results and the objectives of the proposed field study, a preliminary field testing program was designed. Considerable refinement was required based on actual availability of manpower, variations in the operation of the system, and so on. However, the

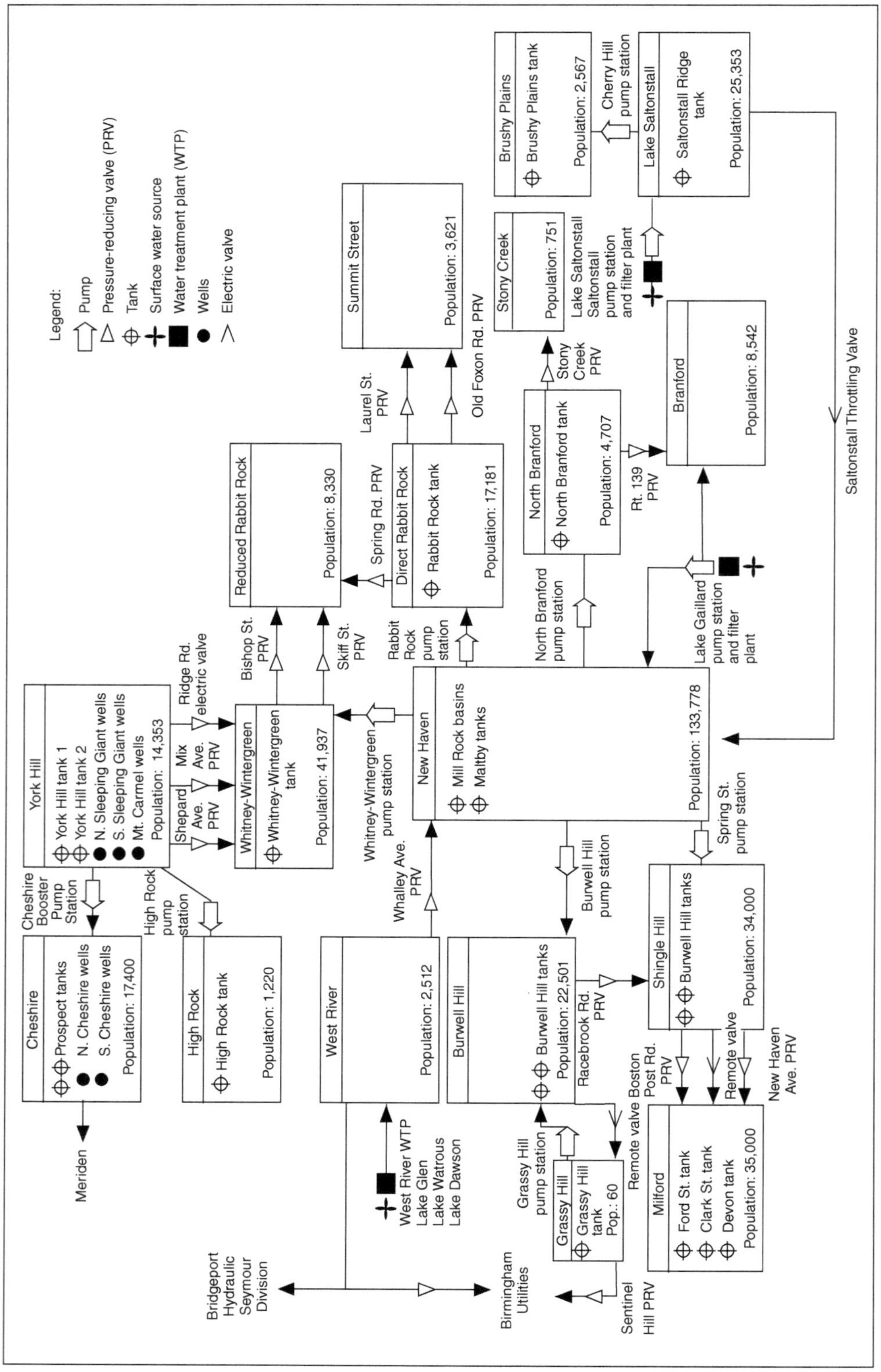

Figure 8-8. South Central Connecticut Regional Water Authority service area schematic

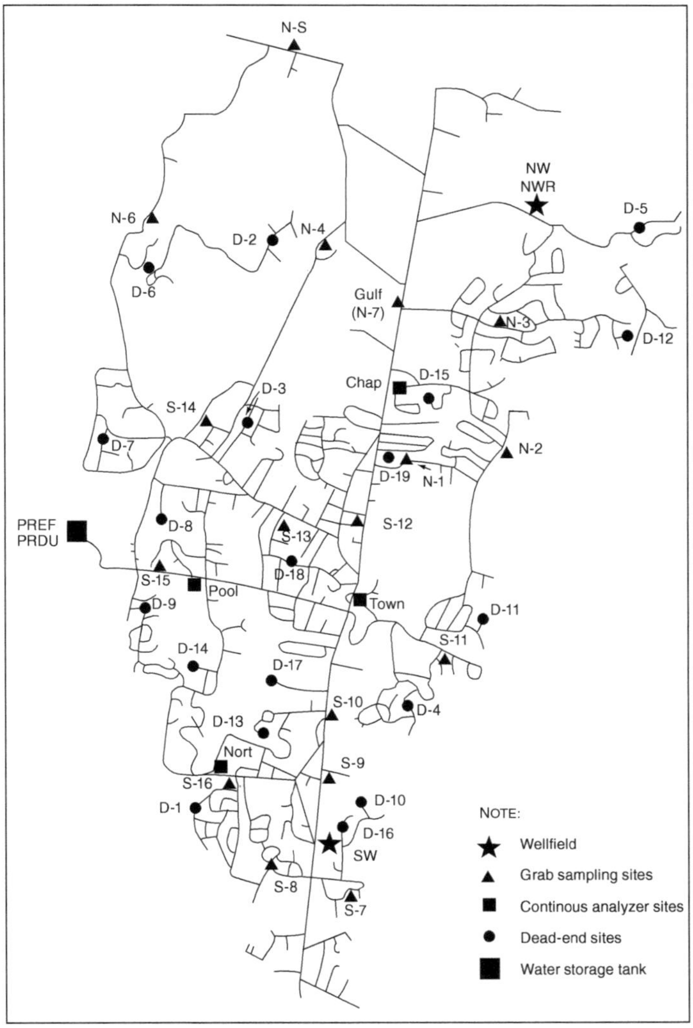

Figure 8-9. Cheshire service area map

plan served as a starting point for final design of the sampling plan.

Fluoride was picked as the tracer for several reasons. It is regularly added to the water at a concentration of approximately 1 mg/L as required by the state Department of Health Services, it does not dissipate from the water, it is easily tested for, and turning off the fluoride feed could be done with no health or aesthetic effects to the water. A fluoride concentration between 0.8 and 1.2 mg/L in public drinking water is required by the state Department of Health Services' Public Health Code. The natural background concentration of fluoride in the Cheshire wells was approximately 0.10 mg/L. By tracing the changes in the fluoride concentration in the distribution system, accurate travel times could be determined by relating the time the fluoride feed was shut down at the well fields to the time it dissipated at the sampling points (Skov et al. 1991).

Preparation

Based on their strategic location in the distribution system, the suggested sites for continuous analyzers were at three elementary schools and the Cheshire town hall (Figure 8-9). Permission to use the sites was obtained via phone contact and official letters to the appropriate town and board of education personnel. Each site was then visited to determine if a viable location in each building could be found to set up a continuous analyzer and chart recorder. Location requirements included an outlet for power, a sample tap at or near the location where the water service entered the building, a drain to accept the analyzed water, and an area large enough to set up equipment.

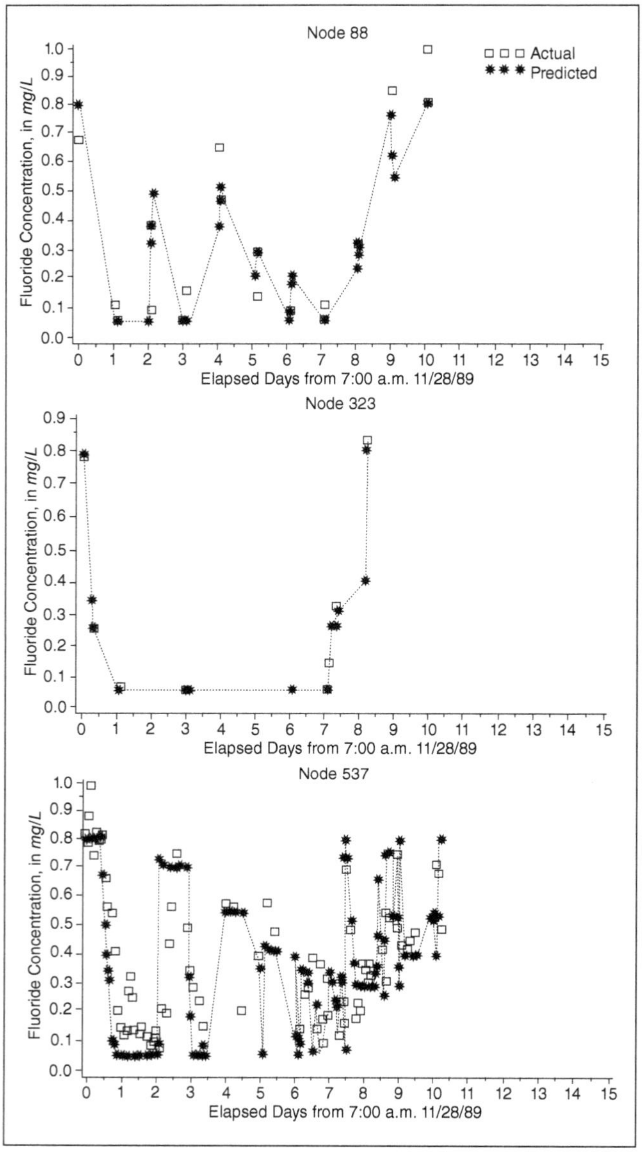

Figure 8-10. Actual versus predicted fluoride concentrations at nodes 88, 323, and 537

Results From the Field Study

Results from the study were very encouraging. The model predictions for the propagation of the nonfluoridated water through the system were very close. Figures 8-10 and 8-11 are comparisons between the model predictions and the field sampling results for nodes 37, 88, 182, 323, 537, and 570 (tank). The general trends predicted were borne out by the field study. As can be seen from the bottom plot in Figure 8-11, the concentration in the tank discharge increased or decreased until it approached the calculated concentration level in the tank. This effect was probably caused by a lack of complete mixing in the tank.

The behavior of the tanks is of particular interest. During the early part of the sampling period, tank level variations were held to a minimum (<3 ft [<0.9 m]). After 2 days, little change in fluoride concentrations was found in the tank and, as a result, the water level was then allowed to vary approximately 8 ft (2.4 m). The wider range in tank level variation had the effect of turning the water over relatively rapidly. Even with the rapid turnover, it took nearly 10 days to fully replace old water with new water in the tanks. It was clear from this analysis that tanks could have a detrimental effect on water quality, particularly as water ages in the tank.

During Aug. 13–15, 1991, another sampling program at the Cherry Hill/Brushy Plains service area began to validate the previously discussed simulation results. The purpose of this sampling program was to gather information to characterize the variation of water quality in the service area and to study the impact of tank operation on water quality.

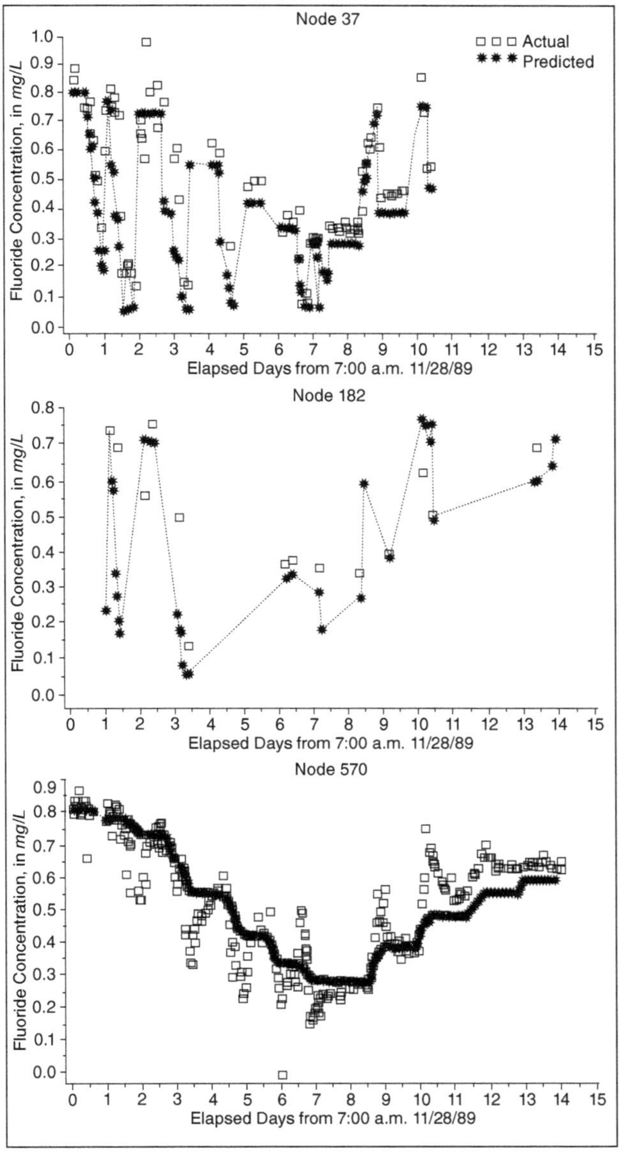

Figure 8-11. Actual versus predicted fluoride concentrations at nodes 37, 182, and 570

Verification Study

The Cherry Hill/Brushy Plains service area covered approximately 2 mi^2 (5 km^2) in the town of Branford in the eastern portion of the SCCRWA area (Clark et al. 1993). The service area was almost entirely residential, containing both single-family homes and apartment and condominium units. Average water use during the sampling period was 0.46 mgd (1,700 m^3/d). The water distribution system consisted of 8-in. (200-mm) and 12-in. (305-mm) mains as shown in the schematic in Figure 8-12. The terrain in the Cherry Hill/Brushy Plains service area was generally moderately sloping, with elevations varying from approximately 50 ft (15.2 m) mean sea level (msl) to 230 ft (70.1 m) msl.

Cherry Hill/Brushy Plains received its water from the Saltonstall system. Water was pumped from the Saltonstall system into Brushy Plains by the Cherry Hill pump station. Within the service area, storage was provided by the Brushy Plains tank. The pump station contained two 4-in. (100-mm) centrifugal pumps with a total capacity of 1.4 mgd (5,300 m^3/d). The operation of the pumps was controlled by water elevation in the tank. Built in 1957, the tank had a capacity of 1.0 mgd (3,800 m^3/d). It had a diameter of 50 ft (15.2 m), a bottom elevation of 193 ft (58.8 m) msl, and a height (to the overflow) of 263 ft (80.2 m) msl. During normal operation, the pumps were set to go on when the water level in the tank dropped to 56 ft (15.2 m) and to turn off when the water level reaches 65 ft (19.8 m).

Presampling Procedures

Prior to the sampling period, the Wadiso hydraulic model (discussed earlier) and the DWQM were applied to establish flow patterns within the service area. Additionally, during the periods of May 21–22, July 1–3, July 8–10, and July 30–Aug. 1, 1991, chlorine residuals

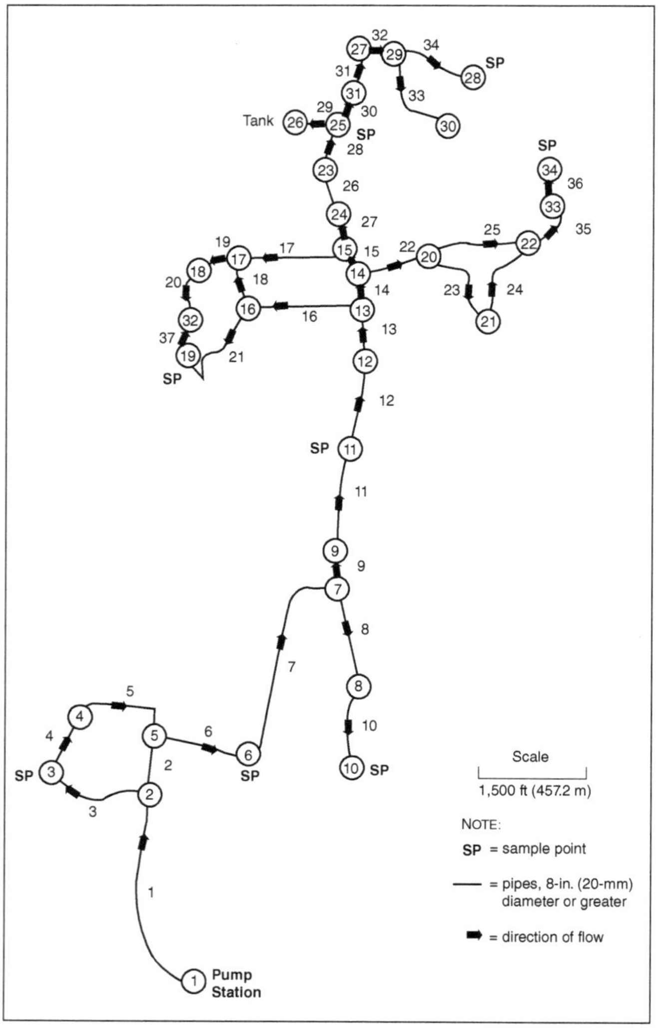

Figure 8-12. Link–node representation of the Cherry Hill/Brushy Plains network with "pumps-on" scenario

were monitored (using a portable free chlorine analyzer and chart recorder) at the tank, and operational patterns (pump records and tank water-level variations) were studied.

Based on these model runs and sampling data, a sampling strategy was adopted. This strategy involved turning the fluoride off at the Saltonstall treatment facility and sampling for both fluoride and chlorine in the Cherry Hill/Brushy Plains service area. Defluoridated water was used as a conservative tracer for the movement of flow through the system and to calibrate the DWQM. The DWQM and a chlorine decay model based on hydrodynamic principles were used to model the dynamics of this substance. Seven sampling sites in the distribution system, in addition to sampling sites at the pump station and tank, were identified as shown in Figure 8-12.

Analysis of Sampling Results

The Wadiso hydraulic model and DWQM water quality model were used to simulate the Cherry Hill/Brushy Plains service area for a 53-hr period from 9:00 a.m. on Aug. 13 to 3:00 p.m. on Aug. 15, 1991. This skeletonization of the distribution system, as shown in Figure 8-12, included all 12-in. (305-mm) mains, major 8-in. (200-mm) mains and loops, and pipes that connected to the sampling sites. Pipe lengths were scaled from maps, actual pipe diameters used, and, in the absence of any other information, a Hazen–Williams roughness coefficient of 100 was assumed for all pipes.

Figures 8-13 and 8-14 show the results of the fluoride sampling study and the modeling efforts at each of the sampling nodes. From these results it is clear that the modeling effort matches the sampling efforts quite well with the exception of the dead ends. The shaded areas at the top of the individual plots in Figures 8-13 and 8-14 show the on-and-off cycle for the pumps. Clearly, the pump cycles heavily influence water quality at several sampling points.

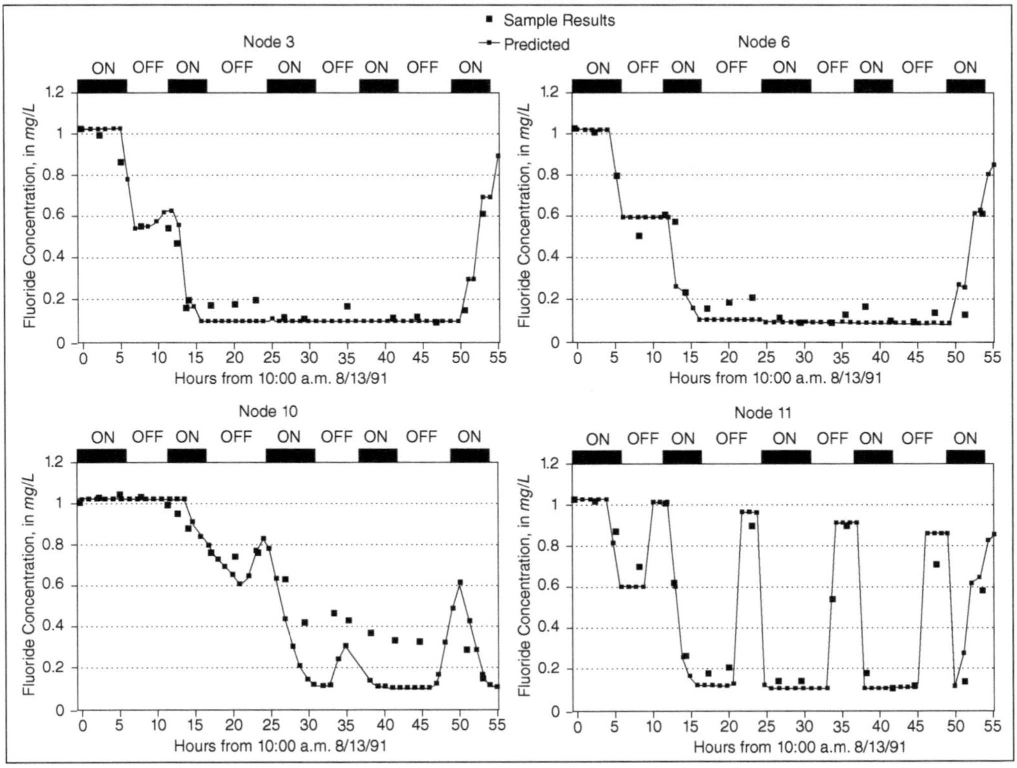

Figure 8-13. Results from fluoride sampling study at nodes 3, 6, 10, and 11

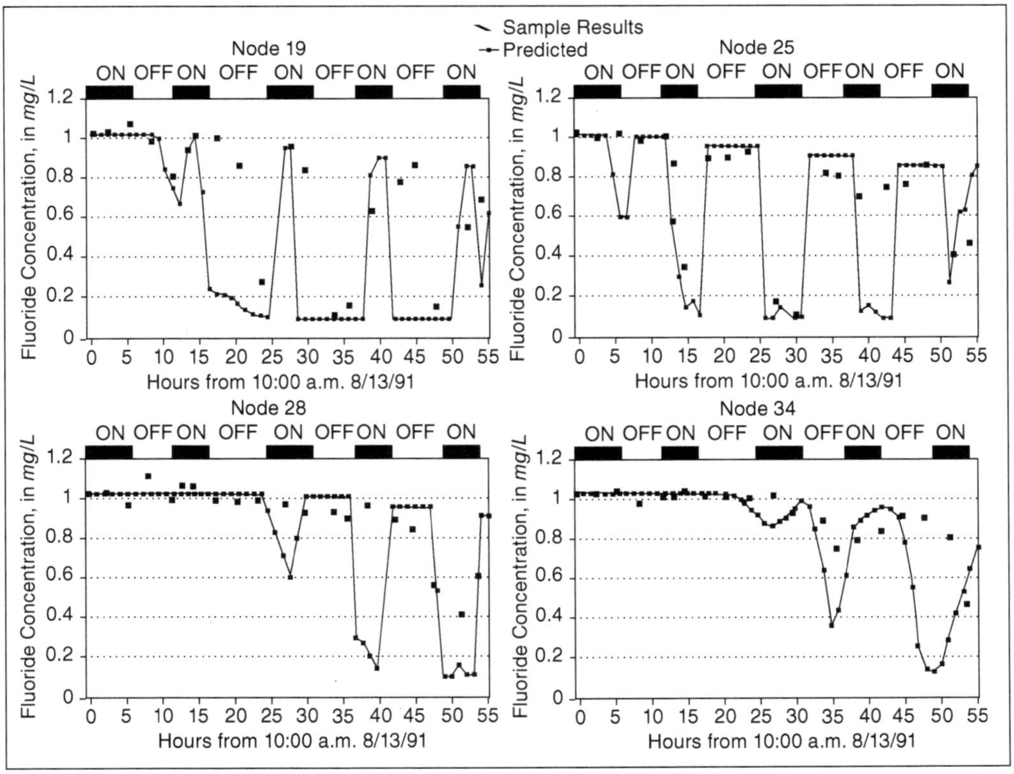

Figure 8-14. Results from fluoride sampling study at nodes 19, 25, 28, and 34

For example, at node 11, during the pumps' on cycle, the fluoridated water is pumped into the system. When the system is being fed from the tank (pumps off), the system is receiving water that had reached an equilibrium concentration of fluoride prior to the stoppage of the fluoride feeders.

Chlorine Residual Modeling

Table 8-1 shows the length of pipe in feet, average velocity in feet per second, residence time in days, and the average chlorine residual at the beginning and ending nodes for various links in the system with the pumps on.

Table 8-1. Hydraulic conditions during "pumps-on" scenario

Link Beginning/ Ending Nodes	Length, *ft (m)*	Residence Time, *days*	Chlorine Concentration Ratio, (upstream/ downstream)	Decay Coefficient, *days*$^{-1}$	Ratio of Pipe to Bulk Decay
1–3	3,700 (1,125)	0.0414	1.08/1.00	1.86	3.38
1–6	4,400 (1,341)	0.0321	1.08/1.00	2.40	4.36
6–11	3,800 (1,158)	0.0286	1.00/0.98	0.71	1.29
6–10	4,900 (1,493)	0.8049	1.00/0.36	1.27	2.30
11–19	5,400 (1,645)	0.1634	0.98/0.16	11.09	20.16
11–25	4,350 (1,325)	0.0424	0.98/0.94	0.98	1.78
11–34	64,000 (19,507)	1.3714	0.98/0.12	1.53	2.78
25–28	2,400 (731)	1.4937	0.94/0.16	1.19	2.16
Tank	—	3.000	0.94/0.16	0.59	1.07

Using the upstream and downstream chlorine concentration and the residence times in the link, the chlorine decay coefficient was calculated for each link. Chlorine demand was calculated based on a first-order assumption as defined by Eq 8-1.

$$C = C_0 e^{-kt} \quad \text{(Eq 8-1)}$$

Where:

C = the concentration at time t, in mg/L
C_0 = initial chlorine concentration, in mg/L
k = decay rate, in min^{-1}
t = time, in min
e = base of natural (Napierian) logarithms

A bench study was conducted in which chlorine demand for the raw water was calculated using Cherry Hill/Brushy Plains water. The chlorine decay rate was calculated as 0.55 day^{-1}. This decay rate might be considered as the bulk decay rate or the decay rate of chlorine in the treated water. As can be seen by the ratio in the sixth column of Table 8-1, the total system demand is much higher than just the bulk decay rate. This additional demand is most likely caused by pipe wall demand, biofilm, and tubercles and is very significant. It is conceivable that some system components may ultimately have to be replaced in order to meet water quality goals. The links resulting in dead ends (which also had the lowest average velocity during the "pumps-on" scenario) also exhibited the highest chlorine demand. It is also clear from the fifth column in Table 8-1 that a single first-order decay rate will not predict chlorine residual adequately.

Results from the SCCRWA study were consistent with the findings of others in showing that water quality varies dramatically from point to point in a drinking water distribution system (Geldreich et al. 1972, Maul et al. 1985a and 1985b, LeChevallier et al. 1987). This study went beyond these earlier studies by presenting a plausible hypothesis as to why this deterioration takes place and demonstrated how system design and specific operating scenarios, such as direct pumping and/or the use of storage tanks, could influence the nature of water quality variation within the system. It also demonstrated the usefulness of water quality propagation models in assessing both the impact and the causative mechanisms for those variations.

The North Marin Case Study

At the time of the study, the NMWD served a suburban population of 53,000 in or near Novato, Calif. The district used two sources of water: Stafford Lake and the North Marin queduct. The North Marin Aqueduct is a year-round source, but Stafford Lake was used only during the warm summer months when precipitation is virtually nonexistent and when demand is high. Novato, the largest population center in the NMWD service area, is located in a warm inland coastal valley with a mean annual rainfall of 27 in. (685 mm). Virtually no precipitation occurs during the growing season from May through September. Eighty-five percent of total water use is residential, and the service area contains 13,200 single-family detached homes, which accounted for 65 percent of all water use (Clark et al. 1994).

The water quality of the two sources differs greatly. Stafford Lake water has a high humic content and is treated with conventional treatment and prechlorination doses of between 5.5 and 6.0 mg/L. The treated water has a residual of 0.5 mg/L when it leaves the treatment plant clearwell. Total trihalomethane formation potential (TTHMFP) levels in the Stafford Lake water were very high. The North Marin Aqueduct water was derived from the Raney Well Field along the Russian River. While technically groundwater, the source water was likely to contain a high proportion of naturally surface-filtered water. Only aqueduct water is disinfected, and it is very low in precursor material with a correspondingly low TTHMFP. Both sources carry a residual chlorine level of approximately 0.5 mg/L when the water enters the system.

Figure 8-15 is an overall map of the entire NMWD service area. Figure 8-16 is a schematic of the distribution network for zone 1, which was the major focus of the study, including both sources, major pipes, major tanks and pumps, and the study sampling points. Major transfers of water out of zone 1 are shown by arrows. Storage tank and pump data for zone 1 are contained in Tables 8-2 and 8-3. As mentioned earlier, depending on the time of year and time of day, water enters the system from either one or both of the sources. The North Marin Aqueduct source operates year-round, 24 hr per day. The Stafford Lake source operated only during the peak demand period from 6:00 a.m. to 10:00 p.m. and generally operated for a period of 16 hr per day. Table 8-4 summarizes NMWD water-use characteristics.

EPANET was applied for the first time to model the system hydraulics, including the relative flow from each source, TTHMs, and chlorine residual propagation (Rossman et al. 1994). The model was based on an earlier network representation made by MWH Global for

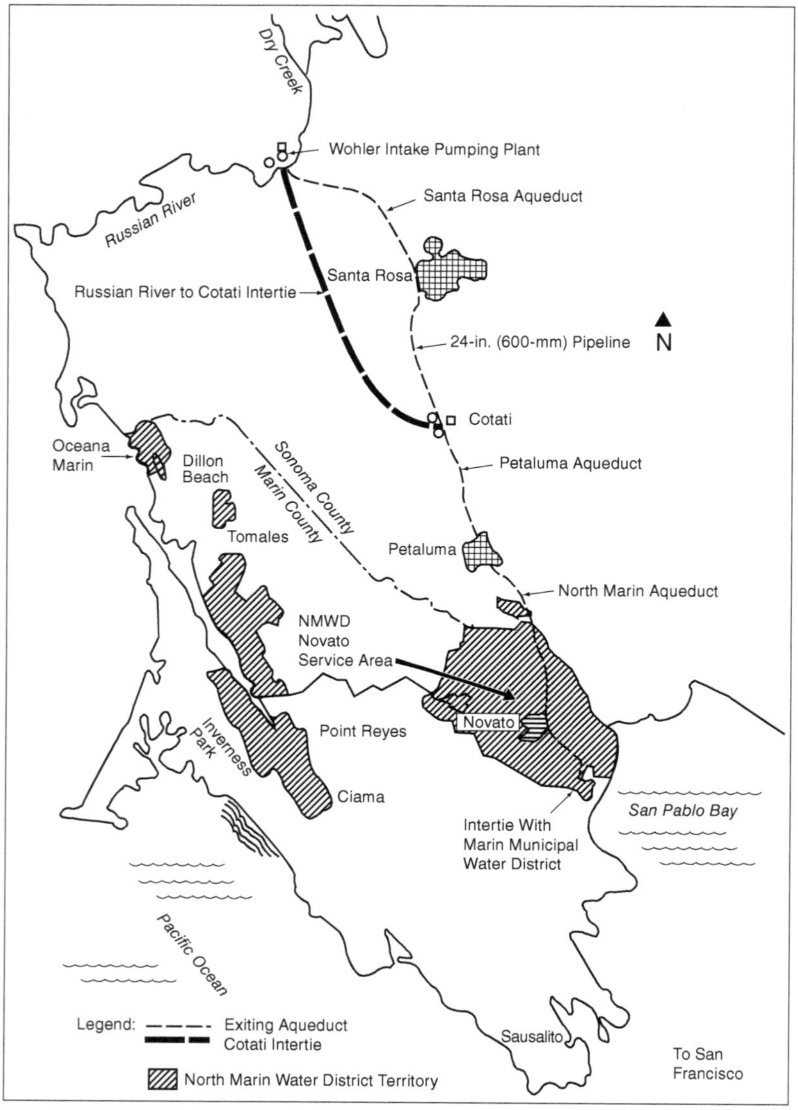

Figure 8-15. North Marin Water District service area

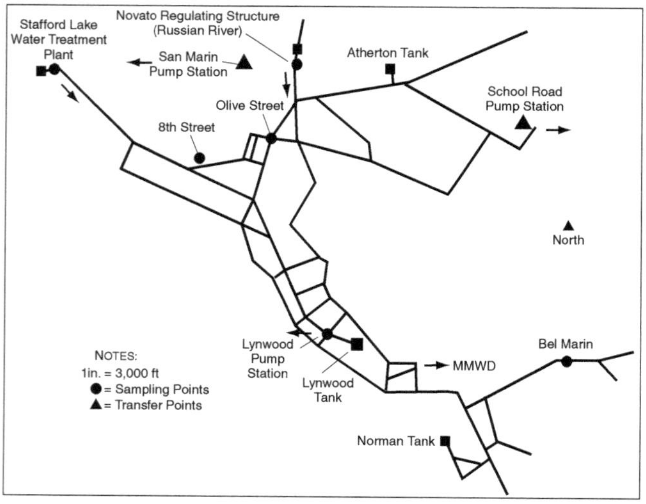

Figure 8-16. Existing zone 1 piping schematic

Table 8-2. Zone 1 storage tank area, NMWD

Tank	Nominal Capacity, *gal*	Inside Diameter, *ft*	Bottom Elevation, *ft* *(msl)*	Overflow Elevation, *ft* *(msl)*	Depth at Overflow, *ft*
Lynwood #1	500,000	53.0	132.7	163.1	30.4
Lynwood #2	850,000	66.0	131.0	164.4	33.4
Norman	500,000	50.0	123.0	156.8	33.8
Atherton	5,000,000	164.4	133.0	164.5	31.5
Plum*	500,000	60.0	134.0	157.7	22.7
Total Storage	7,350,000				

* Out of service during study

msl = mean sea level

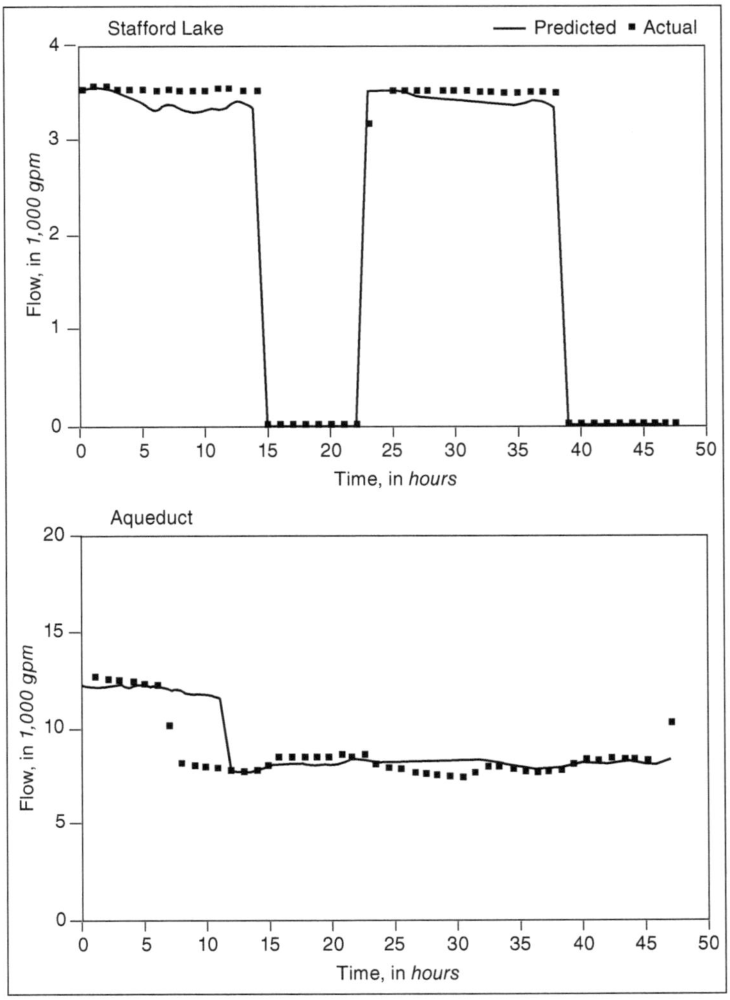

Figure 8-17. Flow from aqueduct and Stafford Lake

NMWD and was calibrated based on a comparison of simulated versus actual tank levels for the May 27–29, 1992, period of operation.

Figure 8-17 compares actual flow data from the aqueduct and Stafford Lake against modeled values. The figure shows a discrepancy between the pump as it actually operated during this study period and the model predictions based on tank level elevations. The model was set to operate on Lynwood tank levels, which, in turn, were based on NMWD supervisory control and data acquisition (SCADA) information. However, during the data collection period, as shown in Figure 8-17, the pump was actually shut

Table 8-3. Zone 1 to zone 2 pump station data, NMWD

Pump Station	Pump Number	Rated Horsepower	Rated Flow, *gpm*	Total Dynamic Head, *ft*	Rated Revolutions, *per min*	Suction Pressure, *psi*	Discharge Pressure, *psi*	Elevation at Pump, *ft*	Controlling Reservoir
San Marin	1	100	1,900	165	1,770	47*	118	48	San Mateo
	2	100	1,800	175	1,770	42*	118	48	
	3	75	2,200	100	1,785	75*	118	48	
Lynwood	1	100	1,900	165	1,770	60	131	8	Control from Central on Pacheco Valle Tank Levels
	2	100	1,900	165	1,770	60	131	8	
	3	100	1,900	165	1,770	60	131	8	
School Road	1	30	400	180	3,500	50	130	32	Crest Tank
	2	30	400	180	3,500	50	130	32	
Stafford Lake WTP Service	1	125	3,500	70	1,770	3	42	147	None
	2	125	3,500	70	1,770	3	42	147	
	3	125	3,500	70	NA	3	38	147	

NOTE: Zone 1 to zone 2 pump stations include Hayden pump station (PS), Hancock PS, Diablo Hills PS, and Woodlands PS. They are not included in the analysis because demands are so small in the areas they serve.

NA = information not available.

* Suction pressure is a function of SCWA Aqueduct pressure (varies 48–30 psi).

Table 8-4. North Marin Water District water-use characteristics

Service Area:	100 mi^2 (259 km^2)
Principal Service Center:	Greater Novato area
Population:	53,000
Character:	Suburban (near San Francisco)
Normal Rainfall:	27.2 in. per year (69.1 cm per year)
Normal Reference Evapotranspiration (ET_0):	44 in. per year (111.8 cm per year)
Applied Water Requirement for Cool Season Grasses:	27.8 in. per year (70.6 cm per year)
Water Supply:	Surface—North Marin Aqueduct 83 percent,
	Stafford Lake 17 percent
All Accounts Metered?	Yes
Overall Average Per Capita Use:	135 gpcpd (511 Lpcpd)*
Distribution of Metered Water Use:	Residential (84.7 percent)
	Commercial (15.3 percent)
Annual Residential Use:	
Single Family (67.3 percent)	144 gpcpd (545 Lpcpd)
Townhouse/Condo (12.7 percent)	115 gpcpd (435 Lpcpd)
Mobile Home (16.6 percent)	77 gpcpd (291 Lpcpd)
Apartment (3.4 percent)	79 gpcpd (299 Lpcpd)
Unaccounted for Water and Water Loss:	5.7 percent

* Gallons/liters per capita per day (based on average household density of 2.7 persons/dwelling).

down 5 hr before normal operating policy would dictate. This slight change in operation led to a prediction of more water in the system than actually occurred, as shown by Figure 8-18 (Norman and Atherton tanks), although the actual and simulated tank levels for the Lynwood tanks were very close. A second simulation was attempted in which the aqueduct pump was operated on a time schedule. However, the tank level predictions for the Lynwood tank did not match the actual data. It was decided to use the first simulation as the basis for the analysis because the Lynwood tank is the controlling tank for the system. Information from the first simulation, coupled with pump operating schedules and demands in the network, dictated the hydraulic conditions used as the basis for this analysis.

The dynamic nature of the system leads to both variable flow conditions and quality in the network. Flow directions frequently reverse within a given portion of the network during a typical operating day. Figure 8-19 (see appendix E) shows the changes in flow direction and the percentage of water from Stafford Lake penetrating the system during a typical operating day. In these figures, N5 is the designation for Stafford Lake water. As can be seen, the system experienced dramatic flow reversals, depending on which source or sources were operating. Consequently, the mix of water ranged from 100 percent Stafford Lake water to 100 percent North Marine Aqueduct water at a given node. Figure 8-20 shows the predicted percentage of water from Stafford Lake at the various sampling points used during the two-day study. A discussion of the consequences of these variable flow patterns for water quality follow.

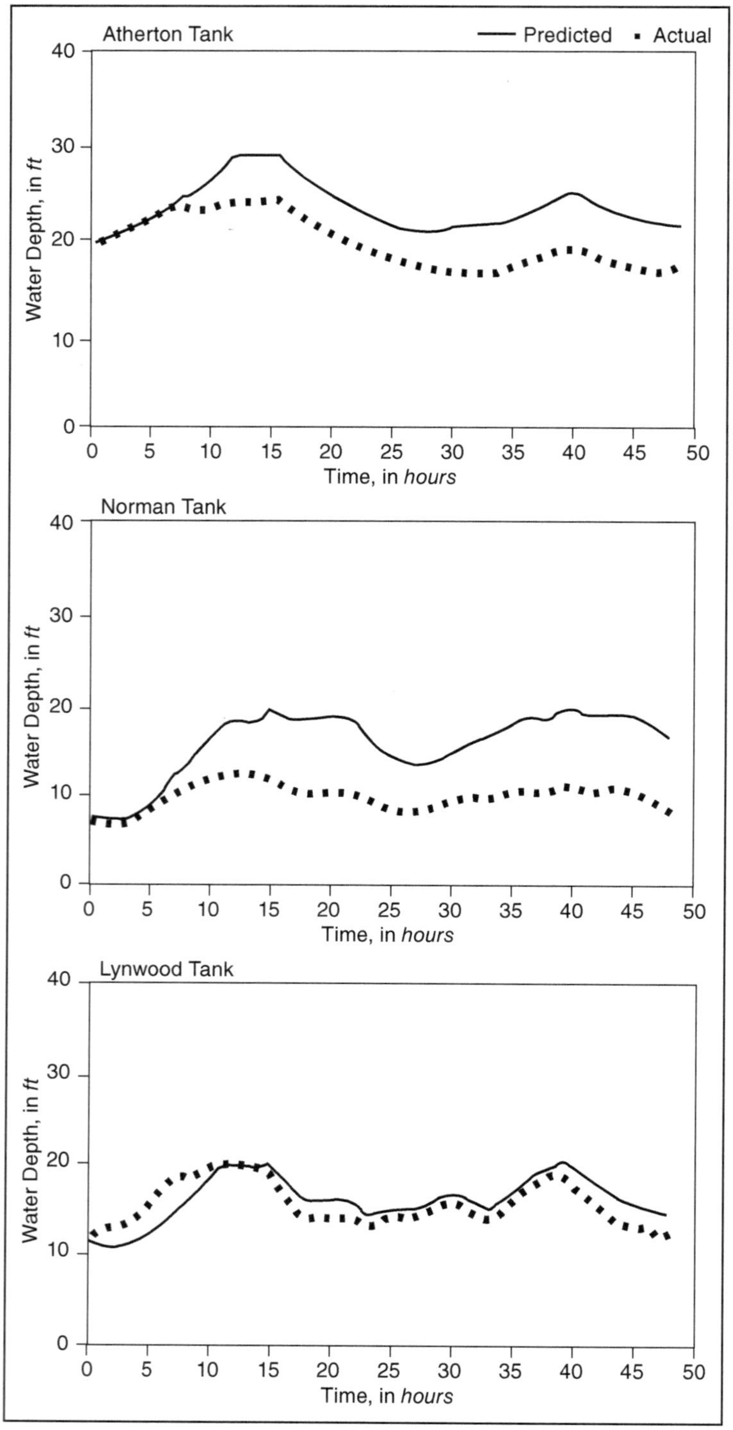

Figure 8-18. Water depths at Lynwood tank, Norman tank, and Atherton tank

Table 8-5. Sampling schedule

Sampling Times	Locations					
	1	2	3	4	5	6
Wednesday (May 27, 1992)						
6:00 a.m.	X	X	X	X	X	X
8:00 a.m.						X
10:00 a.m.						X
12:00 a.m.	X	X	X	X	X	X
2:00 p.m.						X
4:00 p.m.						X
6:00 p.m.	X	X	X	X	X	X
8:00 p.m.						X
10:00 p.m.						X
12:00 p.m.	X	X	X	X		X
Thursday (May 28, 1992)						
2:00 a.m.						X
4:00 a.m.						X
6:00 a.m.	X	X	X	X	X	X
8:00 a.m.						X
10:00 a.m.						X
12:00 a.m.	X	X	X	X	X	X
2:00 p.m.						X
4:00 p.m.						X
6:00 p.m.	X	X	X	X	X	X
8:00 p.m.						X
10:00 p.m.						X
12:00 p.m.	X	X	X	X	X	X

NOTES:
1 = Stafford Lake WTP send-out.
2 = Aqueduct source.
3 = Bel Marin.
4 = Eighth St. area.
5 = 1100 Olive St.
6 = Lynwood tank area.

Water Quality Study

In order to characterize the NMWD water quality, USEPA designed a sampling protocol and sent a team of investigators to work with NMWD and MWH Global staff during the period May 27–29, 1992. Table 8-5 shows the sampling schedule at six sites. Table 8-6 contains the results of a round-the-clock sampling study conducted over a two-day period. Column 1 of Table 8-6 shows the name of the sampling station. Columns 2 through 5 contain values for individual trihalomethane (THM) species. Columns 7 through 10 contain haloacetic acid (HAA) values. Columns 11, 12, 13, 14, and 15 contain values for temperature, free and total chlorine, pH, and sodium, respectively. Figure 8-21 shows the time formation curves for TTHMs and chlorine demand for the two source waters. These figures illustrate the significant difference in water quality from the two sources. As can be seen from Table 8-6, water quality can change dramatically over time in the system. For example, at the Eighth St. sampling point, trichloromethane ($CHCl_3$) levels vary from 38.4 to 129.9 µg/L

Table 8-6. Water quality characteristics of NMWD sources and sampling points

Station	$CHCl_3$,* µg/L	$CHBrCl_2$,* µg/L	$CHBr_2Cl$,* µg/L	$CHBr_3$,* µg/L	UV†	DCAA,‡ ppb	TCAA,‡ ppb	BCAA,‡ ppb	DBAA,‡ ppb	Temp, °F	Cl_2-F§	Cl_2-T§	pH**	Na,†† µg/L
Aqueduct	0.8–3.0	3.1–6.4	7.0–11.4	4.8–7.5	0.011–0.012	<0.80	<0.78	<1.89–2.27	<0.99–7.08	66–70	0.66–0.45	0.32–0.45	7.40	9.50–10.00
Bel Marin	9.6–81.8	6.3–22.0	6.3–8.6	2.9–4.8	0.019–0.063	<0.80–3.97	1.79–22.63	<1.39–2.07	<0.99	68–73	0.02–0.22	0.12–0.23	—	9.20–20.00
Eighth St.	38.4–129.9	12.4–30.7	2.7–8.3	<0.1–4.7	0.032–0.094	<0.80–43.00	9.33–52.56	<1.89–6.50	<0.99	68–73	0.05–0.61	0.17–0.84	—	12.00–29.50
Lynwood	3.6–124.5	4.2–28.9	4.0–8.4	<0.1–5.6	0.015–0.092	<0.80–19.10	0.78–59.06	<1.89–2.94	<0.99	68–73	0.00–0.55	0.60–0.56	7.50–8.36	9.20–27.00
Olive St.	9.0–134.7	5.2–30.2	3.0–10.3	<0.1–6.1	0.015–0.093	<0.80–36.11	1.37–49.69	<1.89–4.46	<0.99	67–72	0.02–0.41	0.14–0.51	—	9.40–29.00
Stafford WTP	108.0–140.5	25.8–32.1	2.4–3.8	<0.1	0.088–0.095	<2.90–53.32	43.54–61.92	5.65–7.63	<0.99	67–71	0.40–0.58	1.08	8.90–9.10	27.00–30.00

* USEPA (1990), method 551.

† APHA et al. (2005), method 5910A0.

‡ USEPA (1990), method 552.

§ APHA et al. (2005), method 409E.

** USEPA (1979), method 150.1.

†† USEPA (1979), method 273.1.

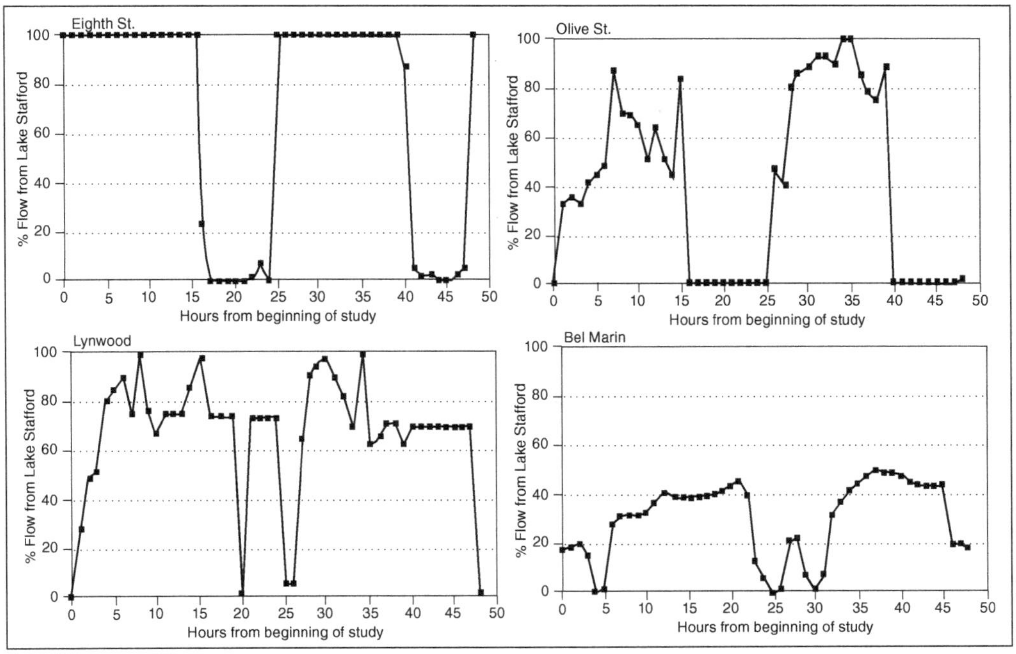

Figure 8-20. Predicted flow from Stafford Lake at Eighth St., Olive St., Lynwood, and Bel Marin sampling sites (percentage)

over the two-day period. This variability is due to the penetration of water from the two different sources. As can be seen from Figure 8-20, a sample taken at Eighth St. may consist of water from the aqueduct, Stafford Lake, or a blend of both. However, over the sampling period, THM levels were relatively constant at the two sources. Given the extreme differences in water quality from the two sources and the variability in percent source water penetrating within the network, it is reasonable to assume that mixing and blending of water are major factors affecting water quality.

Modeling THMs

In order to test the mixing hypothesis, the average values for TTHMs at each of the sources were used. These TTHM levels were treated as conservative, and TTHM levels at each sampling station were calculated based on the percentage of water from a source at that point over time. This assumption was verified with field and bench testing. For example, the average TTHM levels for Stafford Lake water and for the aqueduct were 151.1 and 20.7 μg/L, respectively. If, at a given time and place in the system, 50 percent of the flow is from Stafford Lake and 50 percent is from the aqueduct, then the predicted TTHM values at that point would be 85.9 μg/L. Figure 8-20 shows the results of this assumption at the various sampling points. Some of the differences between actual and predicted values in the figures are likely because of the calibration problems discussed earlier.

It was observed that ultraviolet (UV) radiation absorbance and TTHMs were closely related. In order to establish that UV absorbance is a good predictor for TTHMs, a regression relationship between UV absorbance and TTHMs was developed using the data from the two sources. Data from the first day of the study were used to develop a regression model between TTHMs and UV as follows:

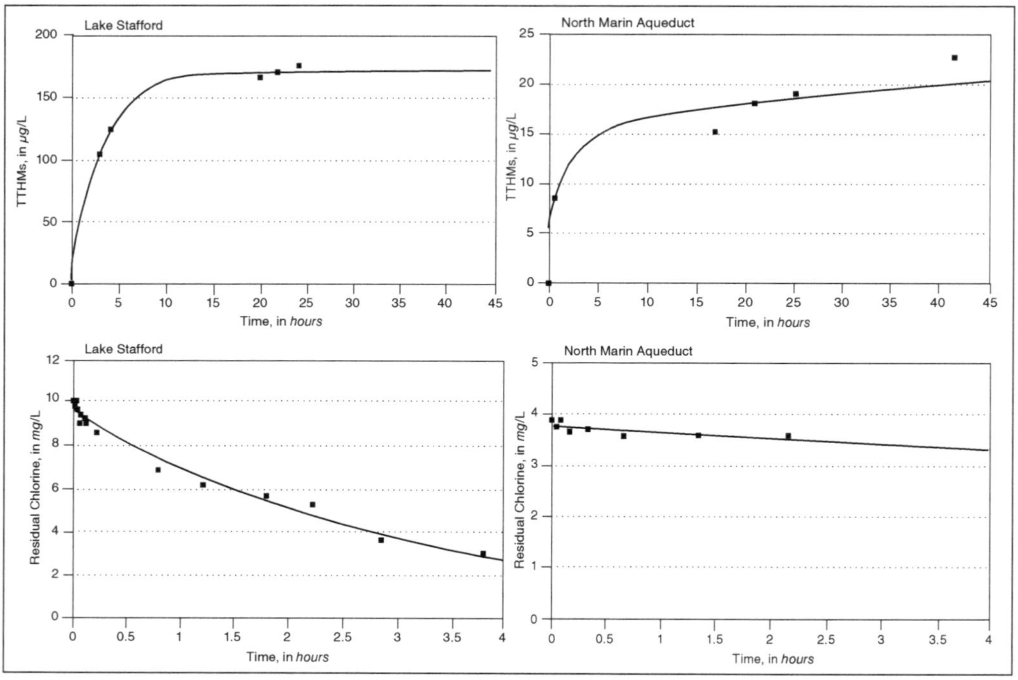

Figure 8-21. TTHM formation potential curves and chlorine demand curves for Stafford Lake and North Marin aqueduct

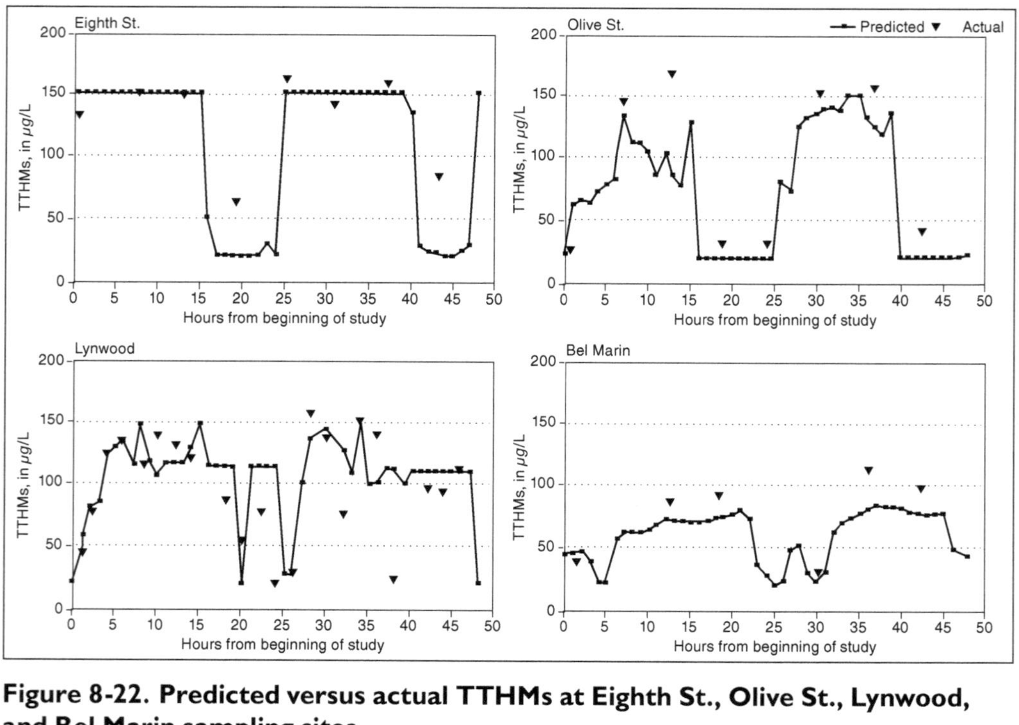

Figure 8-22. Predicted versus actual TTHMs at Eighth St., Olive St., Lynwood, and Bel Marin sampling sites

$$TTHM = 2.056 + 1648.2^{*} UV \quad \text{(Eq 8-2)}$$

Where:

TTHM = total trihalomethanes, in μg/L

UV = ultraviolet absorbance in the 250-nm range, in cm^{-1}

with a p-value = 0.0001, r = 0.9875, and MSE = 11.1. The assumptions of the model (constant variance and normality of error terms) were checked and deemed reasonable. Figure 8-21 shows the data used to develop the regression model. Figure 8-22 shows the use of Eq 8-2 to predict TTHMs at the various sampling points for the second day of the study. As can be seen, the predictions are quite good.

Figure 8-23 shows simulated versus actual UV absorbance data at the four sampling sites based on the same mixing assumptions used for predicting trihalomethane levels. Figure 8-24 compares the actual versus predicted TTHMs. As can be seen, UV can be used effectively to predict TTHMs so that Eq 8-2 can be used to predict THMs using UV absorbance as the independent variable.

Chlorine Demand

Both Stafford Lake and the North Marin aqueduct water generally maintain a 0.5-mg/L chlorine residual level as the treated water enters the system, and it was found that the Stafford Lake water has a much higher chlorine demand than does the North Marin aqueduct water. In order to predict chlorine demand at the various sampling points, a first-order decay relationship was assumed as follows (Clark et al. 1993):

$$C_t = C_0 e^{-kt} \quad \text{(Eq 8-3)}$$

Where:

C_t = chlorine concentration at any time t, μg/L

C_0 = initial chlorine residual, μg/L

k = first-order decay coefficient, hr^{-1}

t = time, hr

e = base of natural (Napierian) logarithms

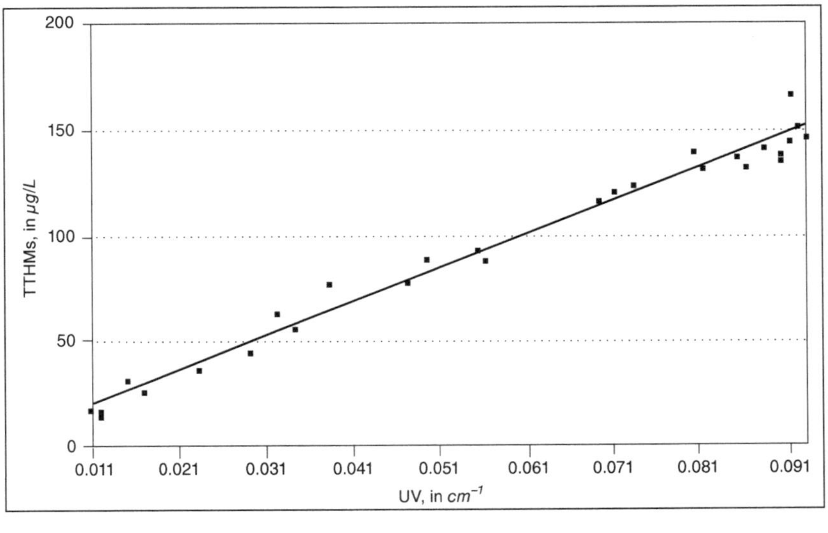

Figure 8-23. Ultraviolet radiation absorbance versus TTHMs from first day of water quality study

In EPANET, chlorine decay is represented by decay in the bulk phase and pipe-wall decay. Based on bulk water calculations, the first-order decay coefficients or bulk demand for the Stafford Lake and North Marin aqueduct sources were 0.31 day^{-1} and 0.03 day^{-1}, respectively. The chlorine residual was estimated using EPANET and the previously assumed hydraulic conditions. Figure 8-25 also shows these predictions based on the four sampling points.

Within EPANET, a mechanism has been provided for considering the loss (or growth) of a substance by reaction as it travels through the distribution system. Reaction is assumed to occur both within the bulk flow and within the pipe-wall material, based on using first-order kinetics. The general expression for chlorine decay in the bulk flow and at the pipe wall is

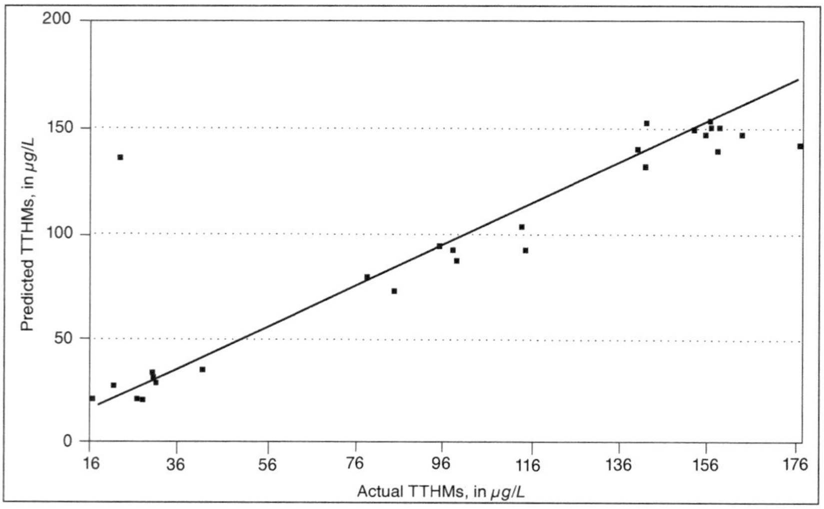

Figure 8-24. Predicted versus actual TTHMs based on Eq 8-2

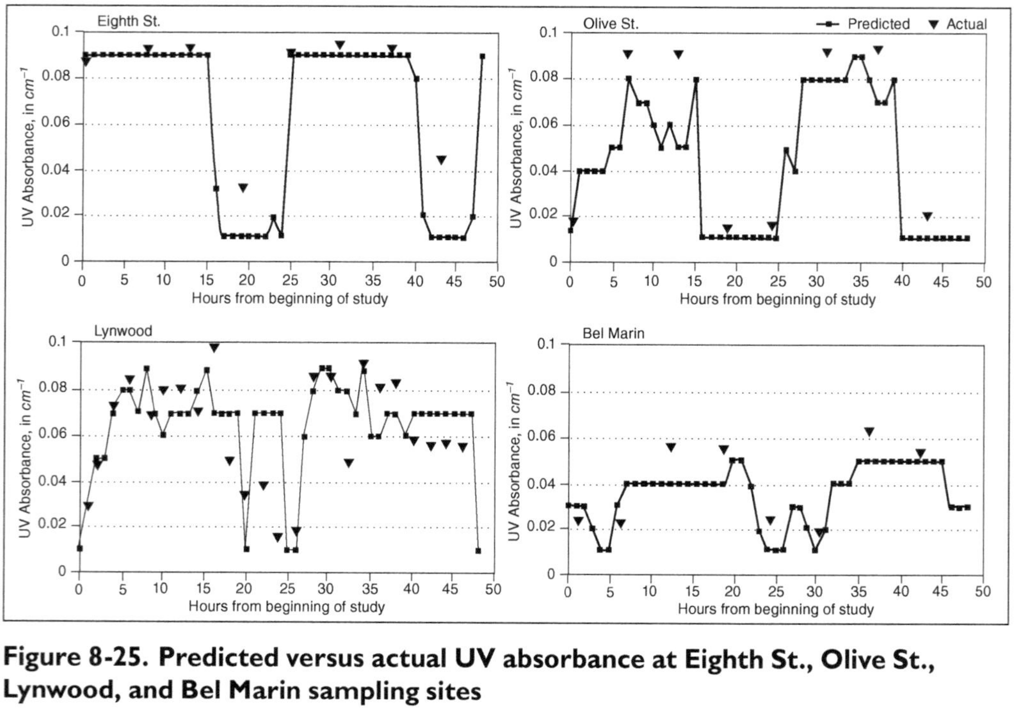

Figure 8-25. Predicted versus actual UV absorbance at Eighth St., Olive St., Lynwood, and Bel Marin sampling sites

$$\frac{dc}{dt} = k_b c - \frac{k_f}{r_h}(c - c_w) \qquad \text{(Eq 8-4)}$$

Where:

dc/dt = rate of chlorine decay, in mg/L/d
c = chlorine concentration in the bulk flow, in mg/L
t = time, in days
k_f = mass transfer coefficient, in ft/d
r_h = hydraulic radius
c_w = chlorine concentration at the pipe wall, in mg/L
k_b = bulk phase chlorine decay coefficient, in day^{-1}

The first term in Eq 8-4 is the bulk flow reaction, while the second term includes c_w, which represents the rate at which material is transported between the bulk flow and reaction sites at the pipe wall. If it is assumed that the rate at the wall is first order with respect to c_w and that it proceeds at the same rate as material is transported to the wall so that no accumulation occurs, the following mass balance at the wall holds:

$$k_f(c - c_w) = k_w c_w \qquad \text{(Eq 8-5)}$$

Where:

k_w = wall reaction constant, in ft/d (m/d)
and other variables as defined previously.

Solving for c_w and substituting into Eq 8-4 yields the following reaction rate expression:

$$\frac{dc}{dt} = -k_b c - \frac{k_w k_f c}{r_h (k_w + k_f)} \qquad \text{(Eq 8-6)}$$

$$k_t = k_b + \frac{k_w k_f}{r_h (k_w + k_f)} \qquad \text{(Eq 8-7)}$$

Substituting Eq 8-6 into Eq 8-7 yields

$$\frac{dc}{dt} = -k_t c \qquad \text{(Eq 8-8)}$$

For steady-state flow conditions in a single pipe, this rate expression yields a first-order decay model for chlorine, i.e.:

$$C_t = C_0 e^{-k_t t} \qquad \text{(Eq 8-9)}$$

Where:

C_t = concentration of chlorine at any time t, in mg/L
C_0 = initial concentration of chlorine, in mg/L

However, the decay constant in this model is now a function of a bulk decay rate constant, a wall decay rate constant, the molecular diffusivity of chlorine, the water's kinematic viscosity, velocity, and the pipe radius (Clark et al. 1995).

With regard to the formation of TTHMs, it has been demonstrated that the rate of formation is a function of time. It has been postulated that DBP formation is governed by the following equation (Clark et al. 1996):

$$\text{DBP} = \text{DBP}_u - \text{DBP}_u^{\,t} \tag{Eq 8-10}$$

Where:

DBP_u = the ultimate formation potential of an individual DBP
$\text{DBP}_u^{\,t}$ = the remaining formation potential at time t

We can also make the first-order assumption

$$\frac{d\text{DBP}_u^t}{dt} = -k\text{DBP}_u^{\,t} \tag{Eq 8-11}$$

Where:

$\text{DBP}_u^{\,t}$ = DBP_u at $t = 0$

Therefore, integrating Eq 8-11 yields

$$\text{DBP} = \text{DBP}_u(1 - e^{-kt}) \tag{Eq 8-12}$$

Eqs 8-9 and 8-12 illustrate that both the loss of chlorine residual and the formation of TTHMs are time-dependent and will vary throughout the system depending on propagation conditions. As mentioned earlier, a study to evaluate these changes was conducted by USEPA in the North Marin Water District (NMWD) and is discussed in the following sections.

Effect of System Demand

It is evident that the pipes in the distribution network can exhibit a demand for chlorine. This demand may come from tubercles, biofilm, and perhaps the pipe-wall material itself (Clark et al. 1993). A comparison between chlorine residuals using the first-order assumptions predicted from EPANET versus actual chlorine residuals provides an excellent illustration of this point.

It is clear that demand exists in the system beyond just bulk water decay. Because EPANET has the capacity to incorporate a wall demand factor in addition to the bulk demand factors for chlorine, the system was simulated again using the bulk demand for the two sources, and trial and error was used to estimate wall demands for four sections of the network as shown in Figure 8-27 (see appendix E). Wall demand k_w, as defined in EPANET, is given in feet per day. One might think of k_w as a penetration velocity for the disinfectant.

Chlorine residuals were reestimated at the four sampling points (Figure 8-26) and, as can be seen, wall demand obviously plays a major role in chlorine residual loss in the NMWD system. Figure 8-27, however, still shows some large differences between predicted and actual chlorine residuals during the middle portion of the study.

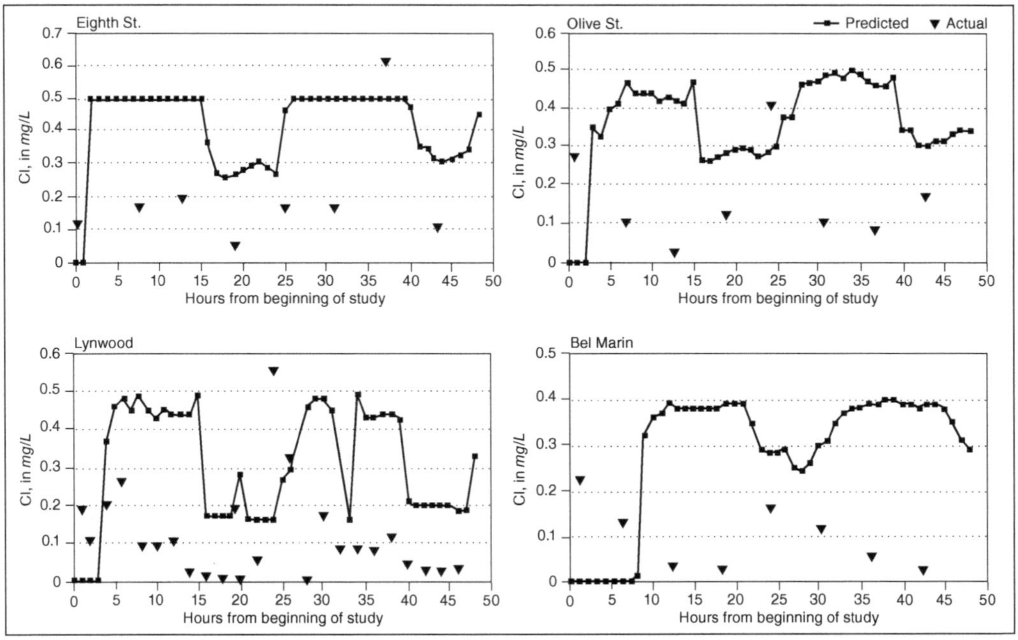

Figure 8-26. Predicted versus actual chlorine residual assuming first-order decay at Eighth St., Olive St., Lynwood, and Bel Marin sampling sites

The wall demand effect may be due to the source or to the age of the system. For example, as illustrated by Figure 8-28, the maximum wall demand was found in those areas served primarily by Stafford Lake. However, these pipes are also the oldest in the system.

AwwaRF/USEPA Study

As a complement to the North Marin study, USEPA and AwwaRF (Vasconcelos et al. 1996) initiated a project that involved several case study utilities and was intended to gain a better understanding of the kinetic relationships describing chlorine decay and TTHM formation in water distribution systems. Several kinetic models were evaluated, tested, and validated using data collected in these field-sampling studies based on the EPANET distribution network model as the framework for analysis.

Another goal was to provide information and guidelines for conducting water quality sampling and modeling studies by water utilities. Chapter 7 presents some of this information as general guidelines for conducting network modeling studies. The specific objectives of the USEPA/AwwaRF study were:

1. Establish protocols for obtaining data to calibrate predictive models for chlorine decay and TTHM formation.
2. Evaluate alternative kinetic models for chlorine decay and TTHM formation in distribution systems.
3. Develop a better understanding of the factors influencing the reactivity of chlorine in distribution and transmission pipes.
4. Make these results available for use in system operations and in planning similar water quality studies.

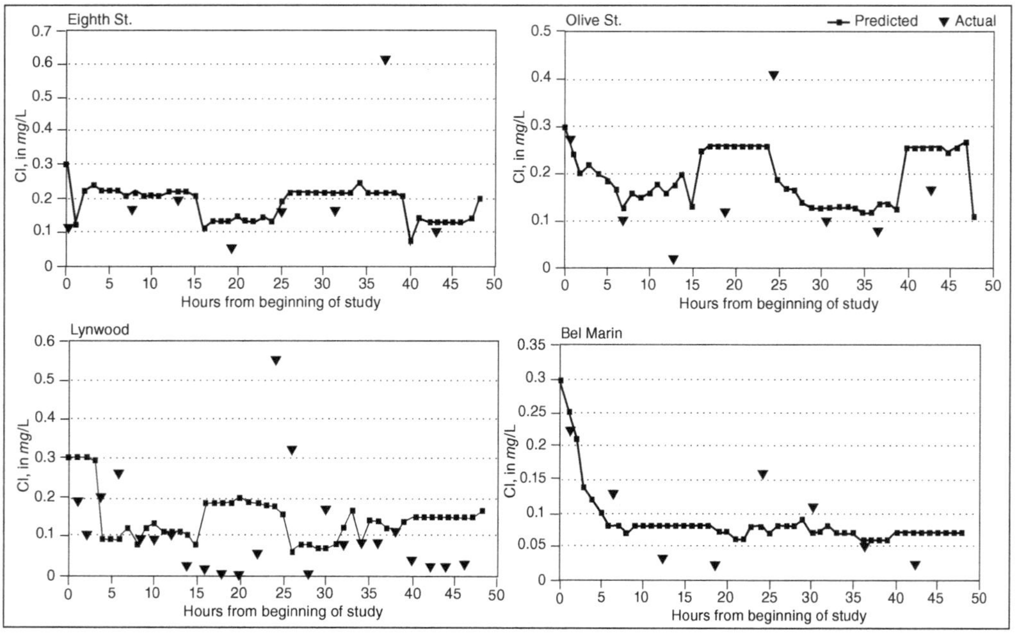

Figure 8-28. Predicted versus actual chlorine residual assuming wall demand factors at Eighth St., Olive St., Lynwood, and Bel Marin sampling sites

Utilities selected for participation in this study represent a wide range of source water qualities. In addition, some systems were included that had small isolated zones or long isolated lines. Systems having significant TTHM formation potential were also selected. The utilities in the study were Bellingham, Wash.

- United Water Corporation (formerly Dauphin Water Company), Harrisburg, Pa.
- Fairfield, Calif.
- North Marin Water District, Novato, Calif.
- North Penn Water Authority, Lansdale, Pa.
- Parisienne des Eaux, France.

Field Sampling Studies

Extensive water quality sampling was conducted at the five participating US utilities. Information was collected on chlorine residuals, TTHM concentrations, and other relevant water quality and hydraulic factors. These data were used to calibrate a hydraulic model and a water quality network model for each test network and to calculate chlorine decay and TTHM formation kinetic constants. Tracer studies were performed to aid in refining the hydraulic calibration for four of the five utilities, and in each utility, a subnetwork and/or a long pipeline was selected for sampling. Sampling plans were prepared that contained detailed descriptions of all aspects of each study, including contingency plans addressing how to deal with unexpected occurrences. A final report was prepared containing all aspects of the fieldwork at the end of each individual study (Vasconcelos et al. 1996).

Evaluation and Testing of Alternative Kinetic Models

The following chlorine decay models were evaluated: (1) first order, (2) *n*th order, (3) limited first order, and (4) parallel first order for bulk decay (Table 8-7). First-order or zero-order

Table 8-7. Mathematical models for alternative decay models

Model	Differential Form ($dC/dt =$)
First order	$-kC$
nth order	$-kC^n$
Limited first order	$-k(C - C^*)$
Parallel first order	$-k_1C_1 - k_2C_2$
	with $C_{1,0} = C_0x$
	and $C_{2,0} = C_0(1 - x)$

C = concentration

C_0 = initial concentration

*k, n, and C** = parameters to be estimated

t = time

decay kinetics was considered for wall decay. Results from this analysis suggested that for TTHM data, limited first-order growth kinetics might be used as the basis for predicting TTHM formation in distribution systems.

Testing Chlorine Kinetic Models. A method similar to that described in the North Marin study was developed for systematically analyzing alternative kinetic models. After determining if wall decay was significant, a series of simulations was performed using alternative wall decay kinetic models in which the wall reaction rate coefficient was systematically adjusted until the best statistical comparison between the observed and simulated chlorine concentrations was obtained. This procedure was repeated for each wall reaction kinetic model tested.

THM Kinetic Model. Trihalomethanes were modeled using a first-order growth, limited reaction-rate model in which the THM formation potential of the source water was assumed to be the limiting concentration that THMs can reach. This simplified model approach is valid only for a system with a single source water quality.

Conclusions

A number of conclusions can be drawn from the AwwaRF/USEPA study and from the other studies conducted by USEPA.

- Chlorine decay in distribution systems can occur because of reactions in the bulk phase and at the pipe wall.
- Chlorine decay bottle tests of a variety of US waters showed that *n*th-order or parallel first-order kinetic models provided better fits to the data than first-order models, although in most cases the difference was minimal.
- Laboratory studies showed that pipe-wall reactions related to corrosion of ferrous pipe materials can consume significantly more chlorine than those related to biofilm.
- The rate of reaction of chlorine at the pipe wall is inversely related to pipe diameter and can be limited by the rate of mass transfer of chlorine to the wall.
- At this time, no method is available for directly determining the kinetics of chlorine decay due to pipe-wall reactions. Field data must be used to calculate these reactions.

- A well-calibrated hydraulic model, one preferably based on a tracer study, is an absolute requirement for a distribution system water quality study.
- The first-order TTHM formation model showed acceptable performance in limited testing in actual distribution systems.
- The enhanced version of the EPANET software developed for this project is an effective tool for calibrating network models for chlorine decay and TTHM formation.

References

American Public Health Association (APHA), American Water Works Association, and Water Environment Federation (WEF). 2005. *Standard Methods for the Examination of Water and Wastewater.* 21st ed. Washington, D.C.: APHA.

Clark, R.M., and R.M. Males. 1985. Simulating Cost and Quality in Water Distribution. *Journal of Water Resources Planning and Management,* 111(4):454–466.

Clark, R.M., and R.M. Males. 1986. Developing and Applying the Water Supply Simulation Model. *Journal AWWA,* 78(8):61–65.

Clark, R.M., W.M. Grayman, R.M. Males, and J. Coyle. 1988. Modeling Contaminant Propagation in Drinking Water Distribution Systems. *Jour. Water Supply: Research and Technology—AQUA,* 3:137–151.

Clark, R.M., and J.A. Goodrich. 1993. Water Quality Modeling in Distribution Systems. In *Strategies and Technologies for Meeting SDWA Requirements.* Lancaster, Pa.: Technomic.

Clark, R.M., J.A. Goodrich, and L.J. Wymer. 1993. Effect of the Distribution System on Drinking Water Quality. *Jour. Water Supply: Research and Technology—AQUA,* 42(1):30–38.

Clark, R.M., W.M. Grayman, R.M. Males, and A.F. Hess. 1993. Modeling Contaminant Propagation in Drinking Water Distribution Systems. *Jour. Environ. Eng.,* 119(2):349–364.

Clark, R.M., G. Smalley, J.A. Goodrich, R. Tull, L.A. Rossman, J.T. Vasconcelos, and P.F. Boulos. 1994. Managing Water Quality in Distribution Systems: Simulating TTHM and Chlorine Residual Propagation. *Jour. Water Supply: Research and Technology—AQUA,* 43(4):182–191.

Clark, R.M., L.A. Rossman, and L.G. Wymer. 1995. Modeling Distribution System Water Quality: Regulatory Implications. *Jour. Water Resources Planning and Management,* 121(6):423–428.

Clark, R.M., H. Pourmoghaddas, L.G. Wymer, and R.C. Dressman. 1996. Modelling the Kinetics of Chlorination By-product Formation: The Effects of Bromide. *Jour. Water Supply: Research and Technology—AQUA,* 45(1):1–8.

Geldreich, E.E., H.D. Nash, D.J. Reasoner, and R.H. Taylor. 1972. The Necessity of Controlling Bacterial Populations in Potable Water: Community Water Supply. *Journal AWWA,* 64(9):596–602.

Gessler, J., and T.M. Walski. 1985. *Water Distribution System Optimization,* TREL-85-11, WES. Vicksburg, Miss.: US Army Corps of Engineers.

LeChevallier, M.W., T.M. Babcock, and R.G. Lee. 1987. Examination and Characterization of Distribution System Biofilms. *Applied and Environmental Microbiology,* 53:2714–2724.

Maul, A., A.H. El-Shaarawi, and J.C. Block. 1985a. Heterotrophic Bacteria in Water Distribution Systems, I. Spatial and Temporal Variation. *The Science of the Total Environment,* 44:201–214.

Maul, A., A.H. El-Shaarawi, and J.C. Block. 1985b. Heterotrophic Bacteria in Water Distribution Systems, II. Sampling Design for Monitoring. *The Science of the Total Environment,* 44:215–224.

Rossman, L.A., R.M. Clark, and W.M. Grayman. 1994. Modeling Chlorine Residuals in Drinking Water Distribution Systems. *Jour. Environ. Eng.,* 120(4):803–820.

Skov, K.R., A.F. Hess, and D.B. Smith. 1991. Field Sampling Procedures for Calibration of a Water Distribution System Hydraulic Model. In *Water Quality Modeling in Distribution Systems.* Denver, Colo.: AwwaRF/USEPA.

USEPA. 1979. *Methods for the Chemical Analysis of Water and Wastes,* EPA/600/4-79-020. Cincinnati, Ohio: USEPA Office of Research and Development.

USEPA. 1990. *Methods for the Determination of Organic Compounds in Drinking Water,* Supplement 1. EPA/600/4-90/020. Cincinnati, Ohio: USEPA.

Vasconcelos, J.J., P.F. Boulos, W.M. Grayman, L. Kiene, O. Wable, P. Biswas, A. Bhari, L.A. Rossman, R.M. Clark, and J.A. Goodrich. 1996. *Characterization and Modeling of Chlorine Decay in Distribution Systems.* Denver, Colo.: AWWA and AwwaRF.

Chapter 9

Case Study Applications

Introduction

As discussed previously, water quality models can provide important insights into the factors that cause the deterioration and/or contamination of drinking water distribution systems. The previous chapter discusses the historical development of water quality modeling. This chapter discusses the application of water quality modeling to several current issues and illustrates the insights that water quality modeling can provide for drinking water utility managers and decision makers. These case studies include the first application of water quality models to waterborne disease outbreaks in Cabool and Gideon, Mo., and an outbreak in Walkerton, Canada. In addition, the reconstruction of a historical contamination event in Dover Township, N.J., and an ongoing epidemiological study at the Marine Corps located at Camp Lejeune, N.C., is presented. In the Dover Township case study, the New Jersey Department of Health determined that the childhood cancer incidence rate in Dover Township was higher than expected. A hypothesis was developed that the higher cancer incident rate was related to the higher exposure to public water supplies with documented contamination. The Agency for Toxic Substances and Disease Registry (ATSDR) developed a water distribution model for the study area using the EPANET software to help test this hypothesis. In another study conducted by ATSDR, it was found that from 1957 through 1987, Marines and their families at Camp Lejeune drank and bathed in water contaminated with toxins. At least 850 former residents have filed claims for nearly $4 billion from the military. No official studies have definitively connected the contamination with illness; however, former residents of Camp Lejeune suffer from a high rate of cancer and other diseases. Children from the Marines have also shown signs of brain cancer and leukemia. Data collected as part of field studies were used to calibrate a hydraulic and water quality model of the Camp Lejeune distribution system (Maslia et al. 2001) to examine this problem.

A case study demonstrating the application of the PipelineNet model, which incorporates both GIS and the EPANET model, is also discussed based on work performed by the East Bay Municipal Utility District (EBMUD). PipelineNet is an EPANET/ArcView-based model and uses the same hydraulic engine as EPANET, which can simulate the fate and transport of contaminants in a water distribution system. The ArcView (or the GIS layer) portion of the model is designed to analyze and display the geospatial distribution of real or simulated contaminants. Finally, the potential application of water quality modeling for regulatory purposes is discussed.

Cabool, Missouri, Case Study

Cabool, located in the southeastern corner of Missouri, experienced a large outbreak of *E. coli* O157:H7 during the winter of 1989–1990. Cabool, a town of approximately 2,100 people, experienced a waterborne disease outbreak in which 243 cases were reported, with 32 hospitalizations and 4 deaths. This was the largest waterborne outbreak of *E. coli* O157:H7 that had been reported in the United States. At the time of the outbreak, the water source was untreated groundwater. Shortly after the outbreak, the USEPA sent a team to conduct a research study to determine the underlying cause of the outbreak.

Exceptionally cold weather prior to the outbreak contributed to two major water system line breaks and 43 water meter replacements throughout the city area. The sewage collection lines in Cabool were generally located away from the drinking water distribution lines but did cross or were near water lines in several locations. At the time of the outbreak, stormwater drained via open ditches along the sides of the streets and roads. During heavy rainfalls, sewage was observed to overflow manhole covers and flow into streets, parking lots, and residential foundations.

The town's water system (untreated groundwater) was implicated in the outbreak. Two of the town's four wells were operating at the time of the outbreak: one was 305 m deep and the other was 396 m deep. Both wells had protected wellheads, and the monitoring data from the 10 years before the outbreak indicated that no coliforms had been detected in either well. Investigation of the outbreak indicated that the distribution system was not well maintained and was vulnerable to sewage contamination at several points. Approximately 35 percent of the total flow was lost in the system—suggesting leaks, inaccurate meters, or unmetered connections. The town sewer system was also in poor condition and operating beyond capacity, resulting in regular sewage backups and overflows.

A number of risk factors apparently contributed to this outbreak. In mid-December 1989, unusually cold weather caused two large water mains and 45 in-ground water meters to fail. Ten cases of bloody diarrhea were reported to the local health department on Jan. 4, 1990. A boil-water order was issued on January 5, and water chlorination was initiated on January 12. Analyses of the temporal distribution of the cases indicated that the first cases occurred seven days before the first water main break (Dec. 23), and the last case occurred three days after the implementation of water chlorination. The early cases may have been caused by leaks and holes that developed prior to the main break. There was a small increase in the incidence of diarrhea after the first main break and a large increase in diarrhea cases about four days after the second main break on Dec. 26.

Hydraulic and water quality models were applied to examine the movement of water and contaminants in the system. Steady-state scenarios were examined and a dynamic analysis of the movement of water and contaminants associated with meter replacement and the aforementioned breaks was conducted. Typical demand patterns were developed from available meter usage for each service connection, and it was found that the water demand was 65 percent of the average well production, indicating inaccurate meters, unmetered uses, and a high water loss in the system.

The modeling effort revealed that the pattern of illness occurrence was consistent with water movement patterns in the distribution system assuming two water line breaks. It was concluded, therefore, that some disturbance in the system, possibly the two line breaks or 43 meter replacements, allowed contamination to enter the water system. Analysis showed that the simulated contaminant movement covered 85 percent of the infected population.

Replacement of the failed water meters may have contributed to contamination of the distribution system. During the replacement of the meters and main break repairs, the lines were subjected to "limited flushing" but were not disinfected by superchlorination, and no water samples were tested for microbial indicators to examine the water quality before bringing the lines back into service. Although sewage overflow into the distribution system via the main breaks and intrusion was believed to be responsible for the outbreak, microbial contamination of the distribution system could not be confirmed because water samples from the distribution system were never collected and analyzed. Hydraulic modeling of the system reinforced the evidence that the second main break had the potential to contaminate a greater portion of the distribution system, including the northern part of the town where 36 percent of the cases occurred (Geldreich et al. 1992).

The model applied to the Cabool outbreak was the dynamic water quality model (DWQM) introduced previously. With the increasing sophistication of water quality propagation modeling, it has become possible to apply these types of models to waterborne disease outbreaks even more easily then during the Cabool outbreak.

Gideon, Missouri, Case Study

In 1993, the town of Gideon, Mo., suffered from an outbreak of salmonellosis that affected more than 650 people and caused seven deaths (Hrudey and Hrudey 2004). At the time of the outbreak, Gideon was a small town (pop. 1,100) in a rural, agricultural area with an unemployment rate greater than 11 percent. Twenty-five percent of the population was living below the poverty level at the time.

The Missouri Department of Health (MDOH) had identified 31 cases of laboratory-confirmed salmonellosis (Clark et al. 1996). The State Public Health Laboratories identified 21 of these isolates as dulcitol negative *Salmonella* serovar *typhimurium*. *Salmonella* is a pathogenic bacterium that has been classified into several serotypes (common set of antigens). *Salmonella* serovar *typhimurium* is among the most common *Salmonella* serovars causing salmonellosis in the United States. Of the 31 laboratory culture-confirmed patients, 15 were hospitalized (including 2 patients hospitalized for other causes and who developed diarrhea while in the hospital). These 15 patients were admitted to 10 different hospitals. Seven nursing home residents exhibiting diarrheal illness died; four of these patients were culture confirmed (the other three were not cultured). Two of the patients had positive blood cultures. Interviews conducted by the MDOH during this period suggested that no food exposures were common to a majority of the patients. However, all of the ill persons, including the culture-confirmed patients, had consumed municipal water, which supported the association. The MDOH reported its suspicion to the Missouri Department of Natural Resources (MDNR).

The Gideon municipal water system was originally constructed in the mid-1930s and obtained water from two adjacent 1,300-ft-deep wells. The well waters were not disinfected at the time of the outbreak. After the outbreak emergency, chlorination was initiated, and later a permanent chlorination system was installed. The distribution system consisted primarily of small-diameter (2-, 4-, and 6-in.) unlined, steel and cast-iron pipe. Tuberculation and corrosion were major problems in the distribution pipes. Raw water temperatures were unusually high for a groundwater supply system (58°F), because the system overlies a geologically active fault. Under low-flow or static conditions, the water pressure was close to 50 psi. However, under high-flow or flushing conditions, the pressure dropped dramatically. These sharp pressure drops were evidence of major problems in the Gideon distribution system.

The municipal system had two elevated tanks. One tank was a 50,000-gal tank (referred to as small tank), and the other was a 100,000-gal tank (referred to as large tank).

In early November, a cold snap caused a thermal inversion in the water storage tanks that resulted in taste and odor problems. In response, the water system was systematically flushed on Nov. 10. The first cases of acute gastroenteritis were reported on Nov. 29 and diagnosed as *Salmonella typhimurium*. However, the outbreak investigation later revealed that diarrhea cases in Gideon started around Nov. 12, with a peak incidence around November 20. By early December, there was a 250 percent increase in absenteeism in the Gideon schools and a 600 percent increase in antidiarrheal medication sales. Over 40 percent of nursing home residents suffered from diarrhea and seven people died (Angulo et al. 1997). However, the outbreak was not linked to the water system until Dec. 15, when the water system samples were reviewed and investigative water sampling was initiated. A boil-water advisory was issued on Dec. 18. On Dec. 22, emergency chlorination was added to the production well, and the two municipal storage tanks were superchlorinated. The last reported cases occurred on Dec. 28.

Water samples collected from the distribution system on Dec. 16, 17, 20, and 21 were positive for total coliforms, and the samples from Dec. 20 and 21 were also positive for fecal coliforms. The outbreak strain of *S. typhimurium* was detected in one large volume sample collected from a fire hydrant. Inspection of the water storage tanks suggested that the outbreak was probably caused by contamination by bird feces in one or more of the tanks. The larger of the municipal tanks was in disrepair and had birds roosting on the roof. The private storage tank had an unscreened overflow pipe and a hole at the top of the tank that was large enough for birds to enter. This private tank had been drained on December 30, but the outbreak strain of *S. typhimurium* was detected in samples of sediment collected on Jan. 5, 1994. The remaining water on the bottom of the tank was described as black and very turbid, with rust, suspended particles, and bird feathers floating on the top. Initially, attention was focused on the private tank as the source of the outbreak as reported by Skala (1994). On Jan. 14, 1994, a USEPA field team, in conjunction with the CDC and the state of Missouri, initiated a field investigation that included a sanitary survey and microbiological analyses of samples collected on site. A system evaluation was also conducted in which EPANET was used to develop various scenarios to explain possible contaminant transport in the Gideon system.

Although the private tank was initially suspected as the cause of outbreak, an in-depth hydraulic analysis of the Gideon system, conducted as part of the outbreak investigation, raised questions about the possibility of the private tank being the source of the outbreak. A subsequent review of as-built drawings of the Gideon system by MDNR personnel revealed that the private tank was separated from the municipal system by a backflow prevention valve, but reports were conflicting about whether this valve was open or closed when the tank was first inspected. In order to test the integrity of the valve, a "fire-pumper" was brought to the site, and it was found that the backflow prevention valve held under a pressure of 100 psi. In a subsequent hydraulic analysis, the private tank was eliminated as a contamination source for the outbreak. The analysis demonstrated that elimination of the private tank yielded results that were consistent with the behavior of the system as observed during the outbreak scenario. This analysis also pointed to the largest municipal tank as the most likely source of the outbreak. A visual inspection of the large municipal tank revealed broken and rusted hatches, bird parts and feathers on the top of the tank, and bird parts and feathers floating on the surface of the tank water. Therefore, the subsequent USEPA field investigations and modeling efforts focused on the two municipal tanks as the source of contamination.

The key analysis was focused on a flushing program conducted earlier by the utility in response to taste and odor complaints. A sequential flushing program was conducted on Nov. 10, 1993, involving all 50 hydrants in the system. The flushing program was started in the morning and continued through the entire day. Each hydrant was flushed for 15 min at an approximate rate of 750 gpm. It was observed that the pump at one of the wells was operating at full capacity during the flushing program (approximately 12 hr), which would indicate that the municipal tanks were discharging during this period.

During the evaluation, it was hypothesized that the taste and odor problems may have resulted from a thermal inversion that had taken place because of a sharp temperature drop prior to the day of the complaint. If stagnant or contaminated water were floating on the top of a tank, a thermal inversion could have caused this water to be mixed throughout the tank and to be discharged into the system, resulting in taste and odor complaints (Fennel et al. 1974). As a consequence, the utility initiated the aforementioned citywide flushing program. Turbulence in the tank from the flushing program could have stirred up the tank sediments that were subsequently transported into the distribution system. It is likely that the bulk water and/or the sediments were contaminated with *Salmonella* serovar *typhimurium*. During the USEPA field visit, a large number of pigeons (bird droppings are known to contain *Salmonella*) were observed roosting on the roof of the 100,000-gal municipal tank.

The USEPA study team evaluated the effects of distribution system design and operation, demand, and hydraulic characteristics on the possible propagation of contaminants in the system. Given the evidence from the lab samples and the results from the valve inspection of the private tank, it was concluded that the most likely contamination source was bird droppings in the large municipal tank. Therefore, the analysis concentrated on propagation of water from the large municipal tank in conjunction with the flushing program. Other possible sources of contamination, such as cross connections, were also studied, but ruled out as a source of the outbreak.

The system layout, demand information, pump characteristic curves, tank geometry, flushing program, and other information needed for the modeling effort were obtained from maps, demographic information, and numerous discussions with consulting engineers and city and MDNR officials. EPANET was used to conduct the contaminant propagation study (Rossman et al. 1994).

The EPANET network model was calibrated by simulating flushing at the hydrants assuming a discharge of 750 gpm for 15 min. The *C* factors (pipe roughness) were adjusted until the head loss in the model matched head losses observed in the field. After the calibration, the hydraulic model was simulated for 48 hr. Thereafter, the flushing program was simulated, starting at 8 a.m. on day 3, by sequentially imposing a 750-gpm demand on each hydrant for 15 min. Employing the TRACE option in EPANET, the percentages of water from both municipal tanks were calculated at each node over a period of 72 hr.

During the simulation of the flushing program, the pump at one of the wells was operated (as previously observed) at full capacity, which was over 800 gpm, and then reverted to cyclic operation. The simulation results showed that the tank elevation fluctuated for both municipal tanks, and both the tanks discharged during the flushing program. At the end of the flushing period, nearly 25 percent of the water from the large municipal tank passed through the small municipal tank where it was again discharged into the system. The model predicted dramatic pressure drops during the flushing program. Based on the information available, it was felt that these modeling results replicated the conditions that existed during the flushing program closely enough to provide a basis for an analysis of water movement in the system.

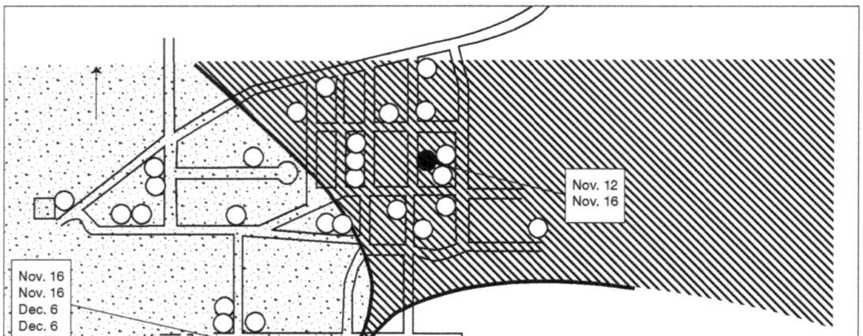

Figure 9-1. Comparison of early confirmed cases of *Salmonella*-positive sample versus the estimated distribution of tank water during the first 6 hr of the flushing program

Data from the simulation study, the microbiological surveillance data, and the outbreak data were used to provide insight into the nature of both general contamination problems in the system and the outbreak itself. The water movement patterns showed the majority of the collected samples that were total coliform—and fecal coliform (FC)—positive occurred at points within the zone of influence of the small and large tanks. During both the flushing program and for large parts of normal operation, these areas were predominantly served by tank water, which confirmed the belief that the tanks are the source of the fecal contamination because positive FC samples were present prior to chlorination. Figure 9-1 shows the comparison of early confirmed cases of *Salmonella*-positive sample versus the estimated distribution of tank water during the first 6 hr of the flushing program.

Walkerton, Canada, Case Study

The first documented outbreak of *Escherichia coli* 0157:H7/*Campylobacter* spp. gastroenteritis associated with a municipal water supply in Canada occurred in the small rural town of Walkerton, Ont., population 1,261, in May 2000 (Grayman et al. 2004). At the time of the outbreak, the town's drinking water was supplied by three wells (wells 5, 6, and 7), which fed a common distribution system. A hydrogeological assessment revealed that two of the three wells (wells 5 and 6) were under the influence of surface water. A microbiological investigation found *E. coli* 0157:H7 and *Campylobacter* spp. on a cattle farm adjacent to Well 5 with identical molecular characteristics to those isolated from outbreak victims.

In order to understand the factors that caused the outbreak, a water quality model of the Walkerton water distribution system was developed. Using a cross-sectional study, it was demonstrated that during the outbreak, residents living in homes connected to the municipal water supply and consuming Walkerton water were 11.7 times more likely to have developed gastroenteritis than those not exposed to Walkerton water.

Modeling of the Walkerton water system involved estimating the following parameters for use in the water quality model:

- Pipe diameter and length, location, age, and composition of all water pipes.
- Size, storage capacity, and active volumes of the two standpipes (water towers) in the system.
- Well pump specifications (including pump curves).
- Pipe friction.

The objective of the water modeling exercise was to recreate the pattern of water flow throughout the town's distribution system immediately before and during the outbreak period. The system was modeled as a network of nodes, which included water sources, tanks, and demand (customer) locations. Individualized temporal water demand patterns were assigned to each commercial node, and residential users were assigned to the nearest residential node. As residential users in Walkerton were not metered, hourly demand was estimated using the daily volume of water supplied to the system after accounting for commercial users and fire events, and literature-based hourly demand patterns. Well pump controls were added to the model, and pump on and off times were set in accordance with the historical pump records. Computerized data from the supervisory control and data acquisition (SCADA) system for the water supply containing 15-min pumpage rates for the three wells were used. Exposure scenarios were created by adding a hypothetical inert contaminant to each well at predetermined times and concentrations and then using the model to follow the movement and relative concentrations of contaminants through the distribution system.

Conditional logistic regression was used to quantitatively evaluate the relationship between exposure to potentially contaminated well water from each of Walkerton's three wells and the likelihood of experiencing infection. A median incubation period of 2–5 days was assumed for both *E. coli* 0157:H7 and *Campylobacter* for water supplied to homes 2–5 days prior to the illness onset date, and potential exposure scenarios were calculated for each case. This was done for each of six exposure scenarios.

The results of this study clearly supported the hypothesis that Well 5 was the primary, if not the only, well involved in the Walkerton *E. coli/Campylobacter* waterborne outbreak. The results also suggested that an extreme rainfall event, which occurred just prior to the peak of the outbreak, may have played a significant role in the propagation of the contaminants.

Dover Township, New Jersey, Case Study

In August 1995, the New Jersey Department of Health (now New Jersey Department of Health and Senior Services [NJDHSS]) determined that the childhood cancer incidence rate in Dover Township (and the Toms River section) was higher than expected. This determination included all malignant cancers and included brain cancer, central nervous system cancer, and leukemia. Consequently, NJDHSS and made a formal request for an evaluation by ATSDR. NJDHSS and ATSDR developed a joint Public Health Response Plan (PHRP) describing actions they would take to investigate this unexpected increase (NJDHSS 2003). The PHRP included several items including the identification of potential environmental exposure pathways associated with two National Priorities List (NPL) sites in Dover Township (Figure 9-2; see appendix E). Two public water supply well fields (Parkway and Holly) located in the vicinity of the NPL sites were identified as potential routes of exposure. These well fields also served areas that had statistically higher childhood cancer rates. Follow-up studies revealed the presence of a previously unidentified compound, styrene acrylonitrile (SAN), in the groundwater at the Parkway well field, which could be traced to the Reich Farm NPL site. A search of historical records also revealed contamination (primarily semivolatile organics [SVOCs]) of the Holly well fields that could be traced to the Ciba-Geigy NPL site.

A hypothesis was developed that the higher cancer incident rate was related to the higher exposure to public water supplies with documented contamination (the Parkway and Holly well fields). ATSDR developed a water distribution model for the study area using the EPANET software to help NJDHSS test this hypothesis. This network model was used to simulate historical characteristics of the water distribution system serving Dover Township from 1962 to 1996. A problem with the approach taken by ATSDR was the lack of historical contaminant-specific data during most of the period covered by the epidemiologic study. Therefore, the modeling effort focused on estimating the percentage of water that an exposed individual might have received from each well that supplied water to the study area.

An extensive study was conducted in 1998 to (1) determine the spatial locations of distribution system facilities (wells, tanks, pump, and hydrants) and (2) equip hydrants with continuous-recording digital data loggers and monitor supply sources (wells, pumps, and tanks). Twenty-five hydrants located throughout the distribution system were equipped with data loggers. System responses during winter demand (March) and summer demand (August) periods were measured (Maslia et al. 2000), including:

- On–off cycling of groundwater wells.
- High-service and booster pump operations.
- Pressure variations, storage tank water-level fluctuations.
- Total water production.

A detailed "all-pipe" hydraulic network model of the study area was developed and calibrated to present-day conditions (1998) using the field investigation results. The model was calibrated through a water quality simulation of the transport of a naturally occurring conservative element (barium) and a comparison of the results with data collected in March and April 1996 at 21 schools and 6 points of entry into the water distribution system. Adjustments were made to the network to reflect growth in the network that had occurred between 1962–1966 and the study period. Some of the data used in this adjustment included water utility pipeline installation records, quarterly billing records, NJDHSS groundwater well records, and annual water utility reports to the state board of public utilities. These data were incorporated into EPANET, and simulations were conducted for each month of the historical period January 1962 through December 1996 (420 simulations). After completing those 35-year/420-month analyses, source-trace analysis simulations were conducted to determine the percentage of water contributed by each well or well field operating during each month for all study subject locations. Grayman et al. (2004) presents a more detailed account of the EPANET model input adjustments.

Simulation Techniques

Using the EPANET network model developed for the Dover Township area, hydraulic modeling was conducted whereby average network conditions were simulated for every month of the historical period (420 simulations). Monthly network hydraulic and water quality simulations (source-trace analysis) were conducted for each water source (point of entry) of the network in order to determine the monthly proportionate contribution of source water at all locations in the Dover Township area serviced by the water distribution system.

Simulation Results and Conclusions

The health scientists using the resulting percentage of water derived from the different sources developed exposure indices for each subject in the study area. The results from the case-control study showed an association between prenatal exposure to contaminated community water and leukemia in female children (NJDHSS 2003). For example, female leukemia cases were found to be five times more likely to have occurred in children born to mothers exposed during the prenatal period to a high percentage of Parkway well water than were control children. The control children were those living in the study area but not exposed to water from the contaminated well fields. Some of the innovations documented by the Dover Township historical reconstruction analysis were:

- Water distribution system modeling and source tracing could be used to quantify exposure on a monthly basis for all locations historically served by the distribution system.
- Sensitivity analyses indicated that operating system changes did not appreciably change the proportionate contribution of water to Dover Township locations.
- The association between exposure and disease was based on being able to integrate distribution system modeling and epidemiologic analyses.

East Bay Municipal Utility District Case Study

The East Bay Municipal Utility District (EBMUD) is located on the eastern side of San Francisco Bay and provides water and sewage treatment to Alameda County and Contra Costa County in California, including the cities of Oakland and Berkeley and several nearby suburbs. The district owns and operated several dams, reservoirs, pipelines, and aqueducts. Water is transported to treatment plants and delivered to local reservoirs and tanks and by gravity to households. The district also has the responsibility for dealing with wet-weather overflow in its service area.

This case study demonstrates the application of the PipelineNet model, which incorporates both GIS and the EPANET model (Panguluri et al. 2005). PipelineNet is an EPANET/ArcView (registered trademark of ESRI)-based model and uses the same hydraulic engine as EPANET, which can simulate the fate and transport of contaminants in a water distribution system. The ArcView (or GIS layer) portion of the model is designed to analyze and display the geospatial distribution of real or simulated contaminants.

The project was jointly funded by AwwaRF and USEPA with the objective of developing techniques to

- Locate monitoring points in a distribution system.
- Determine appropriate timing and frequency of monitoring.
- Establish monitoring techniques and relevant water quality parameters.

The study focused on 16 of the 123 EBMUD pressure zones containing 27 tanks, 748 mi of pipes, 62 pumps, and 17,997 pipe segments with diameters equal to or greater than 2 in. Figure 9-3 shows a partial view of the hypothetical network of pipelines.

Calibration was accomplished by comparing the observed (SCADA data) and simulated (PipelineNet model) water levels in 25 of the tanks located in the study area. The inflow/outflow and water height patterns were matched for a 24-hr time interval using data measured on July 1, 2001. Pump characteristic curves were used to achieve a good comparison between simulated and observed tank levels, and the flow value of the characteristic curve was changed as needed to reflect field conditions. Pumps were operated using time controls.

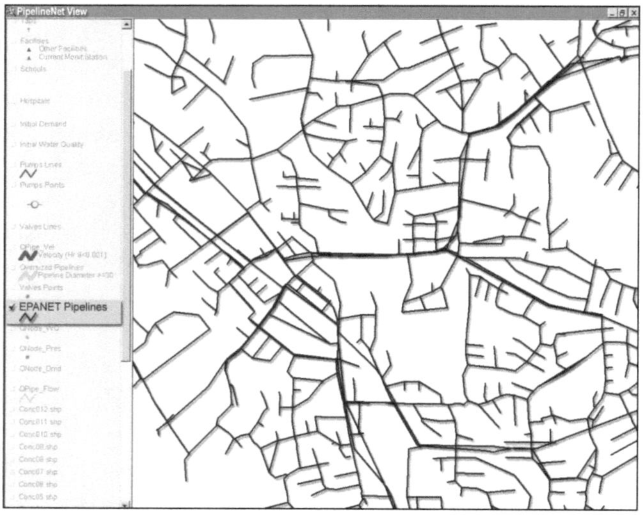

Figure 9-3. Partial view of the pipe network (Panguluri et al. 2005)

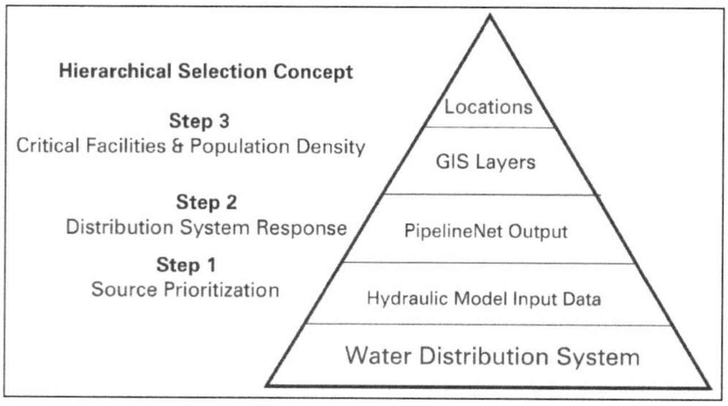

Figure 9-4. Hierarchical selection process (Panguluri et al. 2005)

Monitoring Site Location Methodology

A hierarchical selection process was developed to locate monitoring stations in the distribution system using a three-step approach as described below (Figure 9-4). In the first step all the elements of the water distribution system are assumed to be available for monitoring and are assigned a score of 1. Then all of the nodes that are not amenable or not accessible are assigned a score of 0. This reduced the number of pipes available for monitoring from 17,997 to 14,938.

In the second step (distribution system response factor), PipelineNet was run using the reduced data set to quantify the distribution system response in terms of flow, velocity, and pressure. Each system response parameter was assumed to have equal weight and assigned an initial score of 1 for every pipe. Based on results from the run, scores are reassigned values over a range of 1 and 10, where a score of 10 would indicate a higher level of concern. For each parameter, the user determined the distribution of scores over the parameter range. A score range of 10 to 1, for example, could be distributed over a flow range of 0.001 to 100 gpm (Bahadur et al. 2003):

In the third step, buffer zones (polygons) were created around critical facility locations. Areas of low-, medium-, and high-population density were delineated by polygons. Pipes closest to critical facilities and/or near high-population-density areas were assigned a score of 10.

The total score for each pipe based on Steps 2 and 3 was computed and linked to the GIS pipeline layer. Based on this approach, the user can identify areas where monitoring stations should be placed based on pipes with high scores. Figure 9-5 (see appendix E) shows the pipes in the hypothetical network with scores greater than 27 overlaid with critical facility locations.

Response and Mitigation Tools

Three additional tools were developed in the areas of emergency response, mitigation, and normal operations as follows:

- Consequence assessment tool, which can be used to quickly identify and quantify the population, infrastructure, and resources at risk from a contaminant event including total population at risk, number of taps contaminated, miles of pipe contaminated, total number of hospitals and beds for each hospital, and total number of schools and student population.
- Isolation tool, which provides the ability to change the status of any pipe in the distribution system. The model would then be rerun, under these new hydraulic conditions, and the output examined to determine if this mitigation step was successful in limiting the area of contamination.
- Spatial database display tool, which can overlay the EPANET model output with various geospatial properties. For illustration purposes, Figure 9-6 (see appendix E) shows the display of three criteria: oversized pipes (diameter >30 in.), current monitoring locations, and low-velocity (velocity < 0.001 ft/s) pipes.

The case study demonstrates that PipelineNet model can be used to perform a variety of practical analyses to locate monitoring systems.

Camp Lejeune, North Carolina, Case Study

In April 1941, Camp Lejeune, a 246-mi^2 (640-km^2) training facility for the United States Marine Corps, was constructed. Volatile organic compounds (VOCs) were found in Camp Lejeune's drinking water supply in 1982. These chemicals included trichloroethylene (TCE), a degreaser; perchloroethylene (PCE), a dry cleaning solvent; benzene; and many other chemicals.

As many as 500,000 people may have been exposed to contaminated water at Camp Lejeune over a period of 30 years. Although no official studies have definitively connected the contamination with illness, former residents of Camp Lejeune suffer from a high rate of cancer and other diseases. Children from the Marines have also shown signs of brain cancer and leukemia.

The Agency for Toxic Substances and Disease Registry (ATSDR) initiated a study of the Camp Lejeune drinking water system(s) in order to determine whether an association existed between exposure to contaminated drinking water and birth defects among children of women who lived at Camp Lejeune. The focus of this study was on women who were pregnant between 1968 and 1985. Information collected as part of field studies was used to calibrate a hydraulic and water quality model of the Camp Lejeune distribution system (Maslia et al. 2001).

Camp Lejeune is located in the city of Jacksonville, Onslow County, N.C., bordering the Atlantic Ocean (Figure 9-7). Presently, two operating water treatment plants (WTPs) provide water for the distribution systems of interest to the epidemiologic study: (1) the Holcomb Boulevard WTP that services the Camp Johnson, Tarawa Terrace, and Holcomb Boulevard areas of the distribution system and (2) the Hadnot Point WTP that services the Hadnot Point area of the distribution system. Hadnot Point was the original WTP and at one time serviced the entire base. The Holcomb Boulevard WTP presently services the rest of the military housing areas. A third plant, the Tarawa Terrace WTP, historically serviced the Tarawa Terrace and Camp Johnson areas, but this plant was shut down and replaced by a ground storage tank at Tarawa Terrace that receives water directly from the Holcomb Boulevard WTP.

The topography is very flat, ranging from sea level to less than 40 ft; hydraulic heads range from 140–160 ft, resulting in a very mild hydraulic gradient. Nine elevated storage

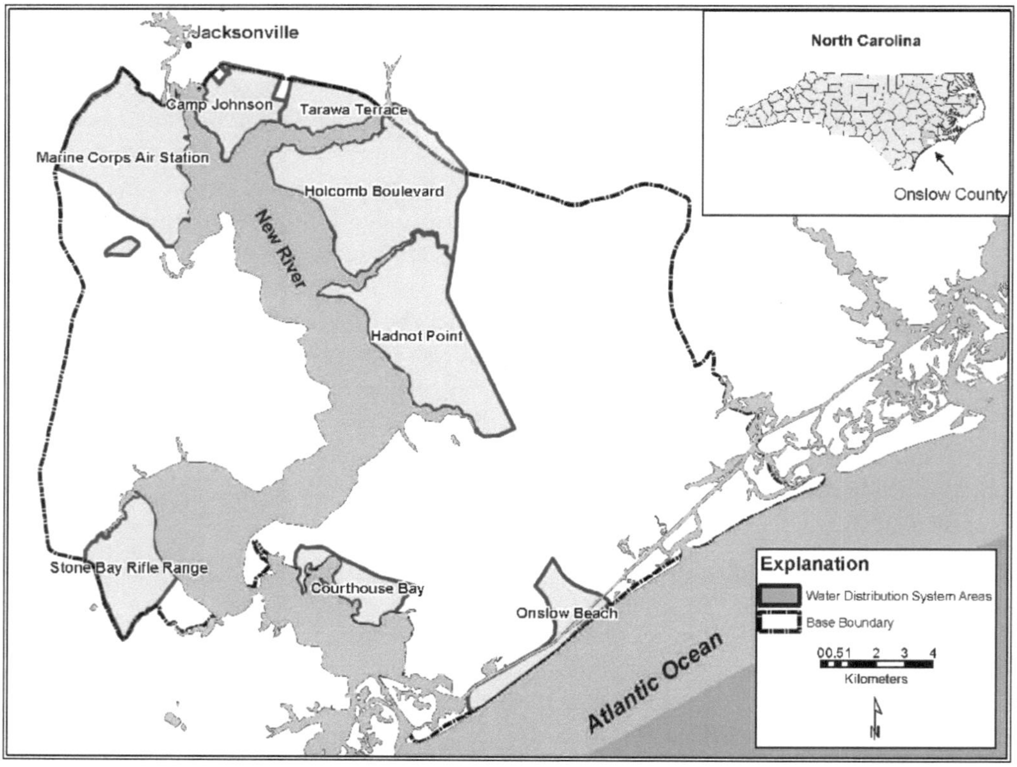

Figure 9-7. Camp Lejeune study area (Panguluri et al. 2005)

tanks are in use in the Holcomb Boulevard and Hadnot Point WTP service areas, and the range in water level fluctuation for the elevated storage tanks is small, generally 1–6 ft. Excellent information is available on the location and characteristics of the water system facilities. SCADA data are available that provide continuous information on plant discharges and tank water levels, but individual buildings and residences are not metered.

Field Work

A variety of field activities were conducted to provide an understanding of the water system operation and to provide information that can be used to calibrate a detailed water distribution system model (Sautner et al. 2005). These activities included:

- Conducting *C*-factor and fire-flow tests
- Recording system pressures and storage tank water levels over time
- Tracer and associated travel time tests
- Recording continuous flow information at key locations

Several of these field activities were performed in tandem in order to provide an integrated understanding of the system operation and performance.

C-Factor and Fire-Flow Tests

C-factor tests and fire-flow tests were conducted in August 2004 at various sites at Camp Lejeune. Continuous pressure loggers attached to hydrants (Figure 9-8) were used to record pressure at 1-min intervals. Standard analog pressure gages were also used as backup during the tests. Hydrant flows were measured using pitot gages installed on hydrants that were flowed during the tests.

Tracer Test and Continuous Measurements

Figure 9-8. Continuous pressure loggers attached to hydrants (Panguluri et al. 2005)

Tracer tests were conducted May 24–27, 2004, in the Hadnot Point (Camp Lejeune) distribution system consisting of four steps: (1) injecting liquid CaCl, 35 percent by weight, into the transmission main on the distribution system side of the WTP to achieve elevated conductance and chloride concentration; (2) injecting sodium fluoride into the transmission main to achieve an elevated fluoride concentration and monitoring fluoride concentration in the distribution system; (3) monitoring distribution system pressures with continuous recording data loggers attached to selected hydrants (Figure 9-8); and (4) monitoring flows and tank water levels from SCADA data. Chloride and fluoride concentrations were monitored at nine hydrants that were equipped with the Horiba W-23XD dual probe ion detector. Conductivity was collected at nine hydrants that were equipped with the Horiba W-21XD single probe ion detector, thus providing a total of 18 monitoring locations for continuously recording conductivity data. Pressure measurements were collected at nine hydrants that were equipped with continuous recording Dixon PR300 pressure data loggers. Grab samples were collected for quality assurance and quality control (QA/QC) purposes. Samples were analyzed at the Hadnot Point WTP by ATSDR staff and then also shipped to the Federal Occupational Health (FOH) environmental laboratory in Chicago for analysis. Twenty-seven hydrants were selected in the Hadnot Point area for monitoring locations.

Results from the chloride injection were used to estimate travel times at different locations throughout the Hadnot Point WTP area. Of special interest are the extremely long arrival times—in excess of 16 hr—in the northwestern part of the of the Hadnot Point WTP area. Post-test field verification by water utility staff confirmed the locations of closed valves.

Additional Test Procedures

A second tracer test was conducted in the Holcomb Boulevard WTP area in September–October 2004. In this test, the normal fluoride feed was turned off for a period of a week and then turned back on and monitored for another week. Nine locations in the distribution system were equipped with Horiba W-23XD continuous recording, dual probe ion detector data loggers.

Case Study Summary

Field tests included: (1) recording system pressures and storage tank water levels, and (2) conducting *C*-factor, fire-flow, tracer, and travel time tests. The test data are being used to assist with hydraulic and water quality model calibration and are the first step in reconstructing historical operations during the period between 1968 and 1985.

Regulatory Studies

The DBPR2 regulation requires that all water utilities that have a disinfectant residual in the distribution system to perform an individual distribution system evaluation (IDSE) unless they obtain a small-system or "40/30" waiver (USEPA 2003). The goal of the IDSE is to identify compliance monitoring sites that may have high DBP levels in distribution systems. Systems that can certify that their total trihalomethanes (TTHMs) and five haloacetic acids (HAA5) are less than or equal to 40 µg/L for TTHM and 30 µg/L for HAA5, respectively, are not required to perform an IDSE. Utilities may choose to perform a System Monitoring Plan (SMP) that involves extensive monitoring, or they may choose to perform a system-specific study (SSS). The SSS uses historical data, distribution system models, or other analyses combined with minimal monitoring to evaluate TTHM and HAA5 levels throughout the distribution system as the basis to select future compliance monitoring sites.

The IDSE guidance manual spells out a series of suggested minimum requirements for the use of a calibrated water distribution system hydraulic model to perform an SSS. In general, for conducting an SSS the water distribution system model should be more comprehensive than models typically used for long-range capital improvement program analysis (e.g., master planning). The guidelines provided in the draft guidance manual are summarized below:

- An EPS model that has been recently calibrated using generally accepted methods.
- An all-pipes model or skeletonized model that includes (a) at least 50 percent of total pipe length in the distribution system, (b) at least 75 percent of the pipe volume in the distribution system, (c) all 12-in.-diameter and larger pipes, (d) all 8-in. and larger pipes that connect major facilities, (e) all 6-in. and larger pipes that connect remote areas of a distribution system, and (f) all active control valves or other system features that could significantly affect the flow of water through the distribution system.
- Water demand should be allocated among the nodes of the model in a manner that reflects the actual spatial distribution of such demand throughout the system.
- A system-specific, diurnal (24-hr) demand pattern should be applied to the overall system demand.
- The model should accurately simulate seasonal system configurations and operational changes.

Once the model is developed, it is then run in EPS mode until a consistent, repeating temporal pattern of water age is established at all nodes of the model. Generally, the model should be run under high DBP formation conditions (typically summer months) to estimate residence times. Based on the modeled water age results, preliminary monitoring sites are identified near locations that satisfy the sampling site requirements. Sampling sites are selected to represent:

- High-TTHM sites, which are generally downstream of storage facilities and in remote locations.
- High-HAA5 sites, which are generally the same as those for high-TTHM sites. However the sites chosen to represent high HAA5 should have a disinfectant residual sufficient to suppress bacteria, which can degrade HAAs.
- Average-residence-time sites, which are sites with residence times close to the flow-weighted mean of all nodal residence times.
- Near-entry-point sites, which are locations near entry points into the water system.

Promulgation of the US Environmental Protection Agency's (USEPA's) Stage 2 Disinfectants/Disinfection Byproducts Rule(D/DBPR2) in 2006 (USEPA 2006) is causing

US drinking water utilities to examine their treatment and distribution system operations with the goal of reducing levels of disinfection by-products (DBPs) in their distribution systems (Yang et al. 2007). Two examples are presented to demonstrate the application of models as part of the IDSE process. One example is based on a hypothetical water distribution system that purchases disinfected groundwater and serves approximately 15,000 people (Panguluri et al. 2005). The second example examines four case-study utilities and discusses how these utilities can or will apply water quality models to comply with the IDSE.

Hypothetical Example

In this example, water is assumed to enter the distribution system from two separate interconnections to a wholesale utility with an average demand of 2.2 mgd. An interconnection in the northern part of the system operates intermittently and provides approximately 80 percent of the demand. An interconnection in the southern part of the system operates continually and provides the remaining 20 percent of the system demand. The system has a 1.5-mil-gal storage tank. The DBPR2 states that a total of six sites are required for a utility of this size using groundwater as follows:

- One site must be near the largest entry.
- One site should represent a site with an average residence time.
- Two sites must represent maximum TTHM conditions.
- Two sites must represent high HAA5 conditions.

It is assumed that the water utility has a distribution system that has been used extensively in the past for both hydraulic and water-quality studies. The model is a skeletonized model that includes all 8-in.-diameter and larger pipes and all major facilities. The pipes in the model are assumed to represent 74 percent of the total length of pipe in the system and 86 percent of the total volume. The model has been previously calibrated based on two tracer studies and has been shown to have excellent agreement with observed field results. Demands have been assigned to nodes based on actual meter readings, and information from the SCADA system has been used to construct a typical diurnal water use pattern. The model is being operated in an EPS mode to simulate a 12-day period and has been calibrated for use in simulating chlorine residuals in the distribution system.

A series of model-runs were conducted to help understand the movement of water and water quality transformations in the system. Specific simulation scenarios included water age and chlorine residual and the results of the water-age model run are shown in Figure 9-9 (see appendix E), which shows the average water age throughout the distribution system over the last 24 hr of the 14 days of the simulation. The last 24-hr period was used in order to avoid the uncertainties associated with the run-in period. The plot shows the nodes in the vicinity of the northern interconnection that receive water with an average age of less than 2 hr. The average water age increases significantly for areas that are farther from the interconnections. The demand flow-weighted average water age for delivered water was calculated to be 27 hr. Water age can, however, also vary quite significantly over the course of a day in a water system, due largely to storage tanks. Figure 9-10 (see appendix E) depicts the minimum water age at each node over the same 24-hr period. Figure 9-11 shows the same effect for Node J-456 in the vicinity of the tank. IDSE guidance does not require utilities to explicitly consider the effects of tanks on diurnal variations in water age. If Node J-456 were selected as representative of high DBP because of its high water age, it would be expected that the DBPs would only be high during the part of the day when water is being discharged from the tank.

The model was also used to display the chlorine residual throughout the system. Figure 9-12 (see appendix E) contains a plot of the minimum chlorine residual. It is important to note sites with long residence time and low residuals that yield high TTHMs and sites with long residence times and high HAA5 levels. Figure 9-13 (see appendix E) also shows a small area in the southwestern portion of the system that the model predicts will potentially experience chlorine residuals less than 0.2 mg/L of chlorine.

Based on the modeling results, various zones were defined in the distribution system that are appropriate for different types of compliance monitoring requirements (Figure 9-13, see appendix E). The zones shown in the plot include:

- Nodes in the vicinity of the north connection with a water age of less than 2 hr.
- Nodes with average water age in the range of 21 to 33 hr.
- Nodes with residence times that exceed twice the average water age (> 54 hr) and having a minimum chlorine residual exceeding 0.20 mg/L (high-HAA5 sites).
- Other nodes with residence times that exceed twice the average water age (> 54 hr) (high-TTHM sites).

As illustrated by the example, if a detailed, calibrated EPS model is available, the model represents an efficient means of defining compliance monitoring locations. Clearly, monitoring site selection would be supplemented by field data as well.

Utility Case Studies

Four utilities were considered in this study ranging in size from 7,000 to 70,000 consumers served (Yang et al. 2007). Their characteristics are summarized in Table 9-1.

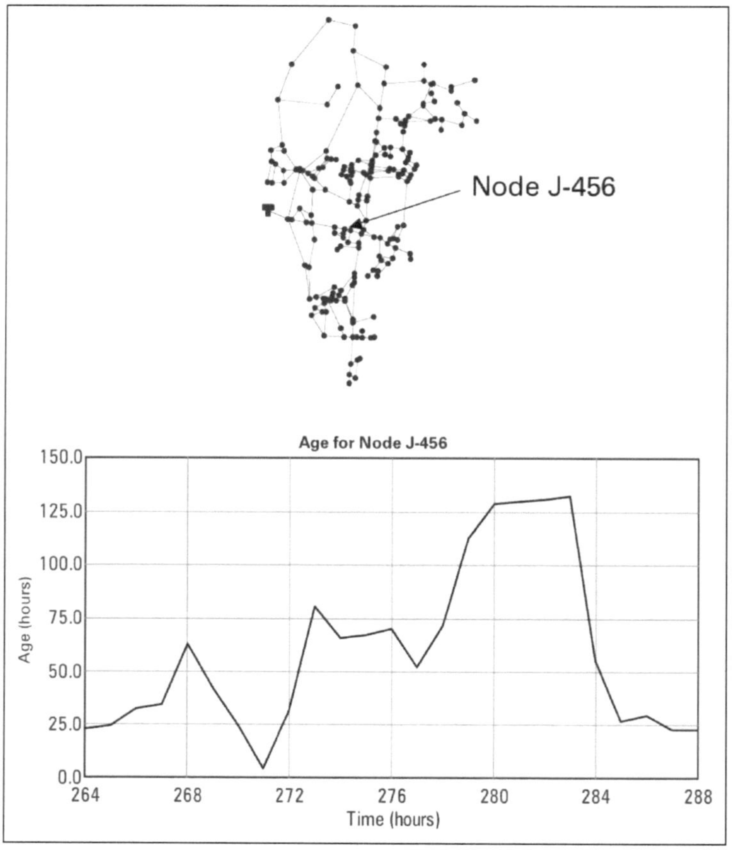

Figure 9-11. Diurnal water age at node J-456 (Panguluri et al. 2005)

Table 9-1. Summary of water utilities selected for water quality modeling (Yang et al. 2007)

Utility	Source Water	Pre/Final Disinfection	Approximate Population Served in 2006	Number of Water Sources	Number of Pressure Zones	Water Main Length, *mi*	Water Quality Modeling Objectives
A	SW	Chlorine/ chloramine (switch to chlorine every spring)	70,000	1	2	357	Stage 2 D/DBPR IDSE modeling SSS
B	GW	Chlorine	7,000	2	1	33	Assist sampling site selection for Stage 2 D/DBPR compliance
C	GW	Chlorine	22,000	3	1	115	Assist sampling site selection for Stage 2 D/DBPR compliance
D	SW	Chlorine	15,000	1	3	62	Identify water quality deficiencies (mainly DBP issues) and conduct improvement planning

DBP—disinfection by-product
D/DBPR—Disinfectants/Disinfection Byproducts Rule
GW—groundwater
IDSE—initial distribution system evaluation
SSS—system-specific study
SW—surface water

For purposes of this study, existing hydraulic models were updated for three of the water systems studied (utilities A, C, and D), but the model for utility B was developed specifically for this study. Hydraulic models were created from existing geographic information system (GIS) data or computer-assisted design drawings, and 24-hr demand patterns were created based on consumption information. The creation of 24-hr demand patterns permitted the use of extended-period simulations (EPS) for water age estimation. Figure 9-14 (see appendix E) shows a schematic of the network for utility D as well as the hydraulic and water quality test locations.

Water quality (chlorine) models were developed by estimating bulk and wall decay coefficients, from a combination of laboratory analysis, field study, and the process of model calibration. Chlorine decay models provide for the prediction of residual levels across the distribution system based on water age and disinfectant residual decay. Utility D "built" a complete DBP model for TTHMs and the sum of its HAA5. Historically monitored or field-measured DBP data was correlated with estimated water age based on the calibrated water system hydraulic model, observed water system chlorine residuals, and other selected water quality parameters. These correlations were used to identify one or more potential parameters that could be a surrogate(s) for predicting DBPs in the water system.

The following calibrations were conducted for the utilities included in the study:

- Steady-state hydraulic calibration was used to establish flow and pressure variations in the modeled water system.

- Extended-period hydraulic calibration was used to compare simulated system pressures, pump discharge rates, and tank levels over a minimum 24-hr period with actual system data.
- Chlorine fate and effect models for utilities B, C, and D were calibrated by estimating chlorine bulk decay and wall decay coefficients.

When simulated values were significantly different from field test results and/or system operational measurements, additional field tests were conducted, and the models were reconciled with the system values.

Water system chlorine modeling can be used to evaluate the spatial variation of chlorine residuals in distribution systems, locate areas with potentially low chlorine residuals, and serve as an indicator of water age or water main conditions. Figure 9-15 (see appendix E) illustrates predicted versus observed chlorine residuals in utility C's distribution system estimated by the calibrated water system chlorine model.

Projected and/or observed chlorine residuals can also be used as an additional surrogate parameter to predict DBP levels in distribution systems when selecting IDSE sampling locations if a significant correlation exists between chlorine residuals and DBP levels. Figures 9-16 and 9-17 show correlations between TTHM/HAA5 levels and total chlorine residuals in the utility A and D distribution systems derived from historical DBP monitoring data and field measurements, respectively. In this study, correlations of TTHM/HAA5 levels with observed water system chlorine residuals were found to be more significant than correlations with water age (data not shown). Applying these correlations to the utility D water quality model allowed prediction of TTHM levels across the distribution system (Figure 9-18; see appendix E). This same information can be helpful in selecting TTHM sampling locations for the IDSE.

Hydraulic and water quality models can be effective tools in developing the IDSE as required under the D/DBPR2. In addition, if properly calibrated, models can assist utilities in addressing a wide range of water quality and system operational issues beyond the focus of the Stage 2 D/DBPR.

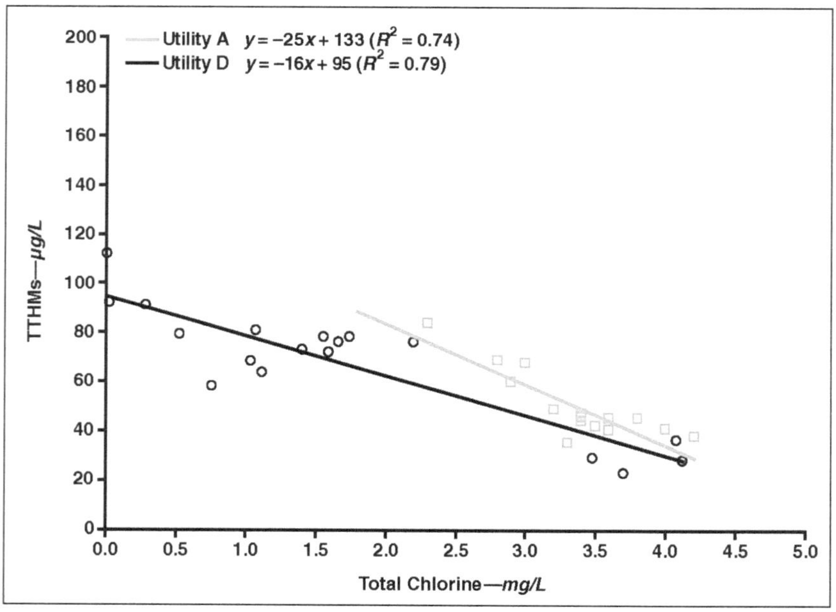

Figure 9-16. Correlations between TTHM and chlorine residuals for utilities A and D (Yang et al. 2007)

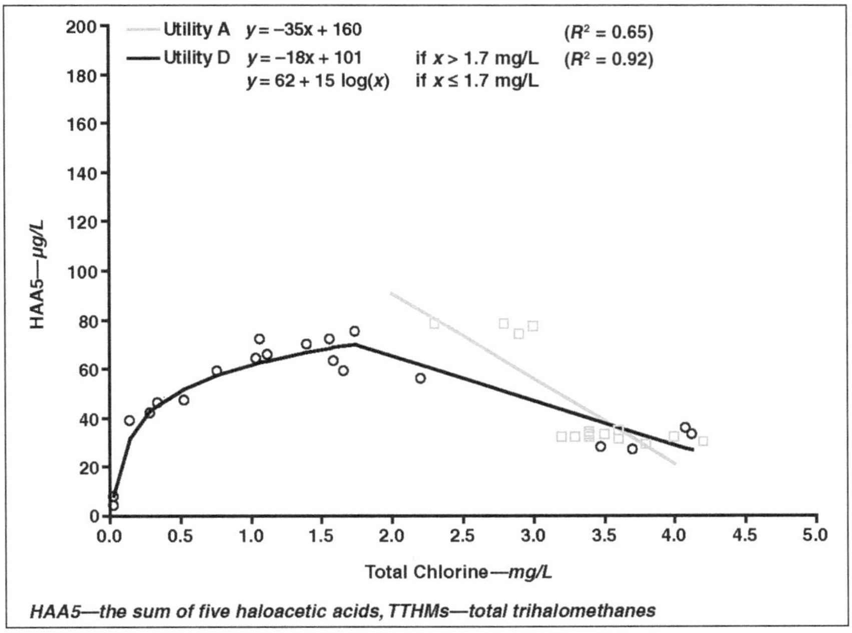

Figure 9-17. Correlations between HAA5 levels and chlorine residuals at utilities A and D (Yang et al. 2007)

References

Angulo, F.J., S. Tippen, D.J. Sharp, B.J. Payne, C. Collier, J.E. Hill, T.J. Barrett, R.M. Clark, E.E. Geldreich, H.D. Donnell, and D.L. Swerdlow. 1997. A Community Waterborne Outbreak of Salmonellosis and the Effectiveness of a Boil Water Order. *American Journal of Public Health*, 87(4):580–584.

Bahadur, R., W.B. Samuels, and J. Pickus. 2003. *Case Study for a Distribution System Emergency Response Tool.* Denver, Colo.: AwwaRF.

Clark, R.M., E.E. Geldreich, K.R. Fox, E.W. Rice, C.H. Johnson, J.A. Goodrich, J.A. Banik, and F. Abdesaken. 1996. Tracking a *Salmonella* Serovar *typhimurium* Outbreak in Gideon, Missouri: Role of Contaminant Propagation Modeling. *Journal of Water Supply: Research and Technology–AQUA*, 45(4):171–183.

Fennel, H., D.B. James, and J. Morris. 1974. Pollution of a Storage Reservoir by Roosting Gulls. *Journal of Society of Water Treatment Exam.*, 23:5–24.

Geldreich, E.E., K.R. Fox, J.A. Goodrich, E.W. Rice, R.M. Clark, and D.L. Swerdlow. 1992. Searching for a Water Supply Connection in the Cabool, Missouri, Disease Outbreak of *Escherichia coli* 0157:H7. *Water Research,* 84:49–55.

Grayman, W.M., R.M. Clark, B.L. Harding, M.L. Maslia, and J. Aramini. 2004. Reconstructing Historical Contamination Events. In *Water Supply Systems Security,* ed. L. Mays. New York: McGraw-Hill, 10.1–10.55.

Hrudey. S.E., and E.J. Hrudey. 2004. *Safe Drinking Water: Lessons From Recent Outbreaks in Affluent Nations.* London: IWA Publishing.

Maslia, M.L., J.B. Sautner, and M.M. Aral. 2000. *Analysis of the 1998 Water-Distribution System Serving the Dover Township Area, New Jersey: Field-Data Collection Activities and Water-Distribution System Modeling.* Atlanta, Ga.: Agency for Toxic Substances and Disease Registry.

Maslia, M.L., J.B. Sautner, M.M. Aral, R.E. Gillig, J.J. Reyes, and R.C. Williams. 2001. *Historical Reconstruction of the Water-Distribution System Serving the Dover Township Area, New Jersey: January 1962–December 1996.* Atlanta, Ga.: Agency for Toxic Substances and Disease Registry.

New Jersey Department of Health and Senior Services (NJDHSS). 2003. *Case-Control Study of Childhood Cancers in Dover Township (Ocean County), New Jersey.* Trenton, N.J.: NJDHSS, Division of Epidemiology, Environmental and Occupational Health.

Panguluri, S., W.M. Grayman, and R.M. Clark. 2005. *Distribution System Water Quality Report: A Guide to the Assessment and Management of Drinking Water Quality in Distribution Systems.* Cincinnati, Ohio: USEPA Office of Research and Development.

Rossman, L.A., R.M. Clark, and W.M. Grayman. 1994. Modeling Chlorine Residuals in Drinking Water Distribution Systems. *Journal of Environmental Engineering*, 120(4):803–820.

Sautner, J.B., M.L. Maslia, C. Valenzuela, W.M. Grayman, M.M. Aral, and J.W. Green Jr. 2005. Field Testing of Water-Distribution Systems at US Marine Corps Base, Camp Lejeune, North Carolina, in Support of an Epidemiologic Study. In *Proc. ASCE-EWRI World Water & Environmental Resources Congress, Anchorage, Alaska*. Reston, Va.: ASCE.

Skala, M.F. 1994. Waterborne *Salmonella* Outbreak in Southeastern Missouri. *Missouri Epidemiologist,* 17(2): 1–2.

US Environmental Protection Agency (USEPA). 2003. *Stage 2 Disinfectants and Disinfection Byproducts Rule, Initial Distribution System Evaluation.* Washington, D.C.: USEPA Office of Water.

USEPA. 2006. Stage 2 Disinfectants and Disinfection Byproducts Rule; Final Rule. *Fed. Reg.*, 71:2:388.

Yang, J., S. Beduhn, and D. Swailes. 2007. IDSE and Beyond: Application of Computer Modeling to Distribution System Water Quality. *Journal AWWA*, 99(7):76–82.

Chapter 10

Advanced Concepts in Water Quality Modeling

Introduction

In the last ten years, important advances have been made in modeling the fate and transport of water quality in drinking water distribution systems. Among these advances are the use of geospatial techniques, modeling the transport of several species simultaneously, the application of modeling for security applications, and transient pressure analysis. This chapter will discuss how water quality models are increasingly being integrated with various geospatial tools such as geographic information systems (GIS). An extension to the original EPANET has been developed that allows it to describe multiple, interacting chemical reactions. This extension is referred to as EPANET-MSX, where MSX stands for Multi-Species eXtension. In addition to a stand-alone executable program, a toolkit library of functions has been developed that can be used to build custom applications.

Water quality models are increasingly being used for modeling security in drinking water distribution systems. USEPA's Threat Ensemble Vulnerability Assessment (TEVA) provides an example of this type of application. Pressure transients have been associated with the deterioration of water quality in pipe networks.

Geospatial Tools

Geospatial tools are of interest to many professionals, and a wide range of geospatial tools can be incorporated into water quality models (Panguluri et al. 2005), such as:

- GIS: Geographic information system.
- CADD: Computer-aided design and drafting tools.
- AM/FM: Automated mapping (or asset management)/facilities management systems.
- CIS: Customer information system.
- DTM: Digital topological model.
- DEM: Digital elevation model.
- GPS: Global positioning system.
- SCADA: Supervisory control and data acquisition systems.
- LIMS: Laboratory information management system.
- RDBMS: Relational database management system.
- SDMS: Spatial data management system.

Each of these tools is discussed briefly below.

Geographic Information System (GIS)

GIS includes the storage, retrieval, manipulation, analysis, and display of geographically referenced data (Mark 1997a). Many of the concepts underlying GIS come from the field of cartography. Figure 10-1 shows the typical inputs and results of current GIS packages (Panguluri et al. 2005). Use of the Geographic Base File/Dual Independent Map Encoding (GBF-DIME) files developed by the US Census Bureau in the 1960s was the first large-scale employment of digital mapping by the government. This development led to the Census Topologically Integrated Geographic Encoding and Referencing (TIGER) files.

Much of the current GIS modeling methodology is an outgrowth of planning approaches that are based on the work of McHarg (1969). McHarg's general approach was to overlay a grid on an area to be studied and then assign different types of attributes to a grid cell. Attributes might include characteristics such as land slope and soil attributes. McHarg created maps of different attributes on transparencies using shading with the darkness proportional to the degree to which that attribute related to a particular use. For example, significantly sloping land could be represented as dark areas because it is difficult to build under these circumstances. The user could physically overlay the transparencies and select the lighter areas that might be most appropriate for development. When computerized, this approach became the basis of the common overlay analysis incorporated into GIS technology. A source of data for GIS technology was computerized photogrammetry and surveying. In the early 1960s, the concept of the digital topological model (DTM) was developed, in which the computer would be used to store a digital database representing the earth's surface.

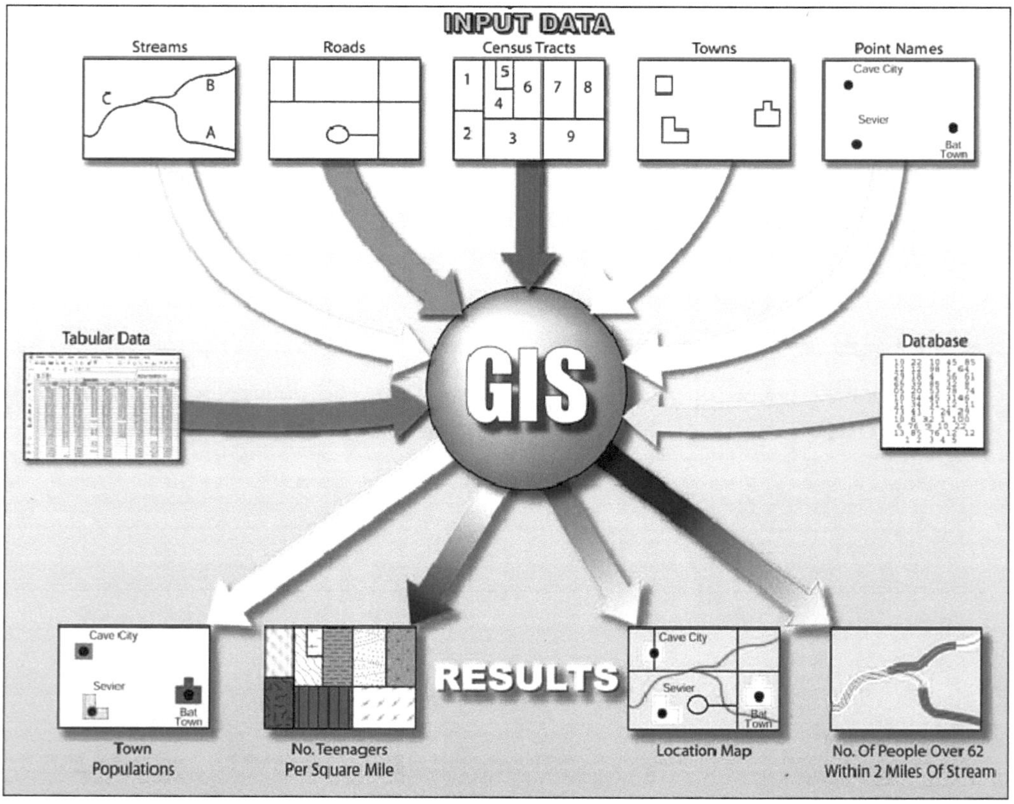

Figure 10-1. Typical inputs and results of current GIS packages (Panguluri et al. 2005)

Computer-Aided Design and Drafting (CADD)

CADD was developed in the 1960s and early 1970s, and several companies were founded that developed and commercialized the basic concepts. Autodesk (maker of AutoCAD) and Bentley Systems (Microstation) were founded and led to the wider availability of CADD on personal computers. Later in the 1980s, Parametric Technology Corp. produced a three-dimensional design system.

Asset Management/Facilities Management (AM/FM) Systems

Asset management and facilities management (AM/FM) systems are based on the use of computer database and mapping technology to store and manage information related to physical assets in a water system. AM/FM systems typically involve the integration of CAD technology and database management technology. AM/FM is used by the electric and gas industry, telecommunications industry, and other industries that maintain physical networks. These systems have been more closely integrated with GIS packages and are now referred to as AM/FM/GIS (Cesario 1995) or even more broadly as geospatial information technology.

Consumer Information System (CIS)

Consumer information systems (CISs) can be used, among other things, for storing and using information on water consumption by customers. Consumers' addresses can be stored as a geographic coordinate. There is increasing interest in integrating automatic meter reading (AMR) systems with CIS. A geographically based CIS provides an excellent mechanism for automatically recording current consumption data to be used in water distribution system models.

Digital Topological Models (DTMs)

An important underlying characteristic of GIS is the concept of topology. Topology may be defined as the locational interrelationship between geographic features. Digital topological models (DTMs) capture relationships such as adjacency, intersection, and connectivity in a database. GIS constructs the topological relationship between individual features, and this capability is used in various analyses and modeling tasks.

Digital Elevation Models (DEMs)

Digital elevation models (DEMs) and DTMs refer to representation of ground surfaces in a computer. Various methods utilizing regular/irregular grids and triangulated irregular networks (TIN) (Mark 1997b) are used to provide efficient representations of surfaces and have become a regular feature of today's GIS packages. DEM databases are readily available for most of the United States from the US Geological Survey and other sources.

Global Positioning System (GPS)

A global positioning system (GPS) is a satellite-based navigation system made up of a network of 24 satellites placed into orbit by the US Department of Defense. GPS was originally intended for military applications, but in the 1980s the system became available for civilian use. GPS works in any weather conditions, anywhere in the world, 24 hr a day, with no subscription fees or setup charges. GPS satellites circle the earth twice a day in a very precise orbit and transmit signal information to earth, where receivers take this information and use triangulation to calculate the user's exact location.

Supervisory Control and Data Acquisition (SCADA) Systems

As has been discussed, supervisory control and data acquisition systems typically have the capability to remotely access information on the state of a water system, to manually or automatically control components such as pumps and valves, and to store and display current or historical time-series data about system operation.

Laboratory Information Management System (LIMS)

Laboratory information management systems (LIMSs) are computerized systems for managing samples in a laboratory and typically include a mechanism for storing, managing, displaying, and tracking samples. Since the origin of a sample must be identified both spatially and temporally, this information provides a means of associating LIMS data with other spatial database management systems such as GIS.

Relational Database Modeling System (RDBMS)

Relational database modeling systems (RDBMS) typically includes attribute data stored in a relational database system that may be part of a GIS or an asset management system. A relational database system is composed of a series of tables that are related through keywords.

Spatial Data Management Systems (SDMSs)

Spatial data management systems are usually composed of two forms of information: geographic coordinates and attribute information. For example, to represent a water main, the start and end of the pipe and any curves or bends in the pipe must be captured. Attribute information may include the pipe diameter, length, material, age, and other data of interest.

Integrating GIS and Water Quality Modeling

A GIS allows for the identification of a group of objects or features that have both a location based on the establishment of a unique position or a point on the earth's surface and characteristics or attributes. These attributes can be stored, updated, and analyzed in a database over time. These systems provide an opportunity and a platform for utilizing spatial data, although they must be arranged in an organized manner. For example, a water storage tank can be identified by its location in terms of latitude–longitude or other coordinate systems and by its characteristics, such as diameter, height, and type of construction. Similarly, a pipe can be described by its route, diameter, length, material, and age. Zeiler (1999) discussed the issues associated with database design, and various industry groups are attempting to define generic data structures for a particular industry (such as the water utility industry) in order to facilitate data transfer and common usage of a GIS in that industry (Grise et al. 2000).

Many GISs include or incorporate CADD packages, AM/FM systems, LIMSs, and DEMs. At many water utilities, GIS capabilities to store, access, and map data are leading to increased usage of GIS in areas such as planning, facilities management, and management of customer and water quality data. Some water utilities share a GIS database with other entities, such as city or county governments, and other utilities, such as gas, electric, and telephone. GIS systems may include a means of storing land property, parcel, and ownership information and geographic descriptions. They provide a mechanism for storing topographical information, and in the past few years, integration of GIS with water distribution system models has been a significant area of research and development.

Geographic features are stored in two general ways: vector and raster. Though a geographic feature can frequently be stored in more than one way (vector and raster), there may be a preferred way to store each piece of information. Under the general area of vector representation, features can be stored as points, lines, or polygons. Points are geographically represented as a single set of coordinates in two or three dimensions. A line can represent either a single straight-line segment, identified by the coordinates of the two end points, or a series of connected line segments to represent a curved line. A polygon is defined by a closed set of line segments and identifies the area contained within the defined outer boundary. Raster data most commonly refer to a set of data that has been defined in terms of a regular square or rectangular grid system that is tied to a geographical coordinate system. Each grid cell can have one or more characteristics assigned to it. Land-use information derived from a satellite represented as a colorized raster database that uses different colors for different uses, such as population centers, agricultural or forested areas, and open space, can provide a base layer for overlaying system information (Panguluri et al. 2005). Within a GIS, features are organized as separate layers in a manner analogous to the original concept developed by McHarg (1969). When viewing GIS data, layers can be turned on or off or moved forward or backward in order to better understand or view the spatial relationships.

The concept of integrating water distribution system modeling with geospatial database management systems has been evolving over the past quarter of a century and continues to be a major focus of development in the water industry today. One of the first attempts to integrate GIS concepts with water quality modeling occurred when the Water Supply Simulation Model (WSSM), discussed earlier, integrated a geobased, link–node system to several models including a hydraulic model, a steady-state water quality and cost allocation model, and various display and editing routines (Clark and Males 1985). WSSM was expanded to include an interface to GIS files using AutoCAD. USGS digital line graph (DLG) files of road networks and DEMs were used within AutoCAD to create a detailed representation of the water distribution system. The resulting database was used to generate an input file for the Wadiso hydraulic model whose engine was a prototype for EPANET. Commercial vendors of network modeling software have led the integration of modeling software and spatial database technology. Recent commercial modeling software packages have released versions that are fully integrated and operate within the GIS environment. These products include WaterGEMS (Haestad Methods/Bentley Systems) and InfoWater (MWH Soft now called Innoyyze).

As an evolving technology, integrating GIS with water distribution system models still presents some issues to be solved. A water utility GIS may have many functions including mapping, facility management, planning, and modeling support and therefore may require a great deal of detail. It may include hydrants, shutoff valves, water meters, household connections, air-release valves, and other appurtenances, but a water distribution system models may not explicitly include many of these components. In the water quality model representation might only inlude only junction nodes and pipes. If the locations of water meters and customer connections are included in the GIS, the disparity between the two representations will increase. Various philosophies may be adopted to deal with this problem and are listed below:

- A very large and detailed water quality/hydraulic model is constructed that includes all of the elements in the GIS. Such a model might have an excessive number of nodes and links.
- Unneeded nodes and links are eliminated when constructing the model representation. Though this approach results in a more appropriate model, it adds an intermediate

step between the GIS and the model. The network is effectively skeletonized and there is no longer a one-to-one correspondence between GIS and model features.
- Multiple representations are established within the GIS in which the detailed representation may be used for facility management while the skeletonized version is used for modeling. This approach requires that the user be aware of the changes in information needed in the different databases.
- The basic link–node network (as used in the model) is maintained as the base case in the GIS in one "layer" and other components (such as hydrants) with links are in different layers. Establishing procedures for updating the GIS is essential, such as authorization of specific users to make changes, and limiting the nature of the changes that might be permanent.

According to Edwards et al. (2009), interest is increasing in integrating geographic information systems with water distribution-system/hydraulic models. Distribution-system models are becoming more complex and increasingly include most of the pipes in a utilities network. There is a trend to integrate the vast amount of data contained in a utility's GIS with the less frequently modified hydraulic model. However, the integration and transfer of data between a GIS and a water distribution-system hydraulic model are challenging. Model integration can be thought of as consisting of the following three levels:
- Interchange, which refers to the lowest level of sophistication for transferring data between a model and a spatial database, such as a GIS. Interchange implies the systems are run separately, and information is extracted from one system and stored in an intermediate file that is subsequently accessed by the other system.
- Interface, which refers to a direct connection between hydraulic model and GIS.
- Integration, which means the two systems work seamlessly together.

Model Structure and Level of Detail

Clearly, different levels of detail are required in the two types of models. GIS is used primarily for managing and mapping of facilities while hydraulic models are dependent on the structure of a network. Figure 10-2 illustrates four basic types of hydraulic modeling structures: all-pipes model, all-pipes reduced model, skeletonized model, and skeletonized reduced model.

The advantages and disadvantages of these models are:
- The all-pipes model preserves the same level of detail as the GIS and maintains a one-to-one relationship between individual GIS and model elements.
- The all-pipes reduced model is similar to an all-pipes model with the exception that intermediate, hydraulically insignificant nodes along pipe reaches of similar diameter, material, and construction date are eliminated. Reduction may decrease the number of pipe segments in an all-pipes model by 50 percent or more.
- The skeletonized model includes all pipe segments greater than a specified diameter, typically 12 to 16 in. (Skeletonized models are sometimes referred to as transmission models because these models represent only larger pipes.)
- The skeletonized reduced model is essentially the same as the skeletonized model except that hydraulically insignificant nodes are reduced.

Variations of the four basic model structures are common. Two of the more common variations are the detailed model and the hybrid model. A detailed model is either a skeletonized or a skeletonized reduced model with some of the smaller-diameter pipes included (typically some of the 6- and 8-in. pipes). A hybrid model combines large transmission mains working and includes many detailed features. Table 10-1 summarizes many of the characteristics of hydraulic models.

Sustainable Modeling Approaches

Figure 10-3 provides a general overview of the development and maintenance process for a sustainable hydraulic model. As can be seen, important interactive steps should be observed in integrating GIS and water quality/hydraulic modeling.

Table 10-2 provides an example of data that would be common in both systems and illustrates how these data might be represented.

In establishing a database that can be used in both a GIS package and a water quality/hydraulic system, certain basic connective and topology rules should be maintained as shown in Table 10-3.

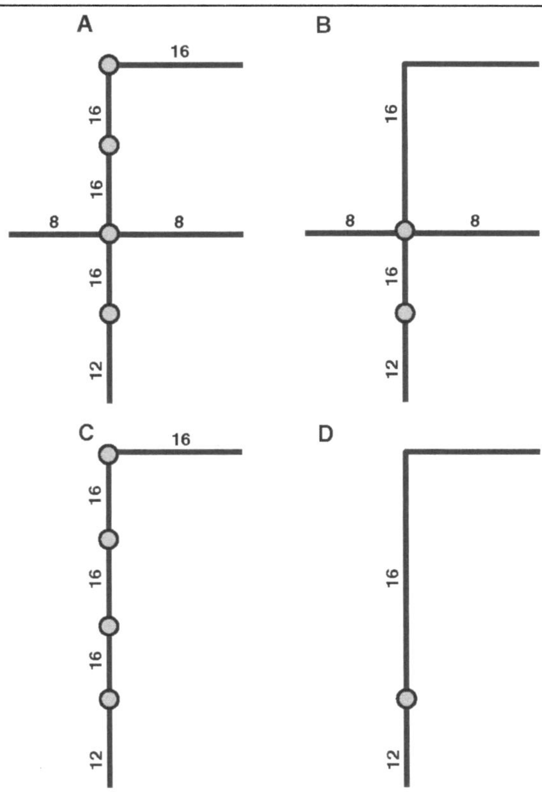

Numbers within the figure represent lengths of pipe in inches. Dots represent hydraulic model junctions and define the endpoints of each pipe segment; where a line does not end with a dot, the pipe segment is not shown in its entirety.

Figure 10-2. Example of four basic hydraulic model structures: (A) all-pipes model, (B) all-pipes reduced model, (C) skeletonized model, and (D) skeletonized reduced model. (Edwards et al. 2009)

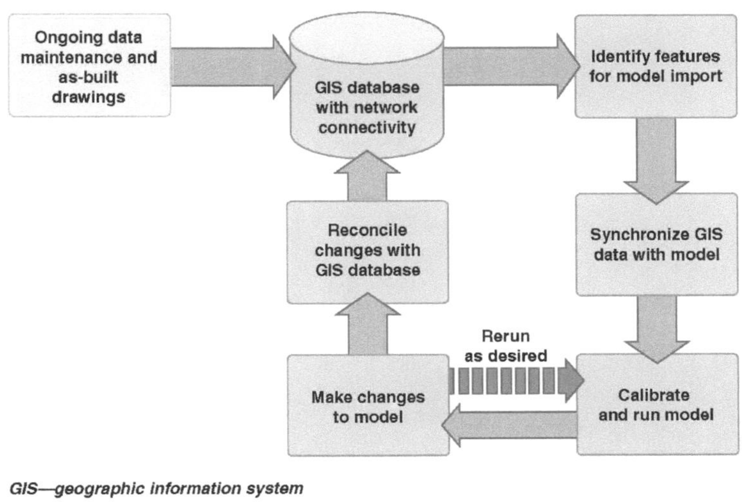

Figure 10-3. A general overview of the development and maintenance process for a sustainable hydraulic model of a water distribution system (Edwards et al. 2009)

Table 10-1. Comparison of hydraulic model characteristics (Edwards et al. 2009)

Characteristics and Considerations	All-Pipes Model	Skeletonized Model	All-Pipes Reduced Model	Skeletonized Reduced Model
General description of model structure	Full system representation (all pipes, valves, hydrants)	Only larger pipes represented (others removed)	Hydraulically insignificant valves and hydrants removed; pipes combined	Hydraulically insignificant facilities removed; pipes combined
Development considerations	Does not require reducing skeletonizing, or creating one-to-many relationships	Requires skeletonizing; does not require creating one-to-many relationships	Requires reducing and creating one-to-many relationships.	Requires skeletonizing, reducing, and creating one-to-many relationships
Maintenance considerations	Easiest-to-use automated maintenance procedures; new GIS pipes can be directly imported into model	Easy-to-use automated maintenance procedures; new GIS pipes can be directly imported into model; skeletonization criteria must be followed	Difficult-to-use automated maintenance procedures; well-defined one-to-many relationships must be maintained	Most-difficult-to-use automated maintenance procedures; alternative manual maintenance procedures may be more cost-effective
Computer processing speed considerations	Slow processing because all pipes and segments are represented	Fast processing because not all pipes are included	Moderate processing because model contains fewer pipe segments	Fastest processing because fewer pipe segments are represented and not all pipes are included
Software-license sizing considerations*	Unlimited pipes typically required	Fewer pipes required	Unlimited pipes typically required	Fewest numbers of pipes required.
Data storage space consideration	Requires maximum amount of disk space	Reduced disk space required	Reduced disk space required; not all pipe segments exist in the model	Minimum disk space required
Ability to conduct Stage 2 D/DBPR sampling site selection evaluations	Yes	Depends on percentage of real pipe length and real pipe volume in model	Yes	Depends on percentage of real pipe length and real pipe volume in model
Ability to conduct unidirectional-flow analyses	Yes	No	No	No
Ability to conduct fire-flow analyses	Yes	Limited	Yes	Limited
Effectiveness in conducting waste quality analyses	Best	Poor	Good	Worst

GIS—geographic information system

Stage 2 D/DBPR—Stage 2 Disinfectants/Disinfection Byproducts Rule

*Hydraulic modeling software is priced based on the maximum number of pipes that can be modeled in any analysis. As the number of pipe elements increases, a larger pipe license may be required with a resulting increase in cost.

Table 10-2. Example data sets and attributes for a hydraulic model and GIS (Edwards et al. 2009)

GIS		Hydraulic Model		
Feature Class	**Feature Type**	**Feature Class**	**Feature Type**	**Recommended Minimum Attributes to Maintain in Both GIS and Hydraulic Model**
Tank, reservoir	Point	Tank, reservoir	Point	Asset identification, name, install date, zone, bottom elevation. overflow elevation, diameter, volume
Fitting, hydrant, nonoperational valve	Point	Junction	Point	Asset identification, install date, retire date, zone, elevation, fitting type
Operational valve, boundary valve	Point	Valve	Point or line	Asset identification, install date, retire date, from zone,* to zone,* elevation, diameter, functional type, body type
Pump	Point	Pump	Point or line	Asset identification, install date, retire date, zone diameter
Pipe	Line	Pipe	Line	Asset identification, install date, retire date, zone, diameter, material, relining date, relining type, from node, to node, length, roughness*

GIS = geographic information system
* Denotes data generally not included in GIS.

Table 10-3. Basic topology rules that should be validated in the initial GIS connectivity checks and established for all GIS data additions (Edwards et al. 2009)

Feature Class	Feature Type	Rule	Possible Errors	Fix or Revision
Fitting, hydrant, pump, valve, water storage	Point	Feature must be coincident with main endpoint	Orphan or duplicate feature	Delete feature
			Feature not snapped to pipe endpoint	Snap feature to main endpoint
Pipe	Line	Must not overlap other line features	Two pipes share identical geometry	Delete duplicate pipe
			Two pipes overlap	Edit pipe geometry
Pipe	Line	Must not overlap itself	Pipe digitized over itself	Edit pipe geometry
Pipe	Line	Must be covered by point feature (fitting, hydrant, pump, valve, water storage)	Pipe endpoint not snapped to point feature	Snap pipe to point feature
			Point feature missing in GIS	Add point feature to GIS

GIS—geographic information system

The complexity of model development requires that sound model development and maintenance procedures be developed and followed. Development of a hydraulic model is a complex task requiring a number of tools and trained personnel, including planners, GIS staff, engineers, distribution system operators, and managers. Maintaining an up-to-date hydraulic model will yield more reliable information for evaluating existing deficiencies, service to potential developments, water quality, and operations. Adoption of GIS standards that support model maintenance will reduce model maintenance costs and facilitate data transfer between both GIS and model users.

GIS Case Studies

A number of drinking water utilities have moved aggressively to incorporate GIS concepts into their water quality modeling programs. We will discuss next the efforts of two utilities, Las Vegas Valley Water District and Denver Water Board, in incorporating GIS into their systems modeling activities.

Las Vegas Valley Water District (LVVWD)

In 1989, the Las Vegas Valley Water District (LVVWD) served 558,000 consumers, but by 2004 the service population had increased to 1,209,000 (Jacobsen and Kamojjala 2005). As a result, the LVVWD service area more than doubled during this period. To address some of the problems caused by this rapid growth, LVVWD integrated its master and operational planning and development review with GIS data, systems modeling, SCADA, and enterprise data (such as CIS, AM/FM, and LIMS) (Jacobsen et al. 2005). LVVWD also developed a one-to-one relationship between its GIS spatial database and its network model. In order to maximize the efficient use of this combined system, LVVWD has taken an "all-pipes capable" approach where the distribution system is segmented by pressure zones and attached to an operational backbone network. Each of the zone models can be attached seamlessly to the backbone network for detailed hydraulic and water quality modeling. Examples of how GIS data are used by LVVWD on a day-to-day basis include:

- *Pressure complaint resolution:* When a pressure complaint is received, the GIS database is searched for parcel and account information, together with modeled and measured pressure data in the vicinity of the complaint. Depending on the results of the search, a crew may be dispatched to trace field pressures and abnormal conditions from the water supply source to the customer location, and to install hydrant recorders to capture dynamic variations in pressure. Modeling and field results are compared to identify possible problems. (Jacobsen and Kamojjala 2005, Panguluri et al. 2005).
- *Water main break analysis:* If a water main break occurs, it is critical to quickly identify distribution system valves that must be closed in order to minimize water loss, potential flooding, or possible contamination. Repairs should be performed quickly so that service can be restored. GIS techniques were employed for identifying the valves to be isolated using GIS and modeling tools. A list of affected customers is generated for appropriate notification, and an analysis is performed to evaluate the impact on existing services (Panguluri et al. 2005). The GIS is also used to compare pressures before and after a shutdown. It was found that response was provided quickly and accurately using integrated GIS and system modeling (Jacobsen and Kamojjala 2005).

Denver Water Board (DWB)

Between January 1997 and December 2000, Denver Water conducted a comprehensive study intended to evaluate the transmission, pumping, and storage capability in its distribution systems. The DWB planned to use the results of this study in order to plan for new facilities and to develop plans for investment in new capital. In the course of this study, Denver Water made extensive use of GIS tools for demand forecasting. Denver Water has been a completely metered system since 1992, and consequently consumption data are available from 1993. Existing demands were spatially allocated using address geocoding (Strasser et al. 2000), and consumption data were extracted from the billing system and imported into an MS Access database. Extracted data included customer address and customer class information (e.g., single family, multifamily, commercial, industrial, and public), and demand was aggregated by customer class. Denver Water used the "address geocoding" feature available within Esri ArcGIS to identify each customer or demand point on a base map (Strasser et al. 2000). The match rate was over 93 percent. Accounts that could not be geocoded, generally master meter accounts and wholesale customers, were entered in the system manually. Results from this geocoding process for one pressure zone is illustrated in Figure10-4 (Panguluri et al. 2005). Each dot in the figure represents addresses that were successfully geocoded. A quality control check was performed on the results of this geocoding process by reconciling consumer demands with Denver Water's annual statistical report (Strasser et al. 2000).

Multispecies Modeling

When the concept of water quality modeling was first developed, it was focused on the fate and transport of a single species (Shang et al. 2008). In order to simulate the propagation of more than one constituent, separate independent "runs" have been made of the model. The case study for the North Marin system illustrates this approach. However, clearly interactions occur among chemical species and between constituents in the bulk phase and on the surface phase.

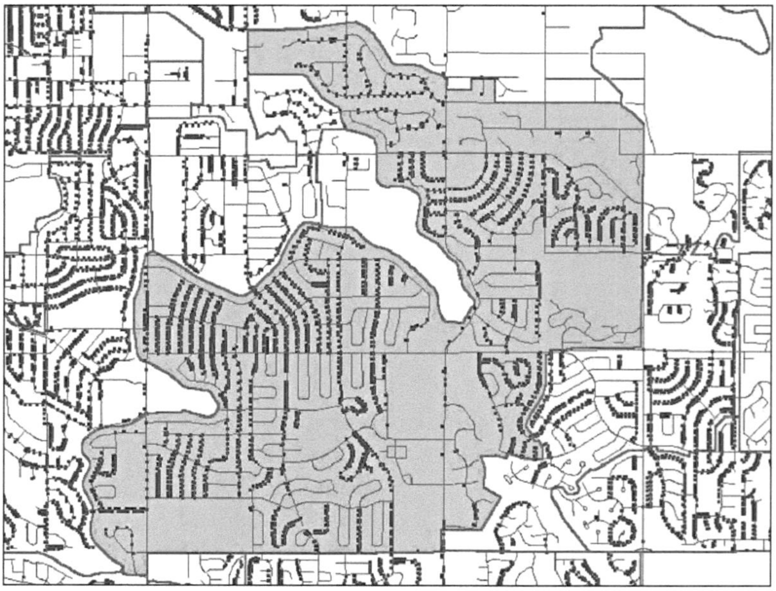

Figure 10-4. GIS geocoding—metered sales demand allocation procedure for Denver Water (Panguluri et al. 2005).

In order to address some these interactions directly, EPANET-MSX was developed. EPANET-MSX assumes a mobile bulk water phase and a fixed pipe surface phase within a water distribution system pipe. The bulk phase species can be chemical or biological components in nature and are transported through the system with the average water velocity. Immobile surface phase species, such as organisms embedded in a biofilm or corrosion by-products, are attached or incorporated into the pipe wall.

Species in the bulk phase can include dissolved constituents (individual compounds or ions) such as HOCI and OCL^-, aggregate components such as TOC, suspended constituents such as bacterial cells and inorganic particulates, and chemicals adsorbed onto particles. Surface species include bacteria incorporated into biofilm, oxidized forms of iron contained within corrosion scale, particulate material that settles out because of gravity or is attached to the pipe wall surface through ionic or molecular (i.e., van der Waals) forces. Other surface compounds are those that can diffuse into or out of plastic pipes or be adsorbed onto or desorbed from iron oxide pipe surfaces. Bacteria and particulates can exist in both the bulk and surface phases and transfer between phases by physical attachment/detachment, chemical adsorption, or molecular diffusion. To deal with this interphase transfer, the component is modeled as two species: one bulk and the other surface. Figure 10-5 illustrates this effect.

Some additional phases might be assumed as existing within a distribution system, including a mobile bed sediment phase and an immobile water phase within the pore

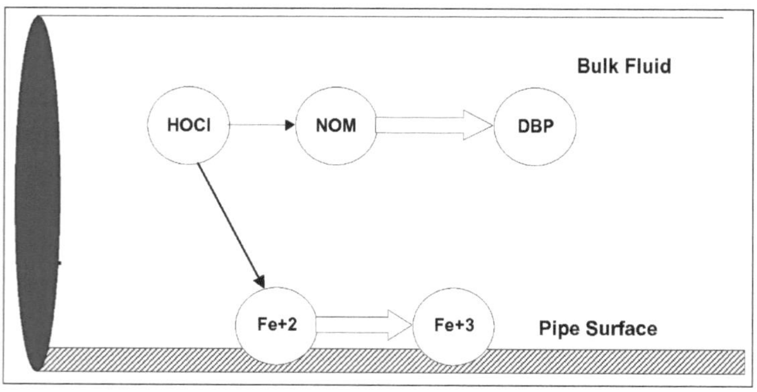

Figure 10-5. Example of reactions in the mobile bulk phase and at the fixed pipe surface in EPANET-MSX (Edwards et al. 2009).

structure of pipe scale. An air phase at the interface of the water surface in a storage tank or standpipe could also be included within this modeling framework.

Some examples of reactions and situations that are amenable to multispecies modeling include:

- The interaction of free chlorine and natural organic matter (NOM).
- The formation of mono-, di-, and trichloramine resulting from the interactions between free chlorine species and ammonia.
- The situation in which more than one water source supplies a distribution system.

EPANET-MSX, like EPANET, assumes a standard link–node format to characterize drinking water distribution systems, as shown in Figure 10-6.

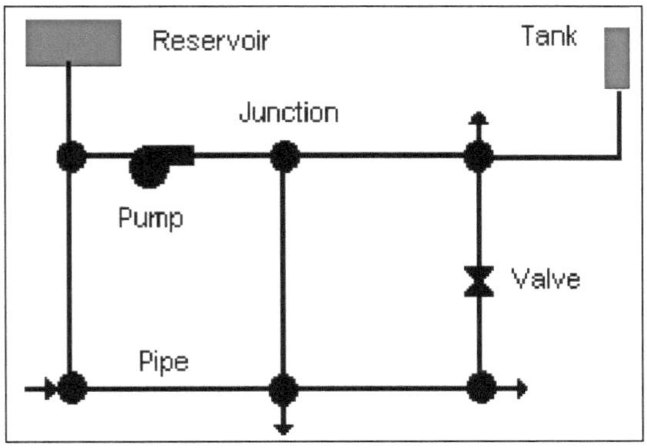

Figure 10-6. Link–node representation of a simple distribution system (Edwards et al. 2009).

Transport Phenomena

Transport phenomena play a major role in the fate and effect of water constituents in a drinking water distribution system pipe, including advection and mixing at nodes and in storage tanks. These are discussed briefly below:

- *Advective transport:* It is assumed that bulk species are transported down the length of a pipe with the same average velocity as the carrier fluid while at the same time reacting with other bulk species and with the pipe wall surface.
- *Mixing at pipe junctions:* All junctions receiving inflow from two or more links are assumed to undergo complete and instantaneous mixing.
- *Mixing in storage tanks:* Inflows to storage tanks and standpipes are assumed to mix completely with the existing contents in storage while these contents are subjected to possible bulk phase reactions (alternative schemes are available for modeling plug flow storage tanks).

Chemical Reactions

Reactions are assumed to be either reversible or irreversible. Reactions that are reversible and fast enough in comparison with the system's other processes are assumed to reach a local equilibrium. Reactions that are not sufficiently fast and/or irreversible are not assumed to reach an equilibrium formulation. In theory, very large backward and forward rate constants (with their ratio equaling the equilibrium constant) can be used to model fast/equilibrium reactions. Therefore, both fast/ equilibrium and slow/kinetic reaction dynamics can be written as a single set of ordinary differential equations (ODEs). These ODEs can be integrated over time to simulate changes in species concentrations. However, this approach can result in reaction rates that may range over several orders of magnitude and lead to such small integration time steps so as to make a numerical solution impractical.

Differential algebraic equations (DAEs) are used in EPANET-MSX to represent the fast/equilibrium reactions and mass conservation. The following DAEs define the interactions between bulk and surface species:

$$\frac{dx_s}{dt}=f(x_b x_s z_b z_s, p) \quad \text{(Eq 10-1)}$$

$$\frac{dx_s}{dt}=g(x_b, x_s z_b z_s, p) \quad \text{(Eq 10-2)}$$

$$0=h(x_b, x_s, z_b, z_s, p) \quad \text{(Eq 10-3)}$$

The vectors of time-varying differential variables x_b and x_s are associated with the bulk water and pipe surface, respectively. The time-varying algebraic variables z_b and z_s are similarly associated, and the model parameters p are assumed time invariant. The algebraic variables are assumed to reach equilibrium in the system within a much smaller timescale compared to the numerical time step used to integrate the ODEs. The dimension of the algebraic equations h must agree with that of the algebraic variables $z = [z_b\ z_s]$, so that the total number of equations in Eq 10-1–Eq 10-3 equals the total number of time-varying species ($[x_b, x_s, z_b, z_s]$).

The oxidation of arsenite (As^{+3}) to arsenate (As^{+5}) by a monochloramine disinfectant residual in the bulk flow and the subsequent adsorption of arsenate onto exposed iron on the pipe wall provide an example of the approach described above (arsenite adsorption is not significant at the pH levels typically found in drinking water). This system consists of four species (arsenite, arsenate, monochloramine in the bulk flow, and sorbed arsenate on the pipe surface). It can be modeled with three differential rate equations and one equilibrium equation:

$$\frac{dAs^{+3}}{dt}=-k_a As^{+3}(NH_2Cl) \quad \text{(Eq 10-4)}$$

$$\frac{dAs^{+3}}{dt}=-k_a As^{+3}(NH_2Cl)-A_v[k_l(S_{max}-As_s^{+5})As^{+5}-k_2 As_s^{+5}] \quad \text{(Eq 10-5)}$$

$$\frac{d(NH_2Cl)}{dt}=-k_b(NH_2Cl) \quad \text{(Eq 10-6)}$$

$$As_s^{+5}=\frac{k_s S_{max} As^{+5}}{1+k_s As^{+5}} \quad \text{(Eq 10-7)}$$

where As^{+3} is the bulk phase concentration of arsenite, As^{+5} is the bulk phase concentration of arsenate, As_s^{+5} is surface phase concentration of arsenate, and NH_2Cl is the bulk phase concentration of monochloramine. The parameters in these equations are as follows: k_a is a rate coefficient for arsenite oxidation, k_b is a monochloramine decay rate coefficient, A_v is the pipe surface area per liter pipe volume, k_1 and k_2 are arsenate adsorption and desorption rate coefficients, S_{max} is the maximum pipe surface concentration of arsenate, and $k_s = k_1/k_2$. In terms of the notation used in Eq 10-1–Eq 10-3, $x_b = \{As^{+3}, As^{+5}, NH_2Cl\}$, $x_s = \{0\}$, $z_b = \{0\}$, $z_s = \{As_s^{+5}\}$, $p = \{k_a, k_b, A_v, k_1 k_2, S_{max}\}$. Example input files for this form of the model are included with the standard EPANET-MSX (Multi-Species eXtension) distribution. The input file for a more complex version of the model is presented in the *EPANET-Multi-Species eXtension User's Manual* (Shang et al. 2008)

Full Network Solution

As discussed earlier, distribution system dynamic water quality models can be classified as either Eulerian or Lagrangian. Eulerian models divide the network into a series of fixed control elements and record the state changes at the boundaries and within these elements, while Lagrangian models track changes of discrete parcels of water as they travel through the network. EPANET-MSX employs the same Lagrangian transport algorithm used by EPANET, which ignores axial dispersion and tracks the movement and reaction of chemicals in discrete water volumes, or segments. These segments are transported through network pipes by the bulk velocity, and they completely mix at junction nodes. This method is relatively efficient, because the number and size of the segments in a pipe can change as hydraulic conditions change.

To summarize, the following steps, depicted in Figure 10-7, are performed for each water quality time step:

1. *React:* Reaction dynamics within each pipe segment and storage tank are calculated over the time step to compute new concentrations throughout the network.
2. *Advect:* Within each pipe, the flow volume is computed and transported over the time step. This amount of volume and its associated bulk species mass from the pipe's leading edge is converted into segments and transported into accumulated mass and volume totals at the downstream node.

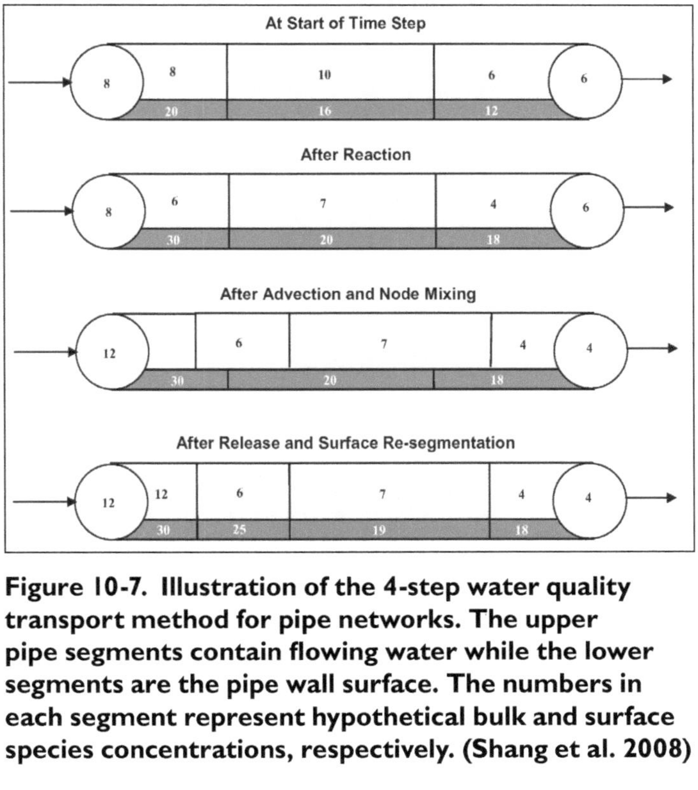

Figure 10-7. Illustration of the 4-step water quality transport method for pipe networks. The upper pipe segments contain flowing water while the lower segments are the pipe wall surface. The numbers in each segment represent hypothetical bulk and surface species concentrations, respectively. (Shang et al. 2008)

3. *Mix:* New bulk species concentrations are calculated at each node based on its accumulated mass and volume inputs from the advection step as well as any external sources.
4. *Release:* A new segment is created at the upstream end of each pipe whose size equals the pipe's flow volume and whose bulk species concentrations equal that of the upstream node. If the difference in quality between the most upstream segment and the upstream node is below some tolerance, the size of the current upstream segment is increased.

Reaction System Solution

The multispecies water quality algorithm modifies the react step (step 1) of the solution scheme shown previously. Reaction dynamics are represented by the system of DAEs within each pipe segment, Eq 10-1–Eq 10-3. The same type of reaction applies for storage tanks, except that the DAEs are modified to consider only bulk reactions. Model parameters p can possibly vary by pipe, and for the equilibrium reactions it is assumed that the Jacobian matrix of h with respect to z, $\frac{\partial h}{\partial z}$, is unique and nonsingular for all t. The implicit functions defined by Eq 10-3

$$z_b = z_b(x_b, x_s, p) \tag{Eq 10-8}$$

$$z_s = z_s(x_b, x_s, p) \tag{Eq 10-9}$$

exist and are continuous and unique, and the partial derivatives are continuously differentiable. These properties, and in particular the ability to numerically evaluate Eq 10-8–Eq 10-9, are critical to the numerical algorithms used for solution of Eq 10-1–Eq 10-3.

Given the implicit functions Eq 10-8–Eq 10-9, the solution of Eq 10-1–Eq 10-3 is performed by substituting Eq 10-8–Eq 10-9 into Eq 10-1–Eq 10-2, and eliminating the algebraic equations Eq 10-3. This leaves a reduced system of ordinary differential equations (ODEs) that can be integrated numerically:

$$\frac{dx_b}{dt} = f(x_b, x_s, z_b(x_b, x_s, p), z_s(x_b, x_s, p), p) = f'(x_b, x_s, p) \tag{Eq 10-10}$$

$$\frac{dx_s}{dt} = g(x_b, x_s, z_b(x_b, x_s, p), z_s(x_b, x_s, p), p) = g'(x_b, x_s, p) \tag{Eq 10-11}$$

The above "substitution" is not performed literally, since Eq 10-8–Eq 10-9 are implicit, as are the reduced trajectories f' and g'. Solving Eq 10-10–Eq 10-11 numerically with an explicit method, such as any of the Runge–Kutta schemes, will require that f' and g' be evaluated at intermediate values of x_b and x_s over the integration time step. Each evaluation will in turn require a solution of the nested set of algebraic equations Eq 10-8–Eq 10-9. Strategies for accomplishing these steps are discussed in the *EPANET Multi Species eXtension User's Manual* (Shang et al. 2008).

In addition to the react step, evaluation of the equilibrium equations also needs to be performed at the mix phase of the overall algorithm, because the blending together of multiple flow streams can result in a new equilibrium condition. This process needs to be performed at each network node, including storage tanks.

Pipe Surface Discretization

The segment bulk water state variables x_b and z_b have moving coordinates due to the nature of the Lagrangian water quality model, which moves with the bulk water velocity. The associated pipe surface variables x_s and z_s have fixed coordinates, because they are associated with the nonmoving pipe. The lack of a common fixed coordinate system for the bulk and surface state variables must be reconciled, because these variables interact through the common pipe–water interface through equations Eq 10-1–Eq 10-3. To resolve this issue, a simple mass-conserving scheme is applied at every water quality time step to update the pipe surface elements to remain consistent with the (advected) water quality segments and redistribute the surface variable mass among the updated elements.

As shown in Figure 10-7, within any single water quality time step, a moving mesh divides each pipe surface into discrete-length elements, such that each shares a common surface/water interface with the water quality segment above it. At the end of the time step the pipe elements will, however, be inconsistent with the water quality segments, due to the bulk phase's advection. In order to remove this inconsistency, the surface species concentration is updated using an interfacial area-weighted average:

$$x_{si}^{new} = \left(\frac{1}{L_i^{new}}\right)\sum_{j=1}^{n}\left(L_i^{new} \cap L_j\right)x_{sj} \quad \text{for } i = 1,....n^{new} \qquad \text{(Eq 10-12)}$$

$$z_{si}^{new} = \left(\frac{1}{L_i^{new}}\right)\sum_{j=1}^{n}\left(L_i^{new} \cap L_j\right)z_{sj} \quad \text{for } i = 1,....n^{new} \qquad \text{(Eq 10-13)}$$

where i is the water quality segment index, and n is the number of water quality segments in the pipe during the most recent react step. L_j is the length of segment j, with corresponding vectors of surface species x_{sj} and z_{sj}, n^{new} is the updated number of water quality segments in the pipe after advection, L_i^{new} is the length of each updated segment, with corresponding updated surface concentrations x_{si}^{new} and z_{si}^{new}. The quantity $(L_i^{new} \cdot Lj)$ is the length of the overlapping intersection between segment j and updated segment i.

Model Implementation

EPANET-MSX (Shang et al. 2008) offers several choices among numerical integration methods for solving the reaction system's ODEs, Eq 10-1 and Eq 10-2. Include among these options are:

- A forward Euler method (as used in EPANET).
- A fifth-order Runge–Kutta method with automatic time step control.
- A second-order Rosenbrock method with automatic time step control.

The EPANET-MSX (Shang et al. 2008) system is supplied as both:

1. A stand-alone console application (epanetmsx.exe) that can run standard water quality analyses without any additional programming effort required,
2. A function library (epanetmsx.d11) that, when used in conjunction with the original EPANET function library (epanet2.d11), can produce customized programming applications.

At this point in time, the extension has not been integrated into the Windows version of EPANET, but integration is expected at some future date. Examples of each type of usage are provided in the *EPANET Multi Species eXtension User's Manual.*

In order to use the multispecies analysis, two input files must be prepared:

- A standard EPANET input file that describes the hydraulic characteristics of the network being analyzed (Rossman et al. 1994). The format of this file is described in the *EPANET User's Manual* (Rossman 2000).
- A second special EPANET-MSX file must be prepared that describes the species being simulated and the chemical reaction/equilibrium model that governs their dynamics.

Distribution System Security

The President's Commission on Critical Infrastructure has categorized infrastructure in the United States as follows: information and communication, physical distribution, banking and finance, and energy and vital human services (Clark and Deininger 2000). Vital human services include community water supply systems, which exist largely at the local and state levels, which are generally governmental in nature and deal chiefly with human needs and safety. Because they tend to be highly localized, failures in one community may have little direct impact on other communities, although the problems and vulnerabilities faced may be similar. Water supply systems are vulnerable to the full range of terrorist threats including physical attack and cyber- and bioterrorism.

The terrorist attacks of September 11, 2001, and the mailing of letters containing *Bacillus anthracis* spores spurred a heightened concern over the vulnerability of critical infrastructure in the United States. The US Environmental Protection Agency (USEPA) is the lead US agency in conducting security-related research for protecting water supply systems. The Homeland Security Strategy Act (2001) and the Public Health Security and Bioterrorism Preparedness and Response Act of 2002 (US Congress 2002) provided the basis for this effort. Additional Homeland Security Presidential Directives (HSPDs) were issued in 2003 and 2004. These directives assigned new responsibilities to the agency for establishing a strong water security science and research program.

Deliberate Intrusions Into a Distribution System

Nilsson et al. (2005) simulated a deliberate biochemical assault on a municipal drinking-water distribution system. The objective of the study was to demonstrate an effective method to characterize potential consumer exposure to contaminants and evaluate system vulnerabilities resulting from a deliberate biochemical assault on a drinking-water distribution system. The basic approach was to link a novel stochastic water-use simulator (PRPsym) with EPANET. The output from these companion codes was then used to generate empirical frequency distributions of mass dose loadings at select nodes in a typical water distribution system. In the Nilsson study, simulation techniques were used to investigate a deliberate biochemical assault on the Cherry Hill/Brushy Plains water distribution system. The attack was modeled as a steady 6-hr injection delivering 3,600 g of a soluble conservative contaminant to a single node on the main line in a small town. Migration of the contaminant plume was tracked for 55 hr throughout the pipe network, and the cumulative mass loading was computed at four target nodes strategically located on looping links and dead-end branches. This exercise was repeated for 1,000 independent trials to establish a baseline distribution of consumer dose exposures at the target nodes. A battery of simulation experiments was then performed to examine the sensitivity of the nodal load distributions to various system characteristics and water-use patterns.

Results show that variability in the total mass load received at a node can be apportioned between the variability in the water-use volume and variability in the mean delivered concentration. Overall, however, the operation of the network storage tank had the greatest influence on the nodal mass loadings. This study demonstrates that Monte Carlo techniques are a useful tool for simulating the dynamic performance of a municipal drinking-water supply system, provided that a calibrated model of realistic network operations is available. Key findings indicate that the relative variability of the contaminant mass received at target nodes over a 55-hr period (measured by the coefficient of variation) was large, ranging from 15 to 50 percent. The variability in nodal loading depended on two factors: variation in the volume of water consumed at the node (affected by stochastic demands) and variation in overall average concentration (affected by initial mixing and flow paths). The behavior of the network storage tank, alternating between a mass sink and a mass source, dictated the overall distribution of contaminant loads delivered to the demand nodes. Simulation results were very sensitive to assumptions about system operation and network configuration. This outcome underscored the vital importance of having an accurate model of the network to achieve realistic results.

Khanal et al. (2006) extended the work of Nilsson et al. (2005) with the intent of examining the sensitivity of network response to the variability in the location, timing, duration, and intensity of a contamination event. This work had two objectives:

- Develop and demonstrate use of an influence map and an exposure index as simple intuitive measures of network response.
- Evaluate the suitability of generalized sensitivity analysis (GSA) as a statistical tool for effectively identifying and ranking the dynamic network variables that are most significant for Monte Carlo simulations of network response to contamination events.

In Part I of the investigation, a deterministic base-case contamination event was applied individually to 89 source nodes in a small municipal pipe network. Consumer exposures were estimated from simulated nodal mass loadings and used to construct a zone of influence (ZOI) map, categorizing individual network nodes on the basis of their potential to reach and expose downstream consumers. A dimensionless exposure index (EI) was introduced as a simple global measure of the average network response expected from a suite of potential contamination events.

In Part II of the investigation, Latin hypercube sampling was used to set up 1,152 randomized contamination events at two injection nodes. A GSA was performed to determine how the network response to simulated contamination events was influenced by changes in four dynamic network variables: base demand, storage capacity, injection mass, and injection duration. The ZOI map for a water distribution system identifies the nodes that, if contaminated, have the greatest impact on population. Water utilities could use this information to better manage security and assist with a host of other decisions, including delineation of sensitive network areas, hardening of water supply infrastructure, deployment of early warning systems, selection of alternate supply routes, development of contamination response plans, and adoption of risk communication strategies. The exposure CDF and its companion metric, the EI, provided simple intuitive ways to interpret results of simulated contamination events. The shape of the exposure CDF offered useful insights about hydraulic topology and network response to contamination.

Comparing exposure CDFs for a network, before and after deployment of in-line sensors, is an objective way to evaluate the effectiveness of proposed early warning systems. In the network example considered here, the sensitivity rankings of the dynamic variables shifted as the threshold value and injection location were changed, but base demand and injection mass were consistently identified as the most significant variables. Using the Kolmogorov–Smirnov d statistic, exposure levels in the example network were found to be most sensitive to variations in base demand and injection mass. Tank storage capacity was important in certain cases, while injection duration tended to be least important. The magnitude of the K–S d statistic quantified the impact of the stochastic input parameters on simulated concentrations, while the sign of the K–S d statistic helped to identify the physical basis for these impacts. The contrary role of the storage tank, acting both as a secondary source of contamination and as an incidental vessel of dilution, was elucidated through the GSA results.

Application of GSA to network simulation results demonstrated both identification and quantification of relationships between inputs and outputs in a stochastic network model.

Shen et al. (2009) has developed a multistage response procedure for identifying possible ingress nodes (PINs) for drinking water networks. The algorithm quantifies the likelihood that a PIN in a given water distribution system is the actual point of ingress. The procedure uses data mining to successively decrease the possible number of PINs based on a preconstructed database in three stages. In each stage, query sentences are executed to locate the PINs, and a Euclidean distance is proposed to estimate the probability of a PIN being the true ingress location. As demonstrated in a case study, the run time of the program to identify the potential location of contaminant ingress at each stage is approximately 2 min, which is rapid and hence reasonable to direct emergency response. The range of PINs is narrowed down in the first three stages, which helps utilities to allocate resources more effectively to the small range of PINs that may be at risk. In addition, the narrower range of PINs is helpful to locate the true ingress node faster in case of false positives.

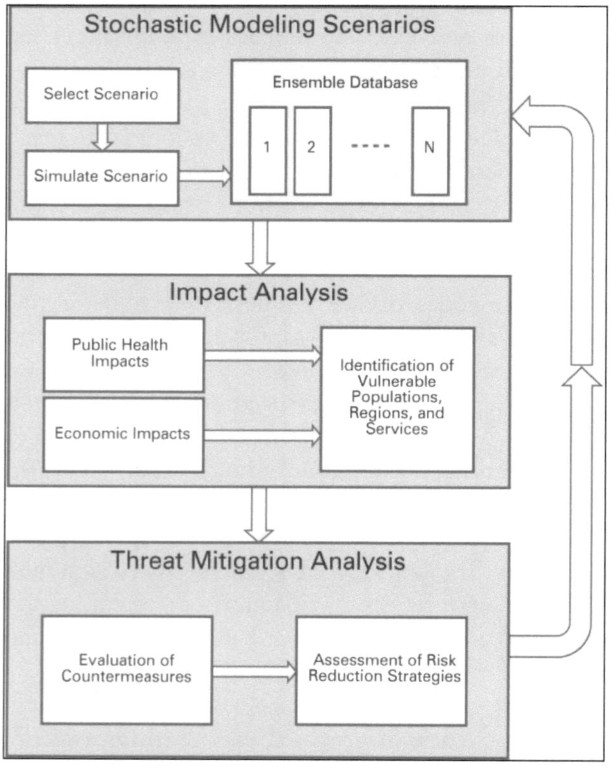

Figure 10-8. Threat Ensemble Vulnerability Assessment Program (Panguluri et al. 2005)

USEPA Research

In response to this increased focus on the vulnerability of drinking water systems to the intentional introduction of chemical, biological, or radiological contaminants, USEPA developed the Threat Ensemble Vulnerability Assessment (TEVA) Program. Some aspects of this program will provide very useful insights into accidental threats to water system integrity as well.

TEVA incorporates a probabilistic framework for analyzing the vulnerability of drinking water distribution systems. Figure 10-8 outlines the major components of the framework as follows:

- Stochastic modeling of scenarios.
- Analysis of potential impacts.
- Assessment of threat mitigation strategies.

These three components present an integrated view of the vulnerability of a distribution system to a wide variety of contamination threats and the potential for a water utility to decrease this vulnerability through a set of mitigation strategies. Without specific intelligence information, it would be difficult to predict how a terrorist group might sabotage a water system. Therefore, TEVA is based on a probabilistic analysis of a large number of likely threat scenarios. TEVA creates a threat ensemble, or a set of contamination scenarios, based on varying the type of contaminant, the amount and concentration of the contaminant, the location of the contaminant introduction into the distribution system, and the duration of the contamination event. System vulnerability is based on an assessment of the entire threat ensemble. A discussion of the three key modeling components follows.

Stochastic Modeling

The stochastic modeling element of TEVA involves the following three steps: selection of the threat ensemble, simulation of the ensemble, and storage of the output in the ensemble database. The threat ensemble is a collection of scenarios to be simulated, for example, the injection of a 55-gal drum containing a biotoxin mixture at a node of a particular distribution system with 1,000 nodes. This threat can be repeated at each of the 1,000 nodes, generating a threat ensemble. Variations in a given scenario could include contaminant type, quantity, concentration, location, or duration. These variations would generate other threat ensembles.

Each scenario in the threat ensemble is simulated using EPANET, including an assumption of first-order contaminant decay of constituents. A recent variation is the possibility of using a recent upgrade to EPANET that allows for the possibility of modeling the fate and transport of multiple dissolved constituents in distribution systems (Uber et al. 2004a). This upgrade permits modeling of reactions at the pipe wall and in the bulk flow and enables the inclusion of chemical reaction products, thereby resulting in more accurate estimates of human exposure and health risk. Results of the stochastic modeling of the threat ensemble are stored in the ensemble database.

Impact Analysis

The impact analysis component uses data stored in the ensemble database to estimate likely public health and economic impacts. Public health impacts include injuries, disease, illness, and deaths. Drinking water consumers can be exposed to contaminants from ingestion, inhalation of volatilized chemicals or particles, and/or dermal exposure. Dose–response models are employed to estimate the various health endpoints. For contagious diseases, dynamic models of disease transmission also must be included in order to accurately assess health impacts. Unfortunately reliable data are lacking for many threat agents, and the ensuing uncertainty in the results must be understood.

Economic impacts include:

- Restoration costs (cleanup, treatment, remediation, and decontamination).
- Denial of service costs (providing alternative sources of water) and other costs, such as medical costs (hospitalization, vaccines).

Although important, psychological costs related to consumers' loss of trust in the water supply system are very difficult to estimate. The distribution of impacts is calculated from the ensemble database, thereby providing an estimate of the expected impacts for the threat scenarios.

Threat Mitigation Analysis

The threat mitigation component explores various mitigation strategies such as the use of:

- Early warning systems (sensors and data analysis tools).
- Operational approaches such as chlorine boosters and backup equipment.
- Emergency response methods, such as isolation of part of the system and public notification.

A range of mitigation strategies can be evaluated with the TEVA simulations using health risk and economic impact analyses to rank and select the best alternative for a set of scenarios (Uber et al. 2004b) to enable a quantitative analysis of the benefits of implementing various strategies for protection.

TEVA (Murray et al. 2004, Uber et al. 2004b, Murray et al. 2005) has been applied, on a case study basis, to three hypothetical distribution systems, each of which differs in population served by approximately an order of magnitude. Results from this case study demonstrated that TEVA has the potential to help water utilities assess the contaminants to which they are most vulnerable, identify the most vulnerable regions of their distribution systems, and select the most appropriate mitigation strategies for their system.

Optimal Monitoring of Water Distribution Systems

Many different approaches to contamination monitoring have been suggested, including using water quality sensors, composite or grab sampling, and various numbers and locations of sensors. Another factor to be considered in any contamination monitoring and surveillance program is the cost, and it is likely that most monitoring programs will be budget constrained. Selecting the best routine sampling sites for detecting intentional contamination and optimizing the placement of a fixed number of monitoring stations are important aspects in the design of a monitoring system. Berry et al. (2005) formulated this problem as an integer linear programming problem in which an objective function is optimized subject to a set of constraints. In the study by Berry et al. (2005), a Greedy heuristic algorithm was used to locate monitoring stations, given a defined budget or number of monitoring stations, in order to minimize health impacts.

Simulation Overview

In order to provide an example of the use of TEVA for evaluating and optimizing the design of a contamination monitoring system for a water distribution system, the following case study was constructed. For the purposes of this analysis, an "all-pipes" EPANET network model was generated to provide a sample distribution system. The system consisted of 1,062 mi of pipe, approximately 12,000 nodes, and an average daily demand of approximately 20 mil gal, serving a population of 130,000. Approximately 6,000 potential sampling locations were randomly selected from the 12,000 nodes assuming each node had a 50 percent probability of inclusion. This selection process was used to make the problem less computationally intensive and to emphasize that an optimal placement of monitoring stations will likely be difficult given legal, financial, or logistical constraints for placing and managing monitoring stations.

To simulate a contamination scenario, many parameters must be specified, including characteristics of the contaminant, the contaminant-introduction scenario, and the consumption patterns of the population. For the purposes of this analysis, a biological agent was considered as the contaminant, and the introduction duration was 24 hr, at a rate of 8.675 L/hr, with a mass rate of 1.45×10^{11} organisms per minute. The agent was assumed to cause infection (50 percent of the time) in an adult when 10,000 or more of the organisms are ingested. The incubation period was assumed to be 24 hr, and the time for effective treatment was 48 hr after the onset of symptoms. After 72 hr, people either recovered or died. A sigmoidal dose–response curve was assumed for the ingestion of organisms with the untreated fatality rate at 16 percent of those infected.

In the TEVA-simulated analysis, the benefits of two composite grab sampling programs were evaluated (daily and every 48 hr) and compared to the benefits provided by a system of real-time (in-line, contaminant-specific) sensors. The benefits of the sampling or monitoring programs were measured by the hypothetical reduction in public health impacts. In this analysis, sample location designs were based on minimizing the mean number of fatalities for two sampling frequencies: 24 hr and 48 hr. Six sampling/sensor station placement scenarios were evaluated in this analysis: 5, 10, 15, 20, 30, and 40 locations for each program. The Greedy algorithm used in this analysis provided an optimal solution for minimizing public health impacts.

For the purposes of this analysis, the grab samples were considered to be filtered samples, which are composited samples that have been collected and concentrated through a filtration device to better enable the collection and analysis of biological organisms. Also, real-time water quality monitors are assumed to detect chemical contaminants or biological organisms through changes in water quality, such as determined by the reduction of chlorine residual or increase in turbidity. These real-time and sampling-based analyses were considered ideal, meaning that detection limits were zero and the biological contaminant was always detected, if present.

TEVA Analysis Approach

This analysis considers attacks at every non-zero demand node (10,000). Only non-zero demand nodes were considered because they represent service connections that are using water from the distribution system on a regular basis. Contaminants were modeled using a mass injection rate, and an assumption of zero volume added, which consequently does not influence the hydraulic properties of the network, i.e., flow increase, flow decrease, or reversal of flow. Hydraulic and water quality simulations were run for 192 hr.

TEVA Analysis Results

Figure 10-9 compares the reduction in mean infections provided by a routine 24-hr filtered sampling program, a routine 48-hr filtered sampling program, and a real-time, continuous monitoring program. Similarly, Figure 10-10 compares the reduction in the maximum number of infections of the same three monitoring programs.

It is assumed that the computer simulations and the monitoring/surveillance programs are successful in reducing public health impacts and that no consumption took place after detection. The results show that there is not a significant difference between daily (24-hr) sampling and 48-hr sampling in terms of reducing health impacts. As expected, the

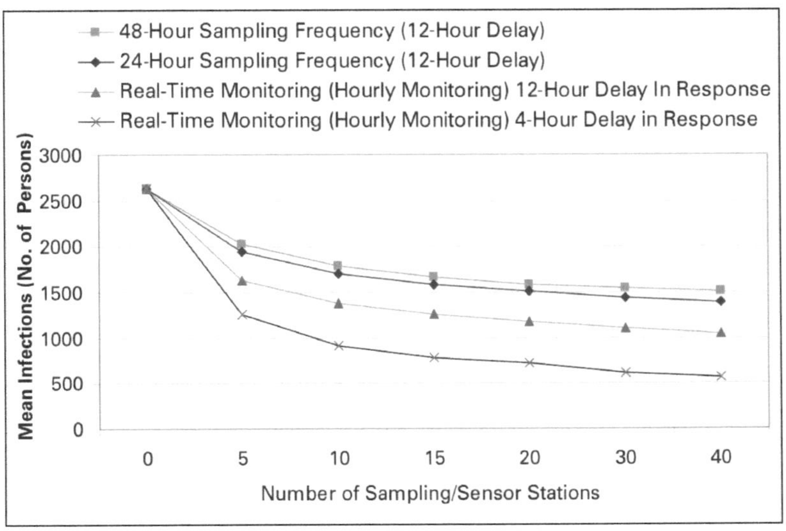

Figure 10-9. Comparison of 24-hr, 48-hr, and real-time, continuous contamination monitoring systems for the reduction in mean infections for a 24-hr contaminant attack (Panguluri et al. 2005)

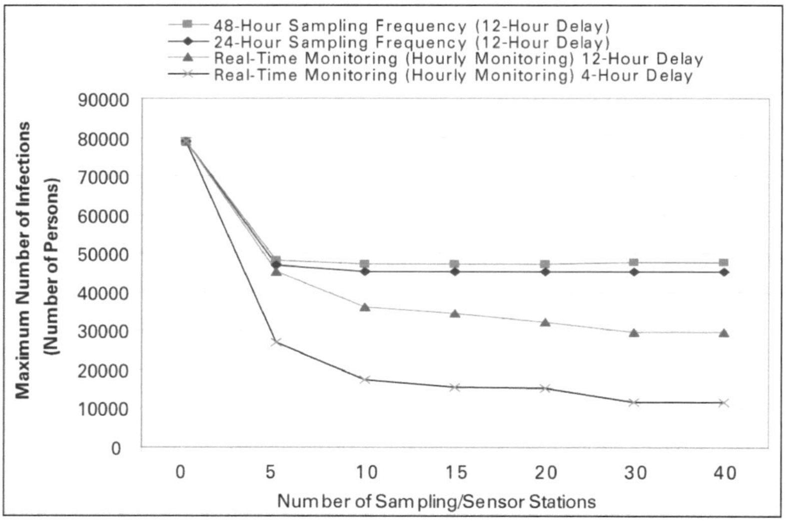

Figure 10-10. Comparison of 24-hr, 48-hr, and real-time, continuous contamination monitoring systems for the reduction in the maximum number of infections for a 24-hr contaminant attack (Panguluri et al. 2005)

continuous monitoring program detects the incident much earlier than the daily sampling program. Figure 10-10 shows an 80 percent reduction in maximum infections is achieved using 40 real-time monitors, as compared to having zero monitors. The results show that it is important that the real-time program be integrated with a response protocol. A comparison of the two real-time, continuous monitoring cases (12-hr delay versus 4-hr delay in notifying the public) illustrates the importance of response time. Shortening the time needed to implement an effective response to reduce further exposure is critical for reducing the number of additional infections.

Statistically analysis of the approximately 10,000 threat scenarios provided an estimate of the hypothetical health impacts in terms of average health impacts (e.g., fatalities or illnesses) and maximum impacts. Results from this case study illustrate that, for this distribution system, strategically placing only 5 or 10 sampling/sensor stations as part of a monitoring and surveillance system can have a significant effect on reducing potential health impacts from intentional contamination.

Water Quality and Pressure Transients

As mentioned previously, a distribution system is a complex system consisting of a complex network of pipes, pumps, valves, reservoirs, and storage tanks. Water is transported from its source or sources to various consumers, and the system is designed to operate consistently and economically and to deliver water in sufficient quantity, of acceptable quality, and at appropriate pressure (Jung et al. 2007).

Transient regimes are inevitable; all systems will, at some time, be started up, switched off, or undergo rapid flow changes such as those caused by hydrant flushing, and they will likely experience the effects of human errors, equipment breakdowns, earthquakes, or other risky disturbances (Boulos et al. 2006, 2005, Wood et al. 2005).

Low-pressure transients may promote the collapse of water mains, leakage into the pipes at joints and seals under subatmospheric conditions, and backsiphonage and potential intrusion of contaminants in the distribution system. LeChevallier et al. (2003) reported the existence of low- and negative-pressure transients in a number of distribution systems. Gullick et al. (2004) studied intrusion occurrences in live distribution systems and observed 15 surge events that resulted in a negative pressure. Most were caused by the sudden shutdown of pumps at a pump station because of either unintentional (e.g., power outages) or intentional (e.g., pump stoppage or startup tests) circumstances. Friedman et al. (2004) confirmed that negative-pressure transients can occur in the distribution system and that the intruded water can travel downstream from the site of entry. Locations with the highest potential for intrusion were sites experiencing leaks and breaks, areas of high water table, and flooded air-vacuum valve vaults.

It has been documented that contaminants can intrude into pipes through leaks from reduced- or negative-pressure transients. In reality, all pipeline systems leak, and hydraulic transients occur more or less continuously in distribution systems, so it is not surprising that low-pressure transients introduce a considerable risk of drawing untreated and possibly hazardous water into a pipeline system.

In fact, soil and water samples were recently collected adjacent to drinking water pipelines and then tested for occurrence of total and fecal coliforms, *Clostridium perfringens*, *Bacillus subtilis*, coliphage, and enteric viruses (Karim et al. 2003). The study found that indicator microorganisms and enteric viruses were detected in more than 50 percent of the samples examined.

These and other results suggest that during negative or low-pressure situations, microorganisms can enter the distribution system directly through pipeline leaks. For these reasons, the designer should not overlook the effects of water hammer or pressure surges in the design and operation of the distribution system or the evaluation of either system performance or ultimate system cost.

Emerging Research Areas

There is much research being conducted in areas that are directly or closely related to hydraulic/water quality modeling. We will briefly discuss these research activities in this section.

Stochastic Characterization of Water Demands

As has been discussed in previous chapters, water demand drives hydraulic/water quality models, and most models assume deterministic or fixed demands. However, realistically, demands are stochastic or have a strong random component. Characterizing water demand from a stochastic perspective requires an extensive database, and recent developments in the installation of automated meter reading programs holds the possibility that such databases may be soon be generally available.

Buchberger and Wu (1995) and Buchberger and Wells (1996) conducted a series of unique experiments and data collection efforts intended to provide a more complete understanding of the factors that influence residential demand. They monitored water demands at four single-family residences for a one-year period. Approximately 8,000 demands per capita were recorded and converted to single equivalent rectangular pulses. Each pulse was classified by type (deterministic or random), location (indoor or outdoor), and day (weekday or weekend). Basic exploratory data analyses were performed to estimate sample statistics, cumulative distributions, and hourly occurrences of the intensity, duration, and volumes of the rectangular pulses according to type, location, and day. In addition, the hypothesis that residential water demands occur as a nonhomogeneous Poisson rectangular pulse process was examined. Results show that the simple rectangular pulse provided a very reasonable approximation of indoor water demands. However, the variance of the daily pulse count appears to be too high for a Poisson process. Potential causes of the high variances in the daily pulse count are discussed. Findings and data from this study can be used to develop and verify fine-resolution network models designed to predict the fate and travel time distribution of substances moving to remote points of consumption in the water distribution system.

The investigators reached the following conclusions:

- Arrivals do follow a nonhomogeneous Poisson process.
- Water use occurs as a rectangular pulse.
- Arrival rate is independent of the number of busy server or water using device.
- Water-use intensity is independent of service duration.
- Water-use intensities are mutually independent.
- Water-use intensities have a common probability distribution.
- Server busy times have a common probability distribution.

Buchberber and Li (2007) extended their research on the stochastic characterization of water demands by developing a model called PRPsym that provides a platform for simulating stochastic water uses and performing aggregation of stochastic and deterministic water demands at network nodes. According to Buchberger and Li (2007), water consumption in a distribution network can be classified into stochastic and deterministic water demands. Deterministic water demands (e.g., some industrial water uses and leaks) can be prescribed by a known occurrence time, demand intensity, and pulse duration, which reflect actual operating conditions. Stochastic water uses (e.g., residential or commercial water demands) are more complex owing to the ever-changing pattern of consumption. As shown in Table 10-4, deterministic water demands such as industrial water uses or leaks can be prescribed by known occurrence time, demand intensity, and duration based on the real operation conditions. Stochastic water uses like residential or commercial water uses are complex owing to the ever-changing pattern of consumption.

Table 10-4. Differences of deterministic and stochastic water demands

Type of Water Uses	Frequency/Occurrence	Intensity	Duration
Deterministic	Known	Known	Known
Stochastic	Random	Random	Random

PRPsym employs the PRP model to simulate random pulses of water demands covering indoor demands and outdoor demands for both network nodes and fixtures. In addition, PRPsym also provides a user interface to input deterministic water demands at desired nodes or fixtures. The aggregation of indoor, outdoor, and deterministic demands at any time gives real-time instantaneous nodal demands for a node or fixture. However, it is not feasible to use instantaneous random flows for simulation of hydraulics and water quality in a large network considering the available modeling techniques. Instead, time-averaged demands time series are useful. If $T \geq 0.5$ minutes, PRPsym provides an efficient method to directly generate the time-averaged demands. However, this efficient method is only applicable to indoor water uses.

PRPsym provides flexible options to control the generation of stochastic water demands and output of results. It can be used to generate a water demand series with any time scale from one second to one day for network nodes or fixtures in a household. It can also be set up to produce files that can be directly used to run EPANET. PRPsym is built with an object-oriented method to provide robust and fast simulation of water demands.

Mass Dispersion and Stochastic Demand

In many peripheral service areas, the flow regime is usually dominated by random intermittent laminar conditions where stagnation is common due to infrequent water use (Buchberger et al. 2003). According to Buchberger and Lee (1999), low flow zones can comprise 25 percent or more of the total infrastructure in the distribution system and tend to service a high percentage of the residential consumer base. In the presence of laminar flow, hydrodynamic dispersion can be an important consideration for simulation of water quality. The advection-dispersion-reaction (ADR) model is widely used to describe miscible constituent transport in water networks, which models the simultaneous processes of advection, dispersion, and reaction. Since analytical solutions for the ADR only exist for limited cases, it is common to use numerical solutions to simulate constituent transport. The traditional numerical methods, such as the finite-element method or finite-difference method (FDM), perform well for dispersion-dominated transport but exhibit a combination of excessive numerical oscillation and numerical dispersion for advection-dominated transport (Liu and Dane 1996). Numerical dispersion is an artificial, grid-dependent smearing of sharp solute concentration fronts. A numerical oscillation is manifested by overshooting and undershooting about the true solution. These problems may be overcome by using extremely fine space and time grids; however, the development of numerical schemes that can handle these numerical difficulties with reasonable computational effort is a challenge (Liu and Dane 1996). Axworthy and Karney (1996) and Lee (2004) both proved that the ADR model significantly underpredicts nonconservative constituents concentration under the laminar flow condition. Therefore, dispersion is important and will most like be incorporated into the next generation of network water quality models.

Tzatchkov et al. (2002) proposed an effective and reliable two-stage Eulerian–Lagrangian numerical solution for ADR model in pipe networks. A technique that uses numerically computed Green's function is proposed to solve network numerical schemes efficiently so as to apply the water quality model to large networks at a reasonable computer cost. The proposed model appears to provide significant improvement in predictions of field-measured fluoride concentration compared to EPANET in low-flow zones in the network.

Lee (2004) studied unsteady dispersion in random intermittent laminar flow. He found that the instantaneous rate of dispersion is the sum of two factors: the dispersion memory from previous pulses and the nonlinear excitation from the current pulse. During periods of stagnation, active dispersion ceases and the memory of previous dispersion decays exponentially. These findings may be used to estimate the actual dynamic dispersion rate in network water quality modeling.

Li et al. (2005) applied numerical models to simulate more realistic network conditions and develop preliminary new guidelines on the conditions under which dispersion is important in pipe networks. Instead of adopting the "usual" one-dimensional advection-dispersion-reaction (lD-ADR) model, a two-dimensional advection diffusion-reaction (2D-ADR) model was selected. This eliminated the error caused by estimation of the dispersion coefficient, an essential parameter of the 1D-ADR model. Numerical results from the 2D-ADR model are compared against numerical results from a 1D-AR (advection-reaction) model to identify differences caused by dispersion-related factors, such as time history of source strength, solute reaction rate, pipe diameter, and flow pattern. Several source load scenarios are simulated using the 2D-ADR numerical code in order to confirm a theoretical order-of-magnitude analysis of each term in the simplified 1D-ADR equation. This comparison highlighted the relative importance of advection, dispersion, and reaction in solute transport during low-flow conditions. The results confirm that dispersion is just as important as advection and, therefore, should not be ignored when modeling water quality in laminar flow dominated zones of a water distribution system.

Li et al. (2006) extended their earlier work by developing and demonstrating a computer program ADRNET for network water quality modeling. The network hydraulic functions in the EPANET toolkit were incorporated and used as a hydraulic engine to simulate the extended-period pipes flows in ADRNET based on the time-averaged stochastic demands. A two-stage Eulerian–Lagrangian scheme was combined with a numerical Green's function technique proposed by Tzatchkov et al. (2002) to solve the ADR equation in network condition. Adjustments and improvements were made in specific difference methods and in estimating spatially averaged dispersion coefficients based on Lee's work (2004) for a more reasonable representation of network conditions. The program was verified against analytical solutions and validated through a case study using field observations in the SCCWRA Cherry Hill/Brushy Plains service network with an all-pipe representation. The authors concluded:

- Their computer program, ADRNET, could be used to simulate water quality in distribution network with ADR model as an extension of EPANET toolkits.
- The Eulerian–Lagrangian numerical scheme combined with numerical Green's function can be applied to solve the ADR model efficiently in network conditions.
- Verification with analytical solution in a pipe illustrated the good performance of convergence and stability for the selected numerical scheme.
- Comparison to the results simulated by EPANET and field observations in the test network showed that the ADRNET provided better results than EPANET in laminar flow–dominated zones and results similar to EPANET in turbulent-dominated flow zones, and the proposed algorithm for dispersion coefficient is appropriate.
- The time scale for averaging pipe flows did not exhibit an obvious effect on the predicted concentration profiles based on the algorithms for the dispersion coefficient. However, different realizations of the stochastic water demands affect concentration profiles by changing transport pathways.

Integrating Real-Time Monitoring and Modeling

It is clear from the North Marin Water District example discussed earlier that modeling can be a very effective tool when used in combination with monitoring data. Hall et al. (2007) conducted an in-depth evaluation of how changes in water quality parameters, which potentially indicate contamination, can be detected by real- or near real-time sensors. The sensors investigated were off-the-shelf commercial products designed to monitor standard drinking water parameters such as pH, free chlorine, oxidation reduction potential (ORP), dissolved oxygen, specific conductance, turbidity, total organic carbon (TOC), chloride, ammonia, and nitrate. The sensors were mounted within a recirculating pipe loop and challenged with contaminants including secondary effluent from a wastewater treatment plant, potassium ferricyanide, a malathion insecticidal formulation, a glyphosate herbicidal formulation, nicotine, arsenic trioxide, aldicarb, and *Escherichia coli* K-12 strain with growth media. Overall, the sensors that responded to most contaminants were those that monitored for free chlorine, TOC, ORP, specific conductance, and chloride. Generally, the technology used in sensor design or the particular manufacturer of the sensor did not affect the response characteristics. The focus of the study was on system vulnerability, but the results may help refine the role of water quality sensors, in conjunction with other data sources such as customer complaints and public health surveillance data. The test was conducted in a pilot-scale distribution system simulator (DSS). To facilitate data collection, a SCADA system was installed at the DSS. This system incorporated the electronic hardware and custom software interfaces necessary for the monitoring devices selected for study. The SCADA system collected, archived, and displayed data from the continuous monitoring devices and the traditional sensors. Software algorithms were used to display historical data and compare any combination of water quality parameters. One important caveat to the results presented in Hall et al. (2007) is that the characteristics of distribution system water used in this type of investigation can significantly affect the results. Thus, while many observations can be made from the data generated by these experiments, the following discussion involves those results that seem most applicable to distribution system waters in general.

No single chemical sensor responded to all of the contaminants used in this study, yet some sensors responded better to the introduction of a larger number of contaminants than did others. Table 10-5 summarizes the response of the sensors in terms of which sensors showed a response of greater magnitude than baseline fluctuations. The table also indicates whether the value of the response was greater than or less than the baseline value. Table 10-5 does not indicate the absolute magnitude of the change. Although the absolute magnitude may be part of a detailed quantitative analysis, qualitative observations were the focus of this study. When performing a quantitative analysis, it is not only the absolute magnitude of the change that is important but also the magnitude relative to the size and fluctuations in the baseline along with the slope of the change (i.e., to determine whether the changes occur over several hours or several minutes). Thus, quantitative evaluation makes use of signal-to-noise principles, which are difficult to generalize about and are location-specific. An examination of Table 10-5 reveals that several sensors responded to a large number of contaminants. These were specific conductivity, TOC, free chlorine, chloride, and ORP. The chlorine sensors appear to respond to all contaminants, although it is well known that some potential contaminants do not react significantly with chlorine. The TOC responded to all the organic (carbon-containing) compounds; the TOC monitor, however, has a higher capital cost when compared with other sensors. The calibration requirements for the sensors in these systems vary from weekly to monthly. It is estimated that a multiparameter monitoring station could have reagent and maintenance costs of several hundred dollars per month.

Kshirsagar et al. (2010) described the development of NetSafe, a network simulation software for predictive simulation of the fate and transport of chemical and biological agents in drinking water distribution systems (DWDS). The system is based on real-time hydraulic and water quality measurements and is capable of identifying the presence of chemical and biological agents and predicting the sources and destinations of such contaminants in DWDS for vulnerability analysis and optimal emergency response. NetSafe provides the based for an integrated model-based detection approach that can be used with both inexpensive chlorine sensors as well as with high-end multispecies sensors. The investigators use the moving average of the difference between the model chlorine concentration and measured chlorine concentration to serve as an indicator that new species in the water are consuming the chlorine. An important aspect of the model is the ability to model chlorine residual transport accurately. That goal is addressed by the technique of recalibrating the model demands continuously using the SCADA information.

The contaminant source identification is based on a particle backtracking approach and determines the possible sources for packets of water starting at each moment in history and arriving at one or more alarm nodes at the current time. Contaminant spread is determined using the EPANET-MSX (Shang et al. 2008) engine and implemented using a conservative tracer. The resulting simulation shows all the nodes that become affected by the contamination event, and that information assists in the vulnerability analysis.

Table 10-5. Sensor response following introduction of contaminant (Hall et al. 2007)

Contaminant	Sensor Response: Increase From Baseline	Sensor Response: Decrease From Baseline
Wastewater	Chloride	Free chlorine
	Specific conductance	ORP
	Turbidity	
	TOC	
Potassium ferricyanide	Free chlorine (DPD)*	
	TOC	
	Chloride	
	Nitrate-nitrogen	
	Ammonia-nitrogen	
	ORP	
Glysphosate formulation	TOC	Free chlorine
	Chloride	ORP
Malathion formulation	TOC	Free chlorine
	Turbidity	ORP
Aldicarb	TOC	Total chlorine
	Turbidity	Free chlorine
		ORP
Escherichia coli in terrific broth	TOC	Total chlorine
	Ammonia-nitrogen	Free chlorine
	Turbidity	
Terrific broth	Turbidity	Total chlorine
	TOC	Free chlorine
		ORP
Arsenic trioxide	Turbidity	Total chlorine
	Ammonia-nitrogen	Free chlorine
		Nitrate-nitrogen
		ORP
Nicotine	TOC	Free chlorine
	Ammonia-nitrogen	Nitrate-nitrogen
	Chloride	ORP

DPD—*N-N*-diethyl-*p*-phenylenedlamine, ORP—oxidation reduction potential, TOC—total organic carbon

*Potassium ferricyanide injected, which was yellow in color, may have masked the DPD colorimetric method, creating an increased reading from the baseline.

Certain contaminants that did not create detectable changes in water quality parameters were omitted for security concerns. Several more-detailed reports of this research are available on the secure WaterISAC website.

Total Water Management

Interest in the concept of total water management is growing, including evaluation of all aspects of water management in urban areas. For example, as noted earlier, in addition to providing drinking water, a major function of most distribution systems is to provide adequate standby fire flow, which must be sustained normally range from 3 to 8 hr. The minimum allowable residual pressure during a fire is normally 20 psi (or 14 m). In order to satisfy this need for adequate standby capacity and pressure, most distribution systems use standpipes, elevated tanks, and large storage reservoirs.

The sizing of water mains is based on a variety of water quantity requirements such as projected demand, pressure requirements, velocity requirements, and pipe carrying capacity, in addition to fire protection requirements. The consequence of designing and operating a system to maintain adequate fire flow and redundant capacity is long transit times between the treatment plant and the consumer, which may be detrimental to meeting drinking water MCLs (Clark and Grayman 1998, Brandt et al. 2004). Snyder et al. (2002) recommended that water systems evaluate existing storage tanks to determine if modification or elimination of the tanks were feasible. Water-efficient fire suppression technologies exist that use less water than conventional standards. In particular, the universal application of automatic sprinkler systems provides the most proven method for reducing loss of life and property due to fire, while at the same time providing faster response to the fire and requiring significantly less water than conventional firefighting techniques. Snyder et al. (2002) also recommended that the universal application of automatic fire sprinklers be adopted by local jurisdictions for homes as well as other buildings.

There is a growing recognition that embedded designs in most urban areas have also resulted in distribution systems that have long residence times due to the large amounts of storage required for firefighting capacity. Almost 20 years ago, Clark and Grayman (1992) expressed concern that long residence times resulting from excess capacity for firefighting and other municipal uses would also provide optimum conditions for the formation of DBPs and the regrowth of microorganisms. They hypothesized that eventually the drinking water industry would be in conflict over protecting public health and protecting public safety. With the promulgation of the Stage 2 D/DBP Rule, this situation has come to pass. Some nonconventional water distribution system designs that might address some of these issues are discussed in the following section, including decentralized treatment, dual distribution systems, and an approach that employs advanced and enhanced treatment to solve distribution system water quality problems.

Decentralized Treatment

Distributed or decentralized treatment systems refer to those in which a centralized treatment plant is augmented with additional treatment units that are located at various key points within the distribution system. Usually, the distributed units provide advanced treatment to meet stringent water quality requirements at consumer endpoints that would otherwise be in violation. Distributed units would be located either at the point-of-use/point-of-entry (POU/POE) of households, for example, or at a location from which aggregates of uses could be served. This might be at the neighborhood or district level, depending on technological and financial requirements.

Research has been conducted to develop guidance as to how the decentralized treatment concept might be implemented in water systems worldwide (Weber 2002, 2004). Weber's approach involves having distributed networks (Distributed Optimal Technologies Networks, or DOT-Nets) in which water supply is optimized by separately treating several components of water and wastewater streams using decentralized treatment units. The approach largely views water supply, treatment, and waste disposal as different aspects of the same integrated system.

The principal trade-off associated with using such systems is the alternative cost associated with upgrading large centralized treatment facilities and distribution networks. Multiple or new pipe networks are a necessary part of the design framework for these satellite systems. That is, new piping would be needed from the advanced water treatment system into the household (or industry), although it would travel a short distance and would be a small percentage of the total plumbing for the building. It is possible that investing in larger satellite systems with separate piping might offer a cost advantage based on economies of scale.

Dual Systems and Water Reclamation

Another option for design and operation of distribution systems is the creation of dual systems (NRC 2006). In these types of systems, reclaimed wastewater or water of subpotable quality may be used for firefighting and other special purposes such as irrigation of lawns, parks, roadway borders and medians; air conditioning and industrial cooling towers; stack gas scrubbing; industrial processing; toilet and urinal flushing; construction; cleansing and maintenance, including vehicle washing; scenic waters and fountains; and environmental and recreational purposes. The design of these systems is such that separate pipe networks are constructed for the potable and nonpotable lines. This differentiates dual systems from most community water supplies, in which one distribution system provides potable water to serve all purposes.

Most dual systems in use today were installed by adding reclaimed water lines alongside (but not connected to) potable water lines already in place. For example, in St. Petersburg, Fla., a reclaimed water distribution system was placed into operation in 1976, and fire protection is provided from both the potable and reclaimed water lines. San Francisco has a nonpotable system, constructed after the 1906 earthquake, that serves the downtown area to augment fire protection. Rouse Hill, Australia, was the first community to plan a dual water system with the reclaimed water lines to serve all nonpotable uses, including fire protection, such that the potable water line can have much smaller pipe diameters. Both systems have service reservoirs for meeting diurnal variations in demand, and if a shortage of water for fire protection occurs, potable water can be transferred to the nonpotable system.

Perhaps the most famous water reuse operation using a dual distribution system is the Irvine Ranch Water District in Irvine, Calif. The Irvine Ranch Water District is one of the first water districts in the United States to practice wastewater reuse. It serves 316,287 people over 133 mi^2 (344.5 km^2), making it about a quarter the size of the Los Angeles Department of Water and Power. It includes 85,500 domestic connections, 3,700 recycled connections, and, as of 2003, 1,075 mi ($1{,}730 \times 10^3$ m) of pipe. The 28 aboveground and below-ground storage tanks range in volume from 0.75 mil gal (0.0028 million m^3) to 16 mil gal (0.061 million m^3) and have a total storage capacity of 131.75 mil gal (0.50 million m^3). The potential advantages of using dual distribution systems include the fact that much smaller volumes of water would need be treated to high standards, which would result in cost savings at the treatment plant. However, the creation of dual distribution systems necessitates the retrofitting of an existing water supply system and reliance on existing pipes to provide nonpotable supply obtained from wastewater or other sources. Its greatest potential may be in new communities where neither type of distribution system currently exists. New communities could better optimize their systems because both types of piping systems could be built simultaneously.

Water Quality/Fire-Flow Modeling Example

In the example discussed here, a hypothetical distribution system was simulated using a series of distribution system models: a combined hydraulic/water quality simulation model modified to search for alternative pipe sizing that will meet fire demands, and a distribution system costing model (Grayman et al. 2001). The purpose of the modeling task was to assess the quantitative effects of providing fire flow on (1) the physical requirements of the distribution system and (2) the resulting water quality. A costing model was then used to determine the cost of the alternative physical systems (Clark et al. 2002).

Four different scenarios were evaluated:

- Existing pipe and storage sizes.
- Modified (downsized) pipes and storage that will still meet the full fire-flow requirements.
- Modified pipes and storage to meet a reduced fire-flow requirement associated with having sprinkler systems in all structures.
- Modified pipe and storage with no fire-flow requirements.

The adequacy of a system was evaluated in terms of its capability to provide fire flow and to meet normal demands. Systems were examined in terms of meeting maximum day conditions with a fire-flow requirement superimposed. The analysis included both a steady-state assessment and an extended-period simulation (EPS) that tested the ability of the system to deliver flow over a typical fire duration (usually 3 hr). Subsequent to the identification of the system configurations under the four design conditions described above, additional modeling was performed to: (1) determine the cost of the resulting networks; and (2) determine the water quality consequences of the reduced pipe diameters and storage requirements. Water quality was assessed in terms of resulting water age using the EPANET model.

The example network contains approximately 44 mi of pipe varying in diameter from 6 in. (15 cm) to 14 in. (36 cm). The average water consumption is approximately 1 mgd (3.785 million L/d). The two tanks in the system have a total volume of 1.37 mil gal capacity (5,185 m^3). The system is fed by a single source.

Nodes have been assigned to five different land-use categories, and demands have been defined in terms of these categories. Table 10-6 lists the average day consumption and fire-flow requirements for each of the categories.

For the assessment, the analysis was done for maximum day conditions. A peaking factor (multiplier times the average day demands) of 1.25 is being used. For water quality assessment, generally lower demand conditions will result in older water. Therefore, a minimum day condition was studied and demands estimated based on the application of a 0.75 multiplier (times average day).

The model and procedure were applied to study the following four scenarios: (1) existing pipe and storage sizes; (2) modified (downsized) pipes and storage that still meet the full fire-flow requirements; (3) modified pipes and storage to meet a reduced fire-flow requirement associated with having sprinkler systems in all structures; and (4) modified pipe and storage with no fire-flow requirements. A modified EPANET model was used to determine the downsized pipe diameters that would meet the fire requirements for scenarios 2 through 4.

The results of the analysis were presented in a series of tables and color-coded maps. Total lengths of pipes of different diameters for the four scenarios are presented in Table 10-7. Required tank volumes are summarized in Table 10-8. Estimated capital costs are presented in Table 10-9.

Water age was used as a surrogate measure of water quality. For each scenario, an extended period simulation was run and the age of the water was calculated. Demand

Table 10-6. Average day and fire-flow requirements by use category

Use Category	Average Day Demand (*gpm*) at Each Node	Fire-flow Requirement (*gpm*)	Reduced Fire Flow (*gpm*)
Low-density residential	3	500	100
High-density residential	4	750	100
Commercial	5	2000	500
Industrial	20	3500	1000
Transmission line	0	0	0

Multiply flows by 0.0631 to convert from gallons per minute to liters/second.

Table 10-7. Results of modeling assessment on pipe lengths by diameter

Pipe Diameter (*in.*)	Base Conditions (*ft of pipe*)	Initial Reductions (*ft of pipe*)	Partial Fire Flow (*ft of pipe*)	No Fire Flow (*ft of pipe*)
14	19,711	12,213	500	500
12	31,839	7,498	11,713	5,324
10	28,029	43,064	5,403	0
8	82,331	66,661	23,385	22.003
6	70,856	64,052	64,610	59,295
4	0	36,328	69,921	41,593
2	0	2,950	57,234	104,051

Multiply diameters by 2.54 to convert from inches to centimeters.
Multiply length by 0.3048 to convert from feet to meters.

Table 10-8. Tank volumes by scenario

Scenario	Tank 1 Volume (*mil gal*)	Tank 3 Volume (*mil gal*)
Base	1.14	0.23
Initial reductions	1.14	0.23
Partial fire flow	0.69	0.12
No fire flow	0.51	0.10

Multiply volumes by 3,785 to convert from million gallons to cubic meters.

Table 10-9. Capital costs (in million dollars) (2002)

Scenario	Base Pipe Costs (*$M*)	Variable Pipe Costs (*$M*)	Tank 1 Costs (*$M*)	Tank 2 Costs (*$M*)	Total Costs (*$M*)
Base	14.5	5.8	1.8	0.5	22.7
Initial reductions	14.5	4.9	1.8	0.5	21.7
Partial fire flow	14.5	2.9	1.1	0.4	18.9
No fire flow	14.5	2.3	0.9	0.4	18.1

weighted average water age was used as an overall measure of the water quality for the scenario. Results are presented in Table 10-10.

As illustrated in the tables, fulfilling the fire-flow requirement had a considerable impact on the resulting network design. In each subsequent scenario, a significant reduction in pipe sizes is possible. The reduction from the first to the second scenario indicated that many of the pipes in the original design are oversized relative to meeting fire-flow requirements. For example, the length of 12-in. and 14-in. (30-cm and 36-cm) pipes dropped from approximately 10 mi (16.1 km) to 4 mi (6.4 km) between the two scenarios. Similarly, as the fire-flow requirement is reduced in the third and fourth scenarios, significant further downsizing is possible. Similarly, tank volume requirements drop as the fire-flow requirements are relaxed in the third and fourth scenarios.

Based on the cost modeling, the cost savings associated with the lower fire-flow requirements (and resulting smaller pipe diameters) are relatively small. For example, the cost reduction from the base case to the no fire-flow scenario is only 20 percent. This level of cost reduction can be explained by the relatively high proportion of the costs that are fixed relative to pipe sizes.

The most significant result is the large reductions in water age associated with reduced fire flows, as indicated in Table 10-10. This suggests a very significant potential improvement in water quality with the smaller pipes and tanks.

This example demonstrates a basic methodology for evaluating the physical, cost, and water quality impacts of providing different levels of fire flow on the distribution system design. Further application to existing distribution systems is planned. Other factors that may be affected by downsizing pipes include the reliability of the system and the impacts of increased velocities due to smaller pipes.

Table 10-10. Demand weighted average water age

Scenario	Age (*hr*)
Base	44.77
Initial reductions	37.99
Partial fire flow	20.97
No fire flow	13.46

Mixing at Junctions

A fundamental assumption underlying most hydraulic/water quality models is complete mixing at junctions. However, recent research has called this assumption into question. Predicting the movement and distribution of contaminants is depended on the assumption of mixing at pipe junctions, where different flow rates and contaminant concentrations can exist. Ho and O'Rear (2009) have reported on experiments and observations of solute mixing in various pipe junction configurations. A major finding was that impinging fluid streams within a junction often do not mix completely. This is contrary to the most common assumption of complete mixing in pipe junctions. The investigators developed analytical models for each configuration, and results were compared with experimental data. The investigators found that if concentrations of two incoming fluid streams differ, they tend to bifurcate and reflect off one another, affecting subsequent solute distribution and mixing of fluid streams.

The authors have proposed a bulk-advective mixing (BAM) model that has been shown to accurately represent this behavior and lead to more accurate water quality assessments. Figure 10-11 (see appendix E) illustrates in an experimental apparatus the mixing that can occur at pipe junctions.

Microbial Fate and Transport

There is growing interest in modeling the fate and transport of microorganisms in drinking water distribution systems. Lu et al. (1995) proposed a comprehensive paper emphasizing the simultaneous decay of chlorine, formation of disinfection by-products, and transport and attachment of microorganisms to pipe walls. Zhang et al. (2004) proposed a mathematical model of bacterial regrowth in distribution systems, which combined hydraulic calculations, inclusive of dispersion, with a description of the following microbial processes: free and attached growth, detachment, endogenous respiration, and inactivation by chlorine. The microbial process description was simplified from previous models based on a sensitivity analysis. The alternating split-operator algorithm was used to solve this model. This method differs from previous approaches by decoupling the transport and reaction processes, allowing a choice among numerical algorithms that is best suited for each part of the model. The proposed solution resolves sharp fronts of a concentration profile more accurately than traditional finite-difference methods. Results were also compared against EPANET to show the importance of accounting for dispersion, as would occur during low water demand conditions when velocity is low. Use of the model to understand the interaction among key parameters affecting bacterial regrowth was illustrated for a simple hypothetical network.

Future of EPANET

EPANET has played a very important role in the application of hydraulic/water quality modeling to drinking water distribution systems. However, some are concerned that USEPA will no longer continue to support the development of EPANET. Rossman and Van Zyl (2010) have proposed that an open source project (OSP) be established for the widely used EPANET pipe network analysis program. It would be an ongoing collaborative effort among a group of geographically dispersed advisors and developers, working on a voluntary basis within an auto-governing organizational structure. Such an approach would allow members of the water distribution system modeling community to participate in the program's development and improvement. Establishing such a project would require justifying the need and benefit of such a project, establishing its membership and organizational structure, identifying the scope of development the project should pursue, and maintaining high quality control standards. A key requirement of an open source EPANET project will be to maintain the program's high degree of dependability and robustness given its widespread use throughout the industry and in academia. The authors suggest the following actions:

- Establish a set of coding standards and conventions that all programmers must follow when submitting code to the project.
- Require that all submitted code be reviewed by one or more project members before being accepted.
- Require that unit testing be done for all new features added to EPANET's engine or user interface.
- Require that regression testing be done against a library of pipe network data sets to ensure that any changes do not "break" the system.
- Pursue a diligent policy of tracking bugs and resolving them in a timely manner.

It will be interesting to follow the development of this idea.

References

Axworthy, D.H., and B.W. Karney. 1996. Modeling Low Velocity/High Dispersion Flow in Water Distribution Systems. *J. Water Resources Planning and Management,* 122(3):218–221.

Berry, J., L. Fleischer, W.E. Hart, C.A. Phillips, and J.P. Watson. 2005. Sensor Placement in Municipal Water Networks. *Journal of Water Planning and Resources Management,* 131(3):237–234.

Boulos, P.F., B.W. Karney, D.J. Wood, and S. Lingireddy. 2005. Hydraulic Transient Guidelines for Protecting Water Distribution Systems. *Journal AWWA,* 97(5):111–124.

Boulos, P.F., K.E. Lansey, and B.W. Karney. 2006. *Comprehensive Water Distribution Systems Analysis Handbook for Engineers and Planners.* 2nd ed. Broomfield, Colo.: MWH Soft, 660 pp.

Brandt, M., J. Clement, J. Powell, R. Casey, D. Holt, N. Harris, and C.T. Ta. 2004. *Managing Distribution Retention Time to Improve Water Quality: Phase I.* Denver, Colo.: AwwaRF.

Buchberger, S.G., and L. Wu. 1995. Model for Instantaneous Residential Water Demands. *Journal of Hydraulic Engineering,* ASCE, 121(3):232–246.

Buchberger, S.G., and G.J. Wells. 1996. Intensity, Duration, and Frequency of Residential Water Demands. *Journal of Water Resources Planning and Management,* 122(1):11–19.

Buchberger, S.G., and Y. Lee. 1999. Evidence Supporting the Poisson Pulse Hypothesis for Residential Water Demand. In *Water Industry Systems: Modeling and Optimization Applications.* Vol. 1, ed. D.Savic and G. Walters. Baldock, UK: Research Studies Press, 89–101.

Buchberger, S.G., J.T. Carter, Y. Lee, and T.G. Schade. 2003. Random Demands, Travel Times, and Water Quality in Dead-Ends. Denver, Colo.: AwwaRF, 467 pp.

Buchberger, S.G., and Z. Li. 2007. PRPsym: A Modeling System for Simulation of Stochastic Water Demands. In *Proc. ASCE-EWRI Conference, Tampa, Fla.* Reston, Va.: ASCE.

Cesario, A.L. 1995. *Modeling, Analysis, and Design of Water Distribution Systems.* Denver, Colo.: AWWA.

Clark, R.M., and R.M. Males. 1985. Simulating Cost and Quality in Water Distribution. *Journal Water Resources Planning and Management,* ASCE, 111(4):454–466.

Clark, R.M., and W.M. Grayman. 1992. Distribution System Quality: A Trade-Off Between Public Health and Public Safety. *Journal AWWA,* 81(7):18.

Clark, R.M., and W.M. Grayman. 1998. *Modeling Water Quality in Drinking Water Distribution Systems.* Denver, Colo.: AWWA.

Clark, R.M., and R.A. Deininger. 2000. Protecting the Nation's Critical Infrastructure: The Vulnerability of US Water Supply Systems. *Journal of Contingencies and Crisis Management,* 8(2):73–80.

Clark, R.M., M. Sivaganesan, A. Selvakumar, and V. Sethi. 2002. Costs Models for Water Supply Distribution Systems. *Journal of Water Resources Planning and Management,* ASCE. 128(5):312–321.

Edwards, J., E. Koval, B. Lendt, and P. Ginther. 2009. GIS and Hydraulic Model Integration: Implementing Cost-Effective Sustainable Modeling Solutions. *Journal AWWA.* 101(11):34–42.

Friedman, M., L. Radder, S. Harrison, D. Howie, M. Britton, G. Boyd, H. Wang, R. Gullick, D. Wood and J. Funk. 2004. *Verification and Control of Pressure Transients and Intrusion in Distribution Systems.* Denver, Colo.: AwwaRF.

Grayman, W.M, R.M. Clark, F. Grablutz, M. Sivaganesan, and T. Schade. 2001. Modeling the Impacts of Fire Flows on Distribution System Water Quality, Design and Operation. In *Proc. World Water and Environmental Resources Congress, May 20–24, 2001, Orlando, Fla.* Reston, Va.: ASCE.

Grise, S., E. Idolyantes, E. Brinton, B. Booth, and M. Zeiler. 2000. *Water Utilities ArcGIS Data Models.* Redlands, Calif.: ESRI Press.

Gullick, R.W., M.W. LeChevallier, R.C. Svindland, and M.J. Friedman. 2004. Occurrence of Transient Low and Negative Pressures in Distribution Systems. *Journal AWWA,* 96(11):52–66.

Hall, J., A.D. Zaffiro, R.B. Marx, P.C. Kefauver, E.R. Krishnan, R.C. Haught, and J.G. Herrmann. 2007. On-Line Water Quality Parameters as Indicators of Distribution System Contamination. *Journal AWWA,* 99(1):66–77.

Ho, C.K, and L. O'Rear Jr. 2009. Evaluation of Solute Mixing in Water Distribution Pipe Junctions. *Journal AWWA,* 101(9):116–127.

Homeland Security Strategy Act of 2001. 2001. HR 1292.

Jacobsen, L., and S. Kamojjala. 2005. Full System Models and GIS Integration. In *Proc. AWWA Annual Conference and Exposition, San Francisco.* Denver, Colo.: AWWA.

Jacobsen L., S. Kamojjala, and M. Fang. 2005. Integrating Hydraulic and Water Quality Models With Other Utility Systems: A Case Study. In *Proc. AWWA Information Management and Technology Conference, Denver.* Denver, Colo.: AWWA.

Jung, B.B., B.W. Karney, P.F. Boulos, and D.J. Wood. 2007. The Need for Comprehensive Transient Analysis of Distribution Systems. *Journal AWWA,* 99(1):112–123.

Karim, M., M. Abbaszadegan, and M.W. LeChevallier. 2003. Potential for Pathogen Intrusion During Pressure Transients. *Journal AWWA,* 95(5):134–146.

Khanal, N., S.G. Buchberger, and S.A. McKenna. 2006. Distribution System Contamination Events: Exposure, Influence, and Sensitivity. *Journal of Water Resources Planning and Management,* 132(4):283–292.

Kshirsagar, S., J. Uber, R. Males, G. Davis, and B. Bedinghaus. 2010. NETSAFE-Enhanced Water Distribution Network Safety Through a Predictive Simulation Software. In *Proc. ASCE Water Distribution System Analysis (WDSA), Tucson, Ariz., Sept. 12–15.* Reston, Va.: ASCE.

LeChevallier, M.W., R.W. Gullick, M.R. Karim, M. Friedman, and J.E. Funk. 2003. The Potential for Health Risks From Intrusion of Contaminants Into Distribution Systems From Pressure Transients. *Jour. Water Health* 1(1):3–14.

Lee, Y. 2004. *Mass Dispersion in Intermittent Laminar Flow.* Ph.D dissertation. University of Cincinnati, Cincinnati, Ohio.

Li, Z., S.G. Buchberger, and V. Tzatchkov. 2005. Importance of Dispersion in Network Water Quality Modeling. In *Proc. ASCE EWRI Conf., Anchorage, Alaska.* Reston, Va.: ASCE.

Li, Z., S.G. Buchberger, and V. Tzatchkov. 2006. Integrating Distribution Network Models With Stochastic Water Demands and Mass Dispersion. In *Proc. ASCE WDSA Conf., Cincinnati, Ohio.* Reston, Va.: ASCE.

Liu, H.H., and J.H. Dane. 1996. An Interpolation-Corrected Modified Method of Characteristics to Solve Advection-Dispersion Equations. *Advances in Water Resources,* 19(6): 359–368.

Lu, C., P. Biswas, and R.M. Clark. 1995. Simultaneous Transport of Substrates, Disinfectants, and Microorganisms in Water Pipes. *Water Research,* 29(3):881–894.

Mark, D.M., N. Chrisman, A.U. Frank, P.H. McHaffie, and J. Pickles. 1997a. *The GIS History Project.* UCGIS Summer Assembly, Bar Harbor, Maine. http://www.geog.buffalo.edu/ncgia/gishist/bar_harbor.html.

Mark, D.M. 1997b. *The History of Geographic Information Systems: Invention and Re-Invention of Triangulated Irregular Networks (TINs).* In *Proc. GIS/LIS '97.* http://www.geog.buffalo.edu/ncgia/gishist/GISLIS97.html.

McHarg, I.L. 1969. *Design With Nature.* Garden City, N.Y.: American Museum of Natural History and Natural History Press.

Murray, R., R. Janke, and J. Uber. 2004. The Threat Ensemble Vulnerability Assessment Program for Drinking Water Distribution System Security. In *Proc. ASCE-EWRI World Water & Environmental Resources Congress, Salt Lake City, Utah.* Reston, Va.: ASCE.

Murray, R., J. Uber, and R. Janke. 2005. A Model for Estimating the Acute Health Impacts Resulting From Consumption of Contaminated Drinking Water. In *Proc. ASCE-EWRI World Water & Environmental Resources Congress, Anchorage, Alaska.* Reston, Va.: ASCE.

National Research Council (NRC). 2006. *Drinking Water Distribution Systems: Assessing and Reducing Risks.* Washington, D.C.: National Academies Press.

Nilsson, K.A., S.G. Buchberqer, and R.M. Clark. 2005. Simulating Exposures to Deliberate Intrusions into Water Distribution Systems. *Journal of Water Resources Planning and Management,* ASCE, 131(3):228–236.

Panguluri, S., W. M. Grayman, and R. M. Clark. 2005. *Distribution System Water Quality Report: A Guide to the Assessment and Management of Drinking Water Quality in Distribution Systems.* Cincinnati, Ohio: USEPA Office of Research and Development.

Public Health Security and Bioterrorism Preparedness and Response Act of 2002. 2002. Public Law (PL) 107-188.

Rossman, L.A., R.M. Clark, and W.M. Grayman. 1994. Modeling Chlorine Residuals in Drinking Water Distribution Systems. *Journal Environmental Engineering,* ASCE, 120(4):803–820.

Rossman, L.A. 2000. *EPANET Version 2 User's Manual.* Cincinnati, Ohio: USEPA Drinking Water Research Division.

Rossman. L.A., and J.E. Van Zyl. 2010. *Proc. Water Distribution System Analysis Symposium, Tucson, Ariz.,* Sept. 12–15. Reston, Va.: ASCE.

Shang, F., J.G. Uber, and L.A. Rossman. 2008. *EPANET Multi Species eXtension User's Manual.* EPA/600/S-07/021. National Risk Management Research Laboratory/National Homeland Security

Research Center, Office of Research and Development. Cincinnati, Ohio: US Environmental Protection Agency.

Shen, H., E.A. McBean, and M. Ghazali. 2009. Multi-Stage Response to Contaminant Ingress Into Water Distribution Systems and Probability Quantification. *Can. J. Civ. Eng.*, 36(11):1764–1772.

Snyder, J.K., A.K. Deb, F.M. Grablutz, S.B. McCammon, W.M. Grayman, R.M. Clark, D.A. Okun, S.M. Tyler, and D. Savic. 2002. *Impacts of Fire Flow on Distribution System Water Quality, Design, and Operation.* Denver, Colo.: AwwaRF.

Strasser, A., N. Diallo, and E.J. Koval. 2000. Development and Calibration of Denver Water's Hydraulic Models for a Treated Water Study. In *Proc. AWWA Annual Conference, Denver, Colo.* Denver, Colo.: AWWA.

Tzatchkov, V.G., A.A. Aldama, and F.I. Arreguin. 2002. Advection-Dispersion-Reaction Modeling in Water Distribution Networks. *J. of Water Resources Planning and Management*, 128(5):334–342.

Uber, J., L. Rossman, and F. Shang. 2004a. Extensions to EPANET for the Fate and Transport of Multiple Interacting Chemical or Biological Components. In *Proc. ASCE-EWRI World Water and Environmental Resources Congress, Salt Lake City, Utah.* Reston, Va.: ASCE.

Uber, J, R. Janke, R. Murray, and P. Meyer. 2004b. A Greedy Heuristics Model for Locating Water Quality Sensors in a Water Distribution System. In *Proc. ASCE-EWRI World Water and Environmental Resources Congress, Salt Lake City, Utah.* Reston, Va.: ASCE.

Weber, W. Jr. 2002. Distributed Optimal Technology Networks: A Concept and Strategy for Potable Water Sustainability. *Water Science and Technology,* 46(6–7):241–246.

Weber, W. Jr. 2004. Optimal Uses of Advanced Technologies for Water and Wastewater Treatment in Urban Environments. *Water Science and Technology: Water Supply,* 4(1):7–12.

Wood, D.J., S. Lingireddy, and P.F. Boulos. 2005. *Pressure Wave Analysis of Transient Flow in Pipe Distribution Systems.* Broomfield, Colo.: MWH Soft, 213 pp.

Zhang, W., C.T. Miller, and F.A. DiGiano. 2004. Bacterial Regrowth Model for Water Distribution Systems Incorporating Alternating Split-Operator Solution Technique. *Jour. Environmental Engineering,* ASCE, 130(9):932–941.

Zeiler, M. 1999. *Modeling Our World.* Redlands, Calif.: ESRI Press.

Chapter 11

Planning and Designing a Water Quality Simulation Study

Introduction

Applying distribution system network modeling requires both an understanding of the underlying fluid mechanics associated with design and operation of the network, and engineering judgment derived from field experience (Clark and Grayman 1998). All models are approximations of the actual systems that are being represented. As discussed, in a network model both the mathematical equations used in the model and the specific model parameters are only numerical approximations. For example, the Hazen–Williams equation used to describe friction head loss is an empirical relationship that was derived based on laboratory experiments. Furthermore, the roughness parameter (*C* factor) used in the Hazen–Williams equation that modelers assign to each pipe is not known with total certainty, because it is not feasible to examine and test every pipe in the system. Therefore, calibration and validation are important aspects of modeling. The goal in calibration is to reduce uncertainty in model parameters to a level such that the accuracy of the model is commensurate with the type of decisions that will be made based on model predictions.

It would be difficult to reduce the use of models to an all-encompassing set of rules and procedures; however, the following general steps are useful as guidelines for those starting a modeling project:

- Definition of scope and problem definition.
- Model selection and model setup.
- Network representation.
- Calibration and validation.
- Model application, analysis, and display.

Each of the above criteria will be discussed in this chapter.

Water distribution system models can be used in a wide variety of applications, including the support of design, planning, and analysis. Because these tasks may result in engineering decisions involving significant investments, it is important that the model be an acceptable representation of the "real world" and that the modeler have confidence in the model's predictions. In order to determine whether a model represents the real world, it is customary to measure various system values (e.g., pressure, flow, storage tank water levels, and chlorine residuals) during field studies and then compare the field results to model predictions. If the model adequately predicts the field measurements under a range of conditions for an extended period of time, the model is considered to be calibrated. If significant discrepancies occur between the measured and modeled data, then further calibration is needed. No general standards exist for defining what is adequate or what is a "significant discrepancy." However, it is recognized that the level of calibration required will depend on the use of the model. A greater degree of calibration is required for models that are used for detailed analysis, such as design and water quality predictions, than for models used for more general planning purposes (e.g., master planning).

Scope and Problem Definition

In planning a modeling study, a clear statement of the study objectives should be developed, even before logistical planning begins (Panguluri et al. 2005). For example, an objective statement might read "determine travel times from Water Treatment Plant A to key locations (transmission mains and representative local mains) in Pressure Zone B under typical summer operation." Such a statement provides a clear understanding of the study's overall goals and objectives. A study objective may also be even more specific and define additional key elements such as tracer material, dosage, and injection duration.

It is important to define the general scope of the modeling effort and attempt to answer the following questions:

- Will the model be used to investigate broad planning alternatives?
- Will the model be used to answer detailed design questions, such as those involved in determining pipe diameters to meet firefighting needs?
- Will the model be applied to a small town or to a major metropolitan area?
- Will water quality modeling be included as part of the effort?
- Does the model need to be integrated with an existing computer-aided drafting (CAD) or design package?

These are only a few of the questions that should be posed in defining the overall scope of the modeling effort. Such questions should always be considered when the modeling process is initiated.

It is useful in planning and designing a study to address and identify specific logistical details. These details might be presented in a comprehensive planning document or work plan that can be reviewed and commented on by parties that may have an interest in the study (e.g., team members, water utility staff and managers, and state regulatory officials). Planning and design-phase elements may include the following:

- Establishing study objectives and timeline.
- Forming a study team.
- Defining study area characteristics.

Depending on the objective, an approximate timeline (schedule) for the study should be formulated. Frequently, external constraints such as weather, system operation, and availability of personnel or equipment may influence this timeline. In other cases, the project timeline may depend on the specific objective of the study. For example, if the maximum community exposure to a contamination event is being studied, the timeline should be consistent with the season and time during which the event is likely to occur. If the study is intended to identify locations in the system where the lowest chlorine residuals are found, the study should be conducted during a period when minimum chlorine residuals occur.

It would be useful to form a study team at the beginning of the project, and depending on the size and scope of the study, the size of the team may vary from as few as 3 members to a sizable group of as many as 20 members. However, the range of functions and responsibilities that must be considered is approximately the same in all types of studies. The team makeup must include members with expertise in planning and carrying out the following activities and functions: understanding study area distribution system and treatment operations; conducting preliminary modeling studies; selecting, acquiring, and installing field equipment; managing and organizing field crews; performing field sampling; conducting laboratory analysis; analyzing and reporting results; and performing communications and notifications.

Study teams may be made up of water utility personnel, consulting engineering firm personnel, contractor staff, students from universities, and in some cases, federal or state governmental agency employees. Specific responsibilities and roles should be assigned to each team member. It is recommended that the study team meet on a regular basis to ensure that the task deadlines are met and the study objectives are achievable. If a tracer study is to be conducted and requires new or never-before-used equipment, training sessions for study team members should be included as part of study timeline and activities.

After the study team is formed, perhaps the first task to be undertaken is to identify the key characteristics of the study area. These characteristics include: the piping system network, pumping and storage operations, inflow and outflow through study area boundaries, temporal and spatial variations in water consumption, presence of large water users that may significantly impact water use patterns, and the study area that could potentially constrain field geography and local features associated with the activities. The effects of distribution system tanks and reservoirs should be considered. A large commercial user such as a golf course in the neighborhood may impact the study events. Several tools and procedures can be applied to improve the team's understanding of the target water distribution system area prior to conducting the tracer study. If a hydraulic model of the distribution system (under study) is available, it would be very helpful to use the model to simulate the tracer study under expected conditions. Examination of documents, such as master plans or operational reports, can also shed light on how the water system behaves. The study team or key members of the study team should also tour the study site with as-built pipe drawings to identify potential locations for safely installing field injection equipment, as well as flow and tracer monitoring equipment.

Model Selection and Setup

Selection of a model as the tool that provides the computational capability to assess design policy alternatives associated with the network is very important. Many models are available that may satisfy a user's needs, and many factors should be considered in the selection process. Selecting a hydraulic–water quality model for a distribution system should be a process of matching the specific needs of the user to the available models.

Selection Criteria

Following is a list of criteria that could be considered when selecting a hydraulic–water quality model for a distribution system:

- Cost
- User interface
- Ease of use
- Components represented
- Water quality modeling characteristics
- Robustness
- Size constraints
- Support
- Documentation

- Speed
- Hardware requirements
- Database/GIS/CAD/SCADA integration
- Flexibility
- Training

Each of these criteria is discussed in more detail in the following sections.

Model Cost

Hydraulic–water quality models range in cost. In some cases, the cost of the model depends on the size of the distribution system that is to be modeled. Some models operate in conjunction with proprietary packages, such as CAD or GIS software, so these costs should also be considered when comparing software packages. For smaller water utilities or consultants, or when modeling is used infrequently, large expenditures for modeling software would generally not be considered prudent, and therefore the cost may be a predominant factor in model selection. However, for larger, more frequent applications, the cost associated with the model acquisition may represent only a small portion of the total budget allocated to distribution system studies. Therefore, the initial cost of a model may only be a minor factor in model selection. EPANET, which is a public sector model, has been used successfully by some small communities and can be used to conduct preliminary evaluations.

User Interface

The look and feel of a model are reflected in the user interface, or the mechanism by which the user interacts with the model. Tremendous advances have been made in user interface development during recent years. Portions of a typical user interface (input and output) from a model in the 1980s are shown in Figure 11-1. By comparison, a graphical user interface

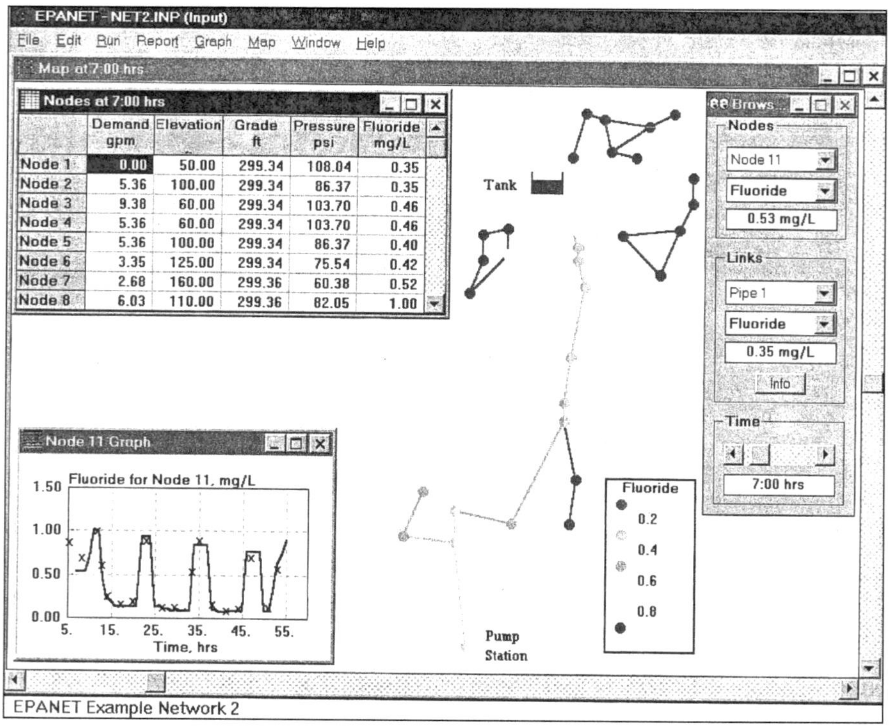

Figure 11-1. Graphical user interface (GUI) from EPANET

(GUI) representative of the current state of the art is illustrated in Figure 11-2. In a few short years, the GUI has progressed from a modeling feature rarely found, to a feature that most users now expect. In water distribution system models, a GUI may include various graphical and tabular options for visualizing results, and graphical methods for entering or editing the characteristics of the distribution system network.

Most currently available models use a Windows-based standard for GUIs. Selecting a user interface is a very personal and subjective process. Demo disks available for many systems provide one mechanism for examining the interface. A preferred method is to examine a full-featured trial package that allows the user to try the model out on small (e.g., 25-link) systems or for a limited time. Such trial packages are available for many models.

A. Partial Input File

```
CHESHIRE SYSTEM—SCENARIO 5 BASED ON WMGSKL2A 6-18-89

        1      1      2      4312      110
        2      3      4      6209      110
        3      5      6      6759      110

ELEVATION (node elevation from DEM)

        1      115.7
        2      210.5
        3      185.1

OUTPUT

        2      32.9775
        5      40.08583
        6      16.6325

TANK 49 22.

RATIO 1.496

ACCURACY .5 .5

END

OC

9
```

B. Partial Output File

```
WADISO—WATER DISTRIBUTION SYSTEM OPTIMIZATION
VERSION: MARCH 19, 1987
PUBLIC DOMAIN PROGRAM DEVELOPED BY GESSLER, SJOSTROM & WALSKI
US ARMY CORPS OF ENGINEERS—WATERWAYS EXPERIMENT STATION
MAX NODE NO = 900 NUMBER OF NODES = 700
MAX PIPE NO = 900 NUMBER OF PIPES = 700

        PIPE NETWORK ANALYSIS AND OPTIMIZATION
        JOB:  CHESHIRE SYSTEM—SCENARIO 5 BASED ON WMGSKL2a 6-18-89

            NODE DATA

            NODE    ELEV.   OUTPUT   E.G.L.   PR. HEAD   PRESSURE
            NO.     FT.     GPM      FT.      FT.        PSI

            1       115.7            430.0    314.3      136.2
            2       210.5   49.      423.0    212.5       92.1
            3       185.1            418.3    233.2      101.0
            4       136.4            419.3    282.9      122.6
            5       252.4   60.      414.9    162.5       70.4
            6       193.9   25.      415.0    221.1       95.8
            7       177.2   81.      415.2    238.0      103.1
            49      393.0   –6.      415.0     22.0        9.5      SUPPLY

            PIPE   NODES      DIAM.   LENGTH    COEF   FLOW     VEL.     HEAD
            NO.    FROM TO    IN.     FT.              GPM      FT/SEC   LOSS

            1      1    2     16.0    4312.0    110.   1404.0   2.2      6.9
            2      4    3     16.0    6209.0    110.   408.     .7       1.0
            3      6    5     12.0    6759.0    110.   44.      .1       .0
            4      7    8      8.0    3684.0     80.   31.      .2       .3
            5      9    10     8.0    5424.0     80.   156.     1.0      7.9
            6      12   11    12.0    3262.0    110.   197.     .6       .6
```

Figure 11-2. Typical user interface (input and output) from a model in the 1980s

Ease of Use

Ease of use is another subjective factor that is important to many users. The trend is toward models that are relatively easy to use. Online help and GUIs' intuitive and standard designs contribute to ease of use. An excellent source of information on this topic is through other users. Since this topic is quite subjective, it is best to communicate with others whose general background and modeling level approximate yours.

Components Represented

In general, a network model should include all primary components that make up the water distribution system. The model should allow flexible representation of tanks, valves, pumps, and control statements and should explicitly represent the full variety of valves (pressure reducing, pressure sustaining, flow control, etc.) that are commonly used in distribution networks. Pump characteristics are generally represented by a pump curve (flow versus head), although there may be subtle differences in the exact method used by the model. *Control statements* refer to the way that the user can tell the model how components should operate under dynamic conditions. For example, all models should permit a pump to be turned on or off by a clock or the water level in a tank. Some models allow compound control statements in which a component is controlled by the state of multiple components (i.e., the use of "and/or" statements). This capability might be critical in a specific situation. When a system contains a variety of components or controls, the ability of the alternative models should be specifically checked before selecting a model for that particular use.

Water Quality Modeling Characteristics

Hydraulic–water quality models vary widely in water quality capabilities. The most useful models have extended-period simulation (EPS) modeling capability. Some variations in water quality modeling characteristics include the capability to model conservative substances, substances that decay (or grow) according to a first-order function, age of water, and coupled constituents (i.e., the rate of change in one constituent depends on the concentration of a second constituent). Disinfection residual kinetics is frequently of interest, and it is important to be able to represent chlorine consumption due to wall demand. The incorporation of the capability to model multiple reactions in the distribution is desirable.

Robustness

Robustness is the ability of a model to find a correct solution under a wide range of situations. Since all dynamic water distribution system models use a numerical scheme, the convergence and stability of the model is important. Convergence problems occur most often in systems with multiple valves (i.e., pressure-reducing valves), in which each valve influences the operation of another valve. For such complex systems, a small-scale test to find which models handle these problems best is suggested prior to selection.

Size Constraints

Most models have some limitation on the number of links, nodes, and components that may be simulated. Such constraints may depend on the memory capacity of the computer being used. In some cases, size may not be an absolute constraint but large systems may result in unacceptably slow performance. The general trend in models is toward supporting simulation of larger and larger systems.

Support

Technical support may be a very important issue for the inexperienced modeler or modeler who is unfamiliar with a newly acquired modeling package. Support requirements may range from none to extensive telephone, e-mail, or on-site support. Continuing support may be part of the initial purchase price of the package, or a separate annual support contract may be required. The type of support available may range from use of the model to problem definition and analysis. The quality of support may also vary among vendors, and discussion of this issue with other model users is recommended.

Documentation

Virtually all models available for public use contain some form of documentation. The trend is toward emphasizing on-line help as a supplement or replacement for traditional printed material. A typical Windows help screen from the EPANET model is shown in Figure 11-3.

Speed

The speed of a model may be measured in terms of the computer time required to run an application of the model and/or the time to graphically display the results. Speed depends on the inherent speed of the computer, but with the trend toward faster computers, this has become less of an issue. If a relatively small system (i.e., less than 500 pipes) is being modeled in a steady-state mode or for short periods (less than 24 hr), most models will run in a few minutes, thus speed may not be an overriding selection factor in any case. However, for larger systems and/or long simulations, run times may be measured in hours, and the inherent speed of the model may then become a very important factor.

Hardware Requirements

Most models operate on a standard personal computer (PC) with moderate random access memory (RAM) and speed. However, some models require more memory or greater speed than others for proper operation. Though the trend in the industry is toward more powerful machines, some models may require users to upgrade their present computers or to purchase new computers.

Database/GIS/CAD/SCADA Integration

Water distribution system models can be used to analyze design and operational scenarios. Frequently the information needed to apply the model or the results generated by the model come from or are used by external packages, such as database management systems, GIS, CAD, or SCADA. The model may be integrated with such packages through tight linkages or by sharing files. Some models operate from within a proprietary CAD or GIS package while others have a stand-alone graphical package provided with the modeling software. Such tight integration may provide for a very smooth interface but requires the user to understand the CAD or GIS package.

Flexibility

Often, some desired feature is not currently supported by a model. In most cases, the user is totally constrained by the capabilities of the existing packages. In some packages, however, some flexibility is possible to modify or expand the analysis and reporting capabilities. Complete flexibility to modify the model is only possible if the model source code is available (generally the case only for public domain models).

Training

Training materials for models may include books, manuals, videos, and hands-on training courses. For many software packages, open training courses are offered by the model distributor. Many of these distributors will also provide special classes either at the distributors' offices or in the office of the client. In most cases, these courses provide some general background on modeling but usually emphasize the specific hands-on training for a specific model.

Input File Section: [TANKS]
Definition Sketch Examples

Purpose:
Describes each storage tank or reservoir in the network

Formats:
node elev (initlevel minlevel maxlevel diam [minvol])

Parameters:

node	node ID
elev	elevation of bottom of tank where water level is zero, ft (m)
initlevel	initial water level above tank bottom, ft (m)
minlevel	lowest allowable water level, ft (m)
maxlevel	highest allowable water level, ft (m)
diam	tank diameter, ft. (m)
minlevel	volume of water below minimum level, cu ft (cu m)

Remarks:
One line should appear for each tank or reservoir.

For reservoirs you need only enter the node ID and elevation. By definition, the water surface in a reservoir remains at a fixed elevation while it varies in a tank as flows enter or leave.

Water surface elevation in tanks equals the bottom elevation plus water level. Tanks are assumed to be cylindrical between their minimum and maximum levels. Non-cylindrical bottom sections can be accommodated by supplying the volume of the section as the last parameter on the line. For non-circular tanks, use a diameter equal to 1.128 times the square root of the area.

A [TANKS] section is required.

Definition of Tank Levels

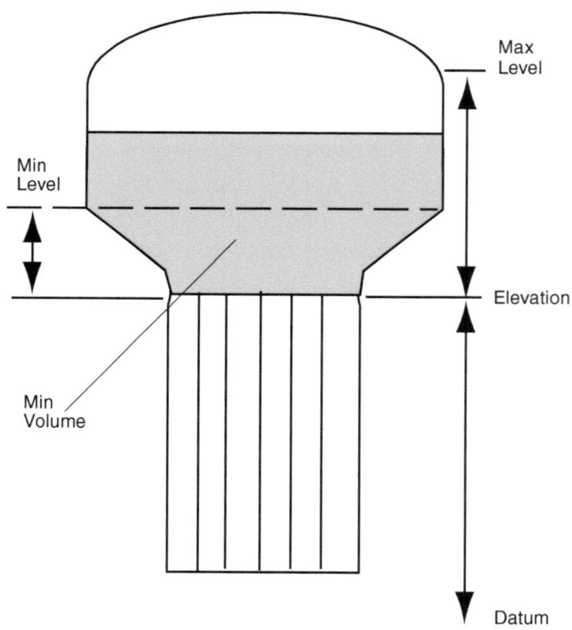

Figure 11-3. Typical help screen from EPANET

Review of Available Models

Many water distribution system models are available for performing hydraulic analysis. Many of these models have been specifically developed or modified to accommodate water quality modeling. Table 11-1 contains a list of water distribution system models available at the time of this writing that have water quality modeling capabilities. The water quality modeling field is growing and new models are continually being introduced.

Table 11-1. Available hydraulic and water quality network modeling software packages (adapted from Panguluri et al. 2005)

Network Modeling Software	Company	EPANET Based	Web Site
AQUIS	7 Technologies		www.7t.dk/aquis
EPANET	EPA	X	www.epa.gov/ord/nrmrl/wswrd/epanet.html
InfoWater H_2ONET/H_2OMAP	Innovyze	X	www.innovyze.com
InfoWorks WS	Innovyze		www.innovyze.com
MikeNet	DHI, Boss International	X	www.dhisoftware.com/mikenet
Pipe2000	University of Kentucky		www.kypipe.com
PipelineNet	SAIC, TSWG	X	www.tswg.gov/tswg/ip/pipelinenettb.htm
STANET	Fisher-Uhrig Engineering		www.stanet.net
SynerGEE Water	Advantica		www.advantica.biz
Wadiso	GLS Engineering Software	X	www.wadiso.com
WaterCAD/WaterGEMS	Haestad Methods	X	www.haestad.com

EPANET

Most of the discussion in this book has focused on the use of EPANET, which is a public sector hydraulic–water quality model. EPANET is a computer program that performs extended-period simulation (EPS) of hydraulic and water quality behavior within drinking water distribution systems. It tracks the flow of water in each pipe, the pressure at each pipe junction, the height of water in each storage tank, and the concentration of a substance throughout a distribution system during an extended-period simulation. In addition to substance concentrations, water age and source tracing can also be performed. The water quality module of EPANET is equipped to model such phenomena as reactions within the bulk flow, reactions at the pipe wall, and mass transport between the bulk flow and the pipe wall. Complete documentation on the use of EPANET on a personal computer under DOS (disk operating software) and Microsoft Windows and the program itself is available through the Internet USEPA home page. The Internet address is http://www.epa.gov/nrmrl/wswrd/dw/epanet.html.

Under Windows, the user is able to edit EPANET input files, run a simulation, display the results on a color-coded map of the distribution system, and generate additional tabular and graphical views of these results. A new extension of EPANET, EPANET-MSX (Multi

Species eXtension), provides very useful flexibility for modeling a wide range of contaminant scenarios.

Setup

Once a model is chosen, the initial step is to install the model on the computer and to become familiar with its capabilities and limitations. Installation is generally straightforward. This familiarization process may be accomplished through formal training courses or trial and error using example problems. However, before full-scale application of the model, the user should become confident in his or her ability to represent the network and to interpret results from the model.

Network Representation

Engineering analysis of water distribution systems is frequently limited to the solution of the hydraulic network problem, i.e., given the physical characteristics of a distribution system modeled as a node–link network and the demands at nodes (junctions where network components connect to one another), the flows in links (a connection of any two nodes) and head at all nodes of the network are determined. This problem is formulated as a set of simultaneous nonlinear equations, and a number of well-known solution methods exist, many of which have been coded as computer programs known as *hydraulic network models* (Cesario 1995).

Mathematical methods for analyzing the flow and pressure in networks have been in use for more than 50 years. They are generally based on well-accepted hydrodynamic equations. Computer-based models for performing this type of analysis were first developed in the 1950s and 1960s and greatly expanded and made more available in the 1970s and 1980s. Currently, dozens of such models are readily available for use on personal computers.

Hydraulic models were developed to simulate flow and pressures in a distribution system either under steady-state conditions or under time-varying demand (extended-period simulation) and operational conditions. Hydraulic models may also incorporate optimization components that aid the user in selecting system parameters that result in the best match between observed system performance and model results.

As mentioned, all water distribution system models represent the network as a series of connected links and nodes. Links are defined by the two end nodes and generally serve as conveyance devices (e.g., pipes), while nodes represent point components (e.g., junctions, tanks, wells, and treatment plants). In most models, pumps and valves are represented as links. Water demands are aggregated and assigned to nodes; an obvious simplification of real-world situations in which individual house taps are distributed along a pipe rather than at junction nodes.

However, network representation is more of an art than a science. Many networks are skeletonized for analysis, which means that they contain only a representation of selected pipes in a service area. A minimal level of skeletonization should include all pipes and features of major concern. Nodes are usually placed at pipe junctions, where pipe characteristics may change in diameter, *C*-value, or material of construction. Nodes may also be placed at locations of known pressure or where pressure valves are desired.

Network components include pipes, reservoirs, pumps, pressure relief valves, flow control valves, altitude valves, check valves, and pressure-reducing valves. For steady-state modeling, a key input is information on water consumption. Usage is assigned to nodes in most models and may be estimated several different ways. Demand may be estimated by a count of structures of different types using a representative consumption per structure, meter readings, and the

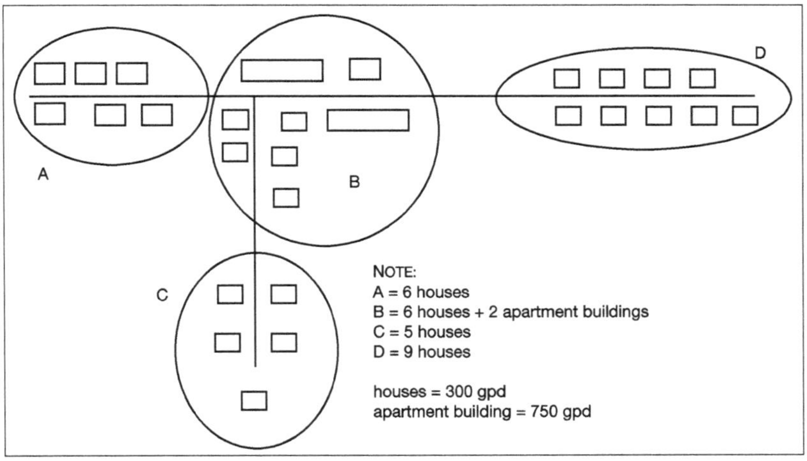

Figure 11-4. Technique for estimating residential water use

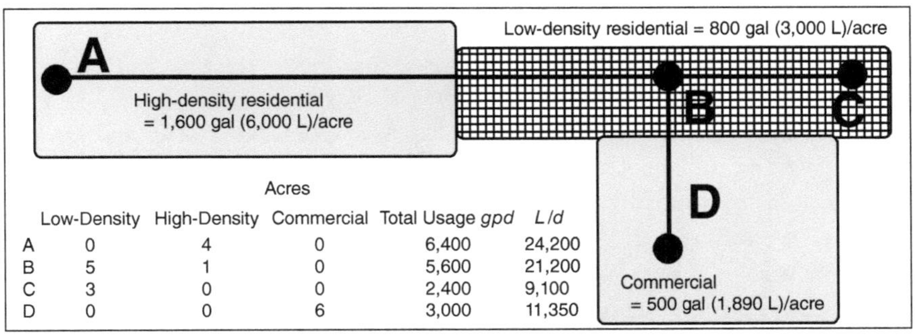

	Acres			Total Usage	
	Low-Density	High-Density	Commercial	*gpd*	*L/d*
A	0	4	0	6,400	24,200
B	5	1	0	5,600	21,200
C	3	0	0	2,400	9,100
D	0	0	6	3,000	11,350

Figure 11-5. Technique for estimating residential, commercial, and residential water demand

assignment of each meter to a node and to general land use. A universal adjustment factor should be used so that total usage in the model corresponds to total production. Figures 11-4 and 11-5 illustrate two methods of estimating water use.

Building a network model, especially if a large number of pipes are involved, is a complex process. The following information is needed to construct a hydraulic model:

- Characteristics of the pipe network components (pipes, pumps, tanks, valves).
- Water use (demands) assigned to nodes (temporal variations required in EPS).
- Topographic information (elevations assigned to nodes).
- Control information that describes how the system is operated (e.g., mode of pump operation).
- Solution parameters (e.g., time steps, tolerances as required by the solution techniques).

Construction of the pipe network and its characteristics may be done manually or through use of existing spatial databases stored in GIS or CAD packages. Most commonly, GIS or CAD packages are used in this process and are described in more detail elsewhere. The initial step in constructing a network model is to identify pipes to be included in the model. Nodes are usually placed at pipe junctions; at major facilities (tanks, pumps, control valves); or where pipe characteristics change in diameter, *C* value, or material of construction. Nodes may also be placed at locations of known pressure, at sampling locations, or at locations where water is used (demand nodes). The required pipe network component information includes the following:

- Pipes (length, diameter, roughness factor).
- Pumps (pump curve).
- Valves (settings).
- Tanks (cross section information, minimum and maximum water levels).

Water consumption or water demand is the driving force behind the operation of a water distribution system. Any location at which water leaves the system can be characterized as a demand on the system. Water demands are aggregated and assigned to nodes, which represent an obvious simplification of real-world situations in which individual house taps are distributed along a pipe rather than at junction nodes. It is important to be able to determine the amount of water being used, where it is being used, and how this usage varies with time (Walski et al. 2003). In order to use a model in the extended-period simulation (EPS) mode, information on temporal variations in water usage over the period being modeled is required. Spatially different temporal patterns can be applied to the individual network nodes. The best available information should be used for developing temporal patterns in order to make EPS most effective. For example, some users may have continuous water metering data, while others may use literature values as a first approximation for estimating residential temporal patterns. Temporal patterns also vary with climate. For example, lawn watering in summer months will cause a spike in usage of water during that time period.

We know that most phenomena and behavior vary over time in the real world. Models that assume no variation over time are referred to as steady-state models, which provide a snapshot at a single point in time. In distribution system models, steady-state models assume that demands are constant and operations are constant (e.g., tank water-level elevations remain constant and pumps are either on or off at a constant speed). Though such assumptions may not be valid over long periods of time, steady-state models can provide some useful information concerning the behavior of a network under various representative conditions, including fire demand, nighttime low demands, and so on.

Models that allow for variations over time are referred to as *temporally dynamic models*. Most network distribution hydraulic models incorporate temporal variation by stringing together a series of steady-state solutions and refer to this method as extended-period simulation (EPS). For example, in the EPS mode, demand at a node may be assumed to be 100 gpm (378 L/min) from 2:00 p.m. to 3:00 p.m. and then 150 gpm (567 L/min) from 3:00 p.m. to 4:00 p.m. Thus, one steady-state solution may be assumed from 2:00 p.m. until 3:00 p.m. and a different solution may be calculated from 3:00 p.m. to 4:00 p.m. Further temporal dynamics may be incorporated by checking the water level in a tank and, if the water level reaches a maximum allowable level at 3:20 p.m., then another steady-state solution is started at 3:20 p.m. with the tank discharging instead of filling. Though the EPS solution does introduce some approximations and totally ignores the transient phenomena resulting from sudden changes, such as a pump being turned on, these more refined assumptions are generally not considered significant for most distribution system studies. Figure 11-6 illustrates systemwide demand variation.

Some of the characteristics of the EPS mode follow:

- Requires information on temporal variations in water usage over the period being modeled.
- Permits temporal patterns to be defined for groups of nodes (spatially different patterns can be applied to a given node).

Figure 11-6. Illustration of system-wide variation

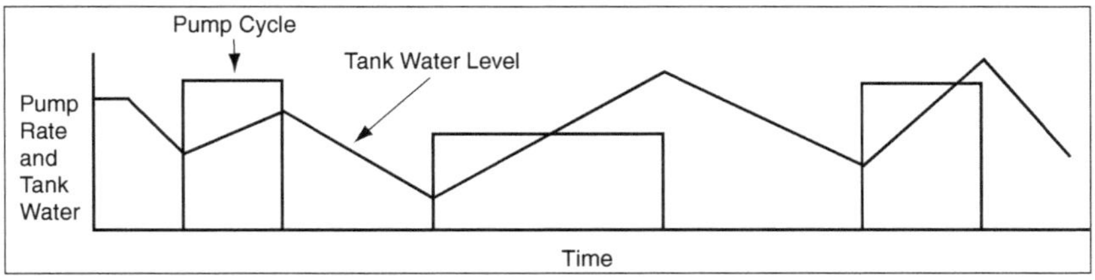

- Uses the best available information on temporal patterns that can be estimated (for example, some users have continuous meters).
- Can sometimes use literature values for a first guess at residential patterns (which may vary with climate).
- Can use analysis of information from supervisory control and data acquisition (SCADA) to estimate systemwide temporal pattern.

In some cases, information from SCADA systems can be used to estimate systemwide temporal patterns. A typical hierarchy for assigning demands includes the following:

- *Baseline demands:* Baseline demands usually correspond to consumer demands and unaccounted-for water associated with average day conditions. This information is often acquired from a water utility's existing records, such as customer meter and billing records. Although the spatial assignment of these demands is extremely important and should include the assignment of customer classes such as industrial, residential, and commercial use, actual metering data should be used when available.
- *Seasonal variation:* Water use typically varies over the course of the year with higher demands occurring in warmer months. When developing a steady-state model, the baseline (average day) demand can be modified by multipliers in order to reflect other conditions such as maximum day demand, peak-hour demand, and minimum day demand.
- *Fire demands:* Water provided for fire services can be the most important consideration in developing design standards for water systems. Typically, a system is modeled corresponding to maximum-use conditions, with needed fire flow added to a single node at a time. It is not uncommon for a requirement that multiple hydrants be flowing simultaneously.
- *Diurnal variation:* All water systems are unsteady because of continuously varying demands. It is important to account for these variations in order to achieve an adequate hydraulic model. Diurnal varying demand curves should be developed for each major consumer class or geographic zones within a service area. For example, diurnal demand curves might be developed for industrial establishments, commercial establishments, and residences.

Large users such as manufacturing facilities may have unique usage patterns. Early software packages limited the number of pipes that could be included due to computer storage restrictions.

This led to the concept of *skeletonizing* a network or including only those pipes that were considered to be the most important. The degree (or level) of skeletonization that is acceptable should depend upon the ultimate use of the model. For example, master plans and energy studies might be based on the use of skeletonized networks. Other applications, such as water quality modeling and designing flushing programs, require a model that includes more pipes. Though no national standard is in place for skeletonization, the USEPA draft guidance issued for modeling to support the IDSE under D/DBPR2 suggests inclusion of the following when developing a network schematic (USEPA 2003):

- At least 50 percent of total pipe length in the distribution system.
- At least 75 percent of the pipe volume in the distribution system.
- All 12-in. diameter and larger pipes.
- All 8-in. and larger pipes that connect pressure zones, influence zones from different sources, storage facilities, major demand areas, pumps, and control valves, or are known or expected to be significant conveyors of water.
- All 6-in. and larger pipes that connect remote areas of a distribution system to the main portion of the system.
- All storage facilities with controls or settings applied to govern the open/closed status of the facility that reflect standard operations.
- All active pump stations with realistic controls or settings applied to govern their on/off status that reflect standard operations.
- All active control valves or other system features that could significantly affect the flow of water through the distribution system (e.g., interconnections with other systems, valving between pressure zones).

Topographical Inputs

Hydraulic models use elevation data to convert heads to pressure. Actual pipe elevations should be used to establish the correct hydraulic grade line. Elevations are assigned to each node in a network where pressure information is required. Various techniques are used to determine elevation information including the following:

- *Topographical maps:* Paper topographical maps produced by the United States Geological Survey (USGS) or other local agencies may be used to manually interpolate elevations for nodes.
- *Digital elevation models (DEMs):* USGS and other agencies produce digital files containing topographical information.
- *Global positioning systems (GPSs) or other field survey methods:* Standard field surveying techniques or modern surveying methods using a GPS satellite can be used to measure elevations at nodes.

Model Control Inputs

In order to apply an EPS model, it is necessary to define a set of rules that tells the model how the water system operates. This may be as simple as specifying that a particular pump operates from 7:00 a.m. to 10:00 a.m. each day or it may be a set of complex "logical controls." These controls may specify which operations such as pump off/on, pump speed, or valve status are controlled using Boolean operators. Factors may be included such as tank water levels, node pressures, system demand, and time of day (Grayman and Rossman 1994).

Calibration/Validation

Calibration and validation are discussed together in this section because they are so closely interrelated. In the calibration step, various model parameters are adjusted so that the results of the model reflect observed field measurements. In some cases, the network representation may be modified in the calibration step. Calibration is required because of uncertainty in the results of any model application due to approximations associated with the formulation of the model and unknowns in some model parameters. For example, the assumption of complete mixing at all nodes represents an uncertainty. The use of nominal pipe diameters that may be in error due to tuberculation, which tends to reduce the actual diameter of pipes over time, may be an approximation. Three important papers on calibration and validation are by Cesario et al. (1996), Walski (1995), and Ormsbee and Lingireddy (1997).

Calibration is an important part of the "art" of modeling water distribution systems. Model calibration is the process of adjusting model input data (or, in some cases, model structure) so that the simulated hydraulic and water quality output sufficiently mirrors observed field data. As mentioned earlier, one way of viewing calibration is to think of a TV screen showing observed and predicted values. Calibration is the process of adjusting the picture so that the predicted value provides the closest estimate for the actual values.

Depending on the degree of accuracy, calibration can be difficult, costly, and time-consuming. The extent and difficulty of calibration are minimized by developing an accurate representation of the network and its components. Several methods may be used for calibrating hydraulic and water quality models. These include both traditional hydraulic field methods and the use of tracers. Traditional methods include extensive measurement of flows and pressures throughout the distribution system and/or measurement of *C* factors for representative pipes. Water quality models are used to predict the movement and transformation of substances, but all distribution system water quality models depend on hydraulic models to provide information on the direction and magnitude of pipe flows. If inaccuracies exist in flow and/or velocity values provided by the hydraulic model, then the water quality predictions will also be inaccurate.

Another calibration method is to use water quality tracers. Naturally occurring or added chemical tracers may be measured in the field and the results used to calibrate hydraulic and water quality models. A commonly used tracer is fluoride. It is conservative (does not degrade), safe, and can usually be added (or normal feed can be curtailed), and the movement can be traced in the system using hand-held analyzers. For conservative tracers, adjustments may be made primarily in the hydraulic model to adequately match the predicted and observed concentrations. Another calibration technique is to measure predicted tank levels derived from computer simulations against actual tank level during a given period of record. For example, using data from SCADA systems or from on-line pressure and tank-level recorders, flows can be adjusted in the simulation model until they match the actual tank-level information. Appendix A describes several extensive tracer studies conducted by the USEPA in conjunction with the Greater Cincinnati Water Works.

Model Application, Analysis, and Display

Three important elements involved in applying water quality models effectively are:

- An understanding of the effect that network configuration has on propagation of substances and contaminants.
- Sampling and analysis procedures.
- Design of the field sampling program.

One of the initial steps in applying a water quality model to a given system is to establish a database for the service area. First, the pipe network is defined. This involves acquiring information on water usage and node elevations and establishing a link–node file that contains various types of information required to model and display the distribution system. The size and configuration of the network are also important in designing a field study.

Three general network configurations that might be studied include the following: isolated pipes; small and isolated portions of a distribution system; and larger pipes in all, or major portions, of a large distribution system (as illustrated in Figure 11-7). No matter

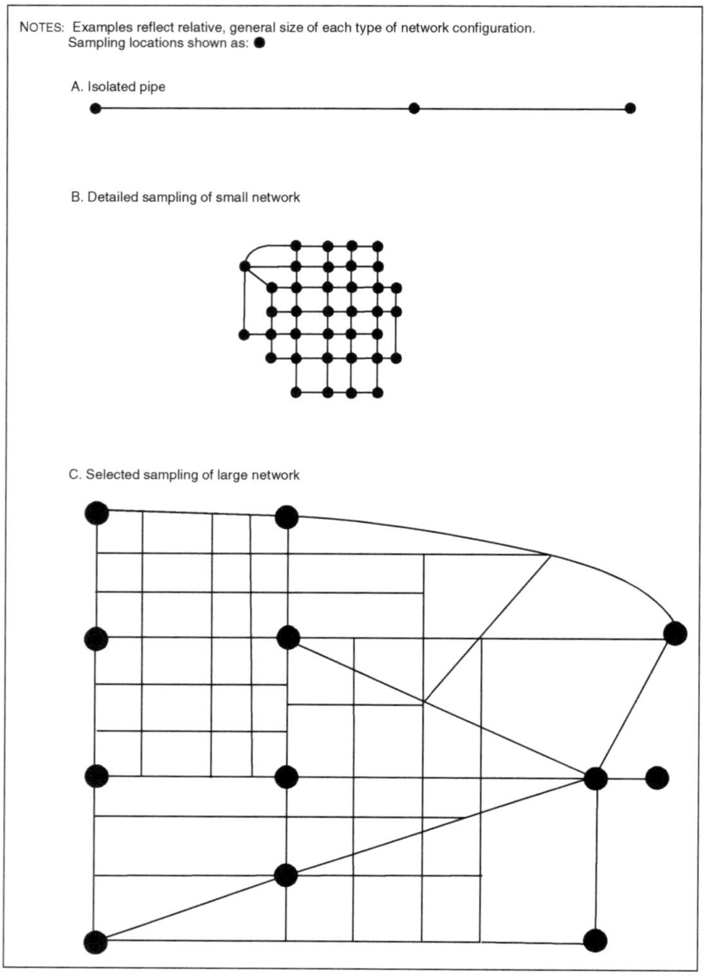

Figure 11-7. General network configurations

the size of the system, the objective is to collect sufficient information so that a model can be calibrated.

If a single pipe is to be studied, it may be viewed as a highly controlled experiment. A pipe of sufficient length should be selected so that variables such as chlorine decay can be measured and so that pipe velocities and characteristics (e.g., diameter, material, and conditions) are known and remain constant over a length of the pipe. Pipes of small diameter (i.e., <12 in. [.305 m]) would be desirable if, for example, the objective is to assess the impact of wall demand. Measurements must be taken at the upstream and downstream ends of the pipe (and, if possible, at intermediate points), and flow rates and pipe diameter must be known in order to determine travel times. By studying these pipes under different flow conditions over several seasons, site-specific changes in such variables as total trihalomethane (TTHM) formation or chlorine decay may be determined.

Frequently, a type of configuration to be considered is a small network or an isolated portion of a larger network where detailed sampling is performed at a large number of locations. Understanding temporal and spatial variations in water usage is very important in understanding the operation of such systems. Therefore, special care must be taken to ensure that good estimates for water usage are available throughout the study area. If sufficient data are available, model variables may be adjusted to reflect field conditions and possibly to develop a relationship between pipe characteristics (such as diameter and material) and decay and/or formation rates for the variable being studied. The level of expertise required for working with a database derived from this network configuration is greater than with an isolated pipe.

Still another configuration to be studied is a large network in which sampling is performed at selected sites within the system. The difference between this configuration and the small network configuration is that the ratio of sampling points to the number of pipes in the network is quite small for the large network study. These data are also useful in adjusting or validating previously developed models.

Overview of Sampling and Analysis Procedures

A general overall approach to designing sampling programs has evolved during the course of the studies described in this book. The effort involves four major steps:

1. Development of a detailed sampling plan.
2. Performance of the field sampling study.
3. Postsampling data analysis and preparation of a report documenting the field study.
4. Analysis of the field data and development (parameterization) of the network model.

Perhaps the most critical step in performing a field study is the development of a sampling plan. The plan should reflect the specific needs and goals of the study and reflect workforce and resource requirements and constraints. Generally, such a plan should err on the side of too much detail but at the same time provide sufficient flexibility to allow for the unexpected events that always occur in field work. A written sampling plan should be developed and reviewed by appropriate personnel.

Planning and Preparing for a Sampling Study

The sampling plan serves as a detailed, written description as to how the sampling study is to be performed. It contains a detailed description of all aspects of the study, including contingency plans addressing unexpected occurrences.

As part of the sampling plan, the network hydraulic–water quality model may be applied to the sampling area using the operating conditions likely to be in effect during the field study. Based on predicted results, sampling locations and sampling frequency can be established. In some cases, alternative operating conditions are simulated in order to determine their effect on the planned sampling strategy. Even though the model may not be fully calibrated and, in fact, may only be an approximation of the actual system, the information gained by predicting the behavior of the system under the expected sampling conditions is frequently helpful. Sampling plans should address the specific issues discussed in the following sections.

Sampling Locations

The actual sampling study should follow the sampling plan as closely as possible. If possible, field data should be examined frequently during the course of the study in order to identify possible problems and to determine deviations from the anticipated results that may require modifications in the plan. Special care should be taken to clearly record all data and to document all aspects of the study.

Specific sampling sites should be carefully and accurately delineated in the plan. Typically, samples can be taken from permanent sampling taps, hydrants, public buildings, businesses, residences, or water utility facilities. For all types of sampling sites, the taps should be allowed to run continuously or care should be taken to be sure that they are adequately flushed prior to each sample to make sure that water is being taken from the main. Each type of sampling location has advantages and disadvantages, as outlined below.

Permanent sampling taps generally provide the best sampling method because they are designed to minimize flushing time, to provide access at all times, and to lessen the chances of sample contamination. However, such devices are relatively expensive to purchase and install and would not be cost-effective for a one-time use. Hydrants provide good accessibility and are widely available. Prior to use, they should be flushed well, and a sampling tap should be installed. Maps and records that describe the system should be consulted prior to use to determine which main the hydrant is connected to and the length and diameter of the connector. Travel times from the main to the hydrant should be calculated based on flow rates, estimated distance, and diameter to determine required flushing time prior to each sample.

Other water-utility-owned facilities can also serve as good sampling locations because of 24-hr accessibility. Public buildings (especially firehouses) are also good candidates because of their general accessibility. Residences and businesses should be considered only as a last resort because of limited accessibility, liability, and inconvenience to the water customer.

Sampling Frequency

Frequency of sampling should be specified in the sampling plan. If samples are to be taken manually, then a "circuit" is generally established in which samples are taken sequentially by a sampling crew. The circuit should be clearly marked on a map. The time required to sample at a single station and over the entire circuit can be estimated by a preliminary test of the circuit or by estimating the times associated with each aspect of the task (e.g., time to flush the sampling tap, time to take and analyze a sample, time to travel between sites, etc.). Adequate rest time (e.g., bathroom stops and meal breaks) or availability of relief crews should be added to sampling time estimates. Additionally, the plan should include contingency plans if the sampling crews fall behind their anticipated schedule. These plans could include the realignment of the circuits, addition of more crew members, or acceptance of less-frequent sample collection.

An alternative to, or variation on, the circuit approach involves sampling along the tracer front. Using this approach, intensive sampling is initiated during the transition period when the tracer is first approaching a sampling station. This approach is logistically more difficult and requires essentially instantaneous analysis of the data so that decisions to move to the next set of stations can be intelligently made.

System Operation

The operation of a water system can significantly affect the movement of a tracer through the system. During a sampling study, system operating conditions should be viewed as a controlled set of variables rather than a random set of occurrences. Generally, the system operation should either reflect normal operating conditions or be set to represent a desired condition. For example, one may wish to study the system under tank-filling conditions so as to eliminate the effects of old water being discharged to the distribution system. For either normal operation or a special condition, preconditioning the system prior to the start of sampling may be necessary so the system operates in the desired fashion. This might require operating the system so that the water levels in selected tanks are at particular levels, ensuring that the system can be operated in the desired manner during the study. Additionally selecting a particular season to perform the study may allow desired operational flexibility. For example, water systems are often stressed during hot summer months, making operational modifications difficult.

Preparation of Sampling Sites

Prior to actual sampling, the site should be adequately prepared. Preparation may include testing of faucets in hydrants, calculation of required flushing time, notification of owners, and marking the sites for easy identification.

Sample Collection Procedures

Procedures for collecting samples should be specified in the plan. These procedures may include required flushing times, methods for filling and marking sample containers, listing reagents or preservatives to be added to selected samples, methods for storing samples, and data logging procedures.

Analysis Procedures

Samples taken in the field may be analyzed at the sampling site, at a field laboratory located in the sampling area, or in a centralized laboratory. Specific procedures to be followed for each type of analysis should be specified in the plan.

Personnel Organization and Schedule

An important part of the sampling plan is a detailed personnel schedule for the sampling study. Ideally, the schedule should include crew assignments and a work schedule for each study participant. Logistical arrangements include lodging for nonresident participants, provision for meals, transportation, and so on. A possible source of assistance during the summer would be college students.

Safety Issues

Many safety issues are associated with sampling studies. These are intensified by round-the-clock sampling and working in unfamiliar areas. The following safety-related concerns should be addressed in the sampling plan: notification of police and other governmental agencies; public notification (e.g., using newspapers or television stations); notification of customers who may be directly affected; issuance of safety equipment, such as flashlights and vests; use of marked vehicles and uniforms identifying the participants as official water utility employees or contractors; issuance of official ID cards or letters explaining their participation in the study; and use of more than one sampler, especially during nighttime or in dangerous areas.

Data Recording

An organized method for recording all data should be devised. A data recording sheet, such as the partial sheet shown in Figure 11-8, should be designed and included as part of the sampling plan. Use of military (24-hr) time is recommended. Data should be recorded neatly in ink and the individual data sheets numbered sequentially. Completed data sheets should not

Study: Name and Location ____________________ **Sheet Number:**__________

MM/ DD	Station ID	Time HHMM	Initials	Temp °C	Chlorine Free	Tot		Sample		FL	Comments

Figure 11-8. Typical data collection form

be kept in the field so as to reduce the chance of loss. Once the data sheets are centrally assembled, they should be copied and the copies kept separately from the originals. Frequently, a sample will be taken and partly analyzed in the field (i.e., for temperature and chlorine residual) and then stored for later analysis (i.e., for fluoride and THMs). Therefore, a method for relating the field and lab samples is needed. Use of date-time-location on both the recording sheet and on the sample container provides a mechanism for identifying samples.

Equipment and Supply Needs

Equipment includes field sampling equipment (e.g., a chlorine meter), safety equipment (vests, rain gear, and flashlights), laboratory equipment, and so forth. Expendable supplies include sampling containers, reagents, and marking pens. As part of the sampling plan, the needs and availability of equipment and supplies should be identified and alternative sources for equipment investigated. Some redundancy in equipment should be planned for, since equipment malfunction or loss is possible.

Training Requirements

Training of sampling crews is essential and should be specified in the sampling plan. Even though crews may be composed of experienced workers, it is important that all crews be trained in a consistent set of procedures. Training topics include sampling locations, sample collection and analysis procedures, data recording procedures, and contingencies. All crew members should read the sampling plan before the training sessions.

Contingency Plans

The old adage "If something can go wrong, then it will" applies to field sampling studies and provides a good basis for contingency planning. Contingency planning should include equipment malfunction, illness of crew members, communication problems, severe weather, unexpected system operation, and customer complaints. All of these events occurred once or more during the sampling studies discussed in this book.

Communications

During a sampling study, key personnel are often distributed throughout the area as follows: samplers at single stations or riding a circuit, laboratory analysts in a permanent or field lab, operations personnel at central control stations, and the study supervisor at any of many locations. In order to coordinate actions during the study and/or to respond to unexpected events, some means of communication is needed. Alternatives include radios (normally available if you are using utility vehicles), cellular phones, walkie-talkies, or the low-tech solution: a person circulating in a vehicle.

Calibration and Review of Analytical Instruments

A specific plan for reviewing and calibrating all instruments is necessary, and this work must be carried out prior to the study. When multiple instruments are used to measure the same parameter (e.g., multiple chlorine meters), each instrument should be calibrated against a standard and compared. A plan to check the calibration of instruments during the study is also essential.

Conducting the Sampling Study

Ideally, conducting a sampling study would simply involve following the detailed sampling plan to completion. In practice, however, unexpected events occur and adjustments must be made in order to meet the study objectives. The most successful studies are well planned and executed in an orderly manner. Ideally, a study should be coordinated through a central location where a study supervisor is located. The supervisor should function as the central coordinator, receiving information from the field, lab, and operations center and then making decisions and disseminating information back to the field. Decisions on changes in sampling schedules or operating procedures should be made from the operating center based on full knowledge of all aspects of the ongoing study. Use of a computerized database system or a manual system for plotting the results of the ongoing study can serve the useful role of showing the behavior of the system and identifying variations to the sampling plan. During a study, it is essential that accurate and complete records be kept at all times. Reliance on memory as to when certain events occurred can lead to inaccurate or incomplete information.

Postsampling Lab Analysis and Preparation of Data Report

During and immediately following the completion of the fieldwork, lab analysis should be completed. When analysis permits, samples can be preserved so that if the resulting data appear inconsistent, tests can be repeated. This is especially true of the fluoride samples, which have a relatively long shelf life. An important step following the fieldwork is the preparation of a data report. The purpose of the data report is to record all pertinent information from the field study and subsequent lab analysis. This includes information on the sampling study, the study area, and results of the lab analysis organized in a usable manner.

All field sampling water quality data should be stored in a computerized database management system (DBMS) in order to facilitate the preparation of a data report and its subsequent use. The data should be used in a relational structure. The data structure is shown graphically in Figure 11-9. In this relational structure, sampling data are uniquely identified by date, time, and sampling station. The primary table holds the key data describing the location, date, time, and sampler, along with the data that is routinely collected and analyzed in the field (chlorine, fluoride, temperature, and so on). Also, a unique identifier is assigned to each sample that undergoes additional laboratory analysis. Using the sample identifier as the link to other tables, the additional analysis is stored in tables corresponding to data analyzed in field laboratories.

Analysis of Data and Development of Model

As soon as possible following completion of the field study, laboratory analysis should be completed and a report prepared containing all aspects of the field work, including all details of the study and the results of all field and laboratory analyses. This report serves as the repository of all information on the field study and, as such, the basis for later analysis.

The final step in the process is the actual analysis of field results and use of the data to develop a water quality model. The results of the analysis might include development of a new model, development of a site-specific relationship between the parameters of an existing model and observed pipe characteristics, or verification of an existing set of parameters or relationships.

A two-step method for performing the analysis, including hydraulic calibration of the network model and development of the water quality model, can be used. Because of the

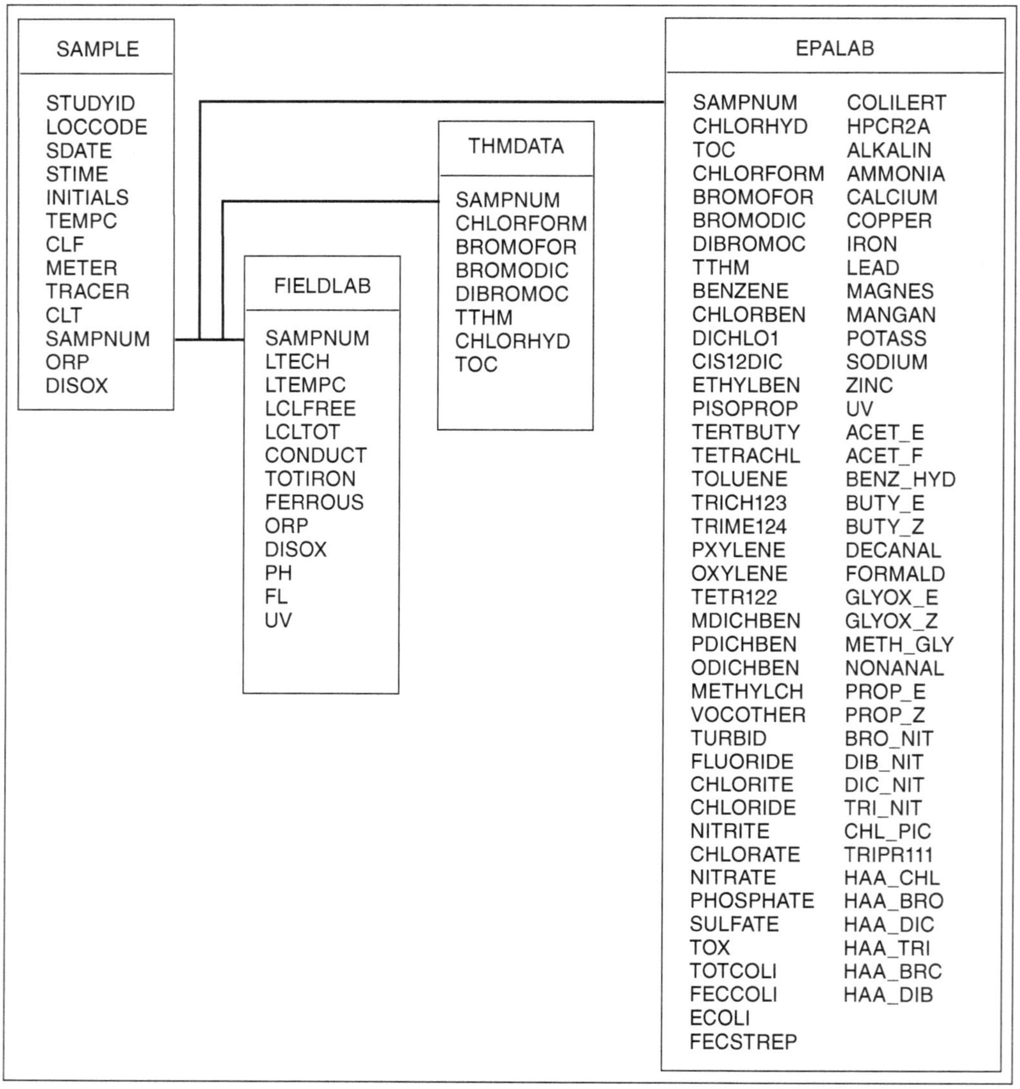

Figure 11-9. Example of relational database structure for water quality sampling data

relative sensitivity of some variables to travel times, it is highly desirable to have a model that has been hydraulically calibrated for the particular sampling study. As discussed earlier, the use of tracers in conjunction with more traditional calibration techniques is preferred because it should provide the most dependable travel time results.

The methods used to adjust model parameters in order to achieve an acceptable match between field tracer data and model results may vary from informal "sensitivity" methods to more formal search methods. Various methods have proven successful in the studies described in this book. Measures of success have included informal visual matches (e.g., "the predicted model results look like they match field results relatively well") to formal numerical measures of goodness of fit. Two formal best-fit measures are illustrated in Figure 11-10. In Figure 11-10, actual data collected are represented by the solid dots, and the best-fit solution is represented by the solid line. The standard error of estimate is a measure of the deviation of predicted from observed results. The "fuzzy" time measure, developed in this study, recognizes

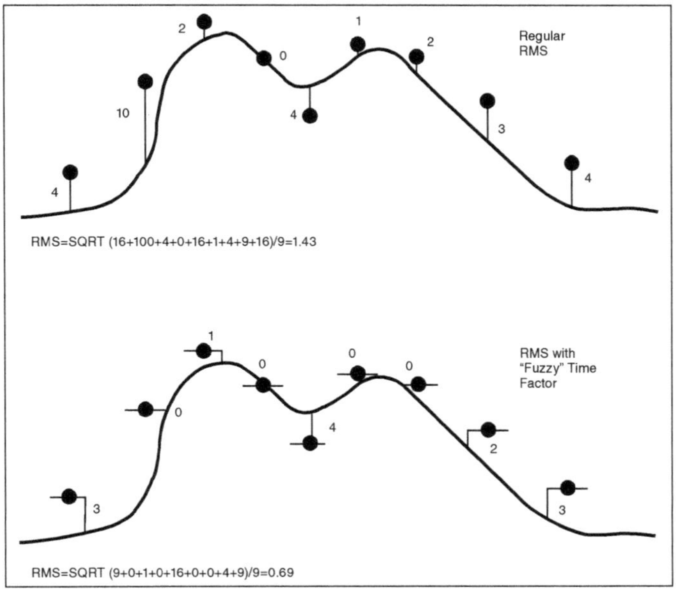

Note: The RMS value of a set of values is the square root of the arithmetic mean (average) of the squares of the original values minus the predicted (or modeled) value

Figure 11-10. Best-fit procedures

that small temporal variations may be acceptable and adjusts for such small temporal shifts prior to the calculation of the standard error of estimate.

In adjusting parameters to achieve an improved match between observed and predicted results, several general rules are followed:

- Parameters should be kept within reasonable ranges. For example, roughness coefficients outside of a range from 50 to 160 should be avoided.
- Adjustment of parameters that have little influence on prediction results is unproductive.
- Parameters should be adjusted, starting with those for which the greatest degree of uncertainty may be expected, and progressing, if needed, to parameters that are assumed to be known with greater certainty.

The following list of model parameters is given in an order that generally approximates increasing levels of certainty. Since there is obviously some subjectivity in defining such an order and variation between sites would be expected, this list should only be viewed as a general guideline:

- *Pipe roughness.* In most applications, pipe roughness is derived from information on pipe material and age. The uncertainty in the estimate of pipe roughness is generally improved when in-situ flow tests have been performed.
- *Temporal water usage patterns.* Most dynamic or EPS models require estimates of average (baseline) water use at each node and then a representation of the temporal water use pattern around the baseline value. Continuous meters are typically unavailable at the point of use, thus water use patterns are estimated based on logic (e.g., schools are in session from 8:00 a.m. until 4:00 p.m., so the heaviest usage is during that period) and/or adjusted so that total water usage by time in the system or subsystem is equal to a known pattern, such as the outflow from a plant. It should be pointed out that as the model representation becomes more detailed to the point that virtually all pipes are included, the uncertainty in temporal patterns at the node level increases because of the stochastic nature of water use at the point of consumption. A possible approach to obtaining demand information would be to use data loggers on a sample of customers to get continuous demand information.
- *Absolute spatial water usage.* Average water use by node is generally estimated using actual meter readings or is based on land-use or dwelling-type estimates. Usually,

significant uncertainty exists in spatial water use that may be adjusted in order to improve predictions. This is especially true in unmetered systems.

- *Pump curves.* The relationship between head and flow through pumps is usually available. However, these curves may reflect the pump condition as installed and can be in error. In-situ pump tests reduce any uncertainty associated with these curves.
- *Valve settings.* Uncertainty exists with regard to specific valve parameters, such as critical pressure settings for pressure-reducing valves. However, more importantly, some valves may not reflect the expected conditions (i.e., valves that are closed or partially closed when they are expected to be open and vice versa). Such occurrences can cause extreme difficulties in calibrating a model, and the user should be ready to try alternative valve settings and to have field crews check existing valve settings.
- *Other system component data.* Other system component data include flows through meters and water level elevations in tanks. Because of malfunctioning or uncalibrated meters, this information, which is routinely considered as correct, may require checking and/or adjustment in order to improve the calibration.
- *Tracer decay characteristics.* A tracer should be as conservative (does not degrade in the system) as possible. However, various chemical and physical reactions can result in the loss of an expected conservative substance. In previous studies, the loss of a fluoride tracer in a system that did not normally fluoridate was observed and substantiated in the laboratory.
- *Pipe diameter.* Most modelers view pipe diameter as a known value that does not need to be modified. However, errors on base maps or encrustation of a pipe can result in significant difficulties in calibrating a model. Examination of pipes that have been replaced can indicate whether or not encrustation is an issue that must be addressed.
- *Tracer concentration measurements.* For systems in which the tracer to be used is not routinely injected, temporary feeding equipment is usually employed. Because such equipment may not function perfectly or the resulting concentrations may vary with flow, frequent concentration measurements should be taken downstream of the injection point and the accuracy of the tracer measurements should be checked at this point and in the system. If uncertainty exists in the measured concentrations, then this parameter may need to be adjusted in the model.
- *Pipe configuration.* Pipe configuration is generally derived from system maps. The primary uncertainty in this case is unknown closed valves or mains that cross but do not intersect or are connected by small-diameter connecting pipes.
- *Elevations.* Nodal elevations are required by models. Generally, the model results are relatively insensitive to minor errors in elevations, thus this parameter should be one of the last to be modified.
- *Pipe length.* Pipe lengths are generally scaled from maps, thus there is minimal uncertainty. Another source might be computerized and database mapping systems.

Once an acceptable level of hydraulic calibration of the model has been achieved, parameter relationships can be established. In previous studies, chlorine decay relationships based on both bulk decay and pipe-wall decay were used, while a first-order growth relationship was used in modeling TTHMs. In establishing chlorine decay values for the bulk decay k_b and the wall demand k_w, various formal and informal search methods were employed.

Best-Fit Values

The following discussion provides some general insights into the procedures used to determine the best-fit parameters, using chlorine decay as an example. Generally, a good starting point for the bulk chlorine decay rate is derived from controlled bottle decay tests. In these tests, samples of the water are stored in headspace-free, darkened bottles at ambient water temperature. Individual bottles are then opened at different times, chlorine residuals measured, and the values plotted against time. For a first-order decay, the resulting curve should be a straight line on semilog paper with the slope representing the decay rate. Frequently, however, the curve is concave, convex, or piecewise linear. Therefore, estimation of the best-fit, first-order decay rate might require some judgment and would be handled by determining a range of decay assumptions to be tested in the model.

Methods for guiding the search for the best-fit values of the bulk and wall decay parameters may be classified as follows:

- *Informal adjustment.* The modeler may informally adjust the parameters relying on experience and judgment to lead to an acceptable set of parameters. This method has been most successful in systems dominated by bulk decay and little wall demand.
- *Relationship between pipe characteristics and wall decay parameters.* It is believed that certain pipe characteristics lead to increased wall demand. For example, pipes of a particular age or material may be more susceptible to chlorine demand at the wall or a relationship may exist between the roughness coefficient and wall demand. Using such postulated relationships, the wall demand parameter may be calculated for each pipe and the resulting predicted chlorine residuals compared to field results.
- *Geographically zoned wall decay parameters.* Pipe location may also serve as a surrogate for various factors that affect wall demand. Methods for grouping pipes and assigning and testing alternative wall demand parameters have been used in many studies (Clark et al. 1994).

Components of a Sampling Study

Field data collection can be characterized according to the following four types: source water data, system data, tracer data, and systems operation data. Each type of data collection technique will be described.

Characterization of Source Waters

A complete characterization of the source water should be performed at each study site. When multiple sources exist, each source is characterized and, if the study area is a significant distance from the source, characterization of samples from one or more locations within the study area should be performed. The constituents analyzed at each site will probably vary slightly based on availability of laboratory facilities and the nature of the water.

Tracer Studies

The formation and decay of contaminants in distribution systems are a function of time. Therefore, velocities in a network must be represented accurately. Traditionally, hydraulic calibration includes matching pressures throughout a network, use of C values derived from field tests, and matching reservoir tank levels and known flows. Though such methods do improve the reliability of velocity predictions in a network model, they do not ensure that the velocities are adequately modeled. For example, it can be shown that if uncertainty exists in actual pipe diameters because of encrustation or poor record keeping, a model that has been calibrated to faithfully reproduce field pressure measurements will not give accurate velocity information.

With tracer methods, a nontoxic conservative constituent is injected into the distribution system over a fixed period, and the movement of the constituent is then monitored throughout the system. The most commonly used conservative tracer is fluoride. In those systems where fluoride is routinely added to the finished water such as the SCCRWA, the fluoride feed was turned off for a period of time, and the resulting fluoride concentrations monitored throughout the study area. The fluoride was subsequently turned on, and the "front" of fluoridated water was then traced through the system. Generally, fluoride was turned off long enough so that the fluoride front would reach the farthest point of the system being studied.

For those systems in which fluoride is not routinely added, provisions can be made to inject fluoride into the water as it enters the study area. Based on experience, it has been observed that for those systems some loss may occur when fluoride is first added to the pipe wall. In order to reduce this effect, it was found that preconditioning of the system by feeding fluoride for a period of a week prior to the study was effective. Fluoride concentrations were measured using electrode methods in the laboratory.

Alternatives to fluoride include constituents normally injected into the system or natural characteristics from multiple sources. In one case (North Marin), sodium hydroxide (NaOH) was added at one water source for pH control. By intermittently turning sodium hydroxide off and on, the movement of water from that source and the other source was monitored. In another system (North Penn), the various water sources had significantly different characteristics (e.g., hardness, THMs, and total to free chlorine ratio). By operating sources at different rates, the water from these systems was traced.

Operating Data

Another type of information required to characterize a water system during a sampling study is system operating data. These data include temporal records for pumping, tank and/or reservoir water level elevations, transfer rates between zones, water usage, and, in some cases, pressures. Information should be collected in either digital or analog form and, in some cases, can be manually read from meters.

Analytical Methods

A key factor in conducting sampling studies is the selection of chemical and physical analytical methods. All methods should be in accordance with accepted standard methods. Some basic field measurements and analytical methods include:

- Chlorine residual measurements
- Colorimetric methods
- Amperometric methods
- Electrode methods
- Trihalomethane measurements
- Liquid–liquid-extraction gas chromatographic method
- Purge-and-trap, gas chromatographic–mass spectrometric (GC–MS) method
- Fluoride measurements
- Sodium 2-(parasulfophenylazo)-1,8-dihydroxy-3,6-naphthalene disulfonate (SPADNS) method (McTigue and Symons 2010)
- Fluoride electrode method

Case Study Applications

Full-scale application of the model may initially involve adjustment of some internal model parameters, such as time step assumptions, and then examination of multiple scenarios. Application of network models to various scenarios can generate large amounts of output. This output is used in the calibration and validation process and to confirm that the model is performing in a reasonable manner. The output is then analyzed to determine how the system responds under the different scenarios to be considered. Many models contain graphic displays and statistical analysis tools or external software packages that may be used in this process.

In order to illustrate some of the calibration methods described earlier and the application issues in this chapter, two examples are discussed. Both examples are based on data taken from a medium-sized Midwestern utility and are discussed in more detail in Appendix A. The two examples are from the same distribution system and part of a skeletonized model extracted from unspecified portions of the distribution system.

The extracted system model was modified and converted to EPANET format. Modifications included: addition of key pipes, updates to consumer demand data, and an interconnection between the case study area and the full system by a fixed grade node (reservoir). These portions of the base model had been previously calibrated using various dynamic calibration methods and were used for routine water utility work. For the purposes of calibration, separate field studies were conducted in each study area.

In both field studies, a food-grade conservative tracer (calcium chloride) was introduced into the system and its movement through the system was monitored by both grab sampling and continuous monitoring (CM) stations installed at key locations in the distribution systems. The CM stations were installed at hydrants that were left partly open for the duration of the study to minimize travel time between the main and sampling location. Each open hydrant was added as a new demand node in the EPANET network model. Additionally, several ultrasonic flowmeters were installed to provide continuous flow measurements at key locations.

Case 1: Small-Suburban Dead-End System

This system is part of a larger pressure zone. It was selected because of the relatively compact size and simple structure, fed by a single supply pipe with no additional storage. As a result, the movement of the tracer was relatively rapid through the system, and it could be monitored with continuous meters placed at several locations. The general layout of this subsystem, the location of the injection site, and the monitoring locations for this study are shown in Figure 11-11.

The calcium chloride tracer was injected as two pulses, a 2-hr pulse, followed by a 2.5-hr period of no injection, and then followed by a higher concentration pulse of 2 hr duration. Chloride concentrations were monitored at a suitable location approximately 100 ft downstream of the injection point with a travel time of approximately 10 min. In preparation for the calibration process, the conductivity readings were converted to chloride concentrations using a relationship developed in the laboratory. Figure 11-12 shows the resulting chloride data set that was used at one location as a basis for evaluating model predictions as part of the calibration process.

The preliminary results indicated some discrepancy between the EPANET-model predicted values and the actual field-measured values, indicating the need for model refinement and recalibration to improve the prediction capability of the EPANET

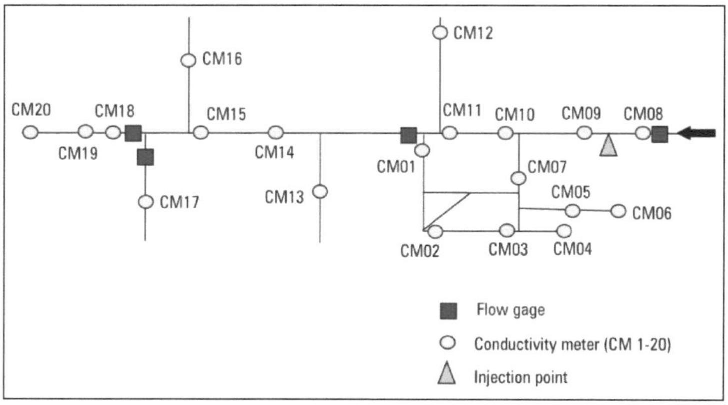

Figure 11-11. Location of injection site and monitoring stations for case study 1 (Panguluri et al. 2005)

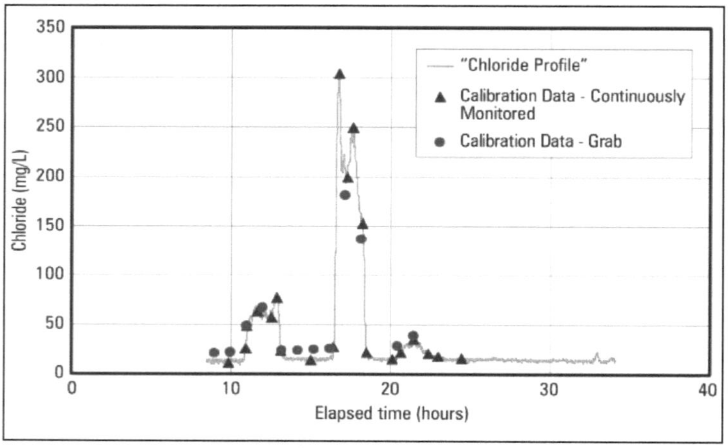

Figure 11-12. Data from chloride data injection study (Panguluri et al. 2005)

model. The results are shown in Figure 11-13 for a continuous monitoring location (CM-18) located on the main feeder pipe. EPANET modeling was performed to evaluate four levels of model refinements. Figure 11-13a (prior to calibration) shows results from a skeletonized EPANET model using the original hourly demand pattern provided by the GCWW and a time-step injection pattern of 60 min. Figure 11-13b shows the same conditions as Figure 11-13a, but a refined 10-min time-step pattern for injection was used along with the conversion of the original hourly demand patterns to 10-min patterns. Figure 11-13c is the same as Figure 11-13b with a refined demand pattern for each node using the field measured flow data, the addition of demand nodes representing water demand of the partially open hydrants, adjustments for a large industrial user of water in the study area (based on data obtained during the study), and residential water billing information provided by the GCWW. Figure 11-13d shows the same conditions as Figure 11-13c with a detailed all-pipe (nonskeletonized) EPANET model. As can seen in Figure 11-13, improvements in the demand estimates and inclusion of the system details in the all-pipe model resulted in a vast improvement in the model's prediction ability for that monitoring location.

Case 2: Large-Suburban Pressure Zone

Similar to the previous example, a field study and calibration exercise were carried out in a large-suburban, pressure zone. This area was selected to demonstrate the application of tracer studies and calibration techniques in a more complex area. The selected area contained multiple pumps and tanks. The selected distribution system area is representative of relatively complex, well-gridded systems found in many larger water systems. The layout of the system, the location of the injection site, and the monitoring locations are shown in Figure 11-14.

Two separate tracer studies were performed in this zone. The first study was used to further calibrate the skeletonized model received from the water utility. The second study served as a validation event to test the veracity of the calibrated model. In the calibration event, the tracer was introduced directly into the main feed line servicing the entire area (characterized by higher flow/higher pressure). In the validation study, the tracer was pulsed. A total of 34 continuous conductivity meters were installed in the system. Four flowmeters were temporarily installed to provide flow measurements at key locations.

During the calibration study, the calcium chloride tracer was injected into the main feed line serving the area for a period of 6 hr. In the validation study, the tracer was pulsed by fill and draw cycles in a storage tank at the same location. In both cases, a target chloride concentration of 190 mg/L or lower was set in order to safely not exceed the 250 mg/L secondary MCL for chloride.

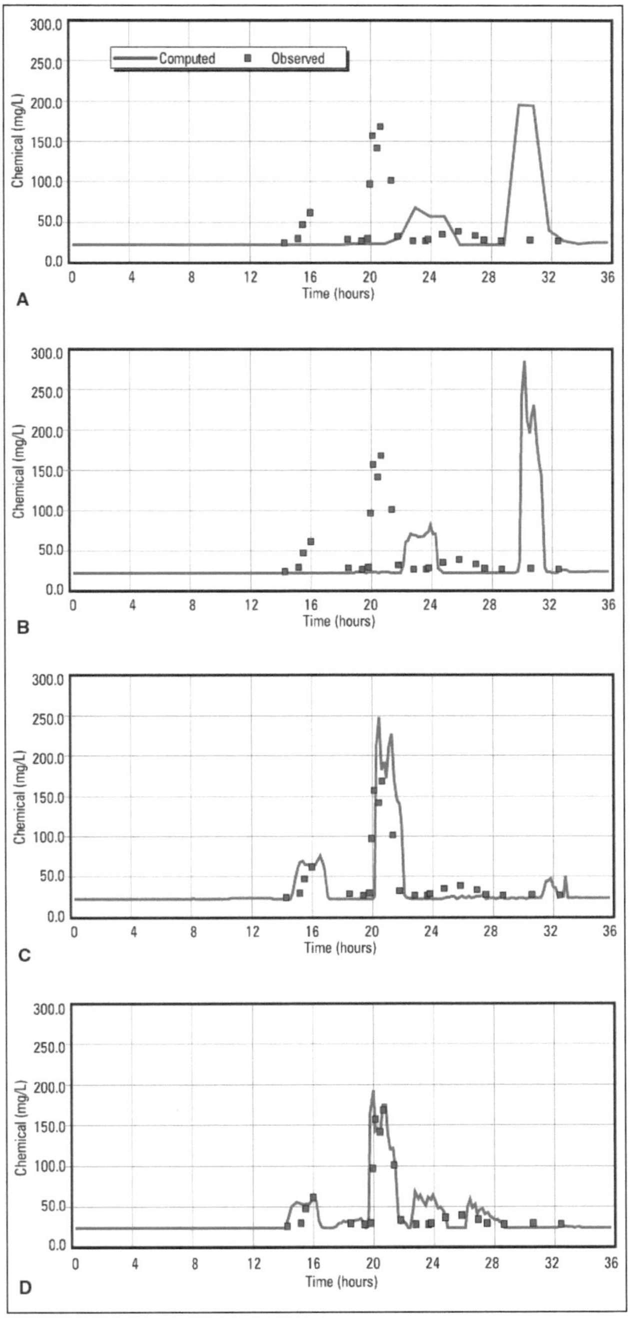

Figure 11-13. Results from a continuous monitoring location (CM-18) located on a main feeder pipe.

During the calibration process, the initial EPANET model simulations were reviewed in detail to determine the flow patterns around various monitoring locations and to attempt to identify causes for discrepancies in the observed and predicted values. A careful examination of the areas of significant discrepancies indicated that these were primarily limited to three geographic subregions within the skeletonized network. Results from Region 2 are discussed here.

In Region 2 (CM52, CM53, CM55 and CM56), the field data indicated that the tracer arrived several hours after the model's prediction. One possible explanation was that this region had lower demand than the average residential demand modeled in this area. The flowmeter data upstream of this location supported this theory as the EPANET predicted flow in this pipe was much higher than the field observed flow (see Figure 11-15). To simulate this possibility, the local demand in this region was reduced by 30 percent in the model. The resultant flow matched the flowmeter data (see Figure 11-16). Also, similar to Region 1, it was found that a potential flow path had again been left out due to skeletonization of the model, which affected CM52. This pipe link was added to the model using the appropriate pipe parameters. Distribution mains between CM55 and CM53 were also found to have been upgraded since the EPANET network model was developed for this area. The EPANET model pipes for this location were updated using the newer information. The model-predicted results improved significantly for this region based on these three adjustments.

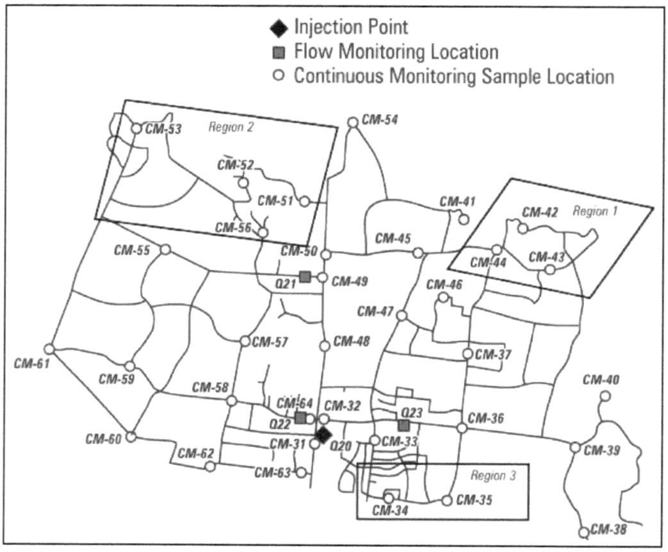

Figure 11-14. System layout for case study 2 (Panguluri et al. 2005)

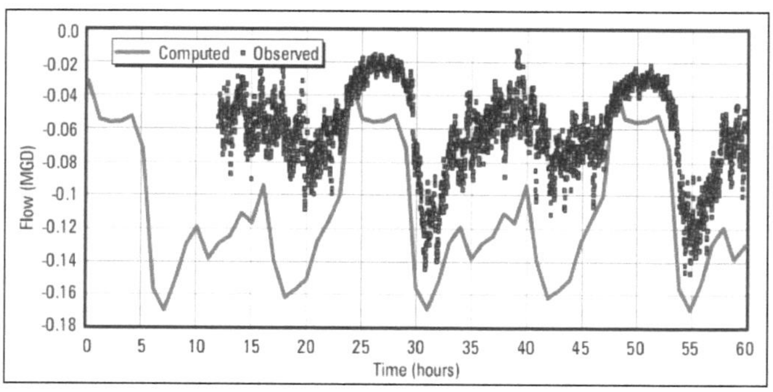

Figure 11-15. EPANET predicted flow versus observed flow (Panguluri et al. 2005)

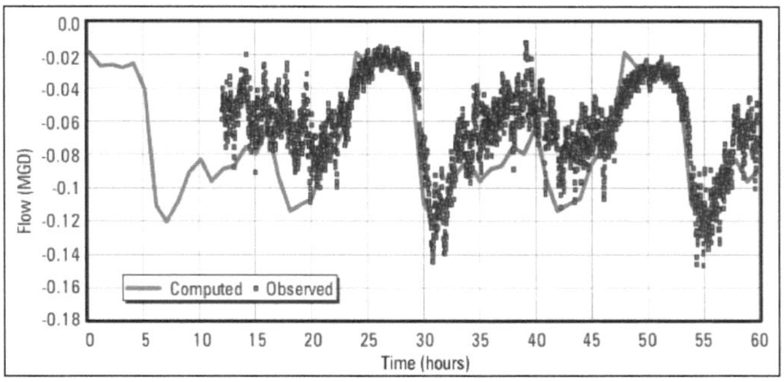

Figure 11-16. EPANET predictions after adjustment (Panguluri et al. 2005)

Typical Model Applications

Appendix B contains specific case study applications of EPANET. A typical example of an EPANET application is presented along with several possible solutions.

References

Cesario, L. 1995. *Modeling, Analysis, and Design of Water Distribution Systems*. Denver, Colo.: AWWA.

Cesario, L., J.R. Kroon, W.M. Grayman, and G. Wright. 1996. New Perspectives on Calibration of Treated Water Distribution Systems. In *Proc. AWWA 1996 Annual Conference*. Denver, Colo.: AWWA.

Clark, R.M., G. Smalley, J.A. Goodrich, R. Tull, L.A. Rossman, J. Vasconcelos, and P. Boulos. 1994. Managing Water Quality in Distribution Systems: Simulating TTHM and Chlorine Residual Propagation. *Jour. Water Supply: Research and Technology—AQUA*, 43(4):182–191.

Clark, R.M., and W.M. Grayman. 1998. *Modeling Water Quality in Drinking Water Distribution Systems*. Denver, Colo.: AWWA.

Grayman, W.M., and L.A. Rossman. 1994. Integrating a Network Model and Expert System. In *Proc. AWWA Computer Conference*. Denver, Colo.: AWWA.

McTigue, N.E., and J.M. Symons, eds. 2010. *The Water Dictionary*, 2nd ed. Denver, Colo.: AWWA.

Ormsbee, L.E., and S. Lingireddy. 1997. Calibrating Hydraulic Network Models. *Journal AWWA*, 89(2), 42–50.

Panguluri, S., W.M. Grayman, and R.M. Clark. 2005. *Water Distribution System Analysis: Field Studies, Modeling, and Management*. EPA/600/R-06/028. Cincinnati, Ohio: US Environmental Protection Agency, Office of Research and Development, National Risk Management Research Laboratory, Water Supply and Water Resources Division.

US Environmental Protection Agency (USEPA). 2003. *Stage 2 DBPR Initial Distribution System Evaluation Guidance Manual*. Washington, D.C.: USEPA Office of Water.

Walski, T.M. 1995. Standards for Model Calibration. In *Proc. 1995 AWWA Computer Conference*. Denver, Colo.: AWWA.

Walski, T.M., D.V. Chase, D.A. Savic, W.M. Grayman, S. Beckwith, and E. Koelle. 2003. *Advanced Water Distribution Modeling and Management*. Waterbury, Conn.: Haestad Press, 268–278.

Chapter 12

Limitations of Modeling

Introduction

As has been discussed previously in this book, water quality modeling is an extremely powerful tool for analyzing the fate and effect of contaminant transport in drinking water distribution systems. However, it is important for modeling practitioners and decision makers to recognize that modeling also has limitations and that models are only representations of reality and not reality itself. An important issue to which modelers should be sensitive is proper model construction and calibration as is discussed in chapter 11 and appendix A. Another issue, and one that is often overlooked by modeling practitioners, is the problem of developing a reasonable characterization of wall demand for inclusion in water quality models. The general mathematical formulation of wall demand is addressed in chapter 6 as part of the development of EPANET. In this chapter, we will discuss results from a research project intended to highlight the fundamental assumptions embedded in EPANET and by extension many of the major proprietary models.

Factors that can influence the quality of water in distribution systems include: chemical and biological quality of source water; effectiveness and efficiency of treatment processes; adequacy of treatment facilities, storage facilities, and distribution system; age, type, design, and maintenance of the distribution network; and quality of treated water (Clark and Coyle 1990). Perhaps one of the most serious public health aspects of water quality deterioration in a network is the loss of disinfectant residual, which can weaken the barrier against microbial contamination. A factor frequently cited as contributing to water quality deterioration and loss of chlorine residual and therefore contributing to wall demand is internal corrosion of the pipe wall material (Clark et al. 2010).

Internal corrosion is the deterioration of the inside wall or lining of a pipe caused by reactions with water, and has been a major area of concern for many years. This deterioration can be the result of physical action that erodes the lining or surface coating of a pipe, chemical dissolution that leaches a pipe's lining or wall material, or electrochemical reactions that remove metal from the wall of the pipe. Many of these reactions contribute to loss of chlorine residuals.

There have been some reports that the loss of chlorine residuals in corroded unlined metallic pipes increases with increasing velocity. This phenomenon has been observed in several studies including research conducted at the University of Birmingham by Powell (1998) and the Detroit Water and Sewerage Department (Grayman et al. 2002, Doshi et al. 2003). However, there is evidence that pipe wall demand may vary significantly depending on pipe wall material. Powell (1998) observed significant differences in wall demand between unlined metallic and PVC pipe. Controlled experimental studies reported in Powell's work also demonstrated a major difference in wall demand between unlined metallic and PVC pipe.

In order to assess the loss of free chlorine in corroded metal and PVC pipe subject to changes in velocity, a study was conducted under controlled conditions in specially constructed pipe loops (distribution system simulator). These loops are located at USEPA's Test and Evaluation (T&E) Facility in Cincinnati (Clark and Haught 2005). The results from these studies will be reported in this chapter.

Previous Studies

Many studies have attempted to determine the factors that cause the deterioration of water quality in networks. Williams (1958) conducted a study intended to identify the cause of sporadic water quality problems in the Brantford, Ont., water system. He found that the treatment plant produced a water that was acceptable from a taste-and-odor point of view. Quality was maintained in high-flow areas of the system but deteriorated dramatically in low-flow areas and in dead ends. Williams concluded that the loss of both chlorine and chloramine residuals in the Brantford system was related to an increase in ferrous iron caused by corrosion in the distribution system pipes.

A study conducted in collaboration with the South Central Connecticut Regional Water Authority (SCCRWA) (Clark et al. 1991) examined the loss of chlorine residuals in a portion of the distribution network. Bench-scale studies were conducted in which the chlorine demand was determined for the treated water (bulk decay rate) and then compared against the total system demand. It was found that the total system decay rate was much higher than the bulk decay rate, and it was concluded that this additional demand was most likely due to pipe wall demand, biofilm, and tubercles.

This same effect was observed in another study conducted in collaboration with the North Marin Water Authority (NMWA) in Northern California (Clark et al. 1994, Vasconcelos et al. 1997). The NMWA uses two sources of water, Stafford Lake and the North Marin Aqueduct. As with the SCCRWA study, the total system chlorine decay rate was considerably higher than the bulk rate in those sections of the system containing older pipe. Figure 8-28 shows the predicted versus actual chlorine residual assuming first-order decay at Eighth St., Olive St., Lynwood, and Bel Marin sampling sites. The effect of wall demand is very clearly shown in this figure.

Benjamin et al. (1996) identified the accumulation of iron corrosion products at the pipe wall and the uptake of these products into the bulk water. The result was a deterioration of water quality.

In a comprehensive study of chlorine residual losses in drinking water distribution systems, Powell (1998) and Powell et al. (2000) found a very large range in wall demand for chlorine. The results suggest that cast-iron pipe rates typically have 10 to 100 times higher wall demand than PVC pipe. An increase in wall demand with velocity, which was most prominent in cast-iron pipe, was also found.

In a comprehensive field study by the Detroit Water and Sewerage Department, a series of 35 chlorine loss tests in pipe segments ranging in length from 153.5 m to 614 m (500 to 2,000 ft), were conducted under varying flow conditions. The majority of the pipes were 70- to 135-year-old unlined cast-iron pipes ranging from 15.24 m to 40.6 cm (6 in. to 16 in.) in diameter. The study showed that higher flow rates resulted in greater residual chlorine loss in the pipe (Grayman et al. 2002, Doshi et al. 2003).

At the University of Illinois, Sarin et al. (2003) conducted a study of iron release from corroded unlined cast-iron pipe using a pipe-loop system constructed from 100-year-old unlined cast-iron pipes from the Massachusetts Water Resources Authority in Boston. They examined the role of pH, alkalinity, and orthophosphates in controlling iron release. The investigators concluded that iron release was closely linked to turbidity and that turbidity is directly proportional to iron concentration. Iron release was found to be sensitive to changes in alkalinity.

In addition to water quality deterioration from corrosion by-products, evidence suggests that corrosion scales support microbiological organisms. Camper et al. (2003) studied the interactions between pipe materials, organic carbon levels, and disinfectants using annular reactors with ductile-iron, PVC, epoxy, and cement-lined coupons at four field sites. The investigators found that, regardless of carbon levels, iron had the highest numbers of bacteria.

Recent corrosion research by the University of Illinois has focused on characterizing the nature of scales in metal pipes (Sarin et al. 2004). This is of special interest because of the large amount of cast-iron pipe used by drinking water utilities in the United States. Sarin et al. (2004) found that corrosion has a number of negative effects on water quality in drinking water distribution systems. It can consume oxidants and disinfectants in the bulk water and create scales that increase the head loss required to deliver water. This results in more energy consumption, biofilm growth, and suspensions of iron particles that create discolored water. They developed a conceptual model that describes the formation and growth of iron scales and their reactions that lead to colored water and reached the following conclusions:

- Corrosion scales govern the interaction of corroded iron in pipes with water.
- When oxidants are in contact with iron scale, the corrosion reaction proceeds continuously at the metal scale interface.
- Stagnant water leads to rapid depletion of oxidants next to the scale. As oxidants are depleted next to the scale, porous scale and anoxic conditions can permit a high iron release rate.
- Flowing water increases the rate of mass transport of oxidants to the surface of the scale and can reduce the rate of iron release. Ferrous iron is oxidized as it diffuses out of the scale, and the iron is precipitated within the scale as a Fe (III) solid.
- When ferrous ion encounters oxidants in the bulk water, it is rapidly oxidized to ferric iron and either deposits on the scale surface or is resuspended by the hydraulic scoring action of flowing water.

Eisnor et al. (2003) studied pipe wall demand for free chlorine on full-scale distribution systems and have shown that free chlorine losses in system pipes were considerably higher than would have been predicted using the bulk water decay rate. The aforementioned studies by Grayman et al. (2002) and Doshi et al. (2003) have suggested that with older unlined metal pipes, chlorine wall demand increases significantly with flow rate.

Clark and Haught (2005) conducted a study to systematically assess the effect of free chlorine loss in corroded unlined metallic pipes subject to changes in velocity. The authors conducted the study under controlled conditions in a specially constructed pipe loop located at the USEPA's T&E Facility in Cincinnati, Ohio. Results from the pipe-loop study supported the concept that the rate of free chlorine residual loss increased with velocity in unlined metallic pipe.

DiGiano and Zhang (2005) developed a bench-scale reactor to test the decay rate of chlorine at the pipe wall. Two 152-mm (6-in.) diameter sections of old cast-iron and new cement-lined ductile-iron pipe were used in their tests. They studied the effects of velocity, corrosion rate, dissolved oxygen, and pH on the chlorine–wall reaction rate. They found that chlorine decay rate was described by zero-order reaction kinetics for cast-iron pipe. The zero-order rate was larger at higher velocities because of higher mass transfer to the pipe surface, although a limit was reached. For the ductile-iron pipe, the chlorine decay kinetics was found to be first-order with respect to chlorine concentrations.

Al-Jasser (2007) attempted to assess the service age of pipes on the effective chlorine wall decay constant. Three hundred and two pipe sections of different sizes and eight different pipe materials were collected and tested for their first-order wall decay constants. It was found that the effect of pipe age was most evident in cast-iron pipes, whereas steel pipes were less affected.

Distribution systems are extremely complex and difficult to study in the field, and mathematical models have proven to be an effective tool for assessing changes in water quality in drinking water distribution systems. In a previous study, two chlorine transport models developed by Rossman et al. (1994) and Biswas et al. (1993) were evaluated (Clark and Haught 2005). These two models have been used to successfully model chlorine transport and decay in distribution system pipes. The Rossman et al. (1994) EPANET model assumes that bulk decay of chlorine can be represented as a first-order reaction. It further assumes that the chlorine decay at the pipe wall is expressed as a first-order function with a uniform concentration over the bulk flow cross section. The Biswas et al. (1993) model assumes that chlorine decay in the bulk phase and chlorine decay at the pipe wall are first-order reactions and neglects axial diffusion. Unlike EPANET, the Biswas et al. (1993) model includes radial diffusion as a factor in chlorine decay.

Recent studies that have attempted to model chlorine residual in drinking water distribution systems have been undertaken by Huang and McBean (2007) and Yang et al. (2007). Huang and McBean suggested a two-component second-order chlorine bulk model decay, and Yang et al. suggested a modified one-dimensional Dankwerts convection–dispersion reaction model.

Data Collection

Data used in this study were collected in pipe loops located at the USEPA's T&E Facility in Cincinnati, Ohio (Rossman et al. 2001). These pipe loops are called the distribution system simulator (DSS). All experimental data were collected according to USEPA Standard Methods. The DSS consists of two sets of three individual 15.24-cm (6-in.) diameter pipe loops (a total of six loops). The three loops in each set are stacked above each other on a project pad. Each loop is 27 m (88 ft) in length and consists of 15.24-cm (6-in.) pipe of various materials. The DSS can be configured to operate as independent loops, collectively as one unit, or in various other experimental configurations. Some of the pipes in the loops have been in service for many years, and there is considerable corrosion and tubercle buildup on the inside surfaces of some of the loops. With no way of measuring the actual internal diameter of the unlined metallic pipe, it was assumed to be the nominal diameter of 15.24 cm (6 in.). In addition, it is likely that biofilm buildup exists on the tuberculated surfaces, which will increase the loss of chlorine residual as water circulates through the loop. However, because there is also no reliable way of measuring the effect of biofilm in an existing system, its effect was ignored in this study.

After filling a pipe loop system, the water is continuously recirculated, using the loop-specific recirculation pump. Sensors are located in 3-in. PVC pipe sections designed for plug-flow sensor contact. They are continuously monitoring, non–water drawing sensors that are read on digital displays along the loop system. The current parameters measured by these sensors are pH, ORP, and DO.

In recirculation tank mode, water is recirculated through a 0.38-m^3 (100-gal) tank. The recirculation tank mode is used when operating the online turbidimeter or online CL-17 chlorine analyzer and when a hydraulic retention time is desired. The CL-17 and turbidimeter require a small amount of loop water for sample analyses. To compensate for the losses through these analyzers, water is added to the loop system from a 0.11-m^3 (30-gal) makeup water tank or from the 1.14-m^3 (300-gal) contact tanks. The makeup water addition rate is set to achieve the desired hydraulic retention time. Any excess water is then removed via the 0.38-m^3 (100-gal) recirculation tank overflow line. This ensures that the loops remain full when operating the sensor loop instruments.

Experimental Protocol

Two loops were used in this study (Loops 4 and 5) in a continuously recirculating "batch" mode (no outflow and no inflow). Loop 4 is unlined ductile-iron pipe, and Loop 5 is composed of PVC. Five experimental runs were conducted in each loop in an attempt to characterize the loss of chlorine in the loop as a function of velocity. The tests were performed at the following constant recirculation flow rates: 0.011 m^3/min (3 gpm), 0.023 m^3/min (6 gpm), 0.038 m^3/min (10 gpm), 0.379 m^3/min (100 gpm), and under no-flow or stagnant conditions. Before each set of experiments, the test loop was flushed with three pipe-volumes of water to ensure that the loop contained water with the appropriate chlorine concentration. The water temperature ranged between 12.439°C (55.6°F) and 26.14°C (74.7°F), and the pH of the water ranged between 8.19 and 9.06. The initial free chlorine concentrations were varied for several of the experiments, but this discussion will focus on the studies in which free chlorine concentrations were held at approximately 2.0 mg/L.

The following parameters were collected as grab samples in the test loops and the reservoir: free chlorine, turbidity, temperature (in-line), and pH (in-line) in the loop; total organic carbon (TOC); THMs; and haloacetic acids (HAAs). The sample schedule varied based on flow rate. No data were used from samples when the free chlorine concentration decreased below 0.05 mg/L. After several experimental runs, it was found that the reservoir parameters changed only slightly during an experimental run, and therefore reservoir sampling was conducted on a reduced schedule. A small volume of deionized water was pumped into the test loop to maintain a constant volume and pressure within the DSS as samples were collected. A dilution factor was calculated based on the deionized water addition and found to be insignificant. The flow rate was recorded at the beginning of each test and monitored during each sampling event. It was not possible to measure head loss across the loop, and hence a *C*-factor could not be calculated. Tables 12-1 and 12-2 summarize the average temperature, pH, turbidity, TOC, HAAs, and THMs for the experiments, although these values varied over time.

Experimental Results

The study concentrated on chlorine residual loss and the effects of the two different pipe surfaces. Figures 12-1 and 12-2 illustrate the types of data collected from the experiments in Loops 4 and 5. These data were collected at a flow rate of 0.038 m^3/min (10 gpm) at a chlorine concentration of 2 mg/L and with ambient TOC. In order to provide an analytical

Table 12-1. Summary of pipe Loop 4 (unlined ductile iron) physical and chemical parameters (chlorine concentration of 2 mg/L)

Flow Rate* m^3/min (gpm)	Facility	Turb† NTU	Temp† °C (°F)	pH†	TOC§¶** mg/L	THMs‡§¶** μg/L	HAAs §¶**
0.0 (0)	Reservoir	0.43	23.52 (74.3)	8.55	0.86	33.21	46.4
	Pipe Loop	0.72	23.72 (74.7)	8.75	1.00	64.7	58.3
0.011 (3)	Reservoir	0.63	22.31 (72.5)	8.53	1.13	33.12	131.7
	Pipe Loop	1.21	20.09 (68.2)	8.84	1.13	57.50	98.9
0.023 (6)	Reservoir	0.61	20.07 (68.1)	8.52	1.09	27.80	13.9
	Pipe Loop	0.67	19.46 (67.0)	8.85	1.15	48.69	26.4
0.038 (10)	Reservoir	0.52	18.67 (65.6)	8.56	0.66	32.63	44.0
	Pipe Loop	0.56	19.28 (66.7)	8.92	0.60	49.41	51.8
0.379 (100)	Reservoir	0.49	13.09 (55.6)	8.19	1.15	24.79	43.3
	Pipe Loop	0.54	17.94 (64.3)	8.81	1.25	46.40	55.6

* The flow rate was recorded at the start of each test and monitored (but not recorded) at each sampling event. Flow rate is monitored by reading an in-line flowmeter.

† pH, temperature, and turbidity were collected via grab sample in Loop 4 and Loop 5 and from the reservoir.

‡ Critical THM components include chloroform, dichlorobromomethane, chlorodibromomethane, and bromoform.

§ USEPA 1983.

¶ USEPA 1990.

** APHA et al. 2005, Method 4500-H.

framework, we will assume a first-order chlorine residual decay over time, although there is evidence that bulk residual decay exhibits second-order kinetics (Clark 1998, Yang et al. 2007, Huang and McBean 2007). For purposes of this analysis, the following is assumed:

$$C_t = C_0 e^{-(k_b + k_r)t} \quad \text{(Eq 12-1)}$$

where C_t = the concentration of chlorine at any time t, C_0= the initial concentration of chlorine, k_b= the bulk decay constant, k_r = the wall decay constant, and t = time in seconds.

Taking the natural log of Eq 12-1 and rearranging yields:

$$\ln\left(\frac{C_t}{C_0}\right) = -(k_b + k_r)t \quad \text{(Eq 12-2)}$$

Figures 12-3 and 12-4 (see appendix E) show the results from applying Eq 12-2 to the data from Loops 4 and 5 uncorrected for bulk demand (time in hours). The slopes of each curve in Figures 12-3 and 12-4 are the sum of k_b and k_r. The effect of mass transfer on total chlorine demand is very clear in Figure 12-3 but is not obvious in Figure 12-4. However, the effect of the two different pipe surfaces on residual chlorine transport is easily seen in these two figures.

Table 12-2. Summary of pipe Loop 5 (PVC) physical and chemical parameters (chlorine concentration of 2 mg/L

Flow Rate* m^3/min (*gpm*)	Facility	Turb† *NTU*	Temp† °C (°F)	pH†	TOC§¶** *mg/L*	THMs ‡§¶** *µg/L*	HAAs §¶**
0.0 (0.0)	Reservoir	0.93	23.55 (74.4)	8.54	1.32	34.13	55.0
	Pipe Loop	0.47	24.29 (75.7)	8.56	1.01	48.9	62.0
0.011 (3)	Reservoir	0.38	22.94 (73.3)	8.58	0.97	35.38	73.0
	Pipe Loop	0.47	20.37 (68.7)	8.58	1.05	42.02	65.3
0.023 (6)	Reservoir	0.39	20.29 (68.5)	8.53	1.12	30.26	21.3
	Pipe Loop	0.38	19.30 (66.7)	8.57	1.12	36.94	17.3
0.038 (10)	Reservoir	0.27	17.78 (64.0)	8.50	0.79	34.07	42.7
	Pipe Loop	0.55	31.94 (89.5)	8.55	0.58	39.47	48.5
0.379 (100)	Reservoir	0.56	15.11 (59.2)	8.39	1.28	30.34	43.3
	Pipe Loop	2.05	18.21 (64.8)	8.45	1.23	40.78	55.6

* The flow rate was recorded at the start of each test and monitored (but not recorded) at each sampling event. Flow rate is monitored by reading an in-line flowmeter.

† pH, temperature, and turbidity were collected via grab sample in Loop 4 and Loop 5 and from the reservoir.

‡ Critical THM components include chloroform, dichlorobromomethane, chlorodibromomethane, and bromoform.

§ USEPA 1983.

¶ USEPA 1990.

** APHA et al. 2005, Method 4500-H.

Figure 12-5 (see appendix E) summarizes the total chlorine demand for Loop 4 and Loop 5. It is obvious that the total chlorine demand ($k_b + k_r$) in the unlined metallic pipe is much greater than the total demand for the PVC pipe. Figure 12-6 shows the calculated bulk water demand (k_b) for all of the experiments based on the tank and reservoir data. In Figure 12-7 (see appendix E), the calculated bulk demand (k_b) is compared to the total chlorine demand ($k_b + k_r$) for the Loop 5 experiments. It can be seen that all of the demand data from Loop 5 falls within one standard deviation of the bulk demand at all velocities and in some cases is lower than the bulk demand. Therefore, it has been concluded that the wall demand for PVC is negligible and is assumed to be zero. Tables 12-3 and 12-4 summarize these calculations.

Model Development

Various approaches have been proposed to model chlorine transport and decay in distribution system pipes. As mentioned previously, successful models have been developed by Rossman et al. (1994) and Biswas et al. (1993). The Rossman et al. (1994) model assumes that bulk decay of chlorine can be represented as a first-order reaction. It further assumes that the

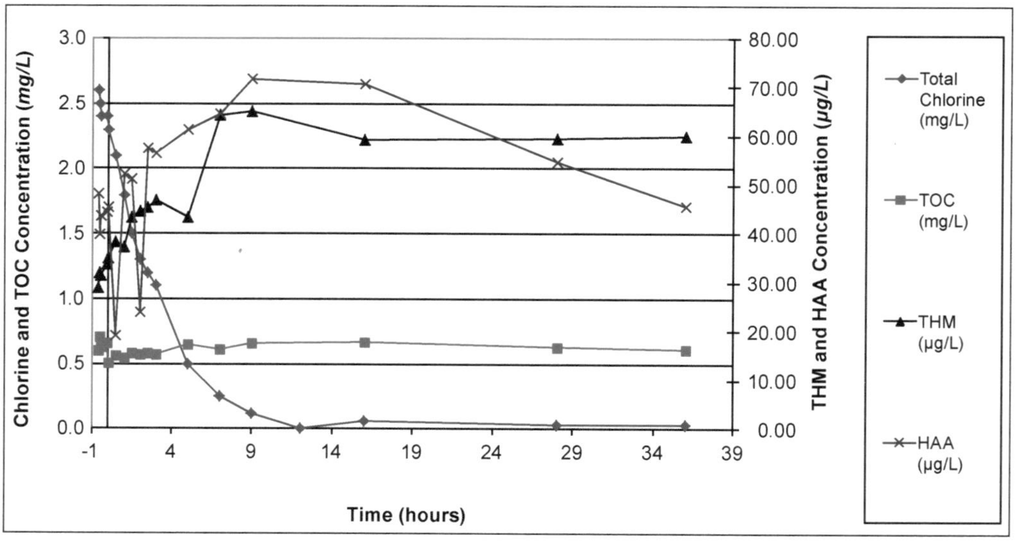

Figure 12-1. Experimental data collected from Loop 4 at 0.038 m³/min (10 gpm), with 2 mg/L Cl, and at ambient TOC (1 mg/L)

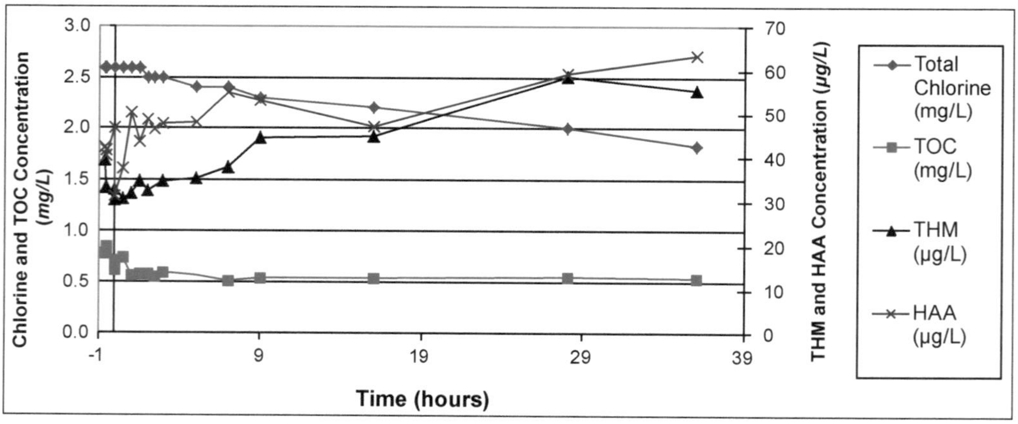

Figure 12-2. Experimental data collected from Loop 5 at 0.038 m³/min (10 gpm), with 2 mg/L Cl, and at ambient TOC (1 mg/L)

Table 12-3. Calculation of chlorine wall demand for ductile-iron pipe

Flow Rate *m³/min* (*gpm*)	Bulk Demand (*hr*⁻¹)	Total Demand (*hr*⁻¹)	Wall Demand (*hr*⁻¹)
0.0 (0.0)	0.042	0.116	0.074
0.011 (3)	0.042	0.172	0.130
0.023 (6)	0.042	0.245	0.203
0.038 (10)	0.042	0.396	0.354
0.379 (100)	0.042	0.454	0.412

Table 12-4. Calculation of chlorine wall demand for PVC pipe

Flow Rate m^3/min (*gpm*)	Bulk Demand (hr^{-1})	Total Demand (hr^{-1})	Wall Demand (hr^{-1})
0.0 (0)	0.042	0.027	−0.015*
0.011 (3)	0.042	0.014	−0.028*
0.023 (6)	0.042	0.011	−0.030*
0.038 (10)	0.042	0.013	−0.029*
0.379 (100)	0.042	0.019	−0.023*

* Assumed as zero

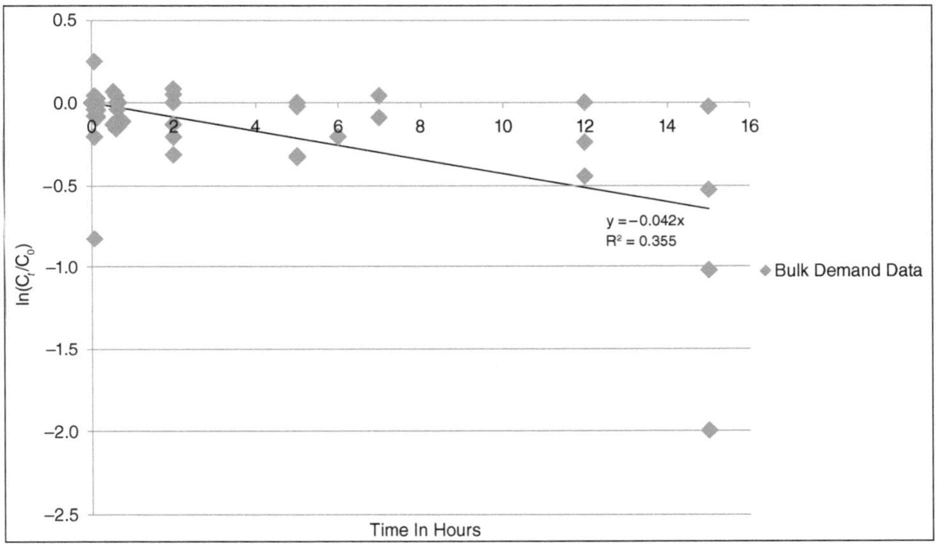

Figure 12-6. Bulk demand based on tank data

chlorine decay at the pipe wall is expressed as a first-order function with a uniform concentration over the bulk flow cross section. The Biswas et al. (1993) model assumes that chlorine decay in the bulk phase and chlorine decay at the pipe wall are first-order reactions and neglects axial diffusion. Unlike EPANET, the Biswas et al. (1993) model includes radial diffusion as a factor in chlorine decay. Both of these models have been discussed previously and were evaluated to characterize the loss of chlorine residuals in metallic pipes (Clark and Haught 2005).

The model developed for use in this study borrows from both the Biswas et al. (1993) model and EPANET (Rossman et al. 1994). A control volume is assumed where the control volume is a cylinder centered on the pipe axis of radius a. The chlorine concentration is assumed to be uniform within the control volume, and thus the concentration gradient occurs over the film adjacent to the pipe wall. The liquid film is assumed to be thin such that the diameter of the pipe and the control volume are assumed to be identical. The total flux of chlorine in the axial direction in the control section is assumed to be a function of the axial position, x (cm) and time t (sec) and is represented by dc/dt. The radial flux to the pipe wall is represented by J_w (mg-cm/L-sec). Assuming the control volume is constant in time, the conservation of mass for the control volume is given as:

$$\frac{dc}{dt} = \frac{J_b(x+\Delta x,t)-J_b(x,t)}{\Delta x} - \frac{2}{a}J_w - r_b \qquad \text{(Eq 12-3)}$$

where r_b is the homogeneous or bulk decay reaction term in mg/L-sec. As the limit of $\Delta x \rightarrow 0$, Eq 12-3 becomes:

$$\frac{dc}{dt} = -\frac{dJ_b}{dx} - \frac{2}{a}J_w - r_b \qquad \text{(Eq 12-4)}$$

If molecular diffusion and turbulent dispersion in the axial direction are assumed to be small, then the Peclet number is large, and the axial flux of chlorine, J_b (mg-cm/L-sec), can be expressed as:

$$J_b = cu \qquad \text{(Eq 12-5)}$$

where u is the cross-sectional average flow velocity in cm/sec. If we assume that the chlorine decay in the bulk flow is first order, the decay term r_b can be given as:

$$r_b = k_b c \qquad \text{(Eq 12-6)}$$

where k_b is the first-order bulk-decay coefficient for chlorine in sec^{-1}. Then, assuming steady-state conditions hold ($dc/dt = 0$ and u is constant), Eq 12-4 is simplified to:

$$u\frac{dc}{dx} + k_b c = -\frac{2}{a}J_w \qquad \text{(Eq 12-7)}$$

Chlorine flux is reaction-limited when chlorine loss at the pipe wall is limited by the rate of the pipe wall reaction. Chlorine flux is mass transport–limited when the chlorine concentration is reduced rapidly enough that the rate of mass transfer is insufficient to restore it. When the rate at which chlorine can be transported to the pipe wall is much lower than the rate at which the chlorine is consumed in the pipe wall reaction, the chlorine concentration vanishes at the pipe wall. Chlorine flux to the pipe wall is bounded below by reaction-limited conditions and on the high side by mass transfer–limited conditions. Under reaction-limited conditions, the rate of chlorine disappearance is limited by the rate of the pipe wall reaction. For example, the concentration of chlorine will become uniform across the pipe cross section when the rate of chlorine decay at the pipe wall is considerably lower than the potential rate of chlorine transport to the pipe wall.

We can define J_w as follows:

$$J_w = k_w c_w \qquad \text{(Eq 12-8)}$$

where k_w is the reaction coefficient at the pipe wall and c_w is the concentration at the pipe wall.

For purposes of this analysis, the following three modeling approaches are examined:

- Wall reaction limited (EPANET).
- Zero-order reaction limited.
- Mass transfer limited.

Wall Reaction Limited (EPANET)

In EPANET, the equation assumed to describe the mass balance of chlorine that is transported to the wall and chlorine consumed at the pipe wall is:

$$J_w = k_{mt}(c - c_w) \quad \text{(Eq 12-9)}$$

where k_{mt} is the mass transfer coefficient, and c is the concentration in the bulk phase.

The expression used to describe the mass balance of chlorine that is transported to the wall and chlorine consumed at the pipe wall is therefore:

$$k_{mt}(c - c_w) = k_w c_w \quad \text{(Eq 12-10)}$$

where k_w is the wall demand coefficient cm/sec. This chlorine consumption at the wall is

$$c_w = [k_{mt}/(k_w + k_{mt})][c] \quad \text{(Eq 12-11)}$$

The governing equation becomes:

$$u\frac{dc}{dx} + k_b c = -\left[\frac{2}{a}\right]\left[k_{mt}k_w/(k_{mt} + k_w)\right]c \quad \text{(Eq 12-12)}$$

Under steady-state conditions, Eq 12-12 can be written as:

$$\frac{dc}{dt} = -k_b c - \left[\frac{2}{a}\right]\left[k_{mt}k_w/(k_{mt} + k_w)\right]c \quad \text{(Eq 12-13)}$$

The total decay of chlorine can therefore be expressed in terms of a single first-order coefficient, K (sec^{-1}) given by:

$$K = k_b + (2/a)(k_{mt}k_w/(k_{mt} + k_w)) \quad \text{(Eq 12-14)}$$

In Eq 12-14, the first term is the bulk decay rate, and the second term represents the wall decay rate and consists of the wall demand coefficient (k_w), the mass transfer coefficient (k_{mt}), and the pipe radius (a). The portion of the first-order decay coefficient k, which is related to the wall decay rate, is therefore

$$r_w = [2/a][k_{mt}k_w/(k_{mt} + k_w)] \quad \text{(Eq 12-15)}$$

where r_w is defined as the wall decay rate (sec^{-1}). If we assume a constant k_w, Eq 12-7 approaches zero as k_{mt} decreases and becomes asymptotic to $2k_w/a$ with increasing k_{mt}. The implications of this property were discussed in detail in Clark and Haught (2005).

Rossman et al. (1994) calculated the mass transfer coefficient as follows:

$$k_{mt} = \mathrm{ShD_{ab}}/2a \quad \text{(Eq 12-16)}$$

where Sh is the Sherwood number and D_{ab} is the molecular diffusion coefficient for chlorine in water (cm^2/sec). Rossman (2000) estimated the Sherwood number under turbulent flow conditions as:

$$\mathrm{Sh} = 0.0149\,\mathrm{Re}^{0.88}\,Sc^{0.333} \text{ for Re} > 2300 \quad \text{(Eq 12-17)}$$

Where:

$$\mathrm{Re} = \mathrm{u}d/\mathrm{v} \quad \text{(Eq 12-18)}$$

$$Sc = \mathrm{v}/\mathrm{D_{ab}} \quad \text{(Eq 12-19)}$$

where d is the pipe diameter and v is the kinematic viscosity of water (cm^2/sec).

For laminar flow, the Sherwood number is given by

$$\mathrm{Sh} = 3.65 + [0.0688(d/L)\,\mathrm{Re}\,Sc]/[1 + 0.04[(d/L)\,\mathrm{Re}\,Sc]^{0.67}] \text{ for Re} < 2{,}300 \quad \text{(Eq 12-20)}$$

where L is the pipe length in cm.

Using the above assumptions, the value for the unlined metallic pipe (Loop 4) wall demand can be "back-calculated." These calculations are summarized in Table 12-5 and Figure 12-8.

As can be seen, the values for k_w with flow are in contrast to an underlying assumption embedded in EPANET: that the wall demand parameter is constant for a given length of pipe. Therefore, based on these data, it is apparent that the wall demand value in EPANET is essentially a fitting parameter, as was explored in the study by Clark and Haught (1995), and has limited predictive capability.

Table 12-5. Summary of EPANET wall demand calculations

Loading (gpm)	Loading (m^3/sec)	Vel (cm/sec)	Re	*Sc*	Sh	K_{mt} (cm/sec)	K_t (/hr^{-1})	K_t (/sec^{-1})	K_w (EPANET)
0.0	0.0	0.0	0.0	822.1	3.7	2.9219E-06	−0.1160	−3.2222E-05	−2.9932E-06
3.0	189.3	1.0	1576.5	822.1	13.3	1.0618E-05	−0.1720	−4.7778E-05	−1.1276E-05
6.0	378.5	2.1	3153.0	822.1	167.0	1.3369E-04	−0.2450	−6.8056E-05	−2.7601E-04
10.0	630.9	3.5	5255.0	822.1	261.8	2.0958E-04	−0.3960	−1.1000E-04	−4.1920E-04
100.0	6309.0	34.6	52549.9	822.1	1985.9	1.5898E-03	−0.4540	−1.2611E-04	6.8860E-04

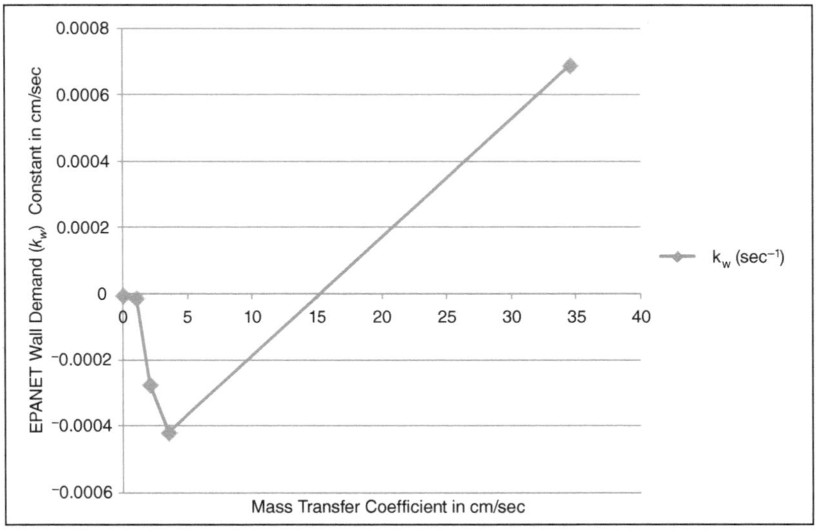

Figure 12-8. EPANET wall demand constant (k_w) versus mass transfer coefficient

Zero-Order Reaction

A model commonly mentioned in the literature is the "zero-order" model (DiGiano and Zhang 2005, Vasconcelos et al. 1997). The zero-order model is based on the assumption that the amount of chlorine (disinfectant) transported to the pipe wall is reaction limited, and the radial flux of the disinfectant must equal the amount of chlorine that reacts at the pipe wall. Therefore, we would assume that the mass transfer equation is

$$k_{mt}(c-c_w)=k_0 \qquad \text{(Eq 12-21)}$$

where k_0 = the reaction constant and $J_0 = k_0$.

The resulting equation is as follows:

$$c = c_0 e^{-k_b t} - \frac{2k_w}{ak_b}\left(1 - e^{-k_{bt} t}\right) \qquad \text{(Eq 12-22)}$$

Assuming k_b = 0.042/hr, Table 12-6 summarizes the solution to Eq 12-22 (k_0) for the various loading rates. Figure 12-9 shows the zero-order reaction rate coefficients versus mass transfer coefficient. These results are consistent with DiGiano and Zhang (2005), in which they found that the reaction coefficient increases with mass transfer but approaches an asymptote as mass transfer increases.

Table 12-6. Zero-order reaction coefficients

Loading m^3/sec (gpm)	Velocity (cm/sec)	Re	Sc	Sh	K_{mt} (cm/sec)	k_b (hr^{-1})	k_0 (mg-cm/L-hrs)
0.0 (0.0)	0	0	822.1311	3.65	2.92192E-06	0.042	0.146
189.3 (3.0)	1.038	1576.498	822.1311	13.26387	1.06181E-05	0.042	0.274
378.5 (6.0)	2.075	3152.996	822.1311	167.0084	0.000133694	0.042	0.604
630.9 (10)	3.459	5254.993	822.1311	261.7973	0.000209575	0.042	0.975
6309.0(100.0)	34.585	52549.93	822.1311	1985.935	0.001589791	0.042	1.016

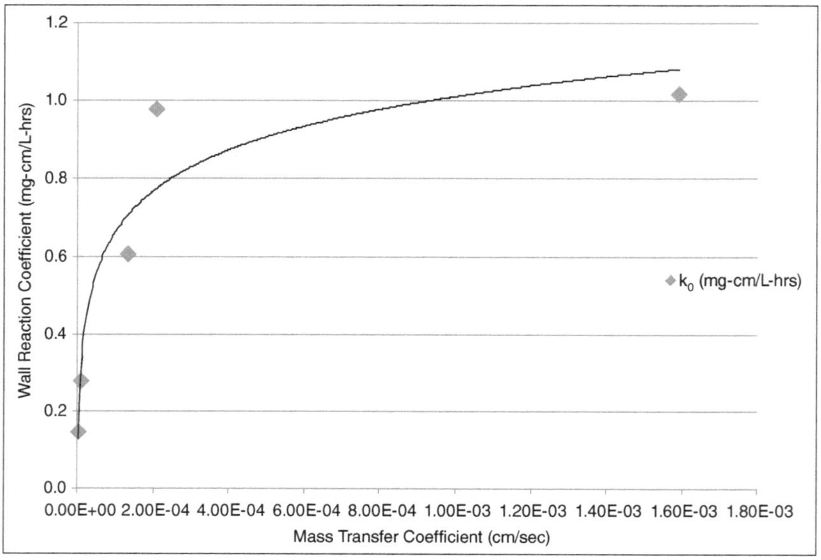

Figure 12-9. Zero-order wall reaction coefficient versus mass transfer coefficient for unlined metallic pipe

Mass Transfer Limited Model

In the mass transfer limited model, it is assumed that the chlorine concentration (c_w) at the pipe wall is zero due to instantaneous reaction. The flux expression then becomes:

$$J_0 = k_{mt} c \quad \text{(Eq 12-23)}$$

When substituted in the general transport equation (Eq 12-7), the mass transfer limited case becomes:

$$c = c_0 e^{-(k_b + \frac{2k_{mt}}{r})t} \quad \text{(Eq 12-24)}$$

However, this model tends to consistently underestimate the actual chlorine demand at low velocities (low mass transfer) and overestimate chlorine demand for higher velocities (high mass transfer) in the unlined metallic pipe. Figures 12-10 and 12-11 illustrate this effect.

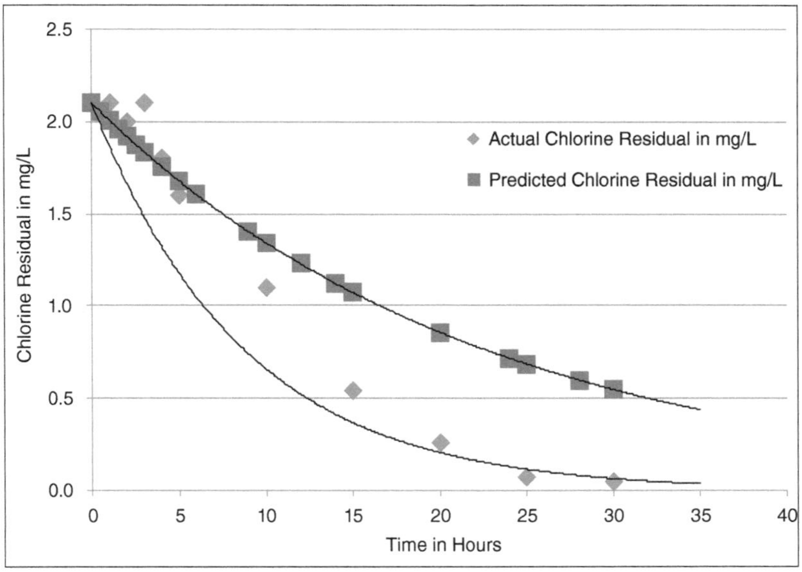

Figure 12-10. Actual and predicted chlorine residual versus time in hours at 0 cm/sec (0 gpm)

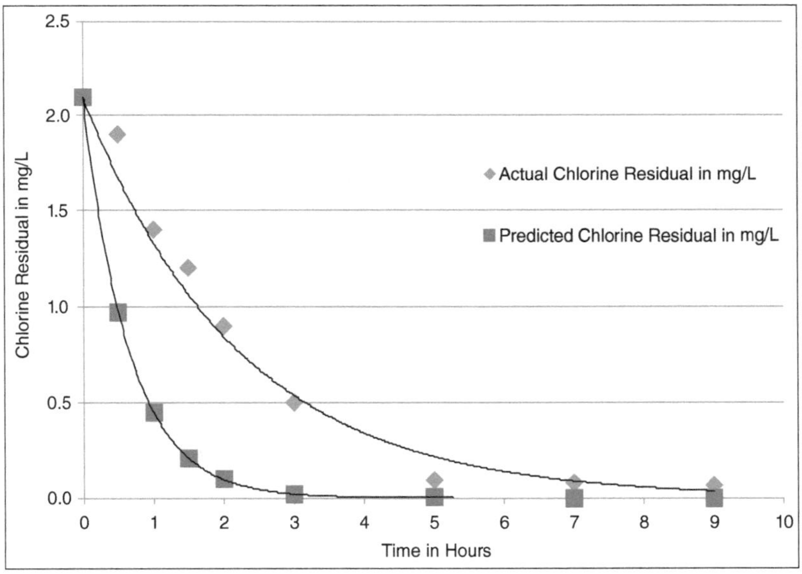

Figure 12-11. Actual and predicted chlroine residual in mg/L at 34.6 cm/sec (10 gpm)

Regulatory Implications

It is clear from the experiments reported above that differences occur in the wall demand characteristics between unlined metallic pipe and PVC. The differences have major regulatory implications. Figure 12-12 illustrates the impact of the two surfaces on the transport of chlorine through the two types of pipe. As can be seen from Figure 12-12, chlorine residual virtually disappeared within 12 hr in the unlined metallic pipe but stayed at a relatively high level in the PVC pipe. However, Figure 12-13 shows the THMs generated in both loops

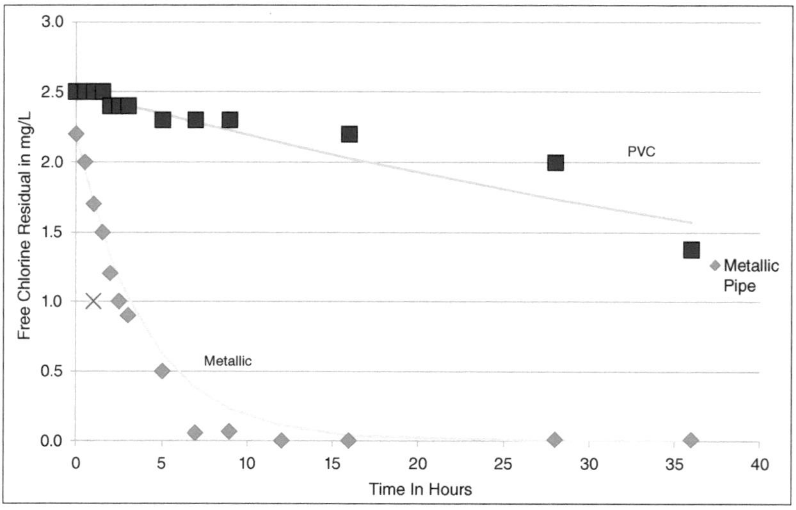

Figure 12-12. Chlorine residual loss for Loop 4 (unlined metallic pipe) and Loop 5 (PVC) at 0.038 m³/min (10 gpm) versus time in hours

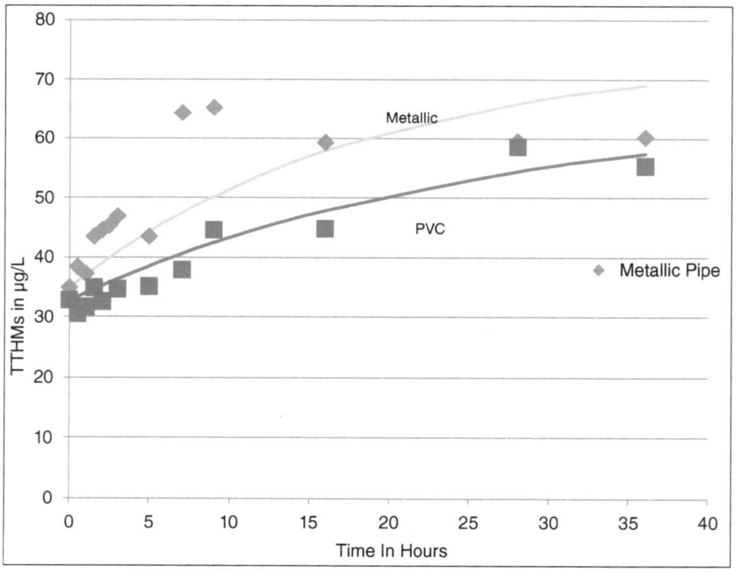

Figure 12-13. TTHMs for Loop 4 (unlined metallic pipe) and Loop 5 (PVC) at 0.038 m³/min (10 gpm) versus time in hours

during the experiment. As can be seen, the THM levels in both loops were very similar, even though chlorine residual transport levels were quite different. For water utilities, these results imply that a change in pipe material can result in more effective maintenance of chlorine residuals.

It has been hypothesized that internal corrosion of metal pipes can result in the deterioration of water quality, including the loss of chlorine residuals. Pipe wall demand for free chlorine has been attributed to a number of factors such as pipe age and material, pipe diameter, temperature, and pipe roughness. Recent studies have demonstrated that in older unlined metal pipes, chlorine wall demand increases with flow rate. DiGiano and Zhang (2005) observed in their experiments that cast-iron pipes exhibited zero-order pipe wall reactions, which increased with velocity due to mass transfer considerations. They found that steel pipes exhibited first-order reactions. Al-Jasser (2007) demonstrated that service age in both steel and cast-iron service age can increase wall demand.

The experimental results reported above demonstrate that surface or pipe wall demand varies significantly among different types of pipe material, and therefore the type of pipe material can influence the transport of free chlorine in drinking water distribution system networks. These characteristics can have regulatory implications. For example, chlorine residuals dissipate rapidly in unlined ductile-iron pipe but are maintained in PVC pipe.

It is also clear from the experimental results that the various models proposed for predicting chlorine transport have limitations. For example, based on the results presented in this chapter the wall demand, which is a fundamental parameter in EPANET, is a fitting parameter and is not necessarily constant under all flow conditions. This effect was also discussed in the paper by Clark and Haught (2005). The zero-reaction model, which is frequently mentioned in the literature, has a similar problem. The zero-reaction coefficient also varies with flow but as pointed out by DiGiano and Zhang (2005) varies in a more consistent and predictable manner. The mass transfer model has similar limitations.

It is also clear that pipe surface material has a major effect on wall demand and that this is a phenomenon that should be studied more thoroughly.

Summary and Conclusions

It has been emphasized several times in this book that, although water quality modeling is an extremely powerful tool for analyzing the fate and effect of contaminant transport in drinking water distribution systems, it also has limitations. Models are only representations of reality and not reality itself. This chapter is intended to stress that point by discussing one of the major unknowns in the water quality modeling field, which is the problem of developing a reasonable characterization of wall demand for inclusion in water quality models. The general mathematical formulation of wall demand was addressed in chapter 6 as part of the development of EPANET.

In this chapter, we have discussed results from a research project intended to highlight the fundamental assumptions embedded in EPANET and by extension many of the major proprietary models. It has been hypothesized that internal corrosion of metal pipes can result in the deterioration of water quality, including the loss of chlorine residuals. Pipe wall demand for free chlorine has been attributed to a number of factors such as pipe age and material, pipe diameter, temperature, and pipe roughness. Recent studies have demonstrated that in older unlined metal pipes, chlorine wall demand increases with flow rate. DiGiano and Zhang (2005) observed, in their experiments, that cast-iron pipes exhibited zero-order pipe wall reactions, which increased with velocity due to mass transfer considerations. They found that steel pipes exhibited first-order reactions. Al-Jasser (2007) demonstrated that service age in both steel and cast-iron pipes can increase wall demand.

To the author's knowledge, the only reported set experiments in which "wall demand" has been determined in parallel for two different types of pipe material simultaneously was conducted by Clark et al. (2010). The experimental results reported in that study, demonstrate that surface or pipe wall demand varies significantly between different types of pipe material and, therefore, the type of pipe material can influence the transport and maintenance of free chlorine in drinking water distribution system networks. These characteristics can have regulatory implications; for example, chlorine residuals dissipate rapidly in unlined ductile-iron pipe but are maintained in PVC pipe. It is clear that pipe surface material has a major effect on wall demand. This is a phenomenon that should be studied more thoroughly and is a major unknown in the field of water quality modeling.

It is also clear, from the experimental results reported by Clark et al. (2010), that the various models proposed for predicting chlorine transport have limitations. For example, based on the results presented in this study, the wall demand parameter, which is a fundamental parameter in EPANET and other water quality models, is essentially a fitting parameter and is not necessarily constant under all flow conditions. This effect was also discussed in Clark and Haught (2005). The zero-reaction model, frequently mentioned in the literature, has a similar problem. The zero-reaction coefficient also varies with flow, however, as pointed out by DiGiano and Zhang (2005), the coefficient varies in a more consistent and predictable manner. The mass transfer limited model has similar limitations. This lack of fidelity seems more serious as wall demand increases.

Despite some of the problems associated with the various types of models evaluated as part of this study, distribution system water quality modeling is an important activity that should be maintained and expanded by water utilities. However, one of the lessons to be learned from the studies cited is that no model is perfect and that it is critical for utilities to maintain a carefully constructed calibration program when distribution system models are used. It is clear that results from distribution system models can provide misleading information and they therefore should be applied with care and understanding.

References

American Public Health Association (APHA), American Water Works Association (AWWA), and Water Environment Federation (WEF). 2005. *Standard Methods for the Examination of Water and Wastewater.* 21st ed. Washington, D.C.: APHA.

Al-Jasser, A.O. 2007. Chlorine Decay in Drinking-Water Transmission and Distribution Systems: Pipe Service Age Effect. *Water Research* 41(2):387–396.

Benjamin, M.M., H. Sontheimer, and P. Leroy. 1996. Corrosion of Iron and Steel. In *Internal Corrosion of Water Distribution Systems.* Denver, Colo.: AwwaRF/DVGW-Technologiezentrum Wasser, 29–77.

Biswas, P., C. Lu, and R.M. Clark. 1993. A Model for Chlorine Concentration Decay in Drinking Water Distribution Pipes. *Water Research,* 27(12):1715–1724.

Camper, A.K., Brastrup, K., Sandvig, A., Clement, J., Spencer, C., and Capuzzi, A.S. 2003. Effect of Distribution System Materials on Bacterial Regrowth. *Journal AWWA,* 95(7):107–121.

Clark, R.M., and J.A. Coyle. 1990. Measuring and Modeling Variations in Distribution System Water Quality. *Journal AWWA,* 82(8):46–53.

Clark, R.M., W.M. Grayman, J.A. Goodrich, R.A. Deininger, and A.F. Hess, 1991. Field Testing Distribution Water Quality Models. *Journal AWWA,* 83(7):67–75.

Clark, R.M., G. Smalley, J.A. Goodrich, R. Tull, L.A. Rossman, J. Vasconcelos, and P. Boulos. 1994. Managing Water Quality in Distribution Systems: Simulating TTHM and Chlorine Residual Propagation. *Journal of Water Supply: Research and Technology–AQUA,* 43(4):182–191.

Clark, R.M. 1998. Chlorine Demand and Trihalomethane Formation Kinetics: A Second-Order Model. *J. Environ. Eng.,* 124(1):16–24.

Clark, R.M., and R.A. Haught. 2005. Characterizing Pipe Wall Demand: Implications for Water Quality. *Journal of Water Resources Planning and Management,* ASCE, 131(3):208–217.

Clark, R.M., Y.J. Yang, C.A. Impellitteri, R.C. Haught, D.A. Schupp, S. Panguluri, and E.R. Krishnan. 2010. Chlorine Fate and Transport in Distribution Systems: Experimental and Modeling Studies. *Journal AWWA,* 102(5):144–155.

DiGiano, F.A., and W. Zhang. 2005. Pipe Section Reactor to Evaluate Chlorine–Wall Reaction. *Journal AWWA*, 97(1):74–85.

Doshi, P.E., W.M. Grayman, and D. Guastella. 2003. Field Testing the Chlorine Wall Demand in Distribution Mains. In *Proc. 2003 Annual Conference of the American Water Works Association*. Denver, Colo.: AWWA, 1–10.

Eisnor, J.D., and G.A. Gagnon. 2003. A Framework for the Implementation and Design of Pilot-Scale Distribution Systems. *Journal of Water Supply: Research and Technology–AQUA*, 52(7):501–519.

Grayman, W.M., L.A. Rossman, M.A. Gill, Y. Li, and D.E. Guastella. 2002. Measuring and Modeling Disinfectant Wall Demand in Metallic Pipes. In *Proc. EWRI Conference on Water Resources Planning and Management.* Reston ,Va.: ASCE.

Huang, J.J., and E.A. McBean. 2007. Using Bayesian Statistics to Estimate the Coefficients of a Two-Component Second-Order Chlorine Bulk Decay Model for a Water Distribution System. *Water Research,* 41:(10)287–294.

Powell, J.C. 1998. *Modeling Chlorine in Water Distribution Networks.* Ph.D. dissertation, University of Birmingham.

Powell, J.C. 2000. Performance of Various Kinetic Models for Chlorine. *Decay. Jour. Water Resources Planning and Management.* ASCE, 126:1:13.

Powell, J.C., J.R. West, N.B. Hallam, C.F. Forster, and J. Simms. 1999. Performance of Various Kinetic Models for Chlorine Decay. *Journal of Water Resources Planning and Management,* 126 (1)3–20.

Powell, J.C., N.B. Hallam, J.R. West, C.F. Forester, and J. Simms. 2000. Performance of Various Kinetic Models for Chlorine Decay. *Journal of Water Resources Planning and Management,* 126(1):13.

Rossman, L.A., R.M. Clark, and W.M. Grayman. 1994. Modeling Chlorine Residuals in Drinking Water Distribution Systems. *Journal of Environmental Engineering,* 120(4):803–820.

Rossman, L.A. 2000. *EPANET Version 2 User's Manual.* Cincinnati, Ohio: USEPA Water Supply and Water Resources Division.

Rossman, L.A., R.A. Brown, P.C. Singer, and J.R. Nuckols. 2001. DBP Formation Kinetics in a Simulated Distribution System. *Water Research,* 35(14):3483–3489.

Sarin, P., J.A. Clement, V.L. Snoeyink, W.W. Kriven. 2003. Iron Release From Corrode Unlined Cast-Iron Pipe. *Journal AWWA,* 95(11)85–86.

Sarin, P., V.L. Snoeyink, D.A. Lytle, and W.M. Kriven. 2004. Iron Corrosion Scales: Models for Scale Growth, Iron Release, and Colored Water Formation. *Journal of Environmental Engineering,* 130(4):364–373.

USEPA. 1983. *Methods for Chemical Analysis of Water and Wastes.* EPA 600/4-79-020. Washington, D.C.: USEPA Office of Research and Development.

USEPA. 1990. *Methods for the Determination of Organic Compounds in Drinking Water Supplement I.* EPA 600/4-90-020. Washington, D.C.: USEPA Office of Research and Development.

Vasconcelos, J.J., L.A. Rossman, W.M. Grayman, P.F. Boulos, and R.M. Clark. 1997. Kinetics of Chlorine Decay. *Journal AWWA,* 89(7):54–65.

Williams, D.B. 1958. Dechlorination Linked to Corrosion in Water Distribution Systems. *Water and Sewage Works,* 106–111.

Yang, Y.J., J.A. Goodrich, R.M. Clark, and S.Y. Li. 2007. Modeling and Testing of Reactive Contaminant Transport in Drinking Water Pipes: Chlorine Response and Implications for Online Contaminant Detection. *Water Research,* 42(6–7):1397–1412.

Appendix A

Calibration/Validation and Tracer Studies

Introduction

Because calibration and validation are critical aspects of water quality modeling, these activities have been examined throughout this book. As part of a study conducted by the USEPA and the Greater Cincinnati Water Works (GCWW), a calibration/validation study was conducted in parts of the GCWW distribution network. This study is discussed in detail in this appendix.

Tracers have been used for decades to determine flow, travel time, and dispersion in surface waters and groundwater. Tracers can be of various types, ranging from a physical object that can be visually detected in a stream or river to dyes or other chemicals whose concentrations can be monitored using special instrumentation. Fluorescent dyes have been used for many years to measure velocity and tidal movement in streams and estuaries. Use of tracers to understand the hydraulic movement in drinking water treatment unit processes or distribution systems is a more recent development. When tracers are used in drinking water, care must be taken to ensure that they will have no adverse health effects and that their use does not result in any violations of primary and/or secondary drinking water maximum contaminant levels (MCLs).

Tracers have been used in drinking water to estimate the travel time through various water treatment unit processes including clearwells (Teefy and Singer 1990, Teefy 1996, DiGiano et al. 2005). Tracer studies have also been conducted in distribution system tanks and reservoirs in an attempt to understand their mixing characteristics (Grayman et al. 1996, Boulos et al. 1996). They have also been used in water distribution networks to provide insight into the complex movement of water in a distribution system, to determine travel times, and to assist in calibration of distribution system hydraulic models (Clark et al. 1993, DeGiano et al. 2005, Vasconcelos et al. 1997, Grayman 2001). For example, Boccelli et al. (2004) and Sautner et al. (2005) have used dual tracers injected into water distribution systems to assess travel time and characterize flow patterns in support of epidemiological investigations. With the recent interest in homeland security issues, tracers are being used to simulate the movement and impacts of accidental or intentional contamination of water distribution systems (Panguluri et al. 2005).

Tracer Studies

Conducting a distribution system tracer study involves (1) injecting the tracer into a pipe upstream of the area to be studied, (2) shutting off or reducing a continuous chemical feed at the water treatment plant, or (3) using a naturally occurring substance in source water. The concentration is measured over time at various locations in the water distribution network as it moves through the study area. To be successful, a tracer study requires careful planning and implementation. This appendix provides information and guidance on planning and conducting tracer studies in drinking water distribution systems.

Tracer studies in distribution systems may provide a wide variety of useful information, including the following:

- Calculating travel time, residence time, or water age in a network.
- Calibrating a hydraulic model.
- Defining zones in a network served by a particular source and/or assessing the degree of blending with water from other sources.
- Determining the impacts of accidental or intentional contamination.
- Identifying appropriate sampling locations within the water distribution network.

Tracer studies may also assist water utilities in complying with various regulatory requirements. For example, the draft *Stage 2 DBPR Initial Distribution System Evaluation Guidance Manual* (USEPA 2003a) recognizes the use of tracers as a means of calibrating models and predicting residence time as a partial substitute for required field monitoring. Several rules and regulations (both existing and proposed) are currently being reviewed, such as the Total Coliform Rule (TCR) and a proposed distribution system rule. Water quality modeling and model calibration are likely to play a role in the development and/or promulgation of these rules.

The scope of a tracer study may vary considerably depending upon the study needs, size, and complexity of the distribution network being evaluated. A study area may consist of a single stretch of pipe, an entire neighborhood, a portion of a large distribution system, a pressure zone, or in some cases the entire distribution network. The resources required to conduct a tracer study will vary with the extent and complexity of the study and the test equipment used. Careful planning and implementation are critical in all cases to ensure meaningful results. This appendix contains information that can be used during the planning phases of a tracer study; a summary of the tasks associated with executing a tracer study; and typical costs associated with conducting a tracer study.

Selecting Tracer Material

Criteria that can influence the selection of a particular tracer include

- Regulatory requirements.
- Analytical methods and instruments available for measuring tracer concentration.
- Injection and storage requirements.
- Chemical reactivity.
- Chemical composition of the finished water.
- Overall cost.
- Public perception.

Ideally, a tracer should be inexpensive, nonreactive with both water and distribution system materials, safe to drink when dissolved in water, easily dispersed in water, aesthetically acceptable to customers, able to meet all drinking water regulations, and inexpensively and accurately monitored in the field by manual and automated methods. There is no one tracer that will meet all of these criteria for a given study. Frequently, trade-offs are made among the criteria listed above that must be assessed when selecting a tracer. The tracer to be used in the study should be determined early in the planning stage and approval for its use received from the water utility and state regulatory agencies.

Tracers may fall into three broad categories: a chemical that is normally added to the water during the treatment process and that may be temporarily shut off during the study; a chemical that is added to the water by the team during the study; or a naturally occurring substance in the source water that may be adjusted in some manner to create a tracer. The most commonly used tracers are fluoride, calcium chloride, and sodium chloride.

Fluoride

Fluoride is frequently added to water supplies because of its health benefits, but it can be turned off for short periods, thereby making the nonfluoridated water a tracer in the system. When fluoridation is not practiced, fluoride can be added to the water system and used as a tracer by injection. It is especially popular with utilities that routinely add fluoride as part of the treatment process because little effort is required to turn the fluoride off and on. When the fluoride feed is shut off, a front of low-fluoride water (or no fluoride if there is no natural background concentration) becomes the tracer. A second tracer test (or a continuation of the initial test) can be performed when the fluoride feed is turned off.

Fluoride can interact with coagulants that have been added during treatment and in some circumstances can interact with pipe walls leading to nonconservative behavior. Thus, when used in systems that do not generally fluoridate, a field test should be performed to determine possible interactions with pipes.

The secondary MCL for fluoride is 4 mg/L. However, if fluoride levels exceed 2 mg/L, customers must be notified. Background levels of fluoride can vary significantly and do exceed the secondary MCL in some geographic areas where fluoride is naturally occurring.

In cases where a utility is not permitted to completely shut off the fluoride feed, it may be feasible to increase the fluoride feed prior to the tracer study and to reduce the fluoride feed during the test. Care should be exercised to avoid exceeding the secondary MCL. However, a sufficient change in the fluoride concentration feed must occur in order to trace the change through the system. Thus, for example, a decrease in feed concentration from 1.2 mg/L to 0.8 mg/L may not be sufficient, but a decrease in concentration from 1.5 mg/L to 0.5 mg/L may be adequate. A change in fluoride dosage may have to be preapproved by state regulators. Depending on the duration of the study, the state agency may choose to allow a temporary shutoff or set a specific lowest-allowable-fluoride-concentration requirement.

In most treatment plants, fluoride is injected prior to a final clearwell. As a result, when the feed is shut off as a part of the tracer study, there is both a time delay and a gradual change in concentration in the clearwell discharge as the nonfluoridated and fluoridated water mix. Therefore, wherever and whenever possible, the clearwell should be operated at minimum water levels during the tracer test in order to achieve a relatively sharp front of nonfluoridated water leaving the treatment plant. The impact of travel through finished water storage reservoirs on the concentration of tracer during the study should also be evaluated. An alternative is to inject fluoride solution (e.g., sodium fluoride) at a point in the main transmission line downstream of the clearwell where both flow and injection rate can be simultaneously monitored and measured.

In a study conducted in the Cheshire service area of the South Central Connecticut Regional Water Authority (SCCRWA) in 1989, the fluoride feed was turned off to provide a tracer to validate a hydraulic and water quality model of its water distribution system

(Clark et al. 1991). SCCRWA normally added fluoride at a level of approximately 1 mg/L. For purposes of the model validation study, the fluoride feed was turned off for a period of seven days and then turned back on with sampling occurring for an additional seven days. This approach yielded, in effect, two tracer fronts. During the study, grab samples were taken every few hours at 16 hydrants, 2 well fields, 1 tank, 4 continuous analyzer sites, and daily at 19 dead-end sites. Additionally, experimental units were installed at a few sheltered sites to automatically measure fluoride concentrations and to take discrete samples for later analysis. A total of 2,150 fluoride grab samples were taken during the study and analyzed in the laboratory.

Note, ion-selective electrodes (ISE) can be used in conjunction with data loggers to provide continuous monitoring capability. At present, however, these instruments are relatively expensive (approximately $5,000 to $10,000 each) and have only been used extensively in large-scale tracer studies (Maslia et al. 2005, Sautner et al. 2005). Generally, grab samples are taken and analysis is performed manually in the field or laboratory.

Under some circumstances, fluoride reacts with pipe-wall material. In one study (Vasconcelos et al. 1996) in a system that did not normally fluoridate, a 13-hr pulse (step input of limited duration) of fluoride was injected into the feed line of a pressure zone. Field measurements of fluoride concentrations in the zone during the study indicated a significant loss of fluoride. It was postulated that some of the fluoride was deposited on the pipe wall. In a follow-up study, this problem was virtually eliminated by injecting fluoride over a period of several days prior to the actual study in order to precondition the pipes.

Calcium Chloride

Calcium chloride ($CaCl_2$) has been used in many tracer studies throughout the United States. It is considered to be safe and relatively easy to handle. Generally, a food-grade substance is required. It can be purchased as a liquid (typically a 30 to 35 percent solution) or as a powder that can be mixed with water to form a solution.

If calcium chloride is chosen as a tracer, the study personnel should be aware of the secondary drinking water MCL for chloride (250 mg/L). A target that is less than the secondary MCL should be set in order to provide a safety factor. Where chloride levels are high, calcium chloride may not be an appropriate choice for a tracer.

Grayman et al. (2000) used calcium chloride as a tracer in two studies of mixing in distribution system tanks. In both studies, the chemical was injected into the inflow pipe of the tank during the fill cycles, and conductivity and chloride were measured at locations within the tank. Calcium chloride has recently been used in several distribution system studies (Panguluri et al. 2005, Maslia et al. 2005, Sautner et al. 2005).

Calcium chloride can be monitored by measuring conductivity or by measuring the calcium or chloride ion (APHA et al. 2005). Conductivity is typically the easiest of these parameters to measure and is most amenable to inexpensive continuous monitors. However, conductivity is not a truly linear parameter (i.e., if a beaker of water of conductivity 100 mS/cm is combined with a like volume of water with a conductivity of 300 mS/cm, the conductivity of the resulting solution will not be exactly 200 mS/cm. As a result, distribution system models (that all assume linearity) can only approximately represent conductivity. Therefore, when using conductivity as the measured parameter, the options are to accept the linear approximation or convert conductivity to a true linear parameter such as chloride or calcium. If the former option is chosen, the amount of resulting error should be established in laboratory tests of waters of varying conductivity. If the latter option is chosen, the relationship between conductivity and chloride (or calcium) must be established in the laboratory.

It should also be noted that most field devices are set up to measure specific conductance instead of conductivity (conductivity is temperature-sensitive, whereas specific conductance is referenced to 25°C). For the purposes of this book, conductivity is assumed to represent specific conductance.

Sodium Chloride

Sodium chloride (NaCl) can be used as a tracer and has many characteristics similar to calcium chloride in that it can be traced by monitoring for conductivity or for the concentration of the chloride or sodium ion. The allowable concentration for sodium chloride is also limited by the secondary MCL for chloride and the potential health impacts of elevated sodium levels. USEPA reports that taste thresholds for sodium vary significantly among individuals, ranging from 30 to 460 mg/L (USEPA 2003b).

In a recent tracer study in Hillsborough County, Fla., two separate tracer chemicals were used to study the movement of water in a large distribution system (Boccelli et al. 2004). Approximately 2,200 gal of a saturated NaCl solution were injected into the finished water of a treatment plant as a series of four pulses ranging in duration from 1 to 3 hr over a 24-hr period. Simultaneously, the normal fluoride feed was shut off at the plant. Continuous conductivity monitors were installed at 14 locations in the distribution system to monitor for the NaCl tracer. Grab samples were taken to monitor the low-fluoride front as it moved through the system and to evaluate water quality changes. The resulting extensive hydraulic and water quality database is being used to calibrate a hydraulic and water quality model of the system (Boccelli and Uber 2005).

Other Chemicals That May be Added as Tracers

Other chemicals added as part of a tracer study include lithium chloride and chlorine. Lithium chloride is a popular tracer in the United Kingdom but is used less frequently in the United States, partly because of the public perception of lithium as a medical pharmaceutical. No field techniques exist for measuring lithium, and it is not easily amenable to automated continuous measurement. Samples must be collected and lithium concentrations measured in the laboratory.

Chlorine is commonly used as a disinfectant in many water systems. Because chlorine is reactive, it will decay over time. Under some circumstances, however, it can be used effectively as a tracer. It is most effective in a water where chlorine is not highly reactive (low decay rate) with either the water or distribution system material, and where the concentration levels can be increased above the normal level to create a front of water with a high chlorine concentration propagating through the system. However, in no case should the chlorine or chloramine be decreased to a level that may affect the disinfection process (Ferguson and DiGiano 2005). Again, any tracer study should first be approved by the state regulators.

Naturally Occurring Tracers

Perhaps the most difficult part of conducting a tracer study is obtaining permission to add a chemical and then injecting the tracer into the system at a concentration consistent with regulations. Much of this effort can be avoided if a natural tracer is available. Natural tracers are generally site-specific, but many options do exist and should be explored. The most common situations are the existence of multiple sources of water with different chemical signatures or cases where a change is planned in the chemical signature at a single source. Examples of these situations are described below.

Some of the chemical signatures that may be used to differentiate between sources include THM concentrations, hardness, conductivity, and treatment coagulant. Sampling in the distribution system for these "tracers" will provide information on zones served by each of the sources and the extent and variation of the mixing that take place in these zones over time. Alternatively, if one water source can be turned off for a period of time until the other source has reached chemical equilibrium throughout the system, the original source can be turned back on and used as a tracer as it propagates through the system. One of the first uses of natural tracers was in the North Penn Water Authority (NPWA) located in Lansdale, Pa. (Clark and Coyle 1990). A field research project was conducted by USEPA and NPWA that resulted in the development of a series of models that were used to study contaminant propagation in the water distribution system. The utility used a combination of groundwater with high levels of hardness and surface water containing higher levels of THMs. This resulted in two sources of water with very different quality characteristics. By monitoring changes in water quality that occurred at selected sampling points in the utility network, it was possible to use hardness and THM concentrations as tracers to validate the model. Making a major change in the incoming water supply such as a change in source water or modifying treatment may provide an opportunity to conduct a tracer test. The increased use of chloramines as a secondary disinfectant, to reduce the formation of DBPs, introduces another potential tracer opportunity. When a water utility switches from chlorine to chloramines (or vice versa), the chemical signature of the water changes and can be monitored by measuring both free and total chlorine. Namely, with chloramination, total chlorine is typically much higher than free chlorine, while with free chlorination, free and total chlorine will typically be very similar. A tracer study can be conducted when a system first adopts chloramination. Alternatively, many water utilities routinely switch back from chloramination to chlorine (e.g., annually for a month) in order to kill ammonia-oxidizing bacteria and thus reduce the chances of nitrification. This provides a recurring opportunity to conduct such a tracer study.

Another case occurred in the North Marin Water District (NMWD) in northern California (Clark et al. 1994), where natural differences in water characteristics were used to serve as a tracer for validation of a water distribution system model. In this USEPA-sponsored study, the utility used two sources of water with dramatically different water quality characteristics. The first source, Stafford Lake, has a very high humic content and thus has a very high THM formation potential. The other source is the North Marin Aqueduct, with a very low humic content and thus a very low THM formation potential. The model was further validated by predicting chlorine residual losses at various points in the network. In a follow-up study supported by AwwaRF (now Water Research Foundation) (Vasconcelos et al. 1997), the investigators used sodium as a tracer to validate the model.

DiGiano and Carter (2001) and DiGiano et al. (2005) traced the flow from two separate treatment plant sources at the same time by simultaneously reducing the fluoride feed at one plant while changing the coagulant added at the other plant. Normally, ferric chloride ($FeCl_3$) was used as a coagulant at both plants. During the tracer study, the coagulant at one plant was changed to aluminum sulfate [$Al_2(SO_4)_3$]. Fluoride, sulfate, and chloride were measured throughout the distribution system. Water utilities should carefully examine their particular system to determine if a natural tracer is available or if source-chemical signatures may be modified to be used as a tracer.

Sweetwater Authority in Southern California took advantage of a normal changeover in source water quality to perform a tracer study in its distribution system (Hatcher et al. 2004). In this case, the utility semiannually changes the primary source of its water supply from local Sweetwater Reservoir raw water to water provided by the California Aqueduct.

These two sources have very different chemical characteristics; most significantly, the organic carbon content (i.e., humic and fulvic acids) of Sweetwater Lake water is much higher than that of the raw aqueduct water. The measurement of molecular organic carbon absorbance at 254 nanometers, employing an ultraviolet–visible (UV–vis) spectrophotometer, is a surrogate measurement for the organic carbon content in water. UV-254 measurements were taken from grab samples at the treatment plant and at 28 sites within the distribution system over the five-day changeover period. The distribution system sites included most of the TCR sampling sites in addition to selected tanks. The resulting database was used to assess the movement of water in the system, the travel time throughout the system, boundary zones in the distribution system between areas served by the surface water plant and secondary sources, and calibration/validation of the distribution system model.

Comparison of Tracers

Teefy (1996) and Panguluri et al. (2005) investigated tracer alternatives for use in studies of residence time in clearwells and described the chemical characteristics of the individual tracers. Table A-1 summarizes various chemical characteristics identified in that report.

Advantages and disadvantages are associated with each of the general types of tracers: conservative (nonreactive) tracers, reactive tracers, chemicals that are normally added to the water but can be turned off, and natural chemical signatures in the finished water. Conservative tracers are more easily modeled than nonconservative tracers. Natural tracers or chemicals that can be turned off are easier to use than injected chemicals. Certain chemicals are more amenable to continuous monitors. These and other factors should all be considered when selecting a tracer for a study.

After investigating tracer options and selecting the most appropriate tracer, the governing state drinking water agency should be contacted. The agency should be provided with the specifics regarding the proposed study including location(s), proposed timeline(s), and selected tracer material. Once agreement has been reached and consent is received, the study team can then proceed with the next steps in the planning process.

Selecting Field Equipment and Procedures

Once a tracer has been selected and approval has been received from the appropriate water utility managers and regulatory agencies, specialized equipment must be identified and procured, including injection pumps, temporary tracer storage tanks, and various flow and tracer monitoring equipment (e.g., tracer chemical, reagents, and/or sample bottles). Vendors should be contacted for technical information, equipment availability, and cost quotations for the required field equipment and analytical instrumentation. The major decisions to be made and the items to be purchased prior to the execution of the study are discussed in the following subsections.

Injection Pump(s)

Pumps that are typically used in drinking water applications can be broadly classified as centrifugal pumps or positive displacement pumps. The centrifugal pumps produce a head and a flow by increasing the velocity of the liquid with the help of a rotating vane impeller. The positive displacement pumps operate by alternating between filling a cavity and displacing the volume of liquid in the cavity. The positive displacement pumps deliver a constant volume of liquid (for a given speed) against varying discharge pressure or head. By design, the positive displacement pumps are better suited to serve as an injection pump for a tracer study.

Table A-1. Tracer characteristics (Panguluri et al. 2005)

	Fluoride	Calcium	Sodium	Lithium	Chloride
Commonly available forms	H_2SiF_6 NaF Na_2SiF_6	$CaCl_2$	NaCl	dry LiCl	$CaCl_2$ NaCl KCl
Analytical methods	IC, ISE, SPADNS method	AA, IC, ICP, EDTA titration, Conductivity	AA, IC, ICP, FEP, Conductivity	AA, IC, ICP, FEP	IC, ISE, $AgNO_3$ titration, $Hg(NO_3)_2$ Titration
Typical chemical cost (2005)	Food-grade H_2SiF_6 \$7.6/100lb–23.97% liquid \$140/55 gal –49% liquid	Food-grade $CaCl_2$ \$150/55 gal –35% liquid	Food-grade NaCl \$12/50 lb \$6/50 lb	Lab-grade LiCl[6] \$22–\$48/ 500 g	Food-grade NaCl \$12/50 lb \$6/50 lb
Typical analytical cost per sample (2005)	\$18 (IC) \$16 (IC) \$12 (ISE) \$25 (IC)	\$10 (ICP) \$12 (ICPMS) \$5 (ICP)	\$10 (ICP) \$12 (ICPMS) \$5 (ICP)	\$12 (ICP) \$12 (ICPMS) \$6 (AA)	\$18 (IC) \$16 (IC) \$12 (USEPA 325.3) \$12 (IC)
Typical background levels in water distribution systems	0–4 mg/L	Varies greatly (1–300 mg/L), use only when low	Varies greatly (1–500 mg/L)	Usually below 5 mg/L	Varies greatly (1–250 mg/L)
Regulatory limits	4 mg/L SDWA MCL, 2 mg/L secondary MCL	None known. See limits for chloride.	20 mg/L for restricted diet (USEPA recommendation)	None known. See limits for chloride.	250 mg/L secondary standard

IC=ion chromatography

ICP=inductively coupled plasma

AA=atomic absorption spectrometry

FEP=flame emission photometric method

EDTA=ethylenediaminetetraacetic acid

ISE=ion selective electrode

SPADNS=Sodium 2-(parasulfophenylazo)-1,8-dihydroxy-3,6-naphthalene disulfonate (SPADNS) (McTigue and Symons 2010)

$AgNO_3$=silver nitrate

Examples of positive displacement pumps include: rotary lobe, progressing cavity, rotary gear, piston, diaphragm, screw, and chemical metering pumps (e.g., bellows, diaphragm, piston, and traveling cylinder).

Selection of the most appropriate positive displacement pump depends on the injection rate, the pressure in the receiving system, the chemical characteristics of the tracer, and local experience and preferences. Two types of positive displacement pumps have generally been used in tracer studies: gear pumps and metering pumps. The final selection depends upon viscosity of the tracer material, variability of pressure in the main, dosage accuracy needs, and other local factors. Furthermore, to control the drive speed (i.e., dosage), these pumps are equipped with alternating current (AC) or direct current (DC) motor. If a pump has an AC motor, frequency is adjusted; if it is equipped with a DC motor, voltage is adjusted to control speed.

USEPA has used gear pumps equipped with variable frequency drives in the past with success for conducting tracer studies. Other studies have reported success with metering pumps with variable speed or variable stroke controllers. The pump should be sized in accordance with the anticipated tracer dosage and pressure range in the main pipe for the selected injection location(s) in the study area. Depending on the location and dosage requirements, more than one size of pump may be needed (excluding backup pumps).

Tracer Storage and Dosage Rate Measurement

Tracers are available in dry or liquid form. If purchased as a powder, provisions for mixing the powder with water must be made. If the tracer is purchased in liquid form, it typically comes in either 55-gal drums or in larger containers such as a 330-gal tote. If only a small amount of tracer is needed, a single 55-gal drum will typically suffice. For greater accuracy, it is recommended that the tracer be transferred from 55-gal drums to a suitably sized day tank with a sight glass (used to periodically monitor the total tracer volume dosed). It is easiest to pump the tracer from a single container rather than having to switch the pump from container to container during the injection process.

If a metering pump is purchased, care must be taken so that the pump flow rate is calibrated for the specific tracer solution (by the vendor). Furthermore, variable area flowmeters (rotameters, with floats contained in an upright conical tube) are relatively inaccurate for measuring tracer dosage even after adjustments are made for density and viscosity. Figure A-1 shows a "flow tube" that can easily be custom fabricated and calibrated to accurately measure the rate of tracer injection. It is recommended that the supply tank also be marked to keep track of the tracer fluid level. Times should be noted at each mark so that it is possible to create a mass balance for the tracer injected during the study.

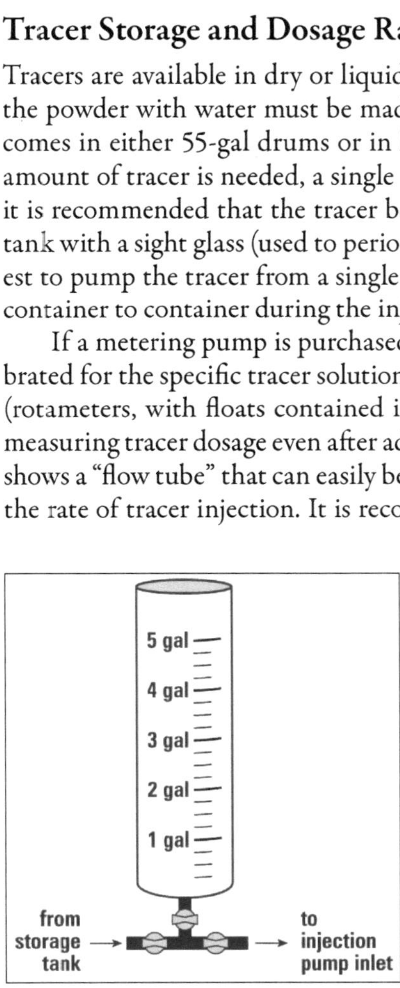

Figure A-1. Flow calibration tube (Panguluri et al. 2005)

During a tracer study when a tracer chemical is being injected into the system, in order to meet water quality regulations and to simplify the modeling, it may be desirable to maintain a constant tracer concentration in the receiving pipe. This can be accomplished by monitoring the resulting concentration in the receiving pipe and manually adjusting the tracer injection rate or through the use of a closed-loop system for automatically controlling the injection rate based on flow in the receiving pipe. The automated process is most effective at a location where the flow in the pipe is varying relatively slowly and where a flowmeter exists. A typical situation is the use of an existing Venturi meter that generates a 4-20 mA signal. This signal can be used as input to a controller that has been calibrated and programmed to control the stroke or speed of a variable stroke or speed injection pump. If the flow in the receiving pipe is varying rapidly over a large flow range, it is difficult for the closed-loop system to respond quickly.

Distribution System Flow Rate Measurement

In order to calculate the concentration of the tracer in the receiving pipe, it is necessary to know the flow rate in the pipe, the injection rate of the tracer, the injected concentration of the tracer, and the background concentration in the water before tracer is added. Flow rate should be measured continuously, because variations in pipe flow rate can affect tracer concentration. These fluctuations in flow can be accommodated by manually adjusting the tracer injection rate in the field or through the use of a flow-paced injection pump that responds to the flow in the receiving pipe.

Placement of additional flowmeters or other flow-measuring devices at various points in the system is recommended. This information will be very useful during the post-tracer modeling studies and is invaluable in calibrating a network hydraulic model. If the existing system does not have an adequate number of flowmeters for purposes of a tracer study, installation of additional meters is recommended.

Various types of flowmeters may be used to measure flow in pipes. They are categorized as either nonintrusive or intrusive meters. Portable ultrasonic flowmeters are nonintrusive and can provide reasonably accurate data if the pipe material is conductive and relatively nontuberculated. The ultrasonic flowmeter requires suitable upstream/downstream straight runs of pipe. Insertion flowmeters are also an option for measuring pipe flow rates. Insertion meters are intrusive, and either may be magnetic (magmeters) that are flange-coupled to the pipe or have propellers that must be inserted through a hole in the pipe. All meters require that the receiving main pipe be exposed (via excavation) or that an existing vault be used. If the injection location is in the vicinity of a reservoir/tank and the water level changes are available in real time, it may, in some instances, serve as a rough surrogate for in-pipe flow measurement. The selected method of flow measurement must be field-tested. Depending on the size of the reservoir/tank and the local demand, the reservoir level changes may not be fast or accurate and precise enough to determine the flow rate in real time.

Field Measurement of Tracer Concentration

Tracer concentration may be measured in the field by using automated monitors that analyze a sample at a preset frequency, by collecting grab samples, or a combination of both. Grab samples can be manually analyzed in the field or in the laboratory.

If grab sampling is used during a tracer study, the sampling team will generally traverse a circuit of several sampling locations. Using such an approach will generally yield a sampling frequency of one sample per station every 1 to 3 hr for an average-sized residential neighborhood (unless multiple crews are used). Some of the factors that will influence sampling frequency include the speed at which the tracer is moving within the distribution system, the number of sampling crews participating in the study, the number of sampling sites selected, the time of the day, and the distance between sampling sites. Equipment requirements for grab sampling are minimal and may include the following: coolers, ice, labeled sample bottles, log books, and temperature blanks. If samples are to be analyzed in the field, the sampling teams will need the appropriate analytical equipment. If samples are to be analyzed in the laboratory, the team will need the means to properly store and transport samples to a central laboratory. The quality assurance project plan (QAPP) may require duplicate or split samples for some or all of the primary samples. When taking a grab sample, care must be taken to flush the tap for a sufficient time to ensure that the sample is representative of the distribution main rather than the service lines.

Reliance solely on grab sampling may be impractical if the study area is large, the tracer front is moving rapidly, or a high frequency of sampling is desired. In these cases, continuous automated monitoring may be the best choice although some grab samples for quality assurance and quality control are recommended. If either calcium chloride or sodium chloride is the tracer selected, an online specific-conductivity meter equipped with an associated data logger is recommended. Automated monitors are available if chlorine residual is used as a tracer. Automated monitors are also available if fluoride is used as a tracer, but there has been relatively limited use under field conditions. Because most automated monitors require a continuous side stream (rather than being inserted directly into a main), the drainage flow from the monitor must be discharged into a sewer, into the street and subsequently into a storm drain, or into a pervious area. This discharge can be an added complication during cold weather when it may freeze. Because this discharge stream is generally chlorinated or chloraminated, regulations may control discharge into natural water courses. Additionally, this discharge flow may have to be accounted for if the data set is being used to calibrate a distribution system model and if the quantity of discharge through a particular meter is significant relative to the demand in the vicinity of the meter. If the total drainage discharge is significant for the purposes of modeling, provisions for continuously or manually measuring the amount of flow being bypassed are needed.

Potential grab and online sampling sites include: dedicated sampling taps, hydrants, pump stations, tank inlet–outlet lines, and faucets located inside or outside of buildings. Figure A-2 depicts an automated monitoring station used by USEPA. This figure illustrates the case where the sampling tap is allowed to run continuously throughout the study with the water going to a drain. The flow rate to or through the sampling tap must be sufficient to minimize the travel time from the main to the monitor.

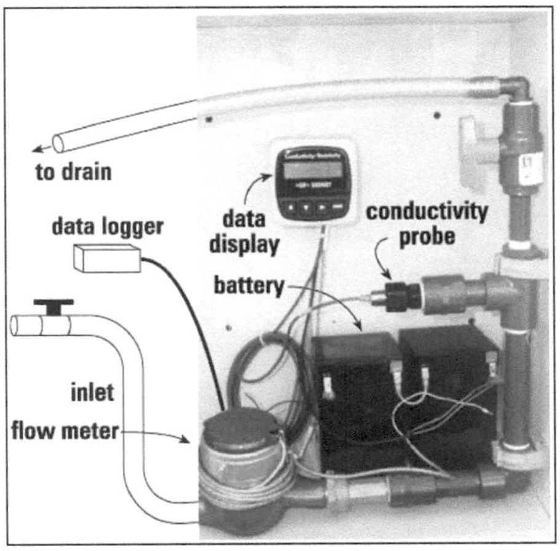

Figure A-2. Automated monitoring station (Panguluri et al. 2005)

USEPA and the Greater Cincinnati Water Works (GCWW) have pioneered the use of online monitors as a central focus for distribution system tracer studies. In a series of field tests, USEPA and GCWW injected calcium chloride tracer into the water system and followed the movement of the tracer using automated conductivity meters strategically placed throughout the study area. Three separate studies were conducted in a large water system representing a small highly urbanized area, a small dead-end suburban area, and a large suburban pressure zone. Based on the success of these studies, similar tracer studies have been conducted using a combination of online monitors and grab samples using both fluoride and sodium chloride as tracers in Hillsborough County, Fla. (Boccelli et al. 2004), and by the Agency for Toxic Substances and Disease Registry (ATSDR) using fluoride and calcium chloride at a large military base in North Carolina (Maslia et al. 2005, Sautner et al. 2005).

Online, automated sampling programs should be complemented with a grab sampling program to add a degree of confidence in measured data and to supplement field data at additional locations or at the automated monitor stations if they fail to record correctly.

Developing a Detailed Study Design

A key element in planning and designing a tracer study is the preparation of a study design document. This document serves as the overall plan for conducting a tracer study and thus the roadmap for execution of the study. Three important study-specific parts of the design plan that may be required before the execution phase are a QAPP, a health and safety project Plan (HSPP), and a contingency plan. The contingency plan describes the actions to be taken if unexpected events occur, for example if distribution system concentrations of the tracer exceed the MCL for chloride or fluoride. The HSPP should at a minimum define the job hazards that might be encountered and the controls, protective equipment, sample handling and work practices, safety review procedures, and emergency procedures to be employed during the study.

The QAPP should clearly define the project objectives, organization, experimental approach, sampling procedures, analytical methods, protocols, instrument calibration requirements, data reporting, data reduction, and data verification procedures.

Addressing Agency and Public Notification

Appropriate agencies, including fire and police departments, should be notified prior to the commencement of field activities. With heightened awareness of security, all people participating should have a valid identification and contact information. A standard statement concerning the study should be developed and provided to all team members in case they receive inquiries at the study site. This same statement should be used by utility personnel to answer any telephone inquires that might be received. A summary information card may be provided to the study participants that could be handed out to the public during the study (if requested). This minimizes the risks of miscommunication.

If the injection site or installation of meters requires excavation, the study team must obtain the necessary permits and approvals. This is especially important if any of the sites are in a residential neighborhood or near a busy street or road. Care should be taken in all cases to provide adequate traffic control. Safety is of paramount consideration.

Executing a Tracer Study

The team should first become familiar with the detailed study design documents. Based on these documents, several tasks then need to be completed during the execution phase of a tracer study. These tasks include:

- Procurement, setup, testing, and disinfection of study equipment (including pumps, storage tanks, chemicals, reagents, tubing, connectors, and continuous tracer monitoring stations).
- Installation of field equipment and testing (both flow and tracer monitoring equipment to confirm study-specific distribution system operation and flow stability).
- Tracer dosage and injection duration calculations.
- Dry runs and planned tracer injection events.
- Real-time field assessments, sampling, and analysis.
- Equipment demobilization, initiation of data collection, reduction, and verification process.

A tailgate safety meeting before commencement of any field work is the best method to increase awareness. These specific execution subtasks are further discussed in the following subsections.

Procurement, Setup, Testing, and Disinfection of Study Equipment

Figure A-3. Tracer injection setup (storage tank, calibration tube, and feed pump) (Panguluri et al. 2005)

Field equipment should be procured on a timeline such that the items arrive several weeks before the planned study date, especially the monitoring and injection equipment that may require assembly. An early arrival will ensure that the equipment can be properly configured and tested before field use.

Unless precalibrated flow-paced injection equipment is purchased (or if the study does not require injection equipment—as in the case of using naturally/normally occurring tracers), the study team should obtain an appropriate injection pump setup. Figure A-3 shows a tracer injection system used by USEPA for field tests. This setup should be calibrated in the lab to compute the speed-specific dosage rate using the tracer solution. If appropriate, a flow-calibration tube should also be fabricated to confirm the flow in the field. Figure A-3 also depicts the flow-tube used by USEPA.

Concurrently, if applicable, the team should initiate the fabrication of the automated tracer monitoring stations. These stations are typically equipped with a probe for measuring the tracer (or a surrogate parameter such as conductivity), associated data logger, and batteries (for powering the probe and the data logger). If accurate measurement of flow through the automated monitoring station is needed, it should be augmented with a household-style water meter and logger. The equipment should be housed in a secure lock box to protect it during the field study. Figure A-2 shows an automated monitoring station used by USEPA and GCWW to conduct a tracer study. The entire setup should be tested in the lab to ensure proper operation and battery capacity to maintain uninterrupted operation.

The grab sampling, laboratory equipment, tracer storage tanks, transportation equipment, and arrangements should be procured and set up. The field equipment hookup, including interconnections between the tracer storage tank, injection pump, and flow-tube, should be leak tested. The equipment used for injection should be properly disinfected and tested prior to field deployment to ensure that no microbiological contamination results from the field tests.

If ultrasonic flowmeters are procured for field deployment, the equipment should be set up in a lab environment to confirm the individual component operation and approximate battery life. The existing flow and data acquisition systems to be used in the field study should be sampled for data accuracy and field communication. During the lab testing phase of the field equipment, the entire field (and backup) crew should familiarize themselves with proper operating procedures for the equipment they are designated to operate.

One procedure for equipment disinfection is to prepare approximately 50 gal of 50 mg/L chlorine disinfectant solution. This solution is then recirculated through the injection pump setup for about 15 min. Thereafter, continuously flush the injection pump using de-ionized water for about 15 min. A water sample should be collected at the end of the flushing cycle and sent for bacteriological analysis (coliform and *E. coli*) to ensure that the disinfection procedure was successful. For the purposes of sampling, sterile sample bottles should be used with a dechlorinating agent (e.g., sodium thiosulfate). The dechlorinating agent is added to remove any residual chlorine or other halogen that may continue the disinfection process in the sample and yield incorrect test results.

Installation of Field Equipment and Testing

Prior to the commencement of field activity each day, conduct a brief tailgate health and safety meeting to remind the crew of potential job hazards. Mobilization of field equipment for excavations (if required for installing main flowmeters) should be initiated to allow for the flow-monitoring devices to be installed prior to the scheduled injection event(s). This time lag will vary according to the needs of the specific study and could range from several days to several weeks. The early installation of flowmeters will allow the study team to capture actual field flow data for performing any revisions to tracer dosage computations and preliminary hydraulic modeling analysis. The flowmeter installation location should meet the manufacturer's recommendations for upstream and downstream straight lengths of undisturbed pipe. The excavations should be performed in accordance with the HSPP. Appropriate drainage for the excavated pits should be arranged in case rain is forecast during the study period.

The measured field flow data should be utilized to confirm the stability and range of flow at the injection location and other major branches of the system where flow is monitored. It may be necessary to operate the distribution system under specified conditions in order to achieve optimum results during the study. The operational changes that may be required include: scheduled cycling of tank levels, pumps, and valves. Time required for the deployment of the automated monitoring stations prior to the start of the tracer tests is dependent on several factors, including the number of monitoring stations, the distances between stations, the ease of attaching the stations to the sampling hydrants, and the effort required to calibrate the monitoring equipment. If feasible and consistent with normal operating policies, the system should be operated to avoid frequent abrupt changes in flow such as would be associated with a pump that was cycling on and off very rapidly.

A day or two prior to the execution of the tracer injection event, the study team should fully deploy the continuous monitoring stations (if used). These stations should be hooked up at the designated sampling locations, and data logs should be checked to ensure data are being collected. Flow through a monitoring station should be sufficient to minimize the time delay in detecting the injected tracer between the main and the sampling location. Experience has shown that 1 to 2 gpm is usually sufficient. The field crew should also test the coverage and reliability of field communication devices (such as cellular phones) in the designated study area.

Tracer Dosage and Injection Duration Calculations

Factors affecting the amount of tracer required for the study include the duration of the injection, the flow rate in the receiving pipe, and the target concentration in the distributed water. This target concentration should be consistent with drinking water standards. For example, if fluoride is being injected (into a system that does not fluoridate) with a secondary MCL of 2 mg/L, a reasonable target concentration level is 80 percent of the MCL, i.e., 1.6 mg/L. The injection rate should be set to meet that goal.

Using the principle of material balance, the resulting tracer concentration in a receiving pipe downstream of the point of injection can be calculated as follows:

$$Q_D = Q_U + Q_T \quad \text{(Eq A-1)}$$

$$C_D = [(C_B \cdot Q_U) + (C_T \cdot Q_T)]/Q_D \quad \text{(Eq A-2)}$$

Where:

Q_D = flow downstream of injection point, L^3/T
Q_U = flow upstream of injection point, L^3/T
Q_T = flow of tracer solution, L^3/T
C = concentration of tracer material downstream of injection point, M/L^3
C_B = background concentration of tracer material in distributed water, M/L^3
C = tracer concentration, M/L^3

For some tracers, the allowable concentration in the distributed water may be controlled by one of the dissolved ions that are part of the tracer. For example, if calcium chloride is the selected tracer, the concentration of the chloride ion in the distributed water controls the amount of tracer that may be injected.

Injection duration depends on the size and complexity of the distribution system and the modeling objectives of the study. A typical duration can range from 1 hr in a small or branched system, to 8 hr or more in a larger, looped system. Some studies have reported success with a series of pulses. However, if the duration of the injection is too short or the series of pulses too close together in time, it is difficult to separate the tracer fronts as they traverse different paths at different velocities through looped systems. The presence of tanks can also impact the needed tracer duration since active filling and drawing can dampen the resulting tracer concentration as it moves through the system.

The injection equipment should be located close to the main in order to minimize the tracer travel time to the main. Alternatively, the travel time should be compensated for during the appropriate phases of the study evaluation.

Dry Runs and Planned Tracer Injection Event(s)

Before the planned full-scale tracer injection event is actually carried out, the project team should consider conducting a smaller duration dry run injection to confirm the system operation and expected levels of tracer concentration. If continuous monitors are to be used in the study, then during the dry run some or all of the monitors should be installed and tested. The timing and duration of the dry run should be such that the injected pulse should be short and clear the system well before the actual event is initiated.

The dry run serves as a final systems check and provides the study team an opportunity to make any necessary last-minute changes prior to the actual study. Thereafter, the actual full-scale injection event should be conducted as planned.

Real-Time Field Assessments, Sampling, and Analysis

While the injection event is ongoing, the study team should carefully monitor the tracer concentration at the immediate downstream location of the injection to ensure that there are no significant deviations in the expected versus observed concentrations in the field. Field crews should communicate directly with the system operations. It is critical that the field personnel are aware of any changes in system operations that may affect the study. Unanticipated changes in water demand may cause the tracer concentration to exceed target concentration levels. In such an event, the field crew should be trained to take measures to minimize any adverse effects. The preventive measures may include lowering (or stopping) the injection rate or achieving appropriate dilution by means of rerouting water through the distribution system (as appropriate). Furthermore, any such tracer concentration exceedances should be confirmed by performing field grab sample analysis to make sure that the exceedance is real and not an instrument anomaly. Until the results are confirmed, it is best to err on the safe side and take preventive measures to maintain water quality.

Periodically, the field crew should take grab samples and inspect the continuous monitoring stations to ensure that the equipment is operating properly. The grab samples should be appropriately handled and analyzed in the field or transported to the laboratory for further analysis. The sampling and monitoring effort should continue well past the conclusion of the injection event until the tracer is expected (and observed) to have moved out of the system. This may take a period of 24 to 48 hr or more after completion of the injection event.

During the course of the sampling event, it is very useful to examine and assess the field data on a near real-time basis. Questions that should be asked include "Are the results reasonable?" "Is the tracer moving through the system at a speed consistent with predictions?" Based on this assessment, modifications may be made in terms of injection rate, grab sampling frequency, or study duration.

Equipment Demobilization, Initiation of Data Collection, Reduction, and Verification Process

After the scheduled injection event(s) are completed, the field crew should download the data (including flow and tracer concentrations) from the various monitoring devices. The data should be spot-checked against field grab sampling data to ensure that there are no time anomalies or gaps in the data log and the readings match relatively well.

After the field sampling events are completed, the crew should demobilize the equipment, remove the automated monitoring stations, refill any excavations, and restore the system operations to their normal conditions. Downloaded data from the field should be processed according to the QAPP and used for further modeling and analysis.

Tracer Study Costs

In general, the cost of conducting a tracer study is proportional to the study area size, number of monitoring sites, study duration, sophistication and amount of equipment, and complexity of poststudy analysis. If a study incorporates an injected tracer and the use of continuous monitors, it can be much more expensive initially than a study using a natural tracer and grab samples. However, the injection equipment and continuous monitoring equipment can be reused at various locations. These are the cost trade-offs between purchase of automated monitoring equipment and labor associated with grab sampling. In some cases, a larger data set derived from an automated monitor is necessary for a detailed analysis. Cost data presented in this section are intended to provide the basis for this type of analysis. For the purposes of this chapter, the overall costs have been broken down into two distinct categories: equipment and labor. Material costs are only a fraction of the total and therefore have been combined and included with equipment costs for simplicity.

Table A-2 lists typical equipment and material costs for those items that may be used in tracer studies. The unit costs are easily scaled to the needs of a specific study. Chemical tracer costs including analytical costs were provided earlier in Table A-1.

Costs may vary widely among studies. For example, if it is necessary to purchase or rent a storage tank or truck, costs will be higher if these types of items are not readily available. If the study team elects to analyze samples in-house rather than using an outside laboratory, the team should balance the costs of labor and additional reagents and chemicals against the cost of performing the analyses at an outside commercial laboratory. Labor costs may be even more variable than equipment and material costs, and labor costs and are a function of the size and complexity of the study. In order to provide an easy basis for comparison, the labor costs are presented in labor hours (Table A-3) and include a combination of engineers and technicians. Labor costs have been estimate for low-, medium-, and high-end studies. These estimates are obtained from actual field studies as described below. This approach should allow utilities to make site-specific cost estimates.

A typical example of a low-end tracer study is provided by the Sweetwater Authority distribution system in Southern California. The Sweetwater system covers a service area of 28 mi^2. The utility was able to take advantage of a naturally occurring tracer and used grab samples taken at 28 existing dedicated sampling sites over a period of 5 days. A study performed in the 21-mi^2 Cheshire service area of the South Central Connecticut Regional Water Authority in 1989 provides an example of a medium-level tracer study. In this case, the normal fluoride feed was shut off for a period of seven days (and then turned back on), and grab samples were taken at intervals of a few hours at 23 sites over a period of 14 days. An example of a high-end study is provided by a two-phased field investigation conducted in two suburban areas of GCWW. The first area is a small (<1 mi^2) dead-end system, and the second area a 12-mi^2 pressure zone. A calcium chloride tracer was injected and monitored using a combination of automated conductivity meters and grab samples. In the smaller area, 20 meters were used and monitoring was conducted over a 24-hr period. In the second area, 33 meters were used and two separate tracer injections were conducted over a period of 5 days. Including both studies, a total of 725 grab samples were taken and analyzed for conductivity, chloride, and calcium. Flow was monitored at 4 locations using ultrasonic flowmeters.

Tracers and tracing techniques have been used for many years in a number of engineering applications to estimate stream velocity and retention time in water and water supply unit processes. More recently, tracers have been used for calibrating drinking water distribution system hydraulic and water quality models. For the purposes of this book, it is assumed that tracer studies are used to calibrate and validate network models. The calibrated and validated

Table A-2. Equipment costs (Panguluri et al. 2005)

Equipment and Material	Unit Cost ($)
Injection pump	$1,000–$5,000
Flowmeter (ultrasonic meter for main pipes)	$7,000–$9,000
Excavation, rigging, and backfill (equipment rental per site)	$1,500
Lab chemicals, batteries, and plumbing supplies (lump sum*)	$1,000–$5,000
Automated monitoring box (self-constructed)	< $200
Online conductivity ISE, meter, and logger	$800–$1,500
Automated monitoring station water flowmeter	$600–$800
Online fluoride meter	$5,000–$10,000
Safety equipment (e.g., vests, first aid kits, raingear, and flashlights)	$500–$1,000
Communication equipment (e.g., radios and GPS)	$500–$1,000
Hydrant equipment (e.g., wrenches, caps, and hoses)	$1,000–$2,000
Transportation (e.g., rental vehicles)	$500–$2,000
Tracer storage tanks (depending on volume and material)	$500–$1,000

Table A-3. Representative labor hours for a range of studies (Panguluri et al. 2005)

Activity	Low-End	Medium	High-End
Planning	27	274	480
Setup	—	150	520
Field study	51	604	370
Laboratory analysis	8	160	120
Poststudy assessment	24	212	740
Total	110	1,400	2,230

network models were then used to estimate other parameters such as water age and travel times. However, the data from a tracer study can be directly used to estimate some specific parameters such as water age (DiGiano et al. 2005). A comprehensive summary of potential uses and regulatory applications for tracer studies is provided in the first subsection of this chapter. Drinking water tracers might include chemicals that are injected into a water distribution pipe, the temporary shutoff of a chemical additive currently being added to treated water (such as fluoride), or significant changes in concentration of disinfectants, DBPs, or natural compounds. The tracer methodology selected would significantly impact the overall costs of the study. Probably the most expensive option would be to inject a chemical tracer, monitor it using leased or purchased online instrumentation, and conduct the study using contractor staff. The least expensive approach would be to take advantage of a natural tracer, monitor the progress of the tracer by grab sampling, and conduct the study using primarily in-house staff. Once a tracer injection methodology has been selected, careful planning and execution will ensure the success of the study.

When planning a tracer study, if the specific steps outlined in this chapter are followed, they should greatly increase the potential for a successful study. These steps include: establishing clear study objectives, forming a study team, defining the study area characteristics, carefully selecting an appropriate tracer, selecting the proper field equipment, developing key planning documents, and ensuring that the public and affected agencies are notified. Application of a distribution and water quality model during the planning stage is highly recommended to simulate the approximate behavior that will be expected during the actual tracer event.

During the execution phase of the study, the following issues should be addressed: procurement of equipment and materials; setup, testing, and disinfection of the procured equipment; availability of analytical instrumentation and laboratory facilities; and, finally, the installation, testing, and operation of field equipment. During the execution phase, it is important to review and understand how tracer dosages and injection duration are to be implemented. Dry runs are highly recommended as a means of debugging the procedures prior to a full study.

Distribution system tracer studies have been conducted for over 15 years, but recent technology developments have improved the efficiency of these studies and provide promise for greatly expanded applications in the future. Specific components that will fuel this expanded use include the following: continuous monitors, which can be easily adapted for use in distribution systems and are being developed and tested, in part in response to water security concerns; automated meter reading (AMR) equipment, which is being installed by many utilities and could provide more detailed temporal and spatial consumption data for hydraulic models; and advanced analysis software, which is evolving and will facilitate the use of large amounts of continuous data in calibrating distribution system models. With increased availability of these technologies, costs are expected to decrease so that larger utilities can afford to purchase and routinely use the equipment, and consulting engineers can affordably offer these services to smaller utilities.

During the field study, it is important that the study team be able to assess the progress of the tracer, in real time, as it propagates through the system. Concise and consistent communications between tracer study team members, test coordinator, and water utility staff are critical at all times during the test.

In the future, it is highly likely that advances currently on the horizon will result in significantly increased use of both online tracer (or water quality) monitors and flow monitoring instrumentation. As the online technology becomes more widely used in drinking water, the use of network water quality models will also be more widely accepted. Online monitoring in conjunction with water quality modeling will provide an in-depth understanding of the manner in which water quality changes can be monitored in a drinking water distribution system. Also, given the current climate of concern over distribution water quality from both regulatory and security viewpoints, it is reasonable to assume that there will be increased interest in applying this type of technology in the water industry.

Hydraulic and Water Quality Model Calibration

Because hydraulic calibration is essential for any modeling study to be meaningful, it is clear that if a hydraulic model is not properly calibrated, the associated water quality model may not make accurate predictions. Water quality modeling is very sensitive to underlying hydraulic models, and a hydraulic model that has been calibrated sufficiently for applications such as master planning may require additional calibration before it is appropriate for use in water quality modeling. The following subsections describe the parameters and techniques employed for hydraulic and water quality model calibration.

Hydraulic Model Calibration

Hydraulic behavior refers to flow conditions in pipes, valves, and pumps, and pressure/head levels at junctions and tanks. Parameters that typically affect hydraulic modeling are pipe roughness factors, minor head losses, demands at nodes, the position of isolation valves (closed or open), control valve settings, pump curves, and demand patterns. When initially establishing and adjusting these parameters, care should be taken to keep the values for the parameters within reasonable bounds. Proper calibration requires that adjustments be made to the correct parameters. It is possible to make compensating errors when adjustments are incorrectly made in one set of parameters in order to match the field results while the parameters that are actually incorrect are left untouched. For example, it might be possible to make an adjustment in pipe roughness factors in order to compensate for a closed isolation valve. In this case, unreasonably low values for the Hazen–Williams roughness coefficients would be selected in order to force a large head loss in the pipes that are actually closed. Though this may result in approximating the pressure measurements made in the field, it will introduce other errors in flow and velocity calculations. Compensating errors may also result from incorrectly adjusting demands or other parameters.

Water Quality Model Calibration

Subsequent to the proper calibration of a hydraulic model, additional calibration of parameters in a water quality model may be required. The following parameters are used by water quality models that may require some degree of calibration:

- *Initial conditions*: Defines the water quality parameter (concentration) at all locations in the distribution system at the start of the simulation.
- *Reaction coefficients*: Describes how water quality may vary over time due to chemical, biological, or physical reactions occurring in the distribution system.
- *Source quality*: Defines the water quality characteristics of the water source over the time period being simulated.

The details of calibration depend on the type and application of the water quality model. Calibration requirements for each type of modeling are described below:

- *Water age*: Water age can be especially sensitive to inflow–outflow rates for tanks, mixing characteristics of tanks, and travel times in dead-end pipes. When modeling a tank, an important parameter is the initial age of the water in the tank at the start of the simulation. This value cannot be measured in the field but can be estimated by dividing the tank volume by the volume of water that is exchanged each day. Frequently, modelers will just assume that the initial age is zero and run the model for a long period until it has reached a dynamic equilibrium. This occurs when the initial water in the tank has been flushed out entirely through the fill and draw process. Figure A-4 shows the effects of the initial water age on the modeled results. As illustrated, a good initial estimate for water age (120 hr in this case) results in a much shorter time period until the dynamic equilibrium is reached. In fact, in this case when the initial age was input as zero hours, the model did not even come close to reaching dynamic equilibrium during the simulation period and would have required a much longer run duration to reach the same point.

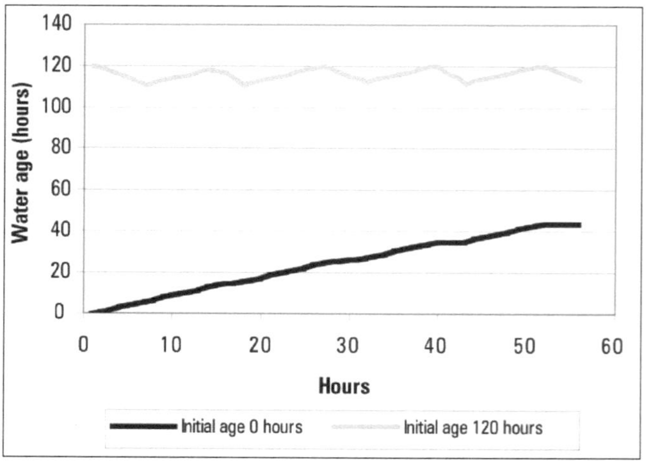

Figure A-4. Effects of the initial water age on the modeled results (Panguluri et al. 2005)

- *Source tracing*: No explicit water quality calibration can be performed because there are no reaction coefficients. Estimates of initial conditions in tanks for percentage of water coming from a source are desirable in order to shorten the length of the simulation. Values for sources are usually set to zero for all sources except for the specific source being traced.
- *Conservative constituents*: No explicit water quality calibration can be performed because there are no reaction coefficients. Estimates of initial conditions in tanks for concentrations of the conservative constituents can usually be determined from field

data and are desirable in order to shorten the length of the simulation. Values for sources are set to the typical concentrations found in the source.

- *Reactive constituents*: For reactive constituents, both the form of the reaction equation and the reaction coefficients must be provided. When modeling chlorine or chloramine decay, the most common formulation is a first-order decay equation including both bulk and wall decay coefficients. Values for these coefficients typically require laboratory and field analysis and calibration in order to match model results to the concentrations measured in the field. Correspondingly, THMs, a group of DBPs formed when water is chlorinated or chloraminated, generally increase in concentration with time (Vasconcelos et al. 1996). This process is frequently represented as a first-order growth function that asymptotically approaches a limiting value representative of maximum concentration reached when all of the NOM has reacted or all of the chlorine has been consumed. Both the limiting value and the rate of growth must be determined in this case. Water quality modeling is very sensitive to the hydraulic representation of the system. To reiterate, hydraulic calibration that may be sufficient for some hydraulic simulation may require additional calibration when used as a basis for water quality modeling.

Static Calibration and Dynamic Calibration

Just as water distribution system models can be run in a steady-state or an extended period mode, calibration can be performed either in a static mode using a steady-state model or in a dynamic mode using an extended-period simulation model. A common approach is to perform a static calibration first followed by EPS, to enhance the static calibration through a dynamic calibration. The options and procedures for these two types of calibration are described below.

Steady-State Calibration Methods

The two most common approaches used in calibrating a steady-state hydraulic model are *C*-factor tests and fire-flow tests. For water quality models of chlorine/chloramines, a test procedure for estimating bulk and wall demand may be employed. In all of these cases, field data are collected under controlled conditions and then applied to determine the model parameters that result in the best fit of the model to the field data.

C-Factor Tests

C-factor tests (sometimes called head loss tests) are performed to estimate the appropriate *C* factors to be used in a hydraulic model. The *C* factor represents the roughness of the pipe in the widely used Hazen–Williams friction equation. Typically, such tests are performed on a set of pipes that are representative of the range of pipe materials, pipe age, and pipe diameters found in the water system that is being studied. The results of the tests are then used to assign *C* factors for other pipes of similar characteristics.

In a field test, a homogeneous section of pipe between 400 and 1,200 ft long is initially isolated. Subsequently, flow, pipe length, and head loss are measured in the field. Typically, nominal pipe diameters are taken from system maps and these values are used along with flow rate to calculate velocity. Two alternative methods exist for determining head loss in the pipe section: a two-gage method (Figure A-5) and a parallel hose method (Figure A-6). These techniques are described in detail in Panguluri et al. (2005).

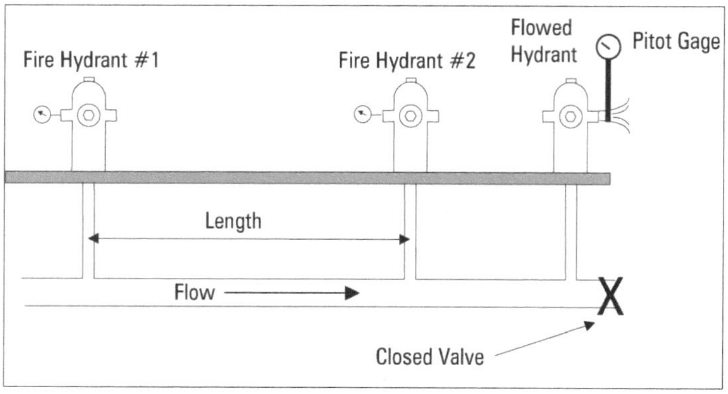

Figure A-5. Schematic of standard two-gage *C*-factor test setup (Panguluri et al. 2005)

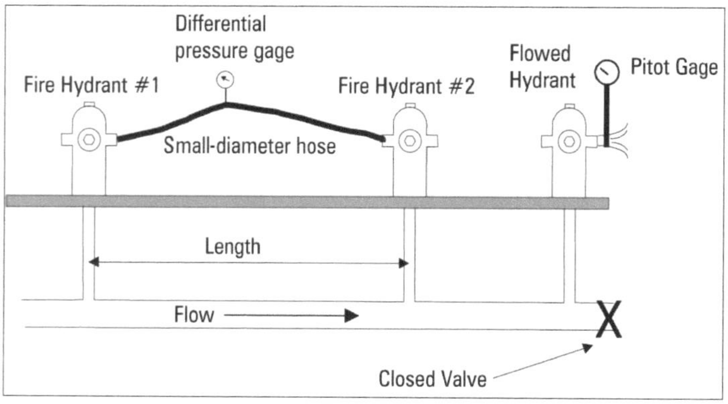

Figure A-6. Schematic of parallel hose *C*-factor test setup (Panguluri et al. 2005)

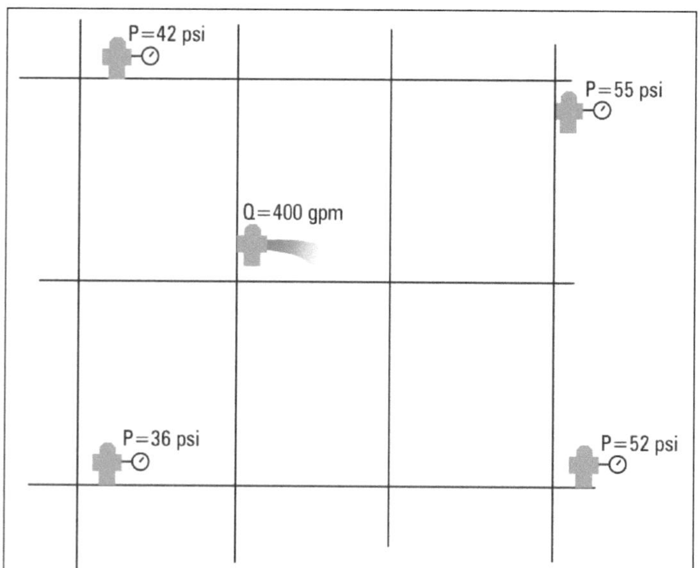

Figure A-7. Fire-flow test setup (Panguluri et al. 2005)

Fire-Flow Tests

Fire-flow tests are routinely performed by water utilities to determine the ability of the system to deliver large flows needed to fight fires. In such a test, fire hydrants are opened, the flow through the hydrants measured, and pressures measured at adjacent hydrants (see Figure A-7). The high demands caused by the open hydrants lead to high flows and increased head loss in pipes in the area around the hydrants. Under these conditions, the system is stressed and the capacity of the system to deliver these flows is very sensitive to the roughness of the pipes.

These fire-flow tests can also be very effective as a calibration methodology. In this case, in addition to the standard information routinely collected as part of a fire-flow test (flows and pressures), information is collected on the general state of the system such as pump and valve operation, tank water levels, and general system demand. The distribution system model is then run under the system conditions observed during the test and adjustments made in roughness factors (or other parameters) so that the model adequately represents the data measured in the field.

Figure A-8 illustrates an example setup for a fire-flow test. The diffuser attached to the hydrant in the figure includes a pitot gage used to measure the flow. The cage diffuses the flow and prevents any objects in the stream from being projected out at high speed. In the case shown in Figure A-9, only a single hydrant is opened, with the flow measured at that hydrant and pressure measurements made at four hydrants. Additional hydrants may be flowed and monitored as part of a fire-flow test for calibration.

Figure A-8. A hydrant being flowed with a diffuser as part of a fire-flow test (Panguluri et al. 2005)

Chlorine Decay Tests

Chlorine bulk reaction and wall reaction (or demand) testing procedures can be used to determine the reaction parameters used in water quality models. Bottle tests measure the rate of chlorine reaction that occurs in the bulk flow independent of wall effects. This procedure is performed by first measuring the chlorine at a representative location such as in the effluent from a water treatment plant. Then several bottles are filled with the same water and kept at a constant temperature. Separate bottles are subsequently opened at intervals of several hours (or days) and the chlorine content is measured. The resulting record of chlorine at different times is used to estimate the bulk reaction rate. *Manual M32* (AWWA 2004) provides a more complete protocol for this test.

The purpose of the chlorine decay field testing procedure is to estimate the chlorine wall demand coefficient for representative pipes in the distribution system. The method described here involves the measurement of chlorine concentrations in a pipe segment under controlled flow conditions and use of the resulting chlorine measurements to determine the wall reaction rate for that pipe segment. The method is designed to be complementary with *C*-factor testing so that it can be conducted in conjunction with a *C*-factor test. The method

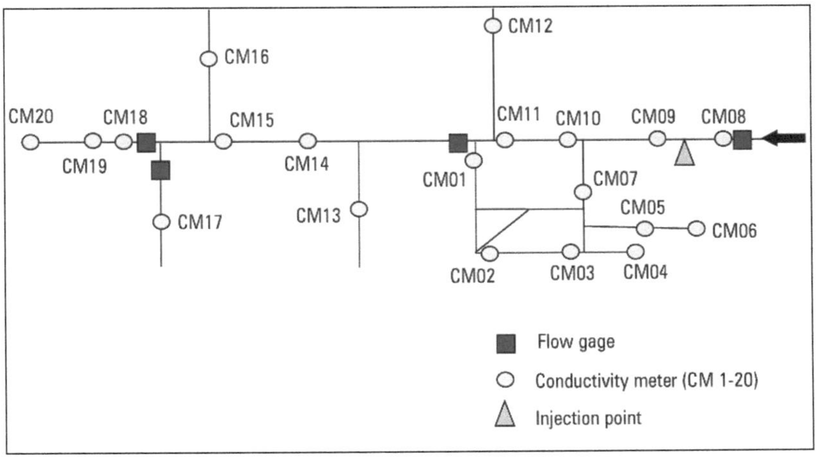

Figure A-9. Schematic representation of small-suburban dead-end system (Panguluri et al. 2005)

is considered to be experimental and feasible only for pipes that are expected to have relatively high wall reaction values, such as smaller-diameter unlined cast-iron pipes. For the smaller-diameter unlined cast-iron pipes, pipe sections with a length in the range of 1,500 to 2,000 ft will be required to estimate wall demand. For other types of pipes that typically have low wall decay factors (e.g., plastic and new pipes), the required length of the pipe may be so long as to make this test impractical. Other factors that should be considered in selecting sites include the following:

- Ability to measure flow in the pipe.
- Ability to valve off the pipe segments.
- Presence of a reasonable chlorine residual (preferably > 0.4 mg/L) at the upstream end of the pipe segment.
- Ability to vary flow in the pipe over a reasonable flow range (e.g., for a 6-in. pipe, a range of flows of 100 to 500 gpm would be desirable).
- Ability to estimate the actual pipe diameter for the pipe segment.

For the selected pipe segment, major lateral(s) and downstream segments should be valved off to control flow in the pipe. Two or three sampling points should be established along the segment of interest (upstream, downstream, and an optional midpoint). Typically, these would be taps on fire hydrants. Prior to the testing, the taps should be run for several minutes to clean out the line. The approximate travel time through the pipe should be calculated and chlorine measurements taken from upstream to downstream so that approximately the same parcel of water is sampled at each station. Flow measurements can be made at any location within the segment.

The test should be repeated for three flow values: a low flow rate, a medium flow rate, and a high flow rate. During each flow test, chlorine residual should be measured at each of the two or three sampling points. Since relatively small variations in chlorine concentration are expected, a good quality field chlorine meter should be employed and three replicates should be taken at each sampling point for each flow test. Following the field analysis, a spreadsheet can be used to back calculate the resulting wall reaction coefficients, or a water distribution model can be used to determine the wall reaction coefficient through trial and error.

Dynamic Calibration Methods

Dynamic calibration methods are associated with the use of an EPS model. The dynamic calibration methods include: (1) comparison of modeled results to measurements made in the field over time and (2) tracer studies. In both cases, model parameters are adjusted so that the model adequately reproduces the observed behavior in the field. Tracer studies are addressed later in this chapter.

Comparison of modeled and measured data can be used for calibration of both hydraulic and water quality models. The most commonly measured hydraulic data are tank water levels, flows, and pressures. Frequently, this information is routinely reported through SCADA systems to a database and can be extracted. In other cases, continuous flowmeters or pressure gages must be installed to collect data during a test period. Generally, tank water level data and flow measurements are the most useful form of data for calibrating an extended period model. Under average water use conditions, temporal variations in pressure measurements typically vary over a relatively small range and then only in response to variations in tank water levels. As a result, they are less useful in calibrating model parameters. If pressure measurements are going to be used for dynamic calibration, the system must be stressed by conducting fire-flow tests during the testing period. The primary model parameters that are adjusted during dynamic calibration are: demand patterns, pump schedules and pump curves, control valve settings, and the position (open or closed) of isolation valves.

Dynamic calibration can also be used for calibrating water quality parameters, such as the wall demand coefficient for computing chlorine residuals. Generally, water quality field studies are performed in conjunction with field hydraulic studies or with a tracer study. For chlorine models, measurements of chlorine are taken at frequent intervals in the field at representative sites. These may include dedicated sampling taps, hydrants, tank inlet/outlets, or other accessible sites. Continuous chlorine meters may also be used. During the model calibration process, the model is first calibrated for hydraulic parameters, and water quality coefficients are subsequently adjusted so that the model results match the field data.

Manual Calibration and Automated Calibration

The aforementioned process of adjusting model parameters so that the model reproduces the hydraulic and/or water quality results measured in the field can involve a significant amount of effort in large or complex systems. As noted earlier in this chapter, many parameters can be adjusted in the model, and the combinations of possible parameter values can sometimes appear to be quite overwhelming. Typically, a manual trial-and-error approach is used. The most influential parameters can be identified based on sensitivity analysis and then adjusted to see if they improve the results. This process is continued until an acceptable level of calibration is achieved or until budgetary constraints dictate closure. It is not unusual for dozens or even hundreds of separate model runs to be made in this process.

An extension to the manual calibration process is an automated approach that allows the computer to search through different combinations of model parameters (with a realm of realistic values) and to select the set of parameters that results in the best match between measured and modeled results. The development of this type of program has been the topic of many studies over the past 25 years (Walski et al. 2003).

Automated methods require a formal definition of an objective function for measuring how good a particular solution is. Generally, the value of a solution is measured by a statistic that reflects the deviation between measured and modeled results in flow and pressure. A commonly used objective function is minimization of the square root of the weighted summation of the squares of the differences between observed and predicted values. The weighting is used to establish a relationship between the errors associated with flow and pressure. For example, the user may choose a 1-psi error in pressure prediction to be equivalent in value to a 10-gpm error in flow.

In most automated methods, the user also groups pipes by common characteristics, such as age, material, and nodes, into common demand characteristics such as residential or commercial. The user then specifies a range of allowable values for pipe roughness factors or a range of multipliers applied to the existing roughness factors. Similarly, a range of allowable demand multipliers is also specified, as are potential pipes where an existing isolation valve may be closed. The optimization routine is then applied and the roughness, demands, and isolation valve positions are selected that result in the minimum error.

Though manual calibration still remains the predominant methodology, automated calibration methods are becoming more available in commercial modeling packages. It is likely that as the automated calibration methods are refined, the technology will expand for routine use with EPS hydraulic and water quality models.

Case Studies

In order to illustrate some of the calibration methods described earlier in this chapter, two case studies are presented in this subsection. The two case studies are similar in general methodology but differ in the overall scale and specifics of the study area. In both cases, the distribution system model that was used as a starting point for the calibration exercise was part of a skeletonized model extracted from unspecified portions of the GCWW distribution system.

Most larger urban water systems generally have at least a skeletonized model of their distribution system. A skeletonized model denotes a model that includes only a major subset of actual pipes rather than all pipes in the distribution system. The extracted system model was modified and converted to EPANET format for use in this project. The modifications included: addition of key pipes, updates to consumer demand data, and an interconnection between the case study area and the full system by a fixed-grade node (reservoir). These portions of the base model had been previously calibrated using various dynamic calibration methods and were used for routine water utility work. For the purposes of calibration, separate field studies were conducted in each study area.

In both field studies, a food-grade conservative tracer (calcium chloride) was introduced into the system, and its movement through the system was monitored by both grab sampling and continuous monitoring (CM) stations installed at key locations in the distribution systems. The CM stations were installed at hydrants that were left partly open for the duration of the study to minimize travel time between the main and sampling location. Each open hydrant was added as a new demand node in the EPANET network model. Additionally, several ultrasonic flowmeters were installed to provide continuous flow measurements at key locations. The general procedures, methodology, and instrumentation used in these field studies are consistent with those presented later.

Case 1: Small-Suburban Dead-End System

The Case 1 system is part of a larger pressure zone. It was selected because of the relatively compact size and simple structure, fed by a single supply pipe with no additional storage. As a result, the movement of the tracer was relatively rapid through the system, and it could be monitored with continuous meters placed at several locations. The general layout of this subsystem, the location of the injection site, and the monitoring locations for this study are shown in Figure A-9.

The calcium chloride tracer was injected as two pulses, a 2-hr pulse followed by a 2.5-hr period of no injection, followed by a higher concentration pulse of 2-hr duration. The injection rate and the resulting concentration of the tracer in the distribution system just downstream of the injection point were carefully monitored to ensure that the resulting chloride concentration did not exceed the secondary maximum contaminant level (MCL) of 250 mg/L for chloride.

Compliance with state and federal regulations during a tracer study is obviously quite important. In order to ensure that the tracer will not exceed allowable levels, it is necessary to monitor information such as the rate of injection of the tracer, the flow in the receiving pipe, and the resulting concentration in the receiving pipe. Frequently, a safety factor for the injection rate is included to account for uncertainty. In this field study, the tracer injection rate was very low and the flowmeter on the injection pump provided approximate values. Chloride concentrations were monitored at a suitable location approximately 100 ft downstream of the injection point with a travel time of approximately 10 min. Due to unexpected variations in flow through the pipe, delay in measurements, and related computations (associated with tracer travel time), chloride values exceeding the target level were experienced for a brief period before the injection rate was adjusted.

The movements of the tracer pulses were monitored by using both manual sampling and continuous conductivity meters located throughout the distribution system. Additionally, four ultrasonic flowmeters were installed in the study area to provide continuous flow measurements at key locations within the distribution system.

In preparation for the calibration process, the conductivity readings were converted to chloride concentrations using a relationship developed in the laboratory. Figure A-10 shows the relationship between conductivity and chloride and the best-fit linear and polynomial relationships between them. This conversion was necessary because conductivity is not a truly linear parameter and, as a result, cannot be simulated exactly in a water distribution system model. The converted continuous concentration readings were then compared to the manually collected data for quality control purposes. Figure A-11 shows the resulting chloride data set that was used at one location as a basis for evaluating model predictions as part of the calibration process.

The preliminary results indicated some discrepancy between the EPANET-model predicted values and the actual field-measured values, indicating the need for model refinement and recalibration to improve the prediction capability of the EPANET model. Therefore, EPANET modeling was performed to evaluate the following four levels of model refinements:

- *Level 1 (prior to calibration)*: A skeletonized EPANET model was used with the original hourly demand pattern provided by GCWW and a time-step injection pattern of 60 min.
- *Level 2*: The same as Level 1, but a refined 10-min time-step pattern for injection was used along with the conversion of the original hourly demand patterns to 10-min patterns.
- *Level 3*: The same as Level 2 with a refined demand pattern for each node using the field-measured flow data, addition of demand nodes representing water demand of the partially open hydrants, adjustment for a large industrial user of water in the study area (based on

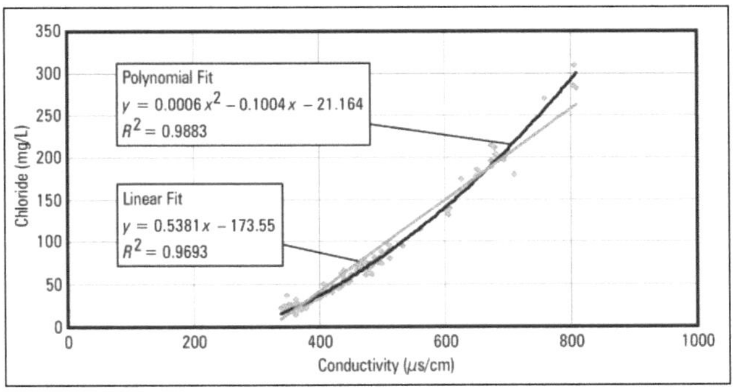

Figure A-10. Empirical relationship between chloride and conductivity (Panguluri et al. 2005)

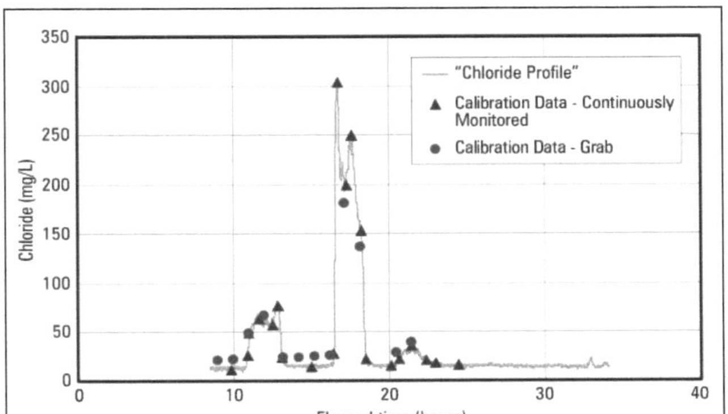

Figure A-11. Sample chloride data used at one station for calibration (Panguluri et al. 2005)

data obtained during the study), and the residential water billing information provided by GCWW.

- *Level 4*: The same as Level 3 with a detailed all-pipe (nonskeletonized) EPANET model.

The results of the four-stage model refinement and calibration process are shown in Figure A-12 for a continuous monitoring location (CM-18) located on the main feeder pipe (see explanation, p. 250). As illustrated, the improvements in the demand estimates and inclusion of the system details in the all-pipe model resulted in a vast improvement in the model's prediction ability for that monitoring location. Similar results were found for most monitoring locations on the main pipe.

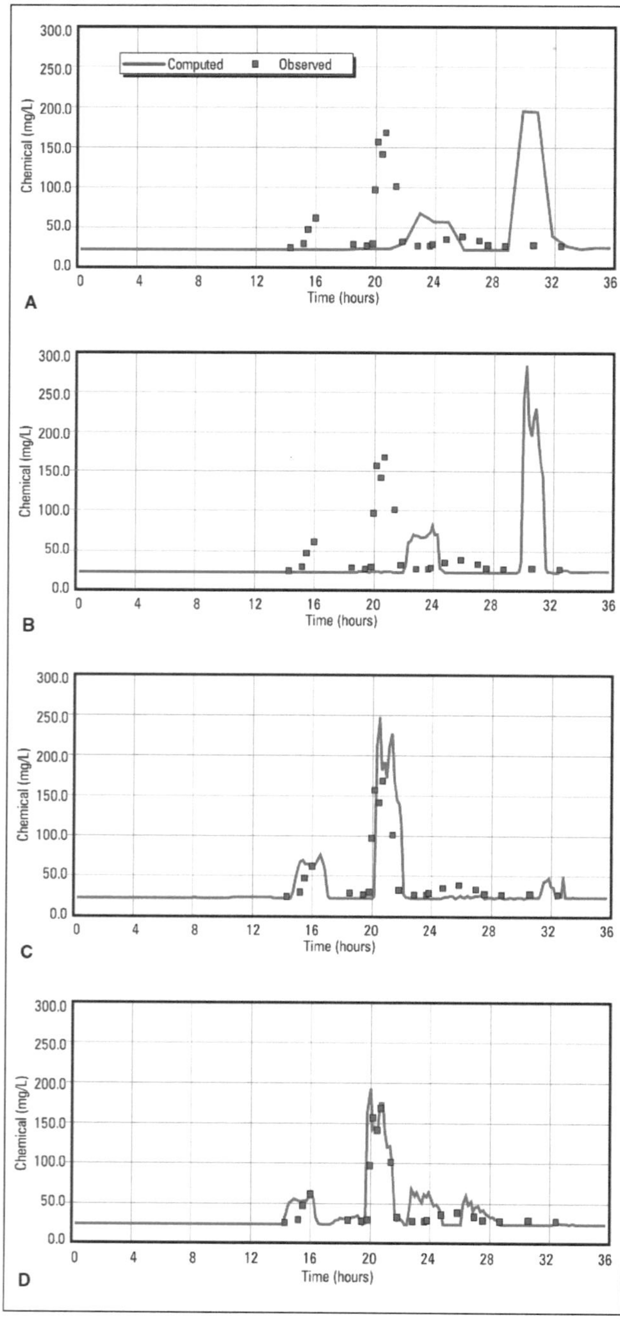

Figure A-12. comparison of model versus field results for continuous monitor location CM-18 at various calibration stages (Panguluri et al. 2005)

During the calibration and refinement process, various model inputs such as flow, demand, and pipe characteristics were adjusted to improve the model prediction. The EPANET model was considered to be calibrated for the area when the field data matched the model-predicted output to an acceptable degree based on visual observation. Depending on the location of the junction (where the model predictions are compared with the field values), both concentration and predicted time of tracer arrival might not be in perfect agreement because of local variation in demands, local flow velocities, and dilution impacts. The sharp tracer fronts observed in this field study made it difficult to employ quantitative statistical measures (e.g., mean error, standard deviation, root mean square error). Therefore, a graphical (visual) approach was considered to be more suitable for model calibration in this application. For example, if the prediction of the arrival time for the tracer differs by even a few minutes from the observed arrival time, use of these standard measures of error could result in a high number, even though the prediction could be viewed graphically as very good.

The calibration of the "looped" portion (referring to the portion of the network on the bottom right side of Figure A-9) of this network proved to be more difficult, and the results for some monitoring locations on the looped piping were less satisfactory. The most problematic were continuous monitoring locations CM-02 and CM-04. Monitoring station CM-02 was located near the confluence of two separate loops, with the actual monitored connection being slightly offset from the junction node. Examination of the model results showed that flow reached that junction from both directions and small variations in the amount of flow in each of the loops resulted in very different travel times. As illustrated in Figure A-13, this complex travel pattern along with the offset location of the monitoring station resulted in poor prediction of travel time to that station. Also, monitoring station CM-04 is located at the end of a dead-end pipe section, and travel to this node is strongly influenced by demands at the very far end of the dead-end section. As illustrated in Figure A-11, this resulted in a poor match of the peak concentration during the second pulse. It is also postulated that dispersion, which is not represented in EPANET, may have had an influence on the peak concentration because of the very low velocities in the dead-end pipe. In some cases, this could also be caused by inaccurate *C*-factors as applied to the distribution system. However (as illustrated in Figure A-13), for monitoring location CM-03 located in the main part of the looping system, the model and field agreement was quite good.

Case 1 data illustrate that, depending on the level of refinement and calibration, a significant variation exists in the capability of a model to accurately represent the system. In general, the parts of the network that are configured as trees (main stem with branches) are more easily calibrated by making adjustments in demands. For looping parts of the system and at dead-ends,

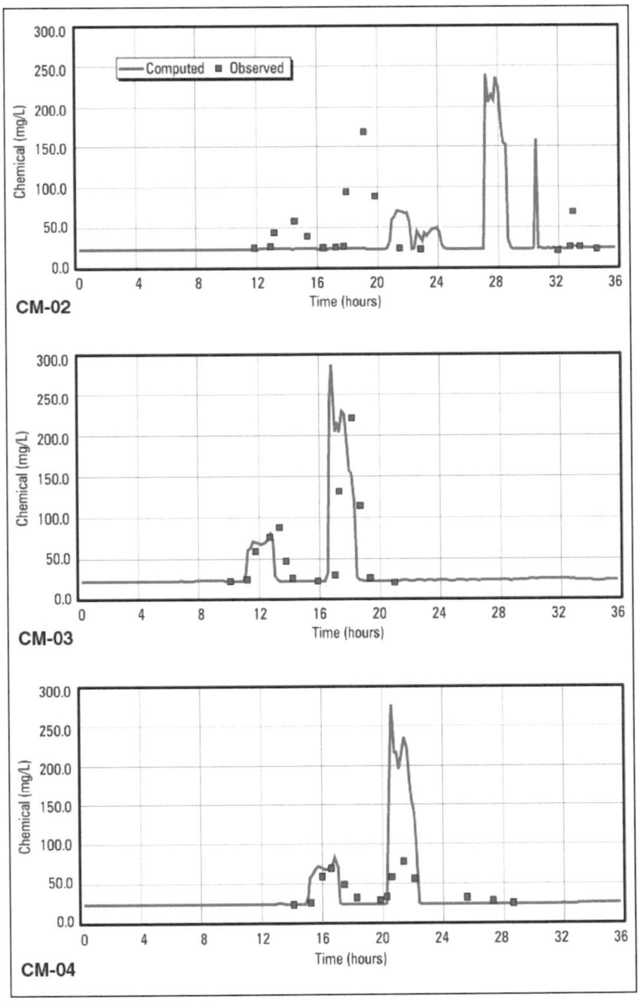

Figure A-13. Calibration of "looped portion" (Panguluri et al. 2005)

results are very sensitive to small variations in demands and system configuration, leading to the possibility of significant prediction errors at some locations. Uncertainty in demand estimates can be a major source of error in the model estimates.

Case 2: Large-Suburban Pressure Zone

Similar to Case 1, a field study and calibration exercise were carried out in a large-suburban pressure zone. This area was selected to demonstrate the application of tracer studies and calibration techniques in a more complex area. The selected area contained multiple pumps and tanks. The selected distribution system area is representative of relatively complex, well-gridded systems found in many larger water systems. The layout of the system, the location of the injection site, and the monitoring locations are shown in Figure A-14.

Two separate tracer studies were performed in this zone. The first study was used to further calibrate the skeletonized model received from the water utility. The second study served as a validation event to test the veracity of the calibrated model. In the calibration event, the tracer was introduced directly into the main feed line servicing the entire area (characterized by higher flow/higher pressure). In the validation study, the tracer was pulsed. A total of 34 continuous conductivity meters were installed in the system. At key locations, 4 flowmeters were temporarily installed to provide flow measurements.

During the calibration study, the calcium chloride tracer was injected into the main feed line serving the area for a period of 6 hr. In the validation study, the tracer was pulsed by fill-and-draw cycles in a storage tank at the same location. In both cases, a target chloride concentration of 190 mg/L or lower was set in order to safely not exceed the 250-mg/L secondary MCL for chloride.

Also during the calibration process, initial EPANET model simulations were reviewed in detail to determine the flow patterns around various monitoring locations and to attempt to identify causes for discrepancies in the observed and predicted values. A careful examination of the areas of significant discrepancies indicated that these were primarily limited to three geographic subregions within the skeletonized network. In addition to these three subregions, there were a few isolated locations where the predicted tracer pattern did not match the observed tracer pattern from the field study. The modeling team carefully examined each of these regions and addressed the zonal issues accordingly. The three subregions are shown in Figure A-14.

In Region 1 (CM42, CM43, and CM44), the field data indicated that the tracer arrived at these continuous monitoring locations several hours before the model's prediction. On closer inspection, it was found that a potential flow path existed that was not included in the skeletonized model. Although the pipe diameter was small, it significantly altered the hydraulic water flow path to that region. This missing pipe link was added to the model, using the appropriate pipe parameters. Furthermore, the modeling team investigated the GIS database to see if any substantial changes occurred in these areas since the time the original water demand patterns were developed five years ago. The updated GIS information indicated a presence of recent housing development in that region. Therefore, additional demand nodes were entered into the EPANET model to accommodate this development. Another possibility for the discrepancy was that the demand in this region was significantly higher than the average residential demand modeled in the area. To simulate this possibility, a sensitivity analysis was performed in which the modeled demand in this region was doubled. The model-predicted results improved significantly for this region based on these three adjustments.

In Region 2 (CM52, CM53, CM55, and CM56), an opposite phenomenon to that in Region 1 was observed. The field data indicated that the tracer arrived several hours after the model's prediction. One possible explanation was that this region had lower demand than the average residential demand modeled in this area. The flowmeter data upstream of this location supported this theory as the EPANET predicted flow in this pipe was much higher than the field observed flow (see Figure A-15a). To simulate this possibility, the local demand in this region was reduced by 30 percent in the model. The resultant flow matched the flowmeter data (see Figure A-15b). Also, similar to Region 1, it was found that a potential flow path had again been left out because of skeletonization of the model, which affected CM52. This pipe link was added to the model using the appropriate pipe parameters. Distribution mains between CM55 and CM53 were also found to have been upgraded since the EPANET network model was developed for this area. The EPANET model pipes for this location were updated using the newer information. The model-predicted results improved significantly for this region based on these three adjustments.

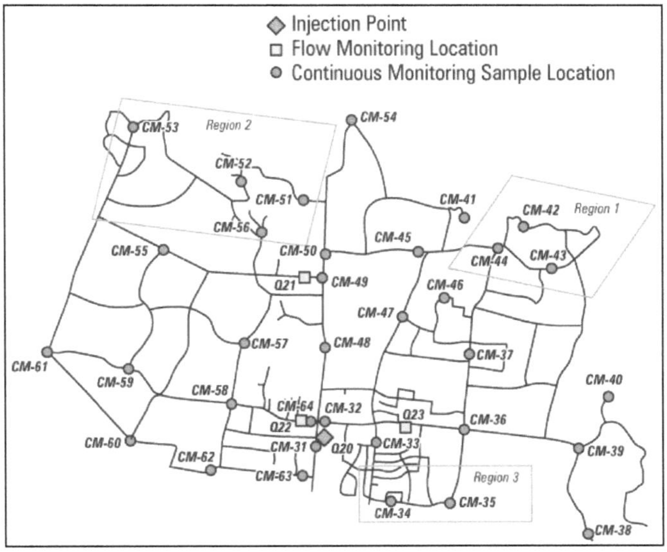

Figure A-14. Schematic representation of Case 2 study location (Panguluri et al. 2005)

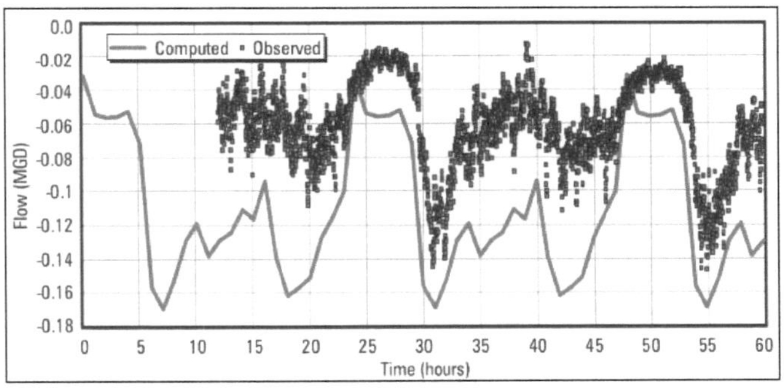

Figure A-15a. Modeled flows compared to measured flows before calibration (Panguluri et al. 2005)

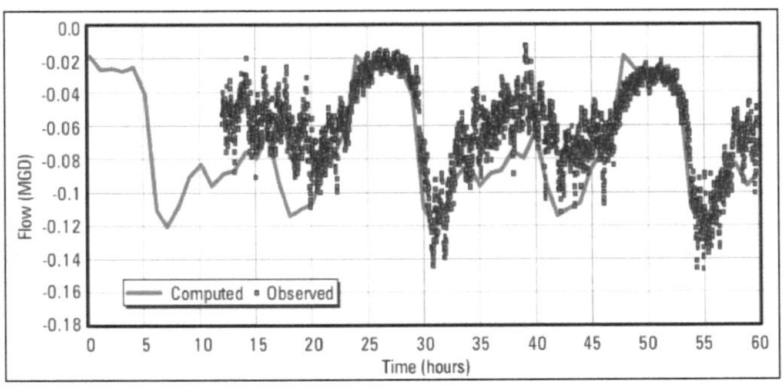

Figure A-15b. Modeled flows compared to measured flows after calibration (Panguluri et al. 2005)

In Region 3 (CM34 and CM35), the field data indicated that the tracer arrived at locations CM34 and CM35 several hours after the model's predicted arrival time. However, the field-verified tracer arrival time matched the predicted tracer arrival time at location CM33, which is slightly upstream of these locations. Also, a review of the water flow pattern in this region indicated that the water traveled from CM33 toward CM34 and CM35 (at all times). Based on the demands in the EPANET model, the pipe lengths, and the regional water flow information, the delay in tracer arrival at CM34 and CM35 could not be explained. A closer inspection of the region revealed a complex grid of interconnected pipes in this region, which were skeletonized as two parallel pipes. This skeletonization eliminated a number of different possible hydraulic flow paths between CM 33 and CM34/CM35. Also, in the EPANET model inputs, it appeared that the demand close to CM34 and CM35 was set artificially higher (to account for the overall demand in the skeletonization process). This model setup resulted in the predicted faster tracer arrival times at CM34/CM35 than those observed in the field. To account for this anomaly, a few pipe segments from the master plan were added to the skeletonized model of this region to better simulate the actual grid demands near CM34 and CM35. This model adjustment resulted in better prediction of the tracer arrival times.

During the calibration process, as demands were adjusted, a mass balance was performed for each hour to ensure that the net water demand in the study area remained the same, i.e., the increase in the demand at certain nodes was balanced by the reduced demand at other nodes to eliminate any net impact on water demand. In the final refinements, a multiplier of 2.0 was used for the base demand in Region 1, and a multiplier of 0.7 for the base demand in Region 2. These refinements showed some improvement in the model's ability to correctly predict the tracer arrival time and concentration. These calibration efforts resulted in a relatively well-calibrated network model. However, some local problems remained, especially in looped areas and areas that were branched off from the main lines.

The substantial changes made to the EPANET skeletonized model representing the large area necessitated a validation process. Therefore, the calibrated EPANET model input file from the first event was used to validate the model's capability to predict the results during the subsequent tracer addition. For the purposes of this validation, the data from the second set of pulsed injections were modeled using the calibrated EPANET network model for the study area to see how the predicted results compared with the continuous monitoring data collected during this event. The modeled and measured concentrations are compared in Figure A-16a for the EPANET calibration. A similar comparison is shown in Figure A-16b for the validation study.

Additionally for the purposes of this analysis, the EPANET predictions from the validation event were compared with the field results for each monitoring site and each site was given a grade as follows:

- Very good match (within ±20 percent of the actual concentration and within ±1 hr of the actual tracer arrival time).
- Moderate match (within ±30 percent of the actual concentration and within ±5 hr of the actual tracer arrival time).
- Poor match (greater than ±30 percent of the actual concentration or greater than ±5 hr of the actual tracer arrival time).

Of the 34 monitoring sites in this study area for the validation event, 15 received a grade of very good match, 14 were in the moderate match category, and 5 received the lowest grade of poor match. In general, it was found that better matches occurred in larger pipes serving large populations, while the poorest matches occurred in more localized loops serving fewer customers. These results are, in general, quite similar to the results obtained for the calibration event, and most problems repeatedly occurred at the same locations for both events. The validation event results confirm the fact that the calibrated EPANET network model can now be used to predict the outcome of a separate event to the same degree of accuracy.

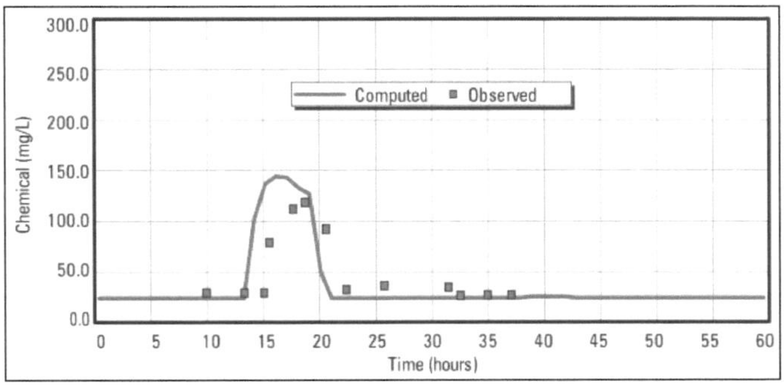

Figure A-16a. Chloride concentration for calibration event at continuous monitor location CM-59 (Panguluri et al. 2005)

Future of Model Calibration

Calibration continues to be a major focus of most modeling efforts. It can provide a model that may be used with greater confidence and produce results that are commensurate with the important decisions that are made based on the application of the model. However, there is significant room for improvements in calibration methodologies and in developing a standardized set of calibration protocols. This has led to an active research program in this area that is expected to continue into the future.

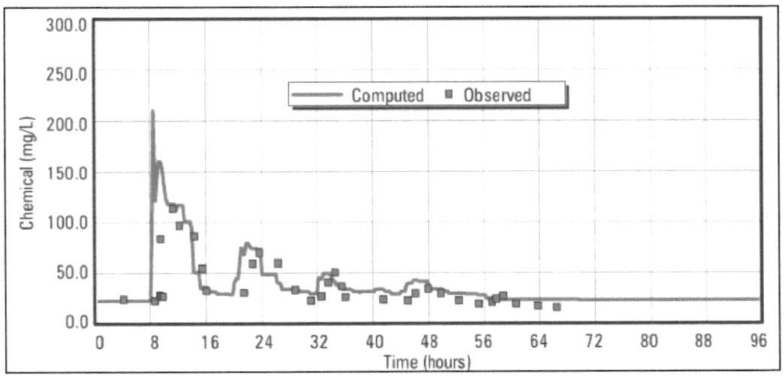

Figure A-16b. Chloride concentration for validation event at continuous monitor location CM-59 (Panguluri et al. 2005)

Calibration Standards

The following issues are raised frequently in the field of distribution system modeling: extent of calibration needed for various applications and standards for calibration. Though these are very reasonable questions, straightforward answers are usually not readily available.

General agreement exists in the modeling profession that the amount and degree of calibration required for a model should depend on the intended use of the model (ECAC 1999). Some applications such as design and water quality analysis typically require a high degree of calibration, while other uses, such as master planning, can be performed with a model that has not been calibrated to such a high standard. However, no universally accepted standards exist.

In the United Kingdom, performance criteria for modeling distribution systems are in place (Water Authorities Association and WRc 1989). These are expressed in terms of the ability to reproduce field-measured flows and pressures within the model, as shown below.

1. Flow:
   - ±5 percent of measured flow when flows are more than ±10 percent of total demand (transmission lines).

- ±10 percent of measured flow when flows are less than ±10 percent of total demand (distribution lines).

2. Pressure:
   - 0.5 m (1.6 ft) or 5 percent of head loss for 85 percent of test measurements.
   - 0.75 m (2.31 ft) or 7.5 percent of head loss for 95 percent of test measurements.
   - 2 m (6.2 ft) or 15 percent of head loss for 100 percent of test measurements.

In 1999, the AWWA Engineering Computer Applications Committee developed and published a set of draft criteria for modeling. These were intended not as true calibration standards but rather as a starting point for discussion on modeling needs.

At this point, no clear movement toward establishing calibration standards has taken place. However, it is likely that the need for further guidance in this area will increase as the extent and sophistication of modeling continue to expand.

Technological Advances

Research is continuing in two areas that strongly influence the likelihood of improved calibration of water distribution systems models: monitoring technology and optimization techniques. The available optimization techniques (and those under development) have been briefly discussed. Active research and development areas include optimization techniques for water quality calibration, EPS models, and use of tracer data. Areas of research, development, and experimental applications in monitoring technology include less expensive meters that can be inserted into pipes in the distribution system and automated monitoring for use in conjunction with tracer studies.

References

American Public Health Association (APHA), American Water Works Association (AWWA), and Water Environment Federation (WEF). 2005. *Standard Methods for the Examination of Water and Wastewater.* 21st ed. Washington, D.C.: APHA, 2-44–2-45.

AWWA. 2004. *Manual M32, Computer Modeling of Water Distribution Systems.* 2nd ed. Denver, Colo.: AWWA.

Boccelli, D.L., F. Shang, J.G. Uber, A. Orcevic, D. Moll, S. Hooper, M. Maslia, J. Sautner, B. Blount, and F. Cardinali. 2004. Tracer Tests for Network Model Calibration. In *Proc. ASCE-EWRI Annual Conference.* Reston, Va.: ASCE.

Boccelli, D., and J. Uber. 2005. Incorporating Spatial Correlation in a Markov Chain Monte Carlo Approach for Network Model Calibration. In *Proc. ASCE-EWRI World Water & Environmental Resources Congress, Anchorage, AK.* Reston, Va.: ASCE.

Boulos, P.F., W.M. Grayman, R.W. Bowcock, J.W. Clapp, L.A. Rossman, R.M. Clark, R.A. Deininger, and A.K. Dhingra. 1996. Hydraulic Mixing and Free Chlorine Residuals in Reservoirs. *Journal AWWA,* 88(7):48–59.

Clark, R.M., and J.A. Coyle. 1990. Measuring and Modeling Variations in Distribution System Water Quality. *Journal AWWA,* 82(8):46–53.

Clark, R.M., W.M. Grayman, J.A. Goodrich, R.A. Deininger, and A.F. Hess. 1991. Field Testing Distribution of Water Quality Models. *Journal AWWA,* 83(7):67–75.

Clark, R.M., W.M. Grayman, R. Males, and A. F. Hess. 1993. Modeling Contaminant Propagation in Drinking Water Distribution Systems. *Journal of Environmental Engineering,* 119(2):349–364.

Clark, R.M., G. Smalley, J.A. Goodrich, R. Tull, L.A. Rossman, J.T. Vasconcelos, and P.F. Boulos. 1994. Managing Water Quality in Distribution Systems: Simulating TTHM and Chlorine Residual Propagation. *Journal of Water Supply: Research and Technology–AQUA,* 43(4):182–191.

DiGiano, F.A., and G. Carter. 2001. Tracer Studies to Measure Water Residence Time in a Distribution System Supplied by Two Water Treatment Plants. In *Proc. AWWA Annual Conference.* Denver, Colo.: AWWA.

DiGiano, F.A., W. Zhang, and A. Travaglia. 2005. Development of the Mean Residence Time From Tracer Studies in Distribution Systems. *Journal of Water Supply: Research and Technology–AQUA,* 54:1–14.

Engineering Computer Applications Committee (ECAC). 1999. Calibration Guidelines for Water Distribution System Modeling. In *Proc. AWWA ImTech Conference, New Orleans, La.* April 18–21. Denver, Colo.: AWWA.

Ferguson, B.A. and F.A. DiGiano. 2005. Impact of Temporary Switches From Monochloramine to Free Chlorine on Water Quality in Distribution Systems. In. *Proc. AWWA Annual Conference.* Denver, Colo.: AWWA.

Grayman, W.M., R.A. Deininger, A. Green, P.F. Boulos, R.W. Bowcock, and C.C. Godwin. 1996. Water Quality and Mixing Models for Tanks and Reservoirs. *Journal AWWA,* 88(7):60–73.

Grayman, W.M., L.A. Rossman, C. Arnold, R.A. Deininger, C. Smith, J.F. Smith, and R. Schnipke. 2000. *Water Quality Modeling of Distribution System Storage Facilities.* Denver, Colo.: AwwaRF and AWWA.

Grayman, W.M. 2001. Use of Tracer Studies and Water Quality Models to Calibrate a Network Hydraulic Model. *Current Methods,* 1(1):38–42.

Hatcher, M.D., W. M. Grayman, C. D. Smith, and M. Mann. 2004. Monitoring and Modeling of the Sweetwater Authority Distribution System to Assess Water Quality. In *Proc. AWWA Annual Conference.* Denver, Colo.: AWWA.

Maslia, M.L., J.B. Sautner, C. Valenzuela, W.M. Grayman, M.M. Aral, and J.W. Green Jr. 2005. Use of Continuous Recording Water-Quality Monitoring Equipment for Conducting Water-Distribution System Tracer Tests: The Good, the Bad and the Ugly. In *Proc. ASCE-EWRI World Water & Environmental Resources Congress, Anchorage, AK.* Reston, Va.: ASCE.

McTigue, N.E., and J.M. Symons, eds. 2010. *The Water Dictionary,* 2nd ed. Denver, Colo.: AWWA.

Panguluri, S., R. Krishnan, L. Garner, C. Patterson, Y. Lee, D. Hartman, W. Grayman, R. Clark, and H. Piao. 2005. Using Continuous Monitors for Conducting Tracer Studies in Water Distribution Systems. In *Proc. ASCE-EWRI World Water & Environmental Resources Congress, Anchorage, AK.* Reston, Va.: ASCE.

Sautner, J.B., M.L. Maslia, C. Valenzuela, W.M. Grayman, M.M. Aral, and J.W. Green Jr. 2005. Field Testing of Water Distribution Systems at U.S. Marine Corps Base, Camp Lejeune, North Carolina, in Support of an Epidemiological Study. In *Proc. ASCE-EWRI World Water & Environmental Resources Congress, Anchorage, AK.* Reston, Va.: ASCE.

Snoeyink, V.L., and I. Wagner. 1996. Principles of Corrosion of Water Distribution Systems. In *Internal Corrosion of Water Distribution Systems.* Denver, Colo.: AwwaRF/DVGW-Technologiezentrum Wasser, 12–13.

Teefy, S.M., and P.C. Singer. 1990. Performance Testing and Analysis of Tracer Tests to Determine Compliance of a Disinfection Scheme With the SWTR. *Journal AWWA,* 82(12):88–98.

Teefy, S.M. 1996. *Tracer Studies in Water Treatment Facilities: A Protocol and Case Studies.* Denver, Colo.: AwwaRF and AWWA.

US Environmental Protection Agency (USEPA). 2003a. *The Stage 2 DBPR Initial Distribution System Evaluation Guidance Manual.*

USEPA. 2003b. Drinking Water Advisory: Consumer Acceptability Advice and Health Effects Analysis on Sodium. Available at: http://www.epa.gov/safewater/ccl/pdfs/ reg_determine1/support_cc1_sodium_dwreport.pdf.

Vasconcelos, J., P. Boulos, W. Grayman, L. Kiene, O. Wable, P. Biswas, A. Bahri, L. Rossman, R. Clark, and J. Goodrich. 1996. *Characterization and Modeling of Chlorine Decay in Distribution Systems.* Denver, Colo.: AwwaRF and AWWA.

Vasconcelos, J.J., L.A. Rossman, W.M. Grayman, P.F. Boulos, and R.M. Clark. 1996. *Characterization and Modeling of Chlorine Decay in Distribution Systems.* Denver, Colo.: AWWA and AwwaRF.

Vasconcelos, J.J., L.A. Rossman, W.M. Grayman, P.F. Boulos, and R.M. Clark. 1997. Kinetics of Chlorine Decay. *Journal AWWA,* 89(7):54–65.

Walski, T.M, D.V. Chase, D.A. Savic, W.M. Grayman, S. Beckwith, and E. Koelle. 2003. *Advanced Water Distribution Modeling and Management.* Waterbury, Conn.: Haestad Press, 268–278.

Water Authorities Association and WRc (Water Research Center). 1989. *Network Analysis: A Code of Practice.* Swindon, UK: WRc.

Williams, G.S., and A. Hazen. 1920. *Hydraulic Tables.* New York: John Wiley & Sons.

Appendix B

Water Quality Modeling Examples

Introduction

As we have seen, water quality modeling tools can provide excellent insight into water system planning and design issues. To illustrate the power of these tools, an example is presented based on the use of EPANET as applied to the Cherry Hill/Brushy Plains service area in the South Central Connecticut Regional Water Authority (SCCRWA). The example was provided by Dr. Steven Buchberger, professor of civil and environmental engineering at the University of Cincinnati. Dr. Buchberger routinely gives this assignment to his graduate course in environmental engineering (CEE 693). Included are the problem statement prepared by Dr. Buchberger and the solved problem sets from three of his students.

In addition, an example illustrating the application of EPANET-MSX to the North Marin Water District Network and a hypothetical biofilm bio-attachment problem are given.

Example Problem Statement

A developer is proposing to connect a large new subdivision to the distribution system for the Cherry Hill/Brushy Plains (CH/BP) water supply network (see Figures B-1 and B-2). The town council is concerned about the ability of the water distribution system to meet peak demands at the subdivision (especially fire flows) and to maintain adequate water quality (chlorine residual) in the distribution system during routine operation as required by the Surface Water Treatment Rule (SWTR). Prior to approval, the council wants proof that the existing water distribution system is adequate to meet increased hydraulic demands and to maintain water quality standards. If the system is not adequate, the developer must propose and pay for improvements. The developer has hired your consulting firm to analyze this water distribution system. Your report must be submitted to the town council prior to their next session (in two weeks). Figure B-1 shows the CH/BP network as it is now, and Figure B-2 shows the CH/BP network as it will be after modifications are made.

The modifed pipe network should look similar to Figure B-2 after the original CH/BP network is modified according to the instructions given in this assignment; see the following discussion for details.

You have received two EPANET files (see Section I and Section II at end of Solution 1 by Wright and McCoy-Simandle for the EPANET input files). Either one can be used to launch the EPANET program. One can be edited with text editor. Table B-1 shows *most* of the modifications that you will make to your EPANET files.

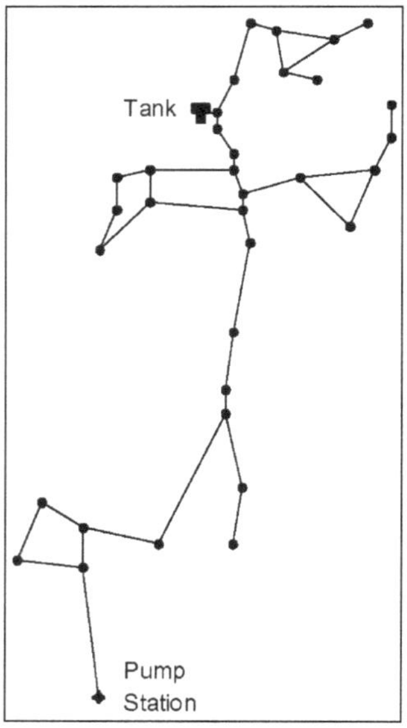

Figure B-1. Original Cherry Hill/ Brushy Plains water distribution system (Rossman et al. 1994)

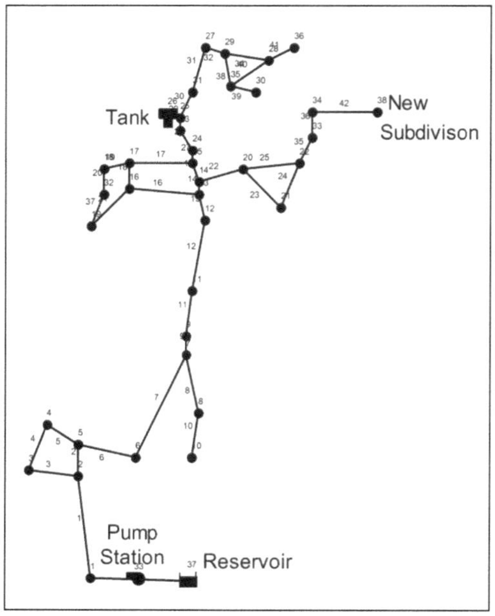

Figure B-2. Revised CH/BP water distribution system with link and node ID numbers

The following information is available for this project:

- The CH/BP network and daily demands have recently been modeled with EPANET. The input data are included in Section I for the Wright/McCoy-Simandle solution.
- The subdivision node is at elevation 200 ft with a base demand of 10 gpm and follows the same hourly demand pattern as the main CH/BP network.
- The subdivision supply line is a 6-in. cast-iron pipe extending 500 ft from Node 34.
- Assume the minimum acceptable pressure at the subdivision node is 40 psi. (25 psi required at fixtures, 10 psi for loss in the subdivision network, and 5 psi for loss in the service line and elevation difference between the main and the second floor of the house).
- Fire-flow requirements are 750 gpm for 2 hr with coincident peak flow. During fire flows, the system pressure is allowed to fall as low as 20 psi.
- To meet SWTR quality standards, the chlorine residual must stay in the range 0.2 to 1.2 mg/L at all nodes during nonemergency operation.
- Design and construction standards for water systems.

Initial township problems to be addressed are:

- *Problem 1: Fire Emergency.* Determine whether or not the CH/BP network provides adequate fire protection to the new subdivision in the event of a fire at Node 34.
- *Problem 2: Disinfectant Residual.* Determine if the CH/BP network satisfies SWTR water quality criterion for chlorine residual throughout the network during routine operation.

Your report to the council should describe the base conditions (hydraulic and water quality) for the water distribution system before and after the new development is connected. If your analysis shows that the new subdivision causes *or exacerbates* problems in the distribution system, you must propose solutions and demonstrate that they will solve the problems.

Table B-1. Suggested revisions to existing network

1. Assign Junction 1 Properties
Base Demand = 0.0 gpm
Demand Pattern = 1
Initial Quality = 0.5 mg/L
Source Quality = 0.0 mg/L

2. New Reservoir and Properties
Add reservoir to map near Junction 1
Total Head = 50 ft
Source Quality = 1.0 mg/L (Concentration)

3. Add a New Pump and Properties
Add pump to map (Reservoir to Junction 1)
Pump Curve = 1

4. Create a Pump Curve
Data Browser: Curves
Use Add Button
Flow = 670 gpm
Head = 260 ft
Enter (to see new pump curve)
OK (to exit)

5. Modify Tank Properties
Minimum Level = 56 ft
Maximum Level = 66 ft
Initial Quality = 1.0 mg/L

6. Setting Pump Controls
(This will tie pump operation to tank level)
Data Browser: Controls, Simple
Simple Controls Editor box, add following lines:
Link 33 Closed if Node 26 Above 66
Link 33 Open if Node 26 Below 56
OK (to exit)

7. New Junction (Subdivision) and Properties
Add a new Junction to map near Junction 34
Elevation = see specifications, page 4
Base Demand = see specifications, page 4
Demand Pattern = 1
Demand Categories (Demands for Junction 38 box)
Row 2 of the "Base Demand" (enter the fire-flow specifications from page 4)
Row 2 Time Pattern = *fire*
OK (to exit)
Initial Quality = 0.5 mg/L

8. Create Fire-Flow Pattern
(See assignment instructions for specifications)
Data Browser: Patterns
Add Button
Pattern Editor
Pattern ID = fire
Description = fire flow
Multipliers = 0 or 1 (**) OK (to exit)

9. Adding a New Pipe Link
Add Link to map (Connect Junctions 34 and 38) Length = see specifications, page 4
Diameter = see specifications, page 4
Roughness: use $C = 100$

10. Adding Text Labels
Place a label box near Junction 38
Add text, "New Subdivision"
Place a label near Node 37
Add text, "WTP Reservoir"

(**) Check demand pattern 1 to find appropriate hours.

Your final report must be a terse professional document (double spaced, 10 pages maximum with graphs). At a few "well chosen" nodes or pipes, show (a) minimum system pressure versus time and (b) minimum chlorine residual versus time. Also plot head in the storage tank versus flow at the pump. Include a summary table showing all proposed changes to the existing CH/BP water distribution system (e.g., new pipe line size and location, information on pump characteristics, number and location of any booster chlorination stations, etc.) Attach EPANET input and output files as an appendix.

Hint: During your water quality analyses, consider running the distribution system for a period of time to eliminate the effect of initial conditions (you determine the required simulation period).

The problem as presented focuses on the current status of the system. However, it might be wise to think about future expansion in the network. For example, nodes 23 and 25 are on the "hill" with the storage tank (look at the node elevations in the junction property box). Hence, you will not be able to increase the pressure at 23 and 25 unless you build a taller storage tank. If you have no other reason to increase the size of the storage tank, then leave things as they are. However, you may have reasons to modify the tank, depending on other aspects of your analysis. Just be aware that it is one of your options as the engineer on this project.

Obviously, cost will be the limiting factor in the real world. In this assignment, do not worry excessively about the economics. Focus on what can be done from a technical standpoint (adding pipes, closing loops, increasing diameters, adding pump stations, increasing pump capacity, increasing tank size, adding new tanks, and so forth). If you identify multiple engineering options, feel free to comment on their economic feasibility.

After you are finished with problems 1 and 2, then address problem 3 as follows.

- *Problem 3: Biochemical Assault.* In recent months, much attention has focused on identifying and fortifying potential weaknesses in the US public infrastructure. One prominent component of the national infrastructure includes municipal water distribution systems. By their very design, these pipe networks serve as efficient delivery devices reaching large segments of the population. If toxic materials were to enter (without warning) a public water distribution system, the health impacts could be devastating.

In this problem you are asked to describe and then analyze a plausible scenario by which a rogue group might attempt to surreptitiously introduce pathogenic or toxic substances into a drinking water distribution system. Obviously, direct contamination of the watershed supply is a possibility. Let us assume, however, that this avenue of sabotage would be detected and corrected at the water treatment plant. Instead, focus on ways that the actual pipe network (somewhere after the treatment plant) might be vulnerable to deliberate intrusions. Once you have identified one (or more) plausible scenarios, use EPANET to simulate an attack on the CH/BP water distribution system. Your analysis should include a brief discussion of the following items:

- Delivery mechanism (How is the substance introduced to the pressurized system?)
- Delivery rates and duration (How much is introduced and for how long?)
- Concentration of substance (What levels of toxicity would be expected in the system?)
- Extent of contamination (How much of the system is affected by the intrusion?)
- Potential impact (Do you think a typical water distribution system is vulnerable to such attacks?)

Some useful hints that might help in addressing problem 3 are follows:

1. Think about the direction of flow and the energy grade line.
2. Nodes can be modeled as demands or as sources.
3. On the node property input dialogue box, go to "Source Quality."
4. Go to HELP, *Source Quality Editor* for information on source types and options.

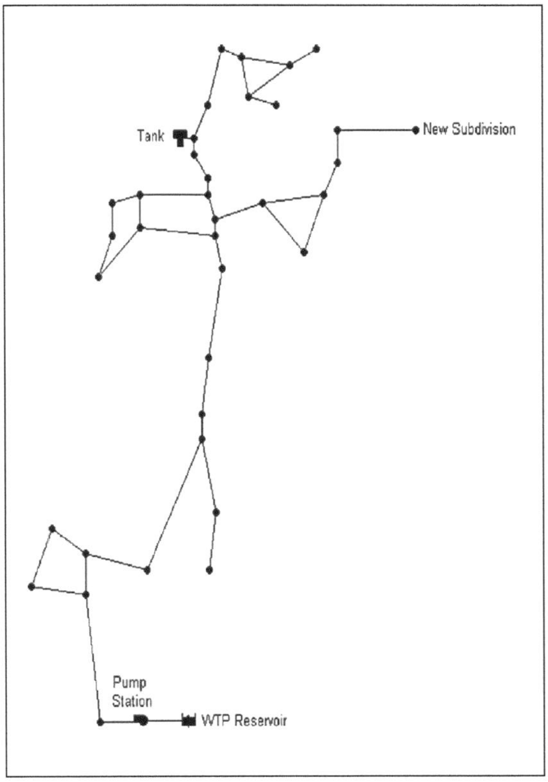

Figure B-3. Cherry Hill/Brushy Plains water distribution system

Solution I

Prepared by Joseph Wright and Chandler McCoy-Simandle, graduate students in CEE 693 Environmental Hydraulics, Department of Civil and Environmental Engineering, University of Cincinnati.

This report has been prepared for the Cherry Hill/Brushy Plains town council in order to provide analysis and insight as to whether or not the existing water distribution system has the capacity to meet peak demands and maintain adequate water quality after the addition of a new subdivision in the northeast section of the network. EPANET (USEPA 2008) was used to model the water distribution network and analyze both flow and water quality throughout the system. Figure B-3 illustrates the Cherry Hill/Brushy Plains water distribution system with the addition of the new subdivision.

Existing Conditions Analysis

Prior to adding the subdivision to the network model, a simulation was run on the water distribution system in order to see whether or not the existing conditions were adequate with respect to flow rate and water quality. A simulation time of 55 hr was chosen to give a complete picture of flow rate, pressure, and water quality and negate effects associated with the initial conditions in the problem statement. Figure B-4 shows the locations of selected nodes within the network.

The point in the system at which the minimum pressure occurs is at the water tank (Node 26). The lowest pressure outside the tank occurs at Node 23 (labeled as Junction 23 in Figure B-4). Figure B-5 (see appendix E) shows how the pressures at each location vary over the simulation time. The minimum pressure in the system occurs after 14 hr where the pressure in the tank (red) is 24.32 pounds per square inch (psi) and the pressure at Junction 23 (green) is 26.31 psi. These low pressures prevail because of the elevation at these locations. The water tank is at an elevation of 235 ft and Junction 23 is at an elevation of 230 ft; these two points are the highest elevations within the network.

According to the Surface Water Treatment Rule (SWTR) quality standards, the chlorine residual must stay within the range of 0.2 to 1.2 milligrams per liter (mg/L) at all nodes during nonemergency operations. The point of lowest chlorine residual is located at Node 34, labeled as Junction 34 in Figure B-6. During the model simulation, there were 16 locations within the distribution system in which the chlorine residual dropped below 0.2 mg/L. Most of these locations were "dead-end nodes" in which the links are not reconnected to the system, and the water in these locations may become stagnant without frequent flushing events. Figure B-7 shows how the chlorine residual at Node 34 varies over the simulation period.

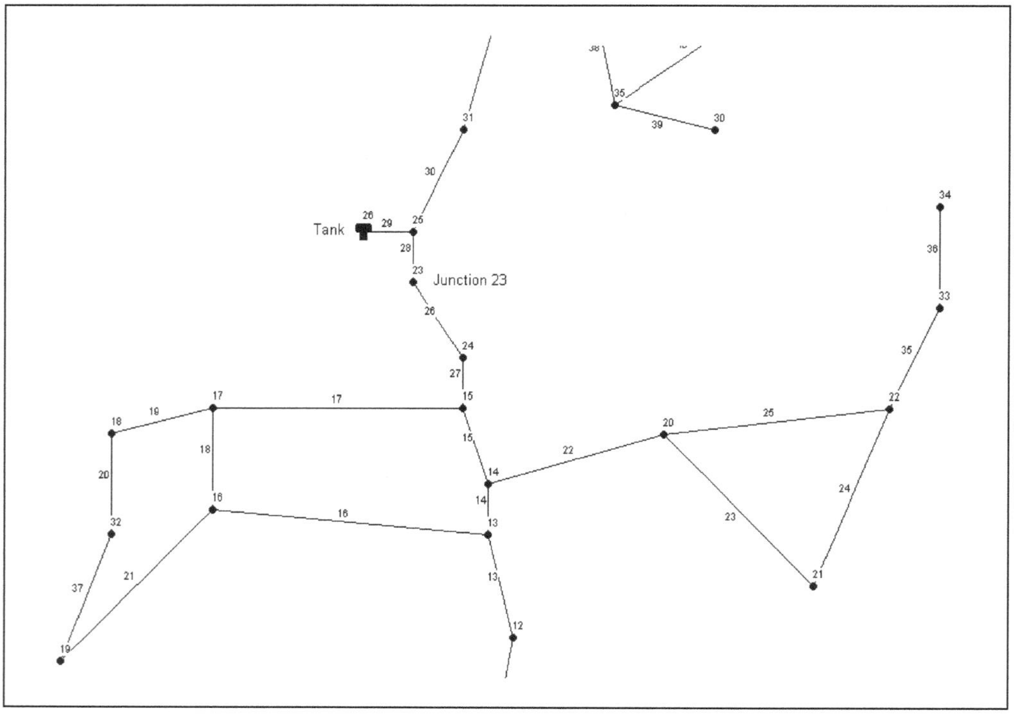

Figure B-4. Locations of Node 23 (Junction 23) and Node 26 (tank)

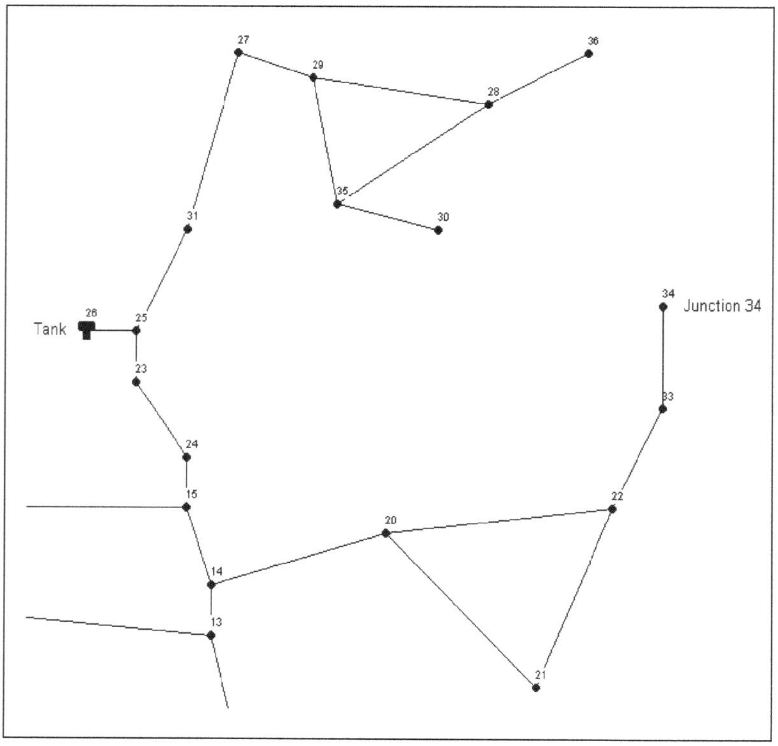

Figure B-6. Location of Node 34 (Junction 34)

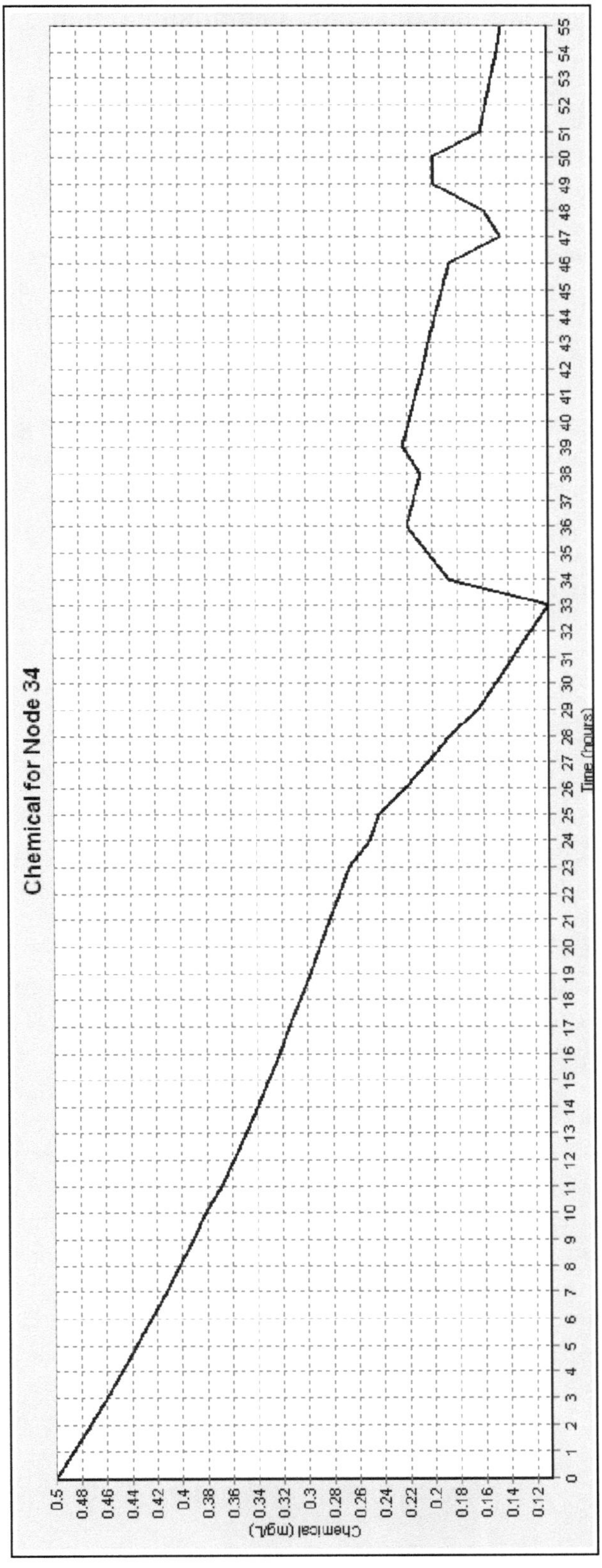

Figure B-7. Minimum chlorine residual versus time for Node 34

The minimum chlorine residual of the system occurs after 33 hr where it is 0.11 mg/L. Moreover, the chlorine residual is under the minimum allowable level of 0.2 mg/L for approximately 20 hr of the 55-hr simulation period. This is primarily because of the fact that Junction 34 is one of the many "dead-end nodes" that do not meet the chlorine residual criteria as stated by the SWTR quality standards. Also, it is important to note that Junction 34 is directly upstream from the location at which the new subdivision would be added. The analysis of the Cherry Hill/Brushy Plains water distribution system provided important information about the existing conditions of the network. While all flow demands are adequately met within the system, the chlorine residuals are not sufficient for nonemergency operations. Therefore, before adding any new construction to the network, the areas in which inadequate residuals exist should be modified in order to meet the SWTR quality standards.

Projected Conditions Analysis

The objective of this study was to observe the water flow rate and chlorine residual within the network and determine whether it was adequate throughout the system after the addition of the new subdivision in the northeast section of the distribution network. In order to perform an accurate simulation for the network including the new subdivision, several assumptions were made concerning flow demands within the subdivision:

- The subdivision is at an elevation of 200 ft with a base demand of 10 gallons per minute (gpm), and it corresponds with the same hourly demand pattern as the existing Cherry Hill/Brushy Plains network.
- It is supplied with a 500-ft 6-in. cast-iron pipe. The minimum acceptable pressure at the subdivision is 40 psi.
- Fire-flow requirements were assumed to be 750 gpm for the 2-hr maximum flow rate of the simulation, during which the system pressure is allowed to fall as low as 20 psi. This is in accordance with fire protection demands, which state that for residences 31 to 100 ft apart, the required fire flow is 750 gpm (Walski et al. 2004); satellite images of the Cherry Hill/Brushy Plains area show that existing houses in the region are usually between 31 and 100 ft apart.

After adding the subdivision to the network model, a simulation of the water distribution system was run in order to see whether adding a new subdivision would cause additional problems and/or exacerbate the existing problems for the flow rate and water quality of the network. The simulation time was kept at 55 hr in order to negate the effects of initial conditions on flow rate and chlorine residual. A plot comparing the head in the water tank to the flow rate provided by the pump is provided in Figure B-8 (see appendix E).

The amount of head in the water tank (shown by the green line) was plotted on the primary (left) vertical axis and the flow rate provided by the pump (shown by the red line) was plotted on the secondary (right) vertical axis. These two plots correspond to the controls of the system: the pump turns on whenever the height of water in the tank drops to 56 ft and turns off when the height of water in the tank reaches 66 ft. Figure B-9 (see appendix E) shows the comparison of the pressure supplied to the new subdivision in red and the node that receives the minimum pressure in the system at Node 23 in green.

The pressure provided to Node 23 was unchanged from the simulations run on the existing conditions. Generally, the pressure supplied to the new subdivision is fairly adequate other than during the emergency fire flow of 750 gpm. It was specified that the pressure was only allowed to fall as low as 20 psi during fire flow; however, the pressure actually drops as low as 1.68 psi after 10 hr.

Since the point of lowest chlorine residual based on the analysis of existing conditions was located at Node 34, another analysis was performed in order to compare the chlorine residual of the new subdivision (represented as Node 38) with that of Node 34, which is located directly upstream. Figure B-10 (see appendix E) illustrates this comparison.

The chlorine residual for Node 34, shown by the red line, is better than it was for the existing conditions simulation. This is due primarily to the emergency fire flow that takes place during hours 10 and 11 of this analysis; by supplying water at a much higher rate to the new subdivision, and therefore to Node 34, there are higher chlorine residuals early on in the simulation. However, the chlorine residual for both locations falls well below the lower limit of 0.2 mg/L after hour 43. While the chlorine residual is better at Node 34, the new subdivision becomes another dead-end node within the water distribution system and will suffer the effect of lower residuals just as Node 34 and other dead-end nodes did from the existing conditions analysis.

The analysis of the effects imposed on the Cherry Hill/Brushy Plains water distribution system with the addition of a new subdivision downstream of Node 34 provided important information about the potential conditions of the network. During normal flow circumstances, the new subdivision has little effect on the network as far as either causing new problems or exacerbating existing problems. However, when analyzing the effect of the subdivision with consideration of fire flows, the distribution network suffers a drastic decrease in overall pressure as a result of the high flow rate demand of 750 gpm during the 2-hr peak flow period. The pressure in the new subdivision drops under 2 psi during the simulation. Overall, negative pressures are induced in the system because the water tank drains to its lower limit and therefore the pump is responsible not only for providing water to the network but also for filling the tank.

Recommendations

As a result of the analyses performed for the Cherry Hill/Brushy Plains water distribution network, several recommendations have been made in order to alleviate problems within the system. Table B-2 lists these recommendations in the suggested order of consideration. The first recommendation that should be considered is the modification of the pump station and water tank controls. The simulation that included the new subdivision showed that inadequate flow rates and negative pressures were induced in the system due to the emergency fire flow. Since the minimum water level in the tank is set at 56 ft and the pump does not turn on until the minimum level is met, the system is totally reliant on the performance of the pump station to provide flow to every junction within the system. To avoid having to upgrade the pump, the controls should be set such that the minimum water level within the

Table B-2. Recommendations for the Cherry Hill/Brushy Plains network

Recommendation	Problem Addressed	Improvement
1. Modify pump tank controls	Low and negative pressure within the system	Provides greater flow rate and increased pressure within the system
2. Install booster chlorination stations	Low chlorine residuals within the system	Provides adequate chlorine to dead-end nodes in order to maintain sufficient water quality
3. Upgrade pump	Low and negative pressure	Provides greater flow fate and increased pressure within

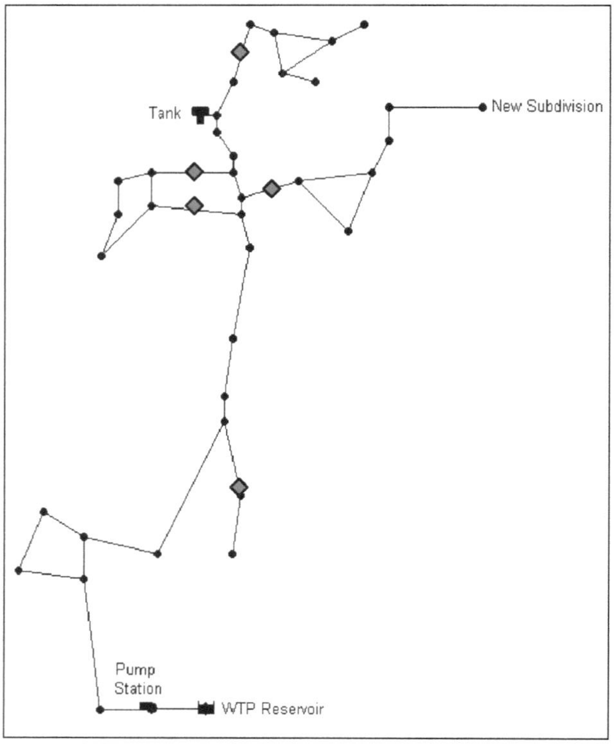

Figure B-11. Location of proposed booster chlorination stations

tank is allowed to drop to 50 ft, and the pump should still turn on at a level of 56 ft. Thus, after the pump turns on, the system is still maintained by the pressure in the tank as well as the pump and adequate flow rates and pressures are sustained throughout the simulation.

However, if any further development is projected to occur within the Cherry Hill/Brushy Plains network, the pump will need to be upgraded in order to meet the standards of the system. Furthermore, booster chlorination stations should be placed along the upstream branches of all dead-end nodes within the system as a means to provide adequate water quality to every junction during nonemergency flows. Figure B-11 indicates these locations in shaded diamond-shaped symbols. It is important to note that all of the recommendations provided here would not include any pipes within the network to be replaced or modified.

The two EPANET files that can be used to launch the EPANET program appear below in Section I and Section II.

Solution 2

Prepared by Matthew D. McCutcheon, graduate student in CEE 693 Environmental Hydraulics, Department of Civil and Environmental Engineering, University of Cincinnati.

The Cherry Hill/Brushy Plains water distribution system is located in the coastal town of Branford, Conn. In 1638, Dutch settlers purchased an area called Totoket from the Mattabeseck Indians, and the Branford settlement established itself in 1644. Growth in the eighteenth and nineteenth centuries led to shoreline communities such as Stony Creek, Indian Neck, and Pine Orchard. Branford was home to industries including Branford Lockworks, Malleable Iron Fittings Company, and the Atlantic Wire Company during the latter part of the nineteenth century. At this time, Branford also became a popular resort area. Shedding itself of the resort image, Branford established many characteristics typically associated with northeastern suburbs in the twentieth century. As of the 2000 census, there were 12,543 households, 7,663 families, and 28,683 people residing in the town (Wikipedia 2011). Recent growth has led town council to considered plans for a new development.

Text continues on p. 327.

Section I: Existing Conditions Simulation

```
                        Existing Conditions Simulation
[TITLE]
EPANET Example Network 2 (CH/BP 1994)
CEE693 Modeling assignment
w/
24 hour simulation demand pattern
chlorine residual

[JUNCTIONS]
;ID            Elev          Demand        Pattern
 1             50            0             1                ;
 2             100           8                              ;
 3             60            14                             ;
 4             60            8                              ;
 5             100           8                              ;
 6             125           5                              ;
 7             160           4                              ;
 8             110           9                              ;
 9             180           14                             ;
 10            130           5                              ;
 11            185           34.78                          ;
 12            210           16                             ;
 13            210           2                              ;
 14            200           2                              ;
 15            190           2                              ;
 16            150           20                             ;
 17            180           20                             ;
 18            100           20                             ;
 19            150           5                              ;
 20            170           19                             ;
 21            150           16                             ;
 22            200           10                             ;
 23            230           8                              ;
 24            190           11                             ;
 25            230           6                              ;
 27            130           8                              ;
 28            110           0                              ;
 29            110           7                              ;
 30            130           3                              ;
 31            190           17                             ;
 32            110           17                             ;
 33            180           1.5                            ;
 34            190           1.5                            ;
 35            110           0                              ;
 36            110           1                              ;

[RESERVOIRS]
;ID            Head          Pattern
 37            50                                  ;

[TANKS]
;ID                    Elevation     InitLevel     MinLevel      MaxLevel
Diameter      MinVol         VolCurve
 26                    235           56.7          56            66
50            0                                    ;

[PIPES]
;ID                    Node1                       Node2                       Length
Diameter      Roughness              MinorLoss     Status
 1                     1                           2                           2400
12            100                    0             Open     ;
 2                     2                           5                           800
12            100                    0             Open     ;
 3                     2                           3                           1300
8              100          0         Open     ;
 4             3                      4                    1200
8              100          0         Open     ;
 5             4                      5                    1000
12             100          0         Open     ;
 6             5                      6                    1200
12             100          0         Open     ;
 7             6                      7                    2700
12             100          0         Open     ;
 8             7                      8                    1200
12             140          0         Open     ;
 9             7                      9                    400
12             100          0         Open     ;
 10            8                      10                   1000
8              140          0         Open     ;
 11            9                      11                   700
12             100          0         Open     ;
 12            11                     12                   1900
12             100          0         Open     ;
 13            12                     13                   600
12             100          0         Open     ;
 14            13                     14                   400
12             100          0         Open     ;
 15            14                     15                   300
12             100          0         Open     ;
 16            13                     16                   1500
8              100          0         Open     ;
 17            15                     17                   1500
8              100          0         Open     ;
 18            16                     17                   600
8              100          0         Open     ;
 19            17                     18                   700
12             100          0         Open     ;
 20            18                     32                   350
12             100          0         Open     ;
 21            16                     19                   1400
8              100          0         Open     ;
 22            14                     20                   1100
12             100          0         Open     ;
 23            20                     21                   1300
8              100          0         Open     ;
 24            21                     22                   1300
8              100          0         Open     ;
 25            20                     22                   1300
8              100          0         Open     ;
 26            24                     23                   600
12             100          0         Open     ;
 27            15                     24                   250
12             100          0         Open     ;
 28            23                     25                   300
12             100          0         Open     ;
 29            25                     26                   200
12             100          0         Open     ;
 30            25                     31                   600
12             100          0         Open     ;
 31            31                     27                   400
8              100          0         Open     ;
 32            27                     29                   400
8              100          0         Open     ;
 34            29                     28                   700
8              100          0         Open     ;
 35            22                     33                   1000
8              100          0         Open     ;
```

Section I: (continued)

```
                        Existing Conditions Simulation
 36                     33                   34                  400
8               100               0          Open     ;
 37                     32                   19                  500
8               100               0          Open     ;
 38                     29                   35                  500
8               100               0          Open     ;
 39                     35                   30                  1000
8               100               0          Open     ;
 40                     28                   35                  700
8               100               0          Open     ;
 41                     28                   36                  300
8               100               0          Open     ;

[PUMPS]
;ID                     Node1                Node2               Parameters
 33                     37                   1                   HEAD 1  ;

[VALVES]
;ID                     Node1                Node2               Diameter
Type    Setting         MinorLoss

[TAGS]

[DEMANDS]
;Junction               Demand         Pattern              Category

[STATUS]
;ID                     Status/Setting

[PATTERNS]
;ID                     Multipliers
;Demand Pattern
 1                      1.26           1.04          .97           .97
.89             1.19
 1                      1.28           .67           .67           1.34
2.46            .97
 1                      .92            .68           1.43          .61
.31             .78
 1                      .37            .67           1.26          1.56
1.19            1.26
;Pump Station Outflow Pattern
 2                      .96            .96           .96           .96
.96             .96
 2                      .62            0             0             0
0               0
 2                      .8             1             1             1
1               .15
 2                      0              0             0             0
0               0
;fire flow
 fire                   0              0             0             0
0               0
 fire                   0              0             0             1
1               0
 fire                   0              0             0             0
0               0
 fire                   0              0             0             0
0               0
 fire                   0              0             0             0
0               0
 fire                   0              0             0             0
0               0
 fire                   0              0             0             0
```

```
                        Existing Conditions Simulation
0               0
 fire                   0              0             0             0
0               0
 fire                   0              0             0             0
0               0
 fire                   0

[CURVES]
;ID                     X-Value        Y-Value
;PUMP: PUMP: PUMP: Pump Curve 1
 1                      670            260

[CONTROLS]
Link 33 Closed if Node 26 Above 66
Link 33 Open if Node 26 Below 56

[RULES]

[ENERGY]
 Global Efficiency      75
 Global Price           0.0
 Demand Charge          0.0

[EMITTERS]
;Junction               Coefficient

[QUALITY]
;Node                   InitQual
 1                      .5
 2                      0.5
 3                      0.5
 4                      0.5
 5                      0.5
 6                      0.5
 7                      0.5
 8                      0.5
 9                      0.5
 10                     0.5
 11                     0.5
 12                     0.5
 13                     0.5
 14                     0.5
 15                     0.5
 16                     0.5
 17                     0.5
 18                     0.5
 19                     0.5
 20                     0.5
 21                     0.5
 22                     0.5
 23                     0.5
```

Section I: (continued)

```
                    Existing Conditions Simulation
 24                     0.5
 25                     0.5
 27                     0.5
 28                     0.5
 29                     0.5
 30                     0.5
 31                     0.5
 32                     0.5
 33                     0.5
 34                     0.5
 35                     0.5
 36                     0.5
 26                     1.0

[SOURCES]
;Node                   Type            Quality         Pattern
 37                     CONCEN          1

[REACTIONS]
;Type          Pipe/Tank                Coefficient


[REACTIONS]
 Order Bulk             1
 Order Tank             1
 Order Wall             1
 Global Bulk            -.5
 Global Wall            -1
 Limiting Potential     0.0
 Roughness Correlation  0.0

[MIXING]
;Tank                   Model

[TIMES]
 Duration               55:00
 Hydraulic Timestep     1:00
 Quality Timestep       0:05
 Pattern Timestep       1:00
 Pattern Start          0:00
 Report Timestep        1:00
 Report Start           0:00
 Start ClockTime        8 am
 Statistic              NONE

[REPORT]
 Status                 No
 Summary                No
 Page                   0

[OPTIONS]
 Units                  GPM
 Headloss               H-W
 Specific Gravity       1.0
 Viscosity              1.0
 Trials                 40
 Accuracy               0.001
 CHECKFREQ              2
 MAXCHECK               10
 DAMPLIMIT              0
 Unbalanced             Continue 10
 Pattern                1
 Demand Multiplier      1.0
```

```
                    Existing Conditions Simulation
 Emitter Exponent       0.5
 Quality                Chemical mg/L
 Diffusivity            1.0
 Tolerance              0.01

[COORDINATES]
;Node                   X-Coord             Y-Coord
 1                      21.00               4.00
 2                      19.00               20.00
 3                      11.00               21.00
 4                      14.00               28.00
 5                      19.00               25.00
 6                      28.00               23.00
 7                      36.00               39.00
 8                      38.00               30.00
 9                      36.00               42.00
 10                     37.00               23.00
 11                     37.00               49.00
 12                     39.00               60.00
 13                     38.00               64.00
 14                     38.00               66.00
 15                     37.00               69.00
 16                     27.00               65.00
 17                     27.00               69.00
 18                     23.00               68.00
 19                     21.00               59.00
 20                     45.00               68.00
 21                     51.00               62.00
 22                     54.00               69.00
 23                     35.00               74.00
 24                     37.00               71.00
 25                     35.00               76.00
 27                     39.00               87.00
 28                     49.00               85.00
 29                     42.00               86.00
 30                     47.00               80.00
 31                     37.00               80.00
 32                     23.00               64.00
 33                     56.00               73.00
 34                     56.00               77.00
 35                     43.00               81.00
 36                     53.00               87.00
 37                     33.86               4.18
 26                     33.00               76.00

[VERTICES]
;Link                   X-Coord             Y-Coord

[LABELS]
;X-Coord            Y-Coord          Label & Anchor Node
 22.59              10.40            "Pump"
 22.45              7.94             "Station"
 26.76              77.42            "Tank"
 36.03              5.63             "WTP Reservoir"

[BACKDROP]
 DIMENSIONS             8.75                -0.15               58.25
         91.15
 UNITS                  None
 FILE
 OFFSET                 0.00                0.00

[END]
```

Section II: Projected Conditions Simulation

```
Projected Conditions Simulation
[TITLE]
EPANET Example Network 2 (CH/BP 1994)
CEE693 Modeling assignment
w/
24 hour simulation demand pattern
chlorine residual

[JUNCTIONS]
;ID          Elev       Demand     Pattern
 1           50         0.0        1
 2           100        8
 3           60         14
 4           60         8
 5           100        8
 6           125        5
 7           160        4
 8           110        9
 9           180        14
 10          130        5
 11          185        34.78
 12          210        16
 13          210        2
 14          200        2
 15          190        2
 16          150        20
 17          180        20
 18          100        20
 19          150        5
 20          170        19
 21          150        16
 22          200        10
 23          230        8
 24          190        11
 25          230        6
 27          130        8
 28          110        0
 29          110        7
 30          130        3
 31          190        17
 32          110        17
 33          180        1.5
 34          190        1.5
 35          110        0
 36          110        1
 38          200        10         1

[RESERVOIRS]
;ID          Head       Pattern
 37          50                          ;

[TANKS]
;ID                     Elevation        InitLevel     MinLevel     MaxLevel
Diameter     MinVol              VolCurve
 26                     235              56.7          56           66
50           0                                         ;

[PIPES]
;ID                     Node1                          Node2                     Length
Diameter     Roughness            MinorLoss            Status
 1                      1                              2                         2400
12           100                  0                    Open    ;
 2                      2                              5                         800
12           100                  0                    Open    ;

Projected Conditions Simulation
 3                      2                              3                         1300
8            100                  0                    Open    ;
 4                      3                              4                         1200
8            100                  0                    Open    ;
 5                      4                              5                         1000
12           100                  0                    Open    ;
 6                      5                              6                         1200
12           100                  0                    Open    ;
 7                      6                              7                         2700
12           100                  0                    Open    ;
 8                      7                              8                         1200
12           140                  0                    Open    ;
 9                      7                              9                         400
12           100                  0                    Open    ;
 10                     8                              10                        1000
8            140                  0                    Open    ;
 11                     9                              11                        700
12           100                  0                    Open    ;
 12                     11                             12                        1900
12           100                  0                    Open    ;
 13                     12                             13                        600
12           100                  0                    Open    ;
 14                     13                             14                        400
12           100                  0                    Open    ;
 15                     14                             15                        300
12           100                  0                    Open    ;
 16                     13                             16                        1500
8            100                  0                    Open    ;
 17                     15                             17                        1500
8            100                  0                    Open    ;
 18                     16                             17                        600
8            100                  0                    Open    ;
 19                     17                             18                        700
12           100                  0                    Open    ;
 20                     18                             32                        350
12           100                  0                    Open    ;
 21                     16                             19                        1400
8            100                  0                    Open    ;
 22                     14                             20                        1100
12           100                  0                    Open    ;
 23                     20                             21                        1300
8            100                  0                    Open    ;
 24                     21                             22                        1300
8            100                  0                    Open    ;
 25                     20                             22                        1300
8            100                  0                    Open    ;
 26                     24                             23                        600
12           100                  0                    Open    ;
 27                     15                             24                        250
12           100                  0                    Open    ;
 28                     23                             25                        300
12           100                  0                    Open    ;
 29                     25                             26                        200
12           100                  0                    Open    ;
 30                     25                             31                        600
12           100                  0                    Open    ;
 31                     31                             27                        400
8            100                  0                    Open    ;
 32                     27                             29                        400
8            100                  0                    Open    ;
 34                     29                             28                        700
8            100                  0                    Open    ;
 35                     22                             33                        1000
```

Section II: (continued)

```
                        Projected Conditions Simulation
8          100                0             Open     ;
 36                  33                     34                 400
8          100                0             Open     ;
 37                  32                     19                 500
8          100                0             Open     ;
 38                  29                     35                 500
8          100                0             Open     ;
 39                  35                     30                 1000
8          100                0             Open     ;
 40                  28                     35                 700
8          100                0             Open     ;
 41                  28                     36                 300
8          100                0             Open     ;
 42                  34                     38                 500
6          100                0             Open     ;

[PUMPS]
;ID                  Node1                  Node2              Parameters
 33                  37                     1                  HEAD 1  ;

[VALVES]
;ID                  Node1                  Node2              Diameter
Type    Setting      MinorLoss

[TAGS]

[DEMANDS]
;Junction            Demand       Pattern                 Category
 38                  10           1                       ;
 38                  750          fire                    ;

[STATUS]
;ID                  Status/Setting

[PATTERNS]
;ID                  Multipliers
;Demand Pattern
 1                   1.26         1.04         .97          .97
.89        1.19
 1                   1.28         .67          .67          1.34
2.46       .97
 1                   .92          .68          1.43         .61
.31        .78
 1                   .37          .67          1.26         1.56
1.19       1.26
;Pump Station Outflow Pattern
 2                   .96          .96          .96          .96
.96        .96
 2                   .62          0            0            0
0          0
 2                   .8           1            1            1
1          .15
 2                   0            0            0            0
0          0
;fire flow
 fire                0            0            0            0
0          0
 fire                0            0            0            1
1          0
 fire                0            0            0            0
0          0
 fire                0            0            0            0
0          0
```

```
                              Projected Conditions Simulation
 fire                 0             0             0             0
0           0
 fire                 0             0             0             0
0           0
 fire                 0             0             0             0
0           0
 fire                 0             0             0             0
0           0
 fire                 0             0             0             0
0           0
 fire                 0             0

[CURVES]
;ID                   X-Value       Y-Value
;PUMP: PUMP:
 1                    670           260

[CONTROLS]
Link 33 Closed if Node 26 Above 66
Link 33 Open if Node 26 Below 56

[RULES]

[ENERGY]
 Global Efficiency    75
 Global Price         0.0
 Demand Charge        0.0

[EMITTERS]
;Junction             Coefficient

[QUALITY]
;Node                 InitQual
 1                    0.5
 2                    0.5
 3                    0.5
 4                    0.5
 5                    0.5
 6                    0.5
 7                    0.5
 8                    0.5
 9                    0.5
 10                   0.5
 11                   0.5
 12                   0.5
 13                   0.5
 14                   0.5
 15                   0.5
 16                   0.5
 17                   0.5
 18                   0.5
 19                   0.5
 20                   0.5
 21                   0.5
 22                   0.5
 23                   0.5
```

Section II: (continued)

```
                         Projected Conditions Simulation
 24                     0.5
 25                     0.5
 27                     0.5
 28                     0.5
 29                     0.5
 30                     0.5
 31                     0.5
 32                     0.5
 33                     0.5
 34                     0.5
 35                     0.5
 36                     0.5
 38                     0.5
 26                     1.0

[SOURCES]
;Node                   Type            Quality          Pattern
 37                     CONCEN          1.0

[REACTIONS]
;Type           Pipe/Tank               Coefficient


[REACTIONS]
 Order Bulk             1
 Order Tank             1
 Order Wall             1
 Global Bulk            -.5
 Global Wall            -1
 Limiting Potential     0.0
 Roughness Correlation  0.0

[MIXING]
;Tank                   Model

[TIMES]
 Duration               55:00
 Hydraulic Timestep     1:00
 Quality Timestep       0:05
 Pattern Timestep       1:00
 Pattern Start          0:00
 Report Timestep        1:00
 Report Start           0:00
 Start ClockTime        8 am
 Statistic              NONE

[REPORT]
 Status                 No
 Summary                No
 Page                   0

[OPTIONS]
 Units                  GPM
 Headloss               H-W
 Specific Gravity       1.0
 Viscosity              1.0
 Trials                 40
 Accuracy               0.001
 CHECKFREQ              2
 MAXCHECK               10
 DAMPLIMIT              0
 Unbalanced             Continue 10
 Pattern                1
```

```
                         Projected Conditions Simulation
 Demand Multiplier       1.0
 Emitter Exponent        0.5
 Quality                 Chemical mg/L
 Diffusivity             1.0
 Tolerance               0.01

[COORDINATES]
;Node                    X-Coord            Y-Coord
 1                       21.00              4.00
 2                       19.00              20.00
 3                       11.00              21.00
 4                       14.00              28.00
 5                       19.00              25.00
 6                       28.00              23.00
 7                       36.00              39.00
 8                       38.00              30.00
 9                       36.00              42.00
 10                      37.00              23.00
 11                      37.00              49.00
 12                      39.00              60.00
 13                      38.00              64.00
 14                      38.00              66.00
 15                      37.00              69.00
 16                      27.00              65.00
 17                      27.00              69.00
 18                      23.00              68.00
 19                      21.00              59.00
 20                      45.00              68.00
 21                      51.00              62.00
 22                      54.00              69.00
 23                      35.00              74.00
 24                      37.00              71.00
 25                      35.00              76.00
 27                      39.00              87.00
 28                      49.00              85.00
 29                      42.00              86.00
 30                      47.00              80.00
 31                      37.00              80.00
 32                      23.00              64.00
 33                      56.00              73.00
 34                      56.00              77.00
 35                      43.00              81.00
 36                      53.00              87.00
 38                      67.38              76.99
 37                      33.86              4.18
 26                      33.00              76.00

[VERTICES]
;Link                    X-Coord            Y-Coord

[LABELS]
;X-Coord            Y-Coord          Label & Anchor Node
 22.88              10.11            "Pump"
 22.74              7.80             "Station"
 26.76              77.42            "Tank"
 68.53              78.29            "New Subdivision"
 35.88              5.34             "WTP Reservoir"

[BACKDROP]
 DIMENSIONS              8.75               -0.15                58.25
         91.15
 UNITS                   None
 FILE
                          Projected Conditions Simulation
 OFFSET                  0.00               0.00

[END]
```

In order for town council to approve development, Matthew McCutcheon Engineering (MME) must perform an engineering analysis of the water distribution system to ensure that the town's water supply can meet demands for the new development. In addition, MME must examine chlorine residual levels to make certain that the town provides a high-quality product suitable for consumption.

Purpose

The town of Branford public safety officials wish to provide an adequate level of fire protection to all areas supplied by the Cherry Hill/Brushy Plains water distribution system (CH/BP WDS). Of particular concern is the proposed subdivision situated on the east side of the CH/BP WDS. To provide sufficient fire protection, officials are asking MME to provide an engineering solution to any and all potential problems within the CH/BP WDS associated with supplying the new single-family units from the CH/BP WDS.

Fire protection is an essential function of any public water distribution system. The amount of water used during a year for extinguishing fires is usually a negligible part of the total annual production; however, the rate of demand during a fire is so high that it becomes the deciding factor in the design of booster stations, storage reservoirs, and distribution mains.

Adequate water quality is also an essential function of any public water distribution system. In order to comply with the Surface Water Treatment Rule (SWTR) quality standards, chlorine residuals must remain at acceptable levels at all nodes during nonemergency operation.

Engineering Methodology

The Insurance Services Office (ISO) has developed a fire suppression–rating schedule for calculating required fire suppression based on the following parameters:

- Type of structure
- Total floor area
- Type of occupancy
- Automatic sprinkler protection
- Exposure distance

A minimum line pressure of 35 to 45 psi is generally required to serve two-story dwellings.

However, fire flows are normally expressed in gallons per minute at a residual line pressure of 20 pounds per square inch (psi) for mobile fire pumping use. The drop in pressure is permissible for emergency situations. The effective size of a public water main determines its ability to supply fire flows. A 6-in. main is designed to support fire flows from a single fire hydrant. Smaller mains should not be equipped with fire hydrants. Significant fire-flow rates are limited primarily to the immediate areas adjoining the 6-in., 8-in., and 12-in. mains within CH/BP WDS.

The ISO uses the above criteria to determine the needed fire flow (NFF) according to the formula:

$$\text{NFF} = 18FA^{0.5}O(X+P) \qquad \text{(Eq B-1)}$$

where F is the class of construction, A is the effective area (square feet), O is the occupancy factor, X is the exposure factor, and P is the communication factor.

Data for determining the construction class and occupancy factor were not furnished; however, the ISO summarizes NFF for residences of two stories or less. The data provided by the ISO are summarized in Table B-3.

Table B-3. NFF for residences two stories or less (from ISO)

Distance Between Buildings (*ft*)	NFF (*gpm*)
More than 100	500
31–100	750
11–30	1,000
Less than 11	1,500

The engineer assumed that a typical side-yard setback for the development would be 20 ft; therefore the minimum distance between any two residences would be 40 ft. The resulting NFF from Table B-3 is 750 gpm.

For design purposes, minor losses at orifices, valves, and fittings were not considered. Minor losses do not contribute a significant amount of energy loss in a pipe network when compared to friction losses in the pipes. Minor losses are particularly important where the direction of flow changes often, such as at pumping stations, treatment plants, and residences. The installation of additional pumps is necessary to provide adequate pressures during fire-flow scenarios.

Results

The engineer used EPANET, a hydraulic modeling program, to simulate the flows and demands on the system with the proposed subdivision included. The results showed pressures at Node 38 below the required minimum pressure of 20 psi. When the engineer added pumps at the pump station, minimum pressures were met throughout the entire system. The pump controls were connected to the water level in the tanks. The controls were such that the startup and shutdown of the pumps were staggered to use only the pumping power necessary to maintain minimum pressures throughout the system. Two additional pumps satisfied the pressure requirements of the system.

When the simulation was run for water quality (represented by chlorine residual), several nodes were below the bottom threshold of 0.20 mg/L of free chlorine. A booster station on the transmission main made providing adequate chlorine residuals possible. The booster station should provide 50 mg/min chlorine to the system at Node 11 during hours 6 through 24 to ensure chlorine residuals are above the lower threshold of 0.20 mg/L. No nodes exceeded the SWTR upper threshold of 1.20 mg/L free chlorine.

Construction Costs

The summary of the financial impact of the Cherry Hill/Brushy Plains Subdivision Expansion is:

500 lin ft of 6-in. Ductile-Iron-Pipe @ $32.50/lin ft	$16,250
2 Pumps @ $5,000	$10,000
1 Chlorine Booster Station @ $6,000	$6,000
Contingencies (10 percent)	$3,250
Total Estimated Construction Cost	$35,500
Related Project Costs (12 percent)	$4,250
TOTAL PROJECT COST	$39,750

In addition to meeting regulated pressures and flows, MME also analyzed the system for susceptibility to terrorist attack. It is important to consider the fate of a biological agent introduced into the system by a rogue organization or individual. The engineer found that an injection of .1 kg/min for 60 min at Node 25 created widespread contamination. The system needed 42 hr of nonemergency flows to "wash out" the contamination.

Summary and Conclusions

The changes that need to be made to the CH/BP WDS in order to satisfy the NFF and SWTR standards are both practical and cost-effective. The additional pumps will provide sufficient flow in the event of a fire during peak demand while also not straining the rest of the system. This option is cost-effective when compared to either adding new tanks or an additional reservoir to the system. Also, the two pumps as well as the booster station can be installed with virtually no interruption to the supply of water throughout the system.

While analyzing the system for a biological attack, many scenarios were considered. While we did find that the system could be infiltrated with a contaminant, it is highly unlikely this will actually occur. The pressure needed to continuously pump the agent back into the system is most likely unreachable at any single node. Also we found that, based on simulated flow direction, it is unlikely that a biological agent would affect the system if introduced at a location other than Node 25.

Overall, the CH/BP WDS should be capable of handling the increased pressure and flow caused by the proposed subdivision. The quality of water in the system will remain high, and flow and pressure will continue to be adequate should the city decide to make the needed improvements to the WDS (see Figure B-12).

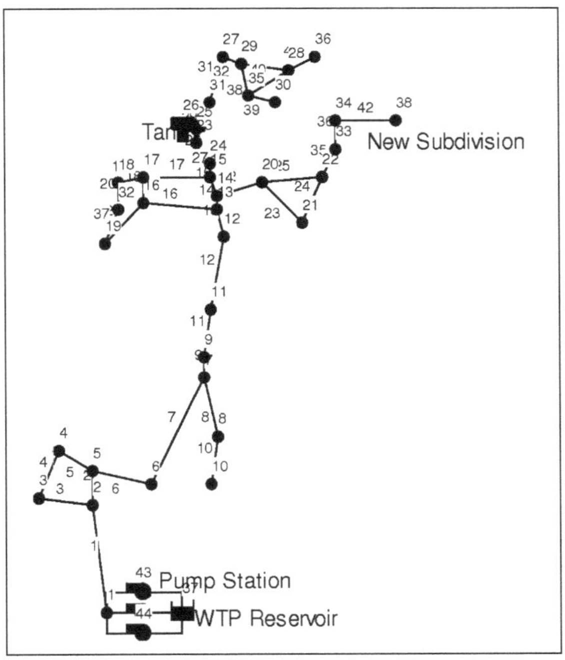

Figure B-12. EPANET example network 2 (CH/BP 1994)

Calculations

This section contains the back-up calculations for Solution 2. Data sheet 1 is the basic EPANET network for the community after the development has been added to the subdivision and additional pumps have been added to the network. Data sheet 2 shows the chlorine residuals at hour 06:00 in the expanded network. Data sheets 3–5 show the fire-flow pressures at node 11 over a 3-hr period. Data sheet 6 depicts the results of a biological attack in the network.

Text continues on p. 337.

Data Sheet I-Basic Network With Added Pumps

Post-Development Additional Pumps at Pump Station										
EPANET Example Network 2 (CH/BP 1994)										
Network Table - Links at 6:00 Hrs										
	Length	Diameter	Roughness	Flow	Velocity	Unit Headloss	Friction Factor	Reaction Rate	Chemical	Status
Link ID	ft	in		GPM	fps	ft/Kft		mg/L/d	mg/L	
Pipe 1	2400	12	100	1576.69	4.47	9.58	0.031	0.97	1	Open
Pipe 2	800	12	100	1318.68	3.74	6.88	0.032	0.77	0.47	Open
Pipe 3	1300	8	100	247.76	1.58	2.24	0.038	1.09	0.47	Open
Pipe 4	1200	8	100	229.84	1.47	1.95	0.039	1.06	0.37	Open
Pipe 5	1000	12	100	219.6	0.62	0.25	0.041	0.67	0.34	Open
Pipe 6	1200	12	100	1528.05	4.33	9.04	0.031	0.74	0.39	Open
Pipe 7	2700	12	100	1521.65	4.32	8.97	0.031	0.73	0.36	Open
Pipe 8	1200	12	140	17.92	0.05	0	0.032	0.28	0.39	Open
Pipe 9	400	12	100	1498.61	4.25	8.72	0.031	0.8	0.35	Open
Pipe 10	1000	8	140	6.4	0.04	0	0.035	0.42	0.44	Open
Pipe 11	700	12	100	1480.69	4.2	8.52	0.031	0.91	0.33	Open
Pipe 12	1900	12	100	1436.17	4.07	8.06	0.031	0.67	0.46	Open
Pipe 13	600	12	100	1415.69	4.02	7.84	0.031	0.66	0.32	Open
Pipe 14	400	12	100	1195.13	3.39	5.73	0.032	1.39	0.32	Open
Pipe 15	300	12	100	1118.33	3.17	5.07	0.032	1.49	0.68	Open
Pipe 16	1500	8	100	218	1.39	1.77	0.039	0.92	0.32	Open
Pipe 17	1500	8	100	-113.04	0.72	0.52	0.043	1.86	0.33	Open
Pipe 18	600	8	100	124.44	0.79	0.63	0.043	1.12	0.31	Open
Pipe 19	700	12	100	-14.2	0.04	0	0.062	0.24	0.33	Open
Pipe 20	350	12	100	-39.8	0.11	0.01	0.053	0.39	0.36	Open
Pipe 21	1400	8	100	67.96	0.43	0.2	0.047	0.85	0.31	Open
Pipe 22	1100	12	100	74.24	0.21	0.03	0.048	0.52	0.55	Open
Pipe 23	1300	8	100	24.71	0.16	0.03	0.054	0.63	0.33	Open
Pipe 24	1300	8	100	4.23	0.03	0	0.071	0.07	0.45	Open
Pipe 25	1300	8	100	25.21	0.16	0.03	0.054	0.63	0.33	Open
Pipe 26	600	12	100	1214.73	3.45	5.91	0.032	1.74	0.78	Open
Pipe 27	250	12	100	1228.81	3.49	6.03	0.032	1.61	0.71	Open
Pipe 28	300	12	100	1204.49	3.42	5.82	0.032	1.87	0.84	Open
Pipe 29	200	12	100	1150.73	3.26	5.34	0.032	1.9	0.91	Open
Pipe 30	600	12	100	46.08	0.13	0.01	0.052	0.99	0.9	Open
Pipe 31	400	8	100	24.32	0.16	0.03	0.054	1.5	0.86	Open
Pipe 32	400	8	100	14.08	0.09	0.01	0.059	1.03	0.79	Open
Pipe 34	700	8	100	2.2	0.01	0	0.076	0.07	0.47	Open
Pipe 35	1000	8	100	16.64	0.11	0.02	0.057	0.56	0.35	Open
Pipe 36	400	8	100	14.72	0.09	0.01	0.058	0.52	0.35	Open
Pipe 37	500	8	100	-61.56	0.39	0.17	0.047	0.96	0.33	Open
Pipe 38	500	8	100	2.92	0.02	0	0.076	0.09	0.46	Open
Pipe 39	1000	8	100	3.84	0.02	0	0.072	0.08	0.48	Open
Pipe 40	700	8	100	0.92	0.01	0	0.108	0.06	0.49	Open
Pipe 41	300	8	100	1.28	0.01	0	0.131	0.08	0.48	Open
Pipe 42	500	6	100	12.8	0.15	0.04	0.057	0.84	0.35	Open
Pump 33	#N/A	#N/A	#N/A	0	0	0	0	0	1	Closed
Pump 43	#N/A	#N/A	#N/A	1576.69	0	-353.28	0	0	1	Open
Pump 44	#N/A	#N/A	#N/A	0	0	0	0	0	1	Closed

Data Sheet 2—Chlorine Residual at Node 11 (06:00 Hrs)

Post-Development Chlorine Booster Station at Node 11				
EPANET Example Network 2 (CH/BP 1994)				
Network Table - Nodes at 6:00 Hrs				
	Demand	Head	Pressure	Chemical
Node ID	GPM	ft	psi	mg/L
Junc 1	0	403.28	153.07	1
Junc 2	10.24	380.3	121.45	0.47
Junc 3	17.92	377.38	137.52	0.37
Junc 4	10.24	375.04	136.51	0.34
Junc 5	10.24	374.79	119.07	0.39
Junc 6	6.4	363.95	103.54	0.36
Junc 7	5.12	339.74	77.88	0.35
Junc 8	11.52	339.74	99.55	0.43
Junc 9	17.92	336.26	67.71	0.33
Junc 10	6.4	339.74	90.88	0.45
Junc 11	44.52	330.29	62.95	0.46
Junc 12	20.48	314.99	45.49	0.32
Junc 13	2.56	310.28	43.45	0.32
Junc 14	2.56	307.99	46.79	0.68
Junc 15	2.56	306.47	50.46	0.71
Junc 16	25.6	307.63	68.3	0.31
Junc 17	25.6	307.25	55.14	0.33
Junc 18	25.6	307.25	89.8	0.35
Junc 19	6.4	307.34	68.18	0.33
Junc 20	24.32	307.95	59.77	0.33
Junc 21	20.48	307.91	68.42	0.33
Junc 22	12.8	307.91	46.76	0.35
Junc 23	10.24	301.41	30.94	0.84
Junc 24	14.08	304.96	49.81	0.78
Junc 25	7.68	299.67	30.19	0.91
Junc 27	10.24	299.65	73.51	0.8
Junc 28	0	299.64	82.17	0.48
Junc 29	8.96	299.64	82.17	0.64
Junc 30	3.84	299.64	73.51	0.48
Junc 31	21.76	299.66	47.52	0.86
Junc 32	21.76	307.26	85.47	0.36
Junc 33	1.92	307.89	55.42	0.35
Junc 34	1.92	307.89	51.08	0.35
Junc 35	0	299.64	82.17	0.48
Junc 36	1.28	299.64	82.17	0.48
Junc 38	12.8	307.87	46.74	0.34
Resvr 37	-1576.69	50	0	1
Tank 26	1150.73	298.6	27.56	0.98

Data Sheet 3—Fire-Flow Pressures at Node 11 (09:00 Hrs)

Post-Development Fire Flow Pressures			
EPANET Example Network 2 (CH/BP 1994)			
Network Table - Nodes at 9:00 Hrs			
	Demand	Head	Pressure
Node ID	GPM	ft	psi
Junc 1	0	289.57	103.81
Junc 2	10.72	289.57	82.14
Junc 3	18.76	289.57	99.47
Junc 4	10.72	289.57	99.47
Junc 5	10.72	289.58	82.14
Junc 6	6.7	289.6	71.32
Junc 7	5.36	289.65	56.18
Junc 8	12.06	289.65	77.84
Junc 9	18.76	289.67	47.52
Junc 10	6.7	289.65	69.18
Junc 11	46.61	289.71	45.37
Junc 12	21.44	289.94	34.64
Junc 13	2.68	290.03	34.68
Junc 14	2.68	290.07	39.03
Junc 15	2.68	291.25	43.87
Junc 16	26.8	290.08	60.7
Junc 17	26.8	290.13	47.72
Junc 18	26.8	290.11	82.37
Junc 19	6.7	290.1	60.7
Junc 20	25.46	286.88	50.64
Junc 21	21.44	281.84	57.12
Junc 22	13.4	277.38	33.53
Junc 23	10.72	295.6	28.43
Junc 24	14.74	292.51	44.42
Junc 25	8.04	297.18	29.11
Junc 27	10.72	297.16	72.43
Junc 28	0	297.15	81.09
Junc 29	9.38	297.15	81.09
Junc 30	4.02	297.15	72.43
Junc 31	22.78	297.17	46.44
Junc 32	22.78	290.11	78.04
Junc 33	2.01	259.19	34.31
Junc 34	752.01	251.95	26.84
Junc 35	0	297.15	81.09
Junc 36	1.34	297.15	81.09
Junc 38	13.4	251.93	22.5
Resvr 37	0	50	0
Tank 26	-1195.93	298.33	27.44

Data Sheet 4—Fire-Flow Pressures at Node 11 (10:00 Hrs)

Post-Development Fire Flow Pressures			
EPANET Example Network 2 (CH/BP 1994)			
Network Table - Nodes at 10:00 Hrs			
	Demand	Head	Pressure
Node ID	GPM	ft	psi
Junc 1	0	386.73	145.9
Junc 2	19.68	363.14	114.02
Junc 3	34.44	360.02	130
Junc 4	19.68	357.82	129.05
Junc 5	19.68	357.61	111.62
Junc 6	12.3	347.06	96.22
Junc 7	9.84	323.68	70.92
Junc 8	22.14	323.68	92.59
Junc 9	34.44	320.41	60.84
Junc 10	12.3	323.67	83.92
Junc 11	85.56	314.92	56.3
Junc 12	39.36	301.67	39.72
Junc 13	4.92	297.71	38
Junc 14	4.92	295.73	41.48
Junc 15	4.92	295.67	45.79
Junc 16	49.2	295.81	63.18
Junc 17	49.2	295.63	50.1
Junc 18	49.2	295.62	84.76
Junc 19	12.3	295.65	63.11
Junc 20	46.74	292.06	52.89
Junc 21	39.36	286.36	59.08
Junc 22	24.6	281.76	35.43
Junc 23	19.68	295.55	28.4
Junc 24	27.06	295.62	45.77
Junc 25	14.76	295.52	28.39
Junc 27	19.68	295.45	71.69
Junc 28	0	295.43	80.35
Junc 29	17.22	295.43	80.35
Junc 30	7.38	295.43	71.68
Junc 31	41.82	295.49	45.71
Junc 32	41.82	295.62	80.43
Junc 33	3.69	262.93	35.94
Junc 34	753.69	255.47	28.37
Junc 35	0	295.43	80.35
Junc 36	2.46	295.43	80.35
Junc 38	24.6	255.4	24.01
Resvr 37	-1599.03	50	0
Tank 26	30.39	295.51	26.22

Data Sheet 5—Fire-Flow Pressure at Node 11 (11:00 Hrs)

Post-Development Fire Flow Pressures			
EPANET Example Network 2 (CH/BP 1994)			
Network Table - Nodes at 11:00 Hrs			
	Demand	Head	Pressure
Node ID	GPM	ft	psi
Junc 1	0	403.46	153.16
Junc 2	7.76	380.49	121.54
Junc 3	13.58	377.61	137.62
Junc 4	7.76	375.22	136.58
Junc 5	7.76	374.96	119.14
Junc 6	4.85	363.96	103.54
Junc 7	3.88	339.37	77.72
Junc 8	8.73	339.37	99.39
Junc 9	13.58	335.8	67.51
Junc 10	4.85	339.37	90.72
Junc 11	33.74	329.66	62.68
Junc 12	15.52	313.69	44.93
Junc 13	1.94	308.74	42.78
Junc 14	1.94	306.3	46.06
Junc 15	1.94	304.62	49.67
Junc 16	19.4	306.13	67.65
Junc 17	19.4	305.74	54.48
Junc 18	19.4	305.74	89.15
Junc 19	4.85	305.84	67.52
Junc 20	18.43	306.27	59.05
Junc 21	15.52	306.25	67.7
Junc 22	9.7	306.25	46.04
Junc 23	7.76	298.88	29.85
Junc 24	10.67	302.92	48.93
Junc 25	5.82	296.89	28.98
Junc 27	7.76	296.88	72.31
Junc 28	0	296.88	80.97
Junc 29	6.79	296.88	80.97
Junc 30	2.91	296.88	72.31
Junc 31	16.49	296.89	46.31
Junc 32	16.49	305.74	84.82
Junc 33	1.45	306.24	54.7
Junc 34	1.45	306.24	50.37
Junc 35	0	296.88	80.97
Junc 36	0.97	296.88	80.97
Junc 38	9.7	306.23	46.03
Resvr 37	-1576.44	50	0
Tank 26	1253.64	295.64	26.27

Data Sheet 6—Water Quality Under a Biological Attack

Biological Attack at Node 25				
EPANET Example Network 2 (CH/BP 1994)				
Network Table - Nodes at 25:00 Hrs				
	Demand	Head	Pressure	Chemical
Node ID	GPM	ft	psi	mg/L
Junc 1	0	298.03	107.47	0
Junc 2	8.32	298.03	85.81	0
Junc 3	14.56	298.03	103.14	0
Junc 4	8.32	298.03	103.14	0
Junc 5	8.32	298.04	85.81	0
Junc 6	5.2	298.05	74.98	0
Junc 7	4.16	298.08	59.83	0
Junc 8	9.36	298.08	81.5	0
Junc 9	14.56	298.09	51.17	5.43
Junc 10	5.2	298.08	72.83	0
Junc 11	36.17	298.12	49.01	13.24
Junc 12	16.64	298.26	38.24	3.33
Junc 13	2.08	298.32	38.27	0.04
Junc 14	2.08	298.37	42.63	0.71
Junc 15	2.08	298.45	46.99	0
Junc 16	20.8	298.25	64.24	0.52
Junc 17	20.8	298.25	51.24	0.37
Junc 18	20.8	298.24	85.9	1.77
Junc 19	5.2	298.24	64.23	0.02
Junc 20	19.76	298.35	55.61	0.01
Junc 21	16.64	298.32	64.27	0.37
Junc 22	10.4	298.32	42.6	0.3
Junc 23	8.32	298.81	29.81	0
Junc 24	11.44	298.55	47.04	0
Junc 25	6.24	298.94	29.87	0
Junc 27	8.32	298.93	73.2	0
Junc 28	0	298.93	81.86	0
Junc 29	7.28	298.93	81.86	0
Junc 30	3.12	298.93	73.2	0
Junc 31	17.68	298.94	47.2	0
Junc 32	17.68	298.24	81.56	16.97
Junc 33	1.56	298.31	51.26	6.67
Junc 34	1.56	298.31	46.93	2.58
Junc 35	0	298.93	81.86	3.87
Junc 36	1.04	298.93	81.86	0
Junc 38	10.4	298.29	42.59	0
Resvr 37	0	50	0	0
Tank 26	-346.09	299.06	27.76	0

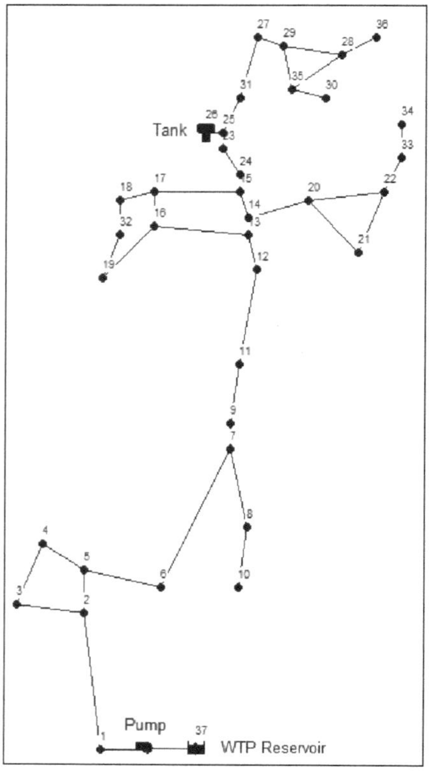

Figure B-13. Original CH/BP WDS

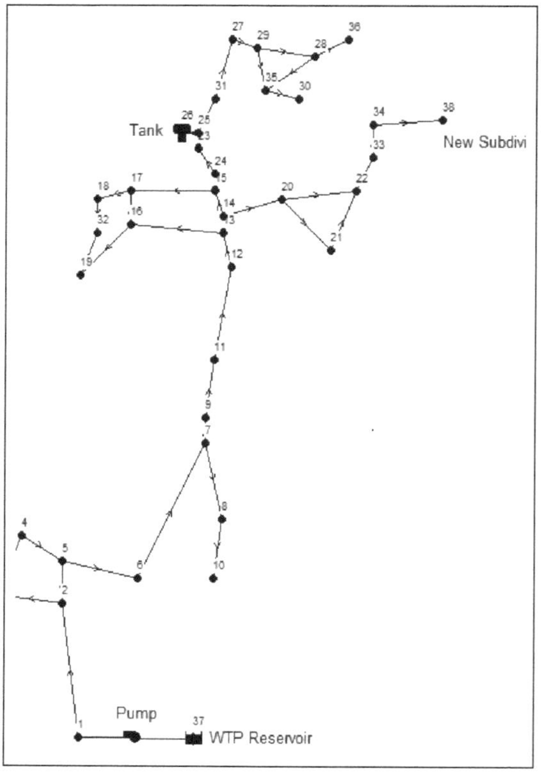

Figure B-14. Revised CH/BP WDS

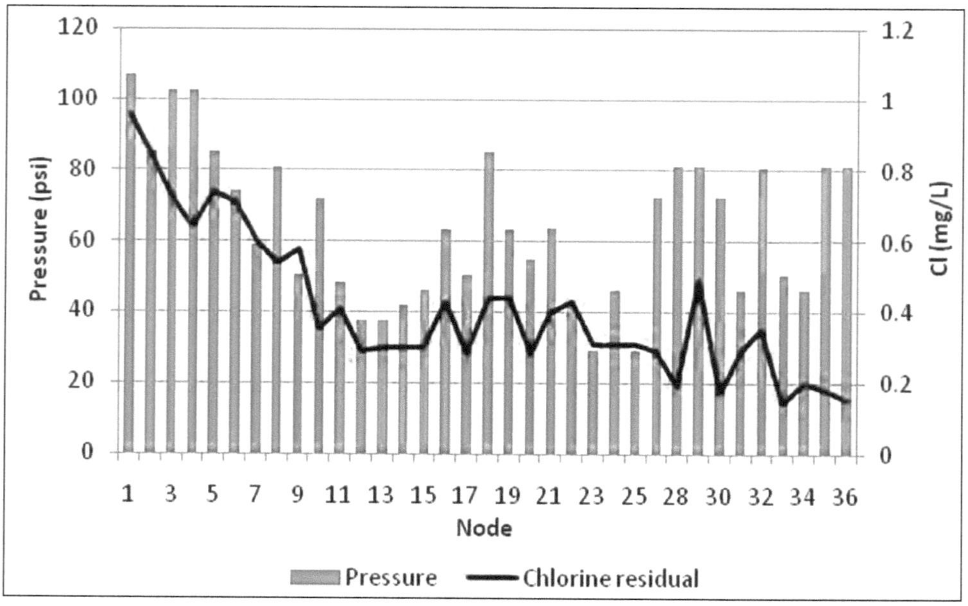

Figure B-15. Pressure and chlorine residual prior to adding new subdivision (59 hr)

Solution 3

Prepared by Nestor Alonso Mancipe Munoz, graduate student in CEE 693 Environmental Hydraulics, Department of Civil and Environmental Engineering, University of Cincinnati.

Cherry Hill/Brushy Plains water distribution system is located in Branford, Conn. This area owned by Mattabeseck Indians (1638) was called Totoket. The Dutch bought it and named the town Branford in 1644. Until the nineteenth century, the town grew and the railroad brought business, industry, and tourism. During the twentieth century, Branford established many characteristics that are associated with northeastern suburbs. It has six historic districts and many recreational facilities including parks and beaches. Based on the census figures from 2000, there were 28,683 people, 12,543 households, and 7,663 families living in Branford (Wikipedia 2011). Continuous growth led the town council to consider plans for further development.

CINGACOL (an environmental engineering firm) conducted a study of the CH/BP water distribution system to ensure that the system will be able to satisfy demands for new development. Pressure and chlorine residual were checked to understand what changes may be necessary to provide reliable service. Fire protection capacity was evaluated to be sure it would be available for any new development. Fire capacity will require large volumes of water during a short period. System design, including disinfection booster stations, storage reservoirs, and distribution mains, should include fire demands.

Base Conditions

- The subdivision node is at elevation 200 ft with a base demand of 10 gpm.
- The subdivision supply line is a 6-in. iron pipe extending 500 ft intersecting at Node 34 . Minimum acceptable pressure at the subdivision node is 40 psi.
- Fire-flow requirements are assumed to be 750 gpm for 2 hr that are coincident with peak flow of pattern 1 (10 a.m. and 11 a.m.). During fire flows, the system pressure is allowed to drop as low as 20 psi. This flow is based on the Fire Protection Rating System (Insurance Services Office 1998). If a 6-in. main is designed to support fire flows from a single fire hydrant and the distance between buildings will range from 30 to 100 ft, the needed fire flow (NFF) is 750 gpm (Walski et al. 2004).
- Based on Policy Statement 129: 40 psi at average day and maximum day demand should be maintained. Also, 20 psi at maximum day demand with maximum applicable fire-flow demand conditions applied at worst-case location should be maintained.
- To meet SWTR quality standards, the chlorine residual must stay in the range 0.2 to 1.2 mg/L at all nodes during no emergency operation.
- To eliminate effect of final conditions during water quality analysis, we consider 72 hr running the distribution system.

Figure B-13 shows the original Cherry Hill/Brushy Plains water distribution system. Figure B-14 shows the proposed new subdivision. Initially, we ran the original CH/BP WDS to see what the system behavior looked like before the new development was connected. Figure B-15 shows results of this simulation at 59 hr (11 a.m. of third day), which corresponds to the peak flow (2.5 times average hourly flow). As we can see, pressure falls (up to 4 nodes out of 36) below 40 psi. Chlorine residual falls (up to 5 of 36 nodes) below 0.2 mg/L. Therefore, the original system is not in fulfillment of the base conditions.

Solutions to Problems 1 and 2

We addressed two problems first:

1. *Fire emergency:* Does the CH/BP network provide adequate fire protection to the new subdivision in the event of a fire in Node 38 (Node 38 is selected because it represents worst-case location for pressure in the network during fire demand)?
2. *Disinfectant residual:* Does the CH/BP network satisfy SWTR water quality throughout the network?

A fire emergency was assumed to have occurred at time 10 a.m. to 11 a.m. on the first day with time simulation of 72 hr. To evaluate results, we selected nodes 13, 25, and 38 (based on results from the original CH/BP WDS for the first two nodes and worst-case location for the third one). Figure B-16 (see appendix E) shows that nodes 13 and 25 are below 40 psi (base conditions) and at Node 38, which is the worst-case location in fire event, pressure fell as low as 2 psi (required pressure 20 psi). Therefore the proposed network is not in fulfillment of the base conditions.

To improve the network pressure operation, some changes were made as shown in Table B-4.

Table B-4. Changes made to CH/BP WDS

Element	Original Set	New Set
Pump Node 46	—	260 ft and 670 gpm (Rule was set to maintain minimum pressures at nodes 13 and 25).
Tank Node 26	66 ft high	95 ft high (Other characteristics are same)
Close loop nodes 36–38	—	Length 800 ft, 6 in. diameter
Close loop nodes 36–34	—	Length 800 ft, 6 in. diameter
Close loop nodes 6–10	—	Length 800 ft, 6 in. diameter
Booster setpoint Node 29	0 mg/L	0.8 mg/L
Booster setpoint Node 20	0 mg/L	0.8 mg/L
Booster setpoint Node 34	0 mg/L	0.5 mg/L
Booster setpoint Tank 26	0 mg/L	1.0 mg/L

Results from simulation with changes except the booster pumps are shown in Figure B-17 (see appendix E). The fire demand at 10 hr dropped pressure in Node 38 (new subdivision). Afterward, the two pumps in the system and the tank maintain pressure above 40 psi for these critical nodes. The configured network will hold a fire event of 2 hr that matches peak flows with 750 gpm while pressures in the system are above 40 psi.

After pressure modifications were set, the chlorine residual was evaluated. Figure B-18 (see appendix E) shows chlorine residual below 0.2 mg/L after hour 36 for nodes 30, 36, and 38 (extreme locations in the network). Booster setpoints were applied as described in Table B-4. Results after simulation with the booster setpoints and prior modifications are shown in Figure B-19 (see appendix E). All nodes are now above 0.2 mg/L except Node 30 at 52 hr. This value is less than 5 percent of the expected 0.2 mg/L during 1 hr. This value does not represent a risk for integrity of the network thus we accept the final configuration. Afterward, the chlorine residual is stable. The final configuration of the network is shown in Figure B-20.

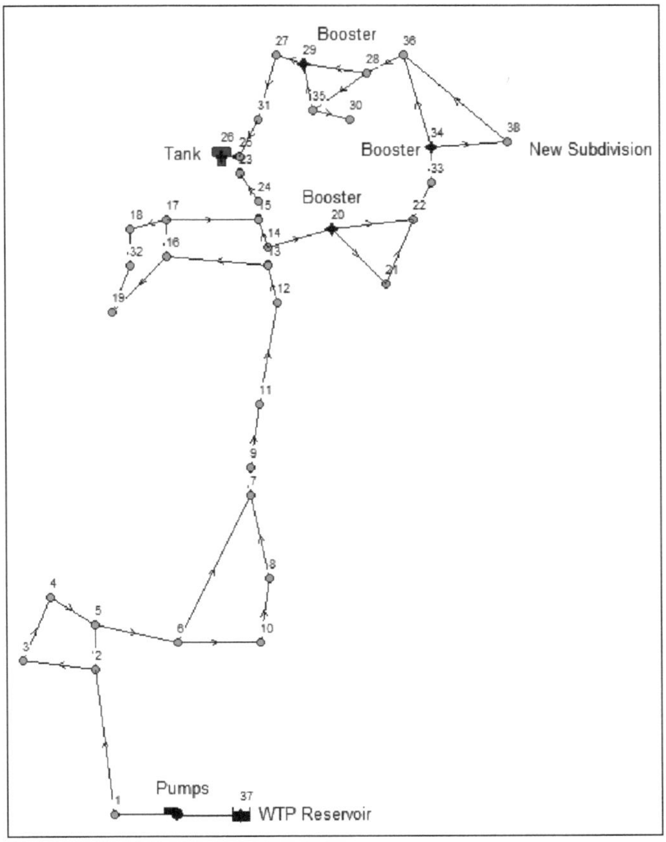

Figure B-20. Final configuration of CH/BP WDS

Solution to Problem 3

The third problem we addressed can be described as follows:

3. *Biochemical assault:* Will the WDS be affected by biochemical attack?

First, the potential pathways of chemical attack to the network were identified. Three options were analyzed as follows:

A. A pump connected to the tank (Node 26) would be an easy way to contaminate the system because this is the highest point of the network, and no additional head will be needed to put the chemical in the water. Fifty-five gallons with 10 percent concentration (100,000 mg/L) of unknown contaminant is applied to the tank during 2 hr at 34 and 35 time simulation. Coefficients of reaction are assumed the same as chlorine, and initial quality was set zero (0 mg/L) in the whole network. Results are shown in contour plots (Figure B-21, see appendix E). Dispersion is going to affect nodes surrounding the tank in the first hour. Afterward, just the nodes 25, 31, 27, 28, 29, 30, and 35 will be affected in following 10 hr after the contamination event occurs. Impact of chemical attack [A] will affect at least one-third of the network.

B. If the attack is made in Node 12 (which is one of the high nodes in the system) with the same characteristics as described above for the option [A], effects are shown in Figure B-22 (see appendix E). A higher impact on water quality will be produced with this scenario than with scenario [A]. At first hour, a good portion of the central area in the network is affected. The impact will remain beyond 10 hr, affecting other areas that were not impacted since the beginning by the attack. Impact of chemical attack [B] will affect at least two-thirds of the network.

C. If the attack is produced in Node 6 (node closer to the water reservoir than the other nodes already analyzed) with the same characteristics as scenario [A], the chemical spread is shown in Figure B-23 (see appendix E). Even though a huge pump will be required to introduce the contaminant into the network, we wanted to show the effects into this node. Clearly the impact of the spill into Node 6 is massive. Almost the whole network will be affected within the first 3 hr. After 10 hr, the concentration starts to decrease. Impact of chemical attack [C] will affect the entire network.

Summary and Conclusions

- The changes needed to be feasible; the operation of the CH/BP WDS is practical but not cost-effective because of tank modifications. Proposed changes will provide enough flow in a fire event during peak demand while fulfillment of standards is effective.
- Although a chemical attack to the CH/BP WDS is highly unlikely to occur, three scenarios were analyzed to see potential impact. When the contaminant is introduced in the initial portion of the network, the impact is higher than if the contaminant is introduced into the tank in middle of the network.

MSX Examples

The examples provided in this section illustrate several different multispecies reaction systems of interest that can be modeled with EPANET-MSX and are taken from the documentation for the EPANET-MSX manual (Shang et al. 2008).

Multisource Chlorine Decay

Multisource networks present problems when modeling a single species, such as free chlorine, when the decay rates observed in the source waters vary quite significantly. As the sources blend differently throughout the network, it becomes difficult to assign a single decay coefficient that accurately reflects the decay rate observed in the blended water. This problem was discussed earlier, in which the North Marin Water District network was solved for each reaction rate independently and a flow-weighted average was used for the solution. As discussed earlier, the network is served by two different sources, as shown in Figure B-24 (see appendix E). The network has been color-coded to show the average fraction of water in each pipe that originates from the North Marin Aqueduct (Source 1).

Assume that free chlorine reacts in the bulk flow along a pipe according to the following first-order rate expression:

$$\frac{dC}{dt} = -kC \qquad \text{(Eq B-2)}$$

where C is the concentration of free chlorine at time t and k is a reaction rate constant. It is assumed that water from the aqueduct has a $k = 1.3$ days, and the water from Stafford Lake (Source 2) has a $k = 17.7$ days. The issue becomes one of determining a k-value for each pipe of the network that will reflect the proper reactivity of the blended water from both sources.

One approach to reconciling the vastly different chlorine decay constants in this example, without introducing a more complex chlorine decay mechanism that attempts to represent the different reactivity of the total organics from the two sources, is to assume that at any time the chlorine decay constant within a pipe is given by a weighted average of the two source values, where the weights are the fraction of each source water present in the pipe. These fractions can be deduced by introducing a fictitious conservative tracer compound at Source 1, denoted as $T1$, whose concentration is fixed at a constant 1.0 mg/L. Then at any point in the network the fraction of water from Source 1 would be the concentration of $T1$ while the fraction from Source 2 would be 1.0 minus that value. The resulting chlorine decay model now consists of two species: a tracer species $T1$ and a free chlorine species C. The first-order decay constant k for any pipe in the system would be given by:

$$k=1.3T1+17.7(1.0-T1) \quad \text{(Eq B-3)}$$

while the system reaction dynamics would be expressed by:

$$\frac{dT1}{dt}=0 \quad \text{(Eq B-4)}$$

$$\frac{dC}{dt}=-(1.3T1+17.7(1.0-T1))C \quad \text{(Eq B-5)}$$

Figure B-25 is the MSX input file that defines this model for a network where the two source nodes are represented as reservoirs with ID names "1" and "2," respectively. Note that it contains no surface species and no equilibrium species, and it assumes that a constant chlorine concentration of 1.2 mg/L is maintained at each source.

Note that the solution presented here is very similar to the original solution presented for the North Marin example.

Bacterial Regrowth With Chlorine Inhibition

This example models bacterial regrowth as affected by chlorine inhibition within a distribution system based on work by Zhang et al. (2004) and includes the following processes depicted in Figure B-26:

a. Both free bacteria in the bulk flow and bacteria attached to the pipe wall use the biodegradable fraction of dissolved organic carbon (BDOC and DOC, respectively) as a growth substrate. Monod kinetics is used to describe this growth with the following rate equations:

$$\left.\frac{dX}{dt}\right|_{growth}=\mu X \quad \text{(Eq B-6)}$$

$$\left.\frac{dS}{dt}\right|=-\mu X/Y \quad \text{(Eq B-7)}$$

where X is the mass concentration of bacterial cells, S is the concentration of BDOC, Y is a yield coefficient (mass of cells produced per unit conversion of BDOC), and μ is a specific growth coefficient. The latter decreases with the BDOC concentration according to:

$$\mu=\frac{\mu_{max}S}{S+K_s} \quad \text{(Eq B-8)}$$

where μ_{max} is the maximum growth rate coefficient and K is the half-saturation constant.

```
[OPTIONS]
  AREA_UNITS  FT2
  RATE_UNITS  HR
  SOLVER      RK5
  TIMESTEP    300

[SPECIES]
  BULK  T1    MG          ;Source 1 tracer
  BULK  CL2   MG          ;Free chlorine

[COEFFICIENTS]
  CONSTANT  k1  1.3       ;Source 1 decay coeff.
  CONSTANT  k2  17.7      ;Source 2 decay coeff.

[PIPES]
  ;T1 is conservative
  RATE  T1   0

  ;CL2 has first order decay
  RATE  CL2  -(k1*T1 + k2*(1-T1))*CL2

[QUALITY]
  ;Initial conditions (= 0 if not specified here)
  NODE   1      T1   1.0
  NODE   1      CL2  1.2
  NODE   2      CL2  1.2
```

Figure B-25. MSX input file for modeling two-source chlorine decay (Shang et al. 2008)

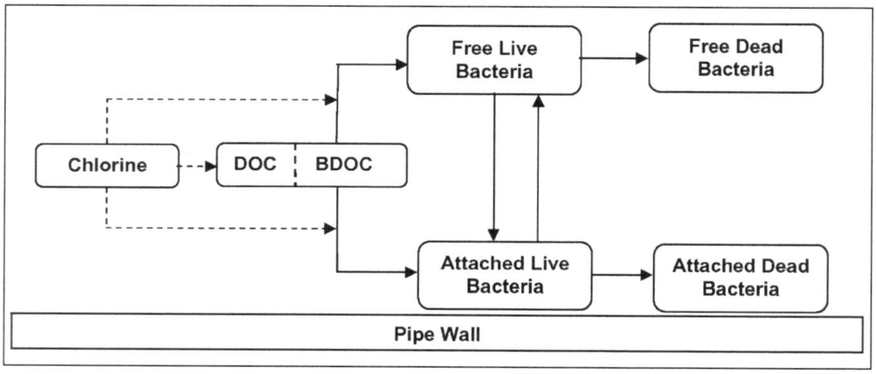

Figure B-26. Conceptual diagram of bacterial regrowth within a pipeline. Dashed arrows represent reactions with chlorine while solid arrows represent transformation processes (Shang et al. 2008)

b. Both free and attached bacteria die at a first-order rate:

$$\left.\frac{dX}{dt}\right|_{decay} = -k_d X \qquad \text{(Eq B-9)}$$

where k_d is a decay rate coefficient.

c. Deposition of free bacterial cells onto the pipe wall is modeled with the following first-order rate process:

$$\left.\frac{dX}{dt}\right|_{deposition} = -k_{dep} X \qquad \text{(Eq B-10)}$$

```
[OPTIONS]
AREA_UNITS  CM2
RATE_UNITS  HR
SOLVER      RK5
TIMESTEP    300

[SPECIES]
BULK   CL2   MG         ;chlorine
BULK   S     MG         ;organic substrate
BULK   Xb    UG         ;mass of free bacteria
WALL   Xa    UG         ;mass of attached bacteria
BULK   Nb    log(N)     ;number of free bacteria
WALL   Na    log(N)     ;number of attached bacteria

[COEFFICIENTS]
CONSTANT  Kb      0.05   ;CL2 decay constant (1/hr)
CONSTANT  CL2C    0.20   ;characteristic CL2 (mg/L)
CONSTANT  CL2Tb   0.03   ;threshold CL2 for Xb (mg/L)
CONSTANT  CL2Ta   0.10   ;threshold CL2 for Xa (mg/L)
CONSTANT  MUMAXb  0.20   ;max. growth rate for Xb (1/hr)
CONSTANT  MUMAXa  0.20   ;max. growth rate for Xa (1/hr)
CONSTANT  Ks      0.40   ;half saturation constant (mg/L)
CONSTANT  Kdet    0.03   ;detachment rate constant (1/hr/(ft/s))
CONSTANT  Kdep    0.08   ;deposition rate constant (1/hr)
CONSTANT  Kd      0.06   ;bacterial decay constant (1/hr)
CONSTANT  Yg      0.15   ;bacterial yield coefficient (mg/mg)

[TERMS]
Ib   EXP(-STEP(CL2-CL2Tb)*(CL2-CL2Tb)/CL2C) ;Xb inhibition coeff.
Ia   EXP(-STEP(CL2-CL2Ta)*(CL2-CL2Ta)/CL2C) ;Xa inhibition coeff.
MUb  MUMAXb*S/(S+Ks)*Ib                     ;Xb growth rate coeff.
MUa  MUMAXa*S/(S+Ks)*Ia                     ;Xa growth rate coeff.

[PIPES]
RATE     CL2    -Kb*CL2
RATE     S      -(MUa*Xa*Av + MUb*Xb)/Yg/1000
RATE     Xb     (MUb-Kd)*Xb + Kdet*Xa*U*Av - Kdep*Xb
RATE     Xa     (MUa-Kd)*Xa - Kdet*Xa*U + Kdep*Xb/Av
FORMULA  Nb     LOG10(1.0e6*Xb)
FORMULA  Na     LOG10(1.0e6*Xa)

[TANKS]
RATE     CL2    -Kb*CL2
RATE     S      -MUb*Xb/Yg/1000
RATE     Xb     (MUb-Kd)*Xb
FORMULA  Nb     LOG10(1.0e6*Xb)

[SOURCES]
CONCEN  SrcNode  CL2  1.2
CONCEN  SrcNode  S    0.4
CONCEN  SrcNode  Xb   0.01
```

Figure B-27. MSX input file for a bacterial regrowth model with chlorine inhibition (Shang et al. 2008)

while detachment of attached cells into the bulk flow also depends on flow velocity:

$$\left.\frac{dX}{dt}\right|_{detachment} = k_{det} XU \qquad \text{(Eq B-11)}$$

where k_{dep} is a deposition rate constant, k_{det} is a detachment rate constant, and U is the bulk flow velocity.

d. The effect of chlorine on limiting the number of viable bacterial cells is modeled by applying an inhibition factor I to the bacterial specific growth rate as:

$$I = exp\left(\frac{-(C-C_t)}{C_c}\right) \quad \text{(Eq B-12)}$$

Here C is the chlorine concentration, C_t is a threshold chlorine concentration below which no inhibition occurs, and C_c is a characteristic chlorine concentration that scales the degree of inhibition. Note that at higher values of C, I becomes smaller, resulting in smaller bacterial growth rates.

e. Chlorine reacts with DOC in the bulk flow to decay at a first-order rate:

$$\left.\frac{dC}{dt}\right| = -k_b C \quad \text{(Eq B-13)}$$

where C is chlorine concentration and k_b is a bulk decay rate coefficient.

The EPANET-MSX specification of the full model is shown in Figure B-27. See the EPANET-MSX manual for further details (Shang et al.).

References

Chow, V.T., D.R. Maidment, and L.W. Mays. 1988. *Applied Hydrology*. New York: McGraw-Hill.

Rossman, L.A. 1994. *EPANET User's Manual*. Cincinnati, Ohio: US Environmental Protection Agency (USEPA).

Rossman, L.A., R.M. Clark, and W.M. Grayman. 1994. Modeling Chlorine Residuals in Drinking Water Distribution Systems. *Journal of Environmental Engineering*, 120(4):803–820.

Shang, F., J.G. Uber, and L.A. Rossman. 2008. *EPANET Multi-Species Extension User's Manual*. EPA/600/S-07/021. National Risk Management Research Laboratory/National Homeland Security Research Center. Office of Research and Development. Cincinnati, Ohio: USEPA.

USEPA. 2008. *EPANET 2*. Cincinnati, Ohio: USEPA.

Walski, T.M. 2004. *Advanced Water Distribution Modeling and Management*. Waterbury, Conn.: Haestad Methods.

Wikipedia. 2011. Branford, Connecticut. http://en.wikipedia.org/wiki/Branford,_Connecticut.

Zhang, W., C.T. Miller, and F.A. DiGiano. 2004. Bacterial Regrowth Model for Water Distribution Systems Incorporating Alternating Split-Operator Solution Technique. *Journal Environmental Engineering*, 130(9): 932–941.

Appendix C

Water Quality Effects of Tanks and Storage

Introduction

A very important aspect of water quality and contaminant propagation in drinking water distribution systems is the effect of system storage. Although direct pumping could maximize water quality by shortening the transport time between source and consumer, it is rarely used today in systems in the United States (AWWA 2008). Most utilities use some type of ground or elevated system storage to equalize demands on supply sources, production works, and transmission and distribution mains. It is then possible to distribute and store water at one or more locations in the service area closest to the user, and the sizes or capacities of these elements may be minimized. Additionally, system flows and pressures are improved and stabilized to better serve the customers throughout the service area. Finally, reserve supplies are provided in the distribution system for emergencies, such as firefighting and power outages.

In most municipal water systems, less than 25 percent of the volume of the storage in tanks is actively used under routine conditions. As the water level drops, the tank controls call for high-service pumps to start in order to satisfy demand and refill the tanks. The remaining water in the tanks (70 to 75 percent) is normally held in reserve as dedicated fire storage.

Storage tanks and reservoirs are the most visible components of a water distribution system but may be the least understood in terms of their effect on water quality. Although these facilities can play a major role in providing hydraulic reliability for firefighting needs and in providing reliable service, they may also serve as vessels for complex chemical and biological changes that may result in the deterioration of water quality. This appendix will examine the issue of the effect of residence time on water quality and will discuss various models that have been developed for describing mixing regimes in tanks.

Mixing Research

P.V. Danckwerts, one of the first investigators to discuss the concept of distribution functions for residence times, explained how mixing and residence time in tanks can be defined and measured in actual systems (Danckwerts 1958). When a fluid flows through a vessel at a constant rate, either "plug flow" (no mixing) or perfect mixing is usually assumed. In practice, many systems do not achieve either condition, thus calculations based on these assumptions may be inaccurate. Danckwerts illustrated the use of distribution functions by showing how they can be used to calculate the efficiencies of reactors and blenders and how models may be used to predict the distribution of residence times in large systems.

A.E. Germeles developed a model based on the concept of forced plumes and mixing of liquids in tanks (Germeles 1975). He considered the mixing between two miscible liquids of slightly different density (<10 percent) when one of them is injected into a tank partially filled with the other. A mathematical model for the mixing of the two liquids was developed from which one can compute tank stratification. The model was also verified experimentally.

Empirical Studies

The impact of storage tanks on water quality was first noted during a study conducted by the US Environmental Protection Agency (USEPA) in conjunction with the South Central Connecticut Regional Water Authority (SCCRWA) (Clark et al. 1991). Some of these studies will be discussed in this appendix.

Several investigators have conducted field studies and attempted to apply relatively simple models to distribution storage tanks. Kennedy et al. (1991and 1993) attempted to assess the effects of storage tank design and operation on mixing regimes and effluent water quality. The influent and effluent flows of three tanks with diameter-to-height ratios ranging from 3.5:1 to 0.4:1 were monitored for chlorine residual. Chlorine levels were also measured within the water columns of each tank. Although chlorine profiles revealed some stratification in tanks with large height-to-diameter ratios, completely mixed models were more accurate than plug–flow models in representing the mixing behavior of all three tanks. These investigators further indicated that the quality of the effluent from completely mixed tanks deteriorated with decreasing volumetric change. The authors found that standpipes were the least desirable tank design with respect to effluent water quality.

Studies conducted by Grayman and Clark (1993) indicated that water quality degrades as a result of long residence times in storage tanks. These studies highlight the importance of tank design, location, and operation on water quality. Computer models, developed to explain some of the mixing and distribution issues associated with tank operation, were used to predict the effect of tank design and operation on various water quality parameters. Because of the diversity of the effects and the wide range of design and environmental conditions, the authors concluded that general design specifications for tanks are unlikely. They also concluded that models will most likely be refined and developed to facilitate site-specific analysis.

Boulos et al. (1996) conducted an extensive sampling study of reservoir water quality at the Ed Heck Reservoir in Azusa, Calif. The Ed Heck Reservoir is an aboveground circular tank made of poured concrete with an inside diameter of 154 ft (46.9 m) and a height of 30 ft (9.1 m). The reservoir had a capacity of 4 mil gal (15×10^9 m^3) and was fed by a 24-in. (0.6-m) inlet pipe that entered near the base of the tank and had a right angle elbow at its end that directed water counterclockwise along the reservoir's perimeter. Energy of the inflowing water was sufficient to cause a visible rotation of the reservoir water in the same counterclockwise direction. Normal operating level for the reservoir varied between 15 and 27 ft (4.6 and 8.2 m) above the reservoir bottom, and the normal mode of operation for the reservoir was simultaneous inflow and outflow. The reservoir has two openings on the roof: a periphery hatch about three quarters of the way around the reservoir from the inlet and a center vent. These access areas permit sampling in both longitudinal and vertical directions. An extensive field sampling study was undertaken, with emphasis on identifying the general free chlorine dynamics in the reservoir. The study consisted of both interior and exterior sampling. Interior sampling was carried out using a specially designed sampling apparatus, and exterior sampling was performed by taking samples through the hatch. The direct interaction between hydraulics and water quality was quantified using a nontoxic conservative tracer (fluoride), and during the study, fluoride, free chlorine residual, and temperature were measured. Sampling data indicate that the Ed Heck Reservoir could be viewed as having two zones:

- Figure C-1, Part A: An annular cylinder extending from the reservoir walls toward the center and a core cylinder. Flow enters the reservoir and, because of the direction of the inlet line, produces a counterclockwise flow pattern. Sampling results from the reservoir hatch indicated that the annular ring is vertically well mixed. Flow is transferred from the annular ring near the top of the reservoir and then gradually moves down the inner core cylinder.
- Figure C-1, Part B: A mass-balance analysis suggests that the volume in the annular ring is significantly greater than the volume in the central core.

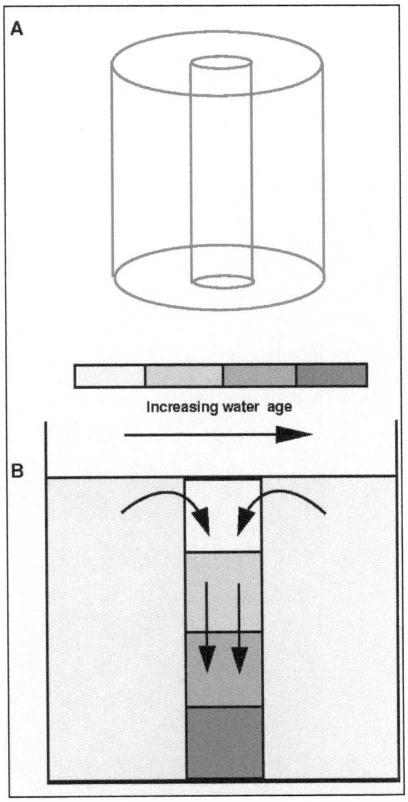

Figure C-1. Inferred zones in reservoir (part A); inferred mixing patterns in reservoir and resulting water age (part B) (Boulos et al. 1996)

The following conclusions were made as a result of the study:

- Fluoride tracer studies were useful for determining both residence time distribution and internal mixing dynamics in storage reservoirs.
- The flow pattern in the Ed Heck Reservoir consisted of a relatively fast rotational flow around a large outer annular ring coupled with a slower downward flow within a smaller central core.
- The short-circuiting between the inlet and outlet lines causes the T_{10} for this reservoir to be less than half that of a true CSTR.

SCCRWA Case Study

As discussed earlier, several case studies involving the effects of tanks and reservoirs were conducted at the SCCRWA. The town of Cheshire, Conn., served by SCCRWA, provides as excellent example of the effects of tank operation on water quality. Primarily a residential area, Cheshire is fed by two separate well fields—the North Cheshire well field, composed of four wells with a combined capacity of 3.6 mgd (13.6 ML/d), and the South well field, with two wells and a capacity of 2.5 mgd (9.5 ML/d). Storage is provided by two adjacent tanks, Prospect tanks 1 and 2, which "float on the system" (i.e., tanks that have little water movement and are designed to maintain system pressure), with 2.5 mil gal (9.5 ML) of storage each. The heights of these tanks are 25 ft (7.5 m), and typically the tanks operate in a range of 14 to 19 ft (4.2 to 5.7 m). The operation of the wells is manually controlled by an operator who responds to the tank water levels. Average daily sendout for Cheshire is approximately

2.2 mgd (8.3 ML/d) during the winter and slightly more in the summer. Water use does not vary significantly by day of the week. Water use is highest during the morning and evening and lowest during the night.

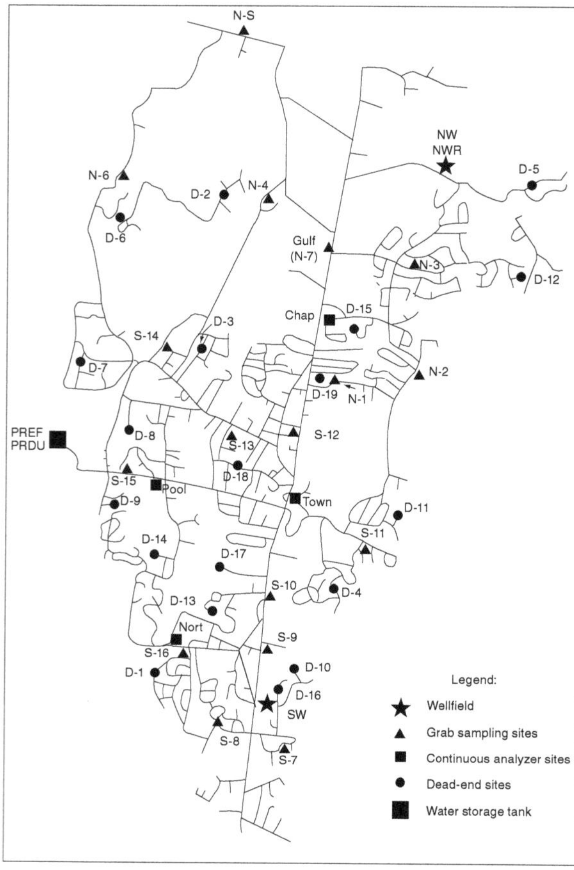

Figure C-2. Cheshire service area map

Field Sampling Program

As part of the overall USEPA-sponsored project discussed earlier, an extensive field sampling program was performed in the Cheshire system during November and December 1989. During this period, the fluoride feed into the water drawn from the two well fields was stopped for a 7-day period during which extensive sampling occurred throughout the system. The fluoride feed was later restarted, and sampling was performed for the next 7-day period. Complete details on the sampling program, described earlier, are provided by Clark et al. (1991) and Skov et al. (1991).

A diagram showing the Cheshire distribution system, wells, the Prospect tanks, and the sampling sites is presented in Figure C-2. Initially during the sampling study, the wells were operated to minimize the variation of water level within the Prospect tanks. After 2 days, this policy was changed to allow a much greater variation of water

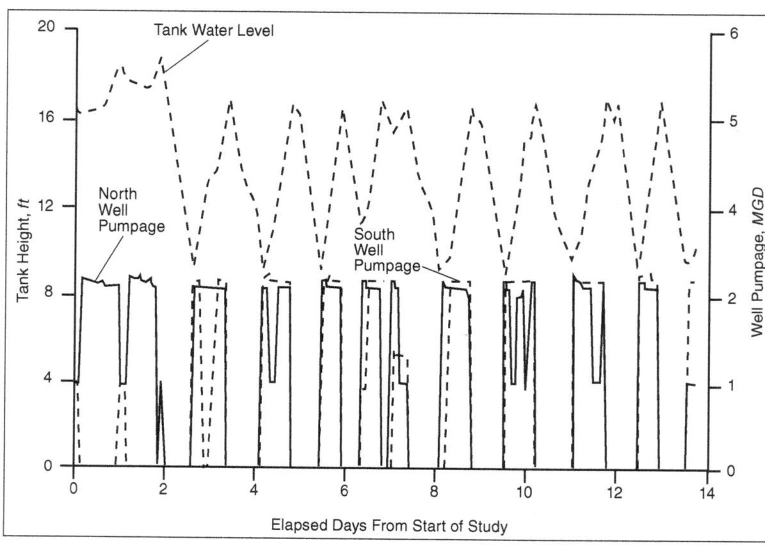

Figure C-3. Well pumpage and tank level sampling period with 9-ft (2.7-m) tank water level variation

level in the tanks. This policy was followed for the remainder of the study. The well pumpage record and tank water level elevations during the study are graphed in Figure C-3.

During the study, water entering and leaving the Prospect tanks was monitored on a frequent basis. An automatic sampler designed to sample at a set interval (generally every 1 or 2 hr) was attached to the outlet line from the tanks. A continuous chart recorder monitored the flow in the combined inlet–outlet line. The resulting water quality profiles are shown in Figure C-4. As illustrated, the variation between fluoride concentration levels entering the tank and leaving the tank was very significant. Only after about 7 days did the levels of fluoride leaving the tank approach the levels entering the tank.

From this information, one could conclude that it would require about 10 days for the tank to reach equilibrium. A second significant conclusion can be reached by examining the fluoride traces in the tank effluent. If the tank was a completely mixed reactor, the effluent traces would be horizontal lines. As illustrated in Figure C-4, during the period when the fluoride concentration levels in the tank exceeded the concentration levels in the system (i.e., during the first seven days of the sampling period when the fluoride was shut off), for each period of discharge, fluoride concentrations increased from a value representative of the inflow concentration to a value representative of the average concentration in the tank. This showed that the tank is not a completely mixed reactor and considerable short circuiting occurred.

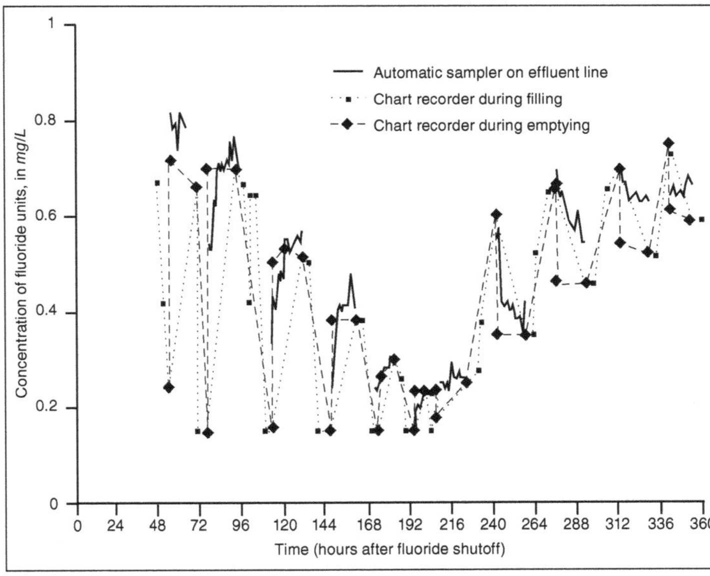

Figure C-4. Fluoride concentrations at Prospect tank

A similar inverse relation was observed during the second 7-day period after the fluoride had been turned back on and the level of fluoride concentration in the system exceeded the concentration level in the tank. During each period of tank discharge, it required approximately 8 hr for the effluent fluoride concentration to reach a constant concentration level, indicating a significant amount of short circuiting.

Effects of Tank Storage Characteristics on Water Quality

The field study showed that storage facilities could have a significant impact on water quality in a distribution system. In order to further investigate the effects of tank location and operation on the water quality in the system, a series of simulations were performed using a hydraulic and water quality model of the Cheshire system.

In the simulations, the effects of the location and operation of the tank were studied (Grayman et al. 1991). Location was studied by simulating the system with the tank located at its actual site (Prospect) and alternatively a tank of similar characteristics located in the

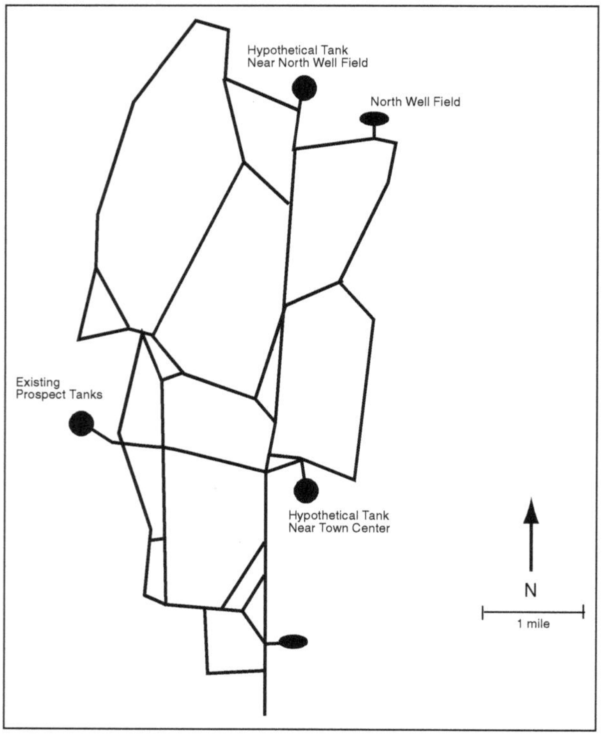

Figure C-5. Skeletonized representation of Cheshire distribution system

vicinity of the North well field and near the center of town. These alternate locations are also shown in Figure C-5. Various operations were simulated by varying the water level elevations in the tank that would activate the wells. A 9-ft (2.7-m) variation (well pumps on when the tank level drops to 9 ft [2.7 m] and off when it reaches 18 ft [5.4 m]), a 6-ft (1.8-m) variation (on at 12 ft [3.6 m] and off at 18 ft [5.4 m]), a 3-ft (1.8-m) variation (on at 15 ft [4.5 m] and off at 18 ft [5.4 m]), and on a limited basis, a 1-ft (0.3-m) variation (on at 17 ft [5.1 m] and off at 18 ft [5.4 m]) were simulated.

Simulating Water Age

Because chlorine reacts with the natural organic matter (NOM) in water, water age is a key factor in the deterioration of water quality. With age, chlorine residuals will decrease and disinfection by-product (DBP) concentrations will increase. Therefore, age serves as a good indicator of water quality.

When simulating water age, an initial age of zero was assumed at all nodes and at the tanks. Water entering the system from wells was considered to be new water (age = 0). For the Prospect location, water level variations of 1, 3, 6, and 9 ft (0.3, 0.9, 1.8, and 2.7 m) were simulated. For the other locations, the minimum water level variation that was examined was 3 ft (1.8 m).

This analysis indicates a significant variation in age based on the different water level variations. For example, the long-term average water age in the Prospect tank is approximately 7 days when the water level in the tank is allowed to vary by 9 ft (2.7 m). However, when the variation in water level is reduced to 1 ft (0.3 m), the long-term average age increases almost threefold to approximately 20 days. A similar pattern is found for the alternate tank location near the town center. For the tank location near the North well, the difference in age is less for the different operating procedures.

This analysis shows the relatively close agreement in results for the Prospect and town center sites, with a significantly lower age when the tank is located in the vicinity of the North well. This phenomenon is easily explained by the relatively short travel time from the well to the North well tank and, as a result, the infusion of primarily new water directly from the well during the time when the tank is filling.

As the water level variation decreases, the spread of older water through the system decreases and the age of the water in the tank and in the vicinity of the tank increases. For example, with a 1-ft (0.3-m) variation in the tank, the age of the water in the vicinity of the tank exceeds 18 days, while with a 9-ft (2.7-m) variation, the age of the water in this area is approximately 7 days. However, with the 1-ft (0.3-m) variation, for 32 of the 48 nodes, the maximum age of the water is less than 2 days, while for the 9-ft (2.7-m) variation, only 13 of the 48 nodes receive water with a maximum age less than 2 days. Thus, when the tank is located at Prospect, one must select between the two objectives of minimizing the age of the water in the vicinity of the tank or minimizing the spread of "old" water through the system.

When the tank is moved to the vicinity of the North well, the results are essentially the opposite. In this configuration, the spread of "old" water is greatest with the 3-ft (0.9-m) variation, and the general age of the water throughout the system is largest (with the exception of the area in the vicinity of the South well, which receives almost all of its flow in the form of "new" water directly from the South well). When the tank is moved to the town center area, a different pattern results, with the 6-ft (1.8-m) variation scenario resulting in the generally youngest water in the vicinity of the tank and throughout the system.

The large variation in results for the various tank locations and operating policies suggests that the age of water is very sensitive to both variables and that modeling is required to assess the likely response of the system to changes in tank location and operation. When water age is coupled with a time-varying substance, such as chlorine residual, the effect of storage on water quality is very apparent.

Chlorine Residual

A time-varying parameter investigated in this study was chlorine residual. Chlorine residual was represented in the model by a first-order decay function. In this study, a value of 1 day^{-1} was used for the first-order decay coefficient. This value is somewhat larger than the observed decay in this distribution system but was useful for illustrative purposes (Characklis 1988). With a decay value of 1, chlorine residual has a half-life (will decay to half of its original value) of approximately 0.7 days. Chlorine residual is an excellent parameter for studying the effects of tanks because it combines the characteristics of a conservative constituent with the effects of age and because of its real-life effects on the potential quality of water as it is delivered to the customer.

In the simulation, a chlorine residual value of 100 (corresponding to 100 percent of the chlorine residual in sources) was assumed at each source. Initially, all nodes (including the tank) were assumed to start with a chlorine residual of zero. The system was then simulated for a period until it reached a dynamic equilibrium (generally less than 10 days), and the variation in chlorine residual throughout the system over the last 48-hour period was noted.

The minimum chlorine residual at each node for the different tank locations and tank water level variations were analyzed. The effects of water level variation at the Prospect tank are shown in Figure C-6 for variations of 1, 3, 6, and 9 ft (0.3, 0.9, 1.8, and 2.7 m), respectively. As with the water age simulations, these effects are very apparent and very significant. As the water level variation increases, the spread of older water through the system increases and results in water with lower chlorine residual throughout the system. For example, with a 1-ft (0.3-m) variation in the tank, the chlorine residual is less than 20 percent (of the chlorine residual in the sources) for only 16 of the 48 nodes in the system. However, when the water level variation is increased to 9 ft (2.7 m), 38 of the 48 nodes have less than 20 percent of the source water chlorine residual. When the tank is moved to the vicinity of the North well, the results are much less dramatic. In this configuration, most of the system receives water that has been stored in the North well tank and consequently has a low chlorine residual for all operating procedures, with the exception of the area in the vicinity of the South well, which receives most of its flow in the form of "new" water directly from the South well.

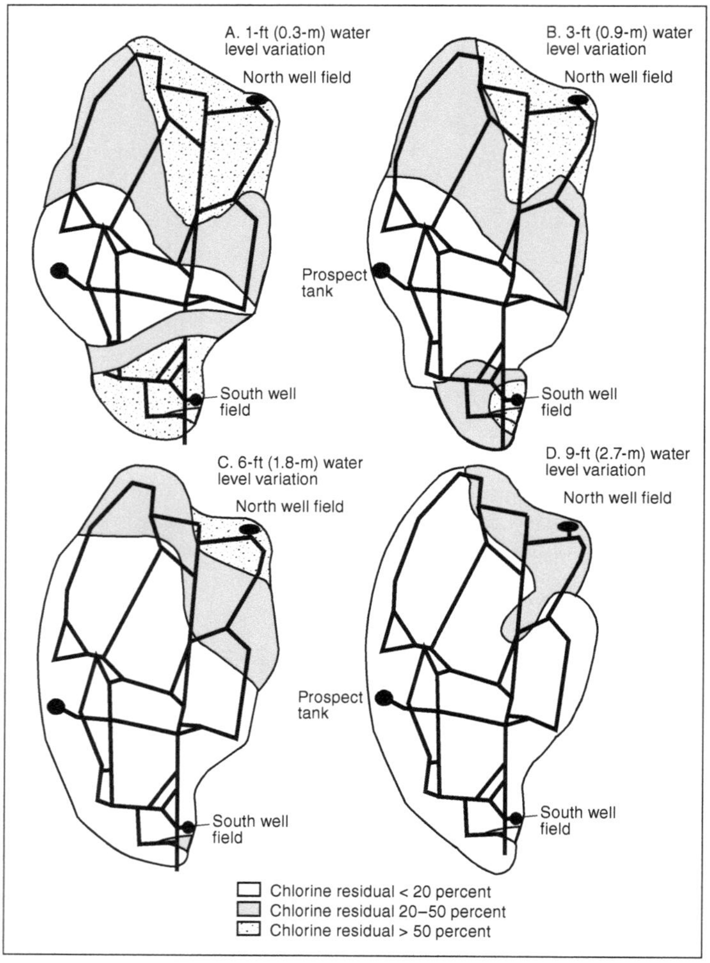

Figure C-6. Minimum chlorine residual at Prospect tank

Storage Tank Models

As indicated earlier, most water quality simulations assume complete mixing of the storage tanks examined. However, sampling data from the Prospect tank and other studies (Kennedy et al. 1991, 1993) indicated that is not necessarily the case. Grayman and Clark (1993) explored the use of compartment modeling to describe tank mixing and to deal with the nonuniform mixing in tanks, as described in the following section.

Tank Mixing Model

A compartmentalization approach will be used to represent different mixing zones in the water column within a tank. For purposes of this analysis, a maximum of three compartments is assumed. Figure C-7 illustrates the inflow and outflow compartmental configurations associated with classical inflow–outflow tank systems. One- (completely mixed), two-, and three-compartment models are developed for the inflow–outflow and continuously flowing tank in the following sections (Clark et al. 1996).

Of the various configurations considered in this study, some basic formulas are common to all of them. The general formula for the change in concentration of a conservative constituent, for example, fluoride, in the tank is

$$\frac{d(C_T V_T)}{dt} = F \cdot C \qquad \text{(Eq C-1)}$$

Where:

F = flow, in ft/sec (m/sec)
C = concentration, in mg/L
d = differential
C_T = concentration of fluoride in tank, in mg/L

V_T = volume of tank, in ft³ (m³)
dt = rate of change in time

or equivalently,

$$V_T \frac{dC_T}{dt} = C_T \frac{dV_T}{dt} = F{\cdot}C \qquad \text{(Eq C-2)}$$

Change in the water volume in the tank is described by

$$\frac{dV_T}{dt} \qquad \text{(Eq C-3)}$$

Solving Eq C-3 numerically, using Euler's method, we obtain:

$$V_T^{(n+1)} = V_T^{(n)} + h{\cdot}F^{(n)} \qquad \text{(Eq C-4)}$$

Where:
h = integration step size
$n + 1$ = the state of the "next" iteration

and other variables are as defined previously.

Inflow–Outflow Tank Model

The general equations for the one-, two-, and three-compartment models are presented in the following sections.

Inflow conditions for one-compartment model: Water is assumed to enter the tank at a flow rate of Q_{in} and fluoride concentration of C_{in} (Clark et al. 1996). The value of C_T is the concentration of fluoride in the tank and V_T is the volume of water in the tank. Therefore, the net concentration of fluoride in the tank during the inflow period would be $(C_{in} - C_T)$. There is no outflow during this stage. The initial conditions in

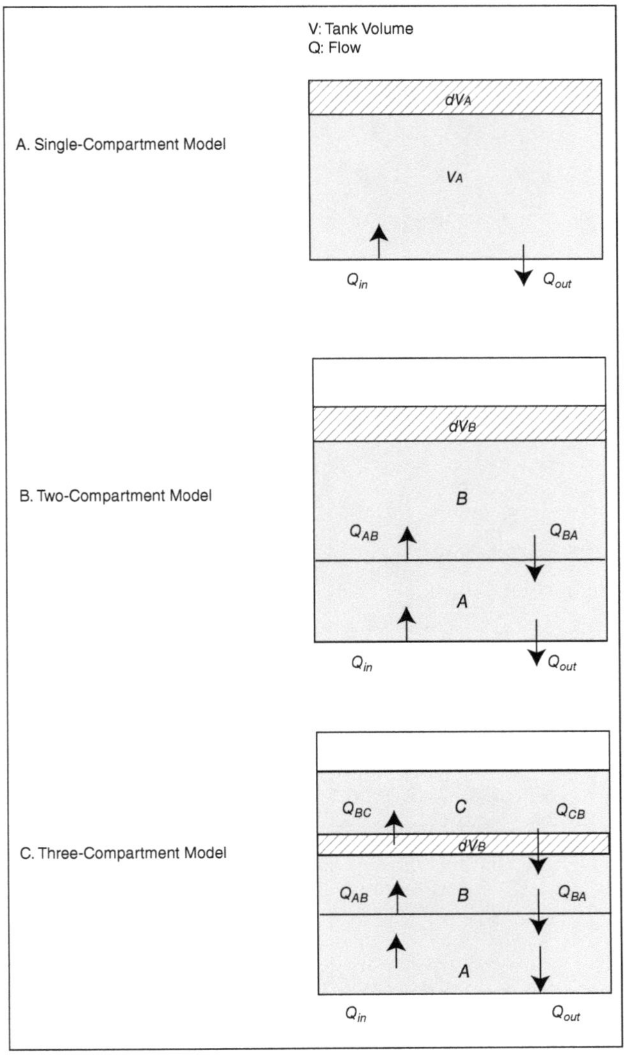

Figure C-7. Compartment model configurations

the tank are $C_T(t = 0) = C_{T0}$ and $V_T(t = 0) = V_{T0}$. By substituting these values in Eq C-1 and C-2, we obtain

$$\frac{dC_T}{dt} = \frac{Q_{in}(C_{in} - C_T)}{V_T} \tag{C-5}$$

Outflow conditions for one-compartment model: Water flows out of the tank at a rate of $-Q_{out}$ and there is no inflow to the tank. Therefore, by substituting the appropriate values for the flow and fluoride concentration in Eq C-1 and C-2, we will have

$$\frac{dC_T}{dt} = 0 \tag{Eq C-6}$$

Inflow conditions for two-compartment model: The volume of water in compartment A, V_A, is fixed. Compartment B has a variable water volume, V_B. Water enters compartment A with a flow of Q_{in} gallons per minute and with a fluoride concentration of C_{in}. Flow rate from compartment A to B is indicated by Q_{AB}, and C_B indicates the fluoride concentration in tank B. The initial conditions for compartment A are $C_A(t = 0) = C_{A0}$ and $V_A(t = 0) = V_{A0}$. Since compartment A has a fixed water volume, then $\frac{dV_A}{dt} = 0$ and $Q_{in} = Q_{AB}$. By substituting these conditions in Eq C-3 and C-4, we will have

$$\frac{dC_A}{dt} = \frac{Q_{in}(C_{in} - C_A)}{V_A} \tag{Eq C-7}$$

Inflow conditions for compartment B with $\frac{dV_B}{dt} = Q_{AB}C_A$ result in

$$\frac{dC_B}{dt} = \frac{Q_{in}(C_A - C_B)}{V_B^{(n)}} \tag{Eq C-8}$$

Outflow conditions for two-compartment model: Water flows out of the compartment B at a rate of $Q_{AB} = Q_{out}$ and a fluoride concentration of C_B. By substituting these values into Eq C-1 and C-2, we have

$$\frac{dC_B}{dt} = \frac{Q_{AB}(C_B - C_B)}{V_B^{(n)}} \tag{Eq C-9}$$

The mass balance equations for outflow from compartment A, with an outflow rate of $Q_{out} = Q_{BA}$, will imply

$$\frac{dC_A}{dt} = \frac{Q_{out}(C_B - C_A)}{V_A} \tag{Eq C-10}$$

Inflow conditions for three-compartment mode: For this model it is assumed that the volumes of compartments A and C are fixed, the volume of compartment B is variable and a function of time t, and there is a flow exchange between compartments B and C, represented by Q_{BC} and Q_{CB}. This flow rate can be adjusted to a fraction of the inflow or outflow, depending on the condition of the tank at a given time.

Since compartments A and C have fixed volumes, then $Q_{in} = Q_{AB}$, $Q_{BC} = Q_C$, and $Q_{BA} = Q_{out}$. The mass balance for inflow for compartment A will result in

$$\frac{dC_A}{dt} = \frac{Q_{in}(C_{in} - C_A)}{V_A} \quad \text{(Eq C-11)}$$

By substituting the net flow values for compartments B and C into Eq C-1 and C-2, we obtain

$$\frac{dC_B}{dt} = \frac{Q_{in}(C_A - C_B) + Q_{BC}(C_C - C_B)}{V_B^{(n)}} \quad \text{(Eq C- 12)}$$

$$\frac{dC_C}{dt} = \frac{Q_{BC}(C_B - C_C)}{V_C} \quad \text{(Eq C-13)}$$

Outflow conditions for three-compartment model: By substituting the outflow conditions for compartment C, with a fixed water volume, and compartments B and A, we have the following:

$$\frac{dC_C}{dt} = \frac{Q_{BC}(C_B - C_C)}{V_C} \quad \text{(Eq C-14)}$$

$$\frac{dC_B}{dt} = \frac{Q_{BC}(C_C - C_B)}{V_B^{(n)}} \quad \text{(Eq C-15)}$$

$$\frac{dC_A}{dt} = \frac{Q_{out}(C_B - C_A)}{V_A} \quad \text{(Eq C-16)}$$

Flow-Through Tank Model

As with the classical tank model, one compartment, two compartments, and three compartments were used to represent the mixing regime in the flow-through tank model. Figure C-8 illustrates the flow conditions associated with these assumptions.

One-compartment model for flow-through model: In this case the tank is filling and emptying at the same time. Therefore, the change in the volume is a function of the difference between inflow and outflow, i.e., $\frac{dV}{dt} = Q_{in} - Q_{out}$. After appropriate adjustments in Eq C-1 and C-2, we have

$$\frac{dC_A}{dt} = \frac{Q_{in}(C_{in} - C_A)}{V_A^{(n)}} \quad \text{(Eq C-17)}$$

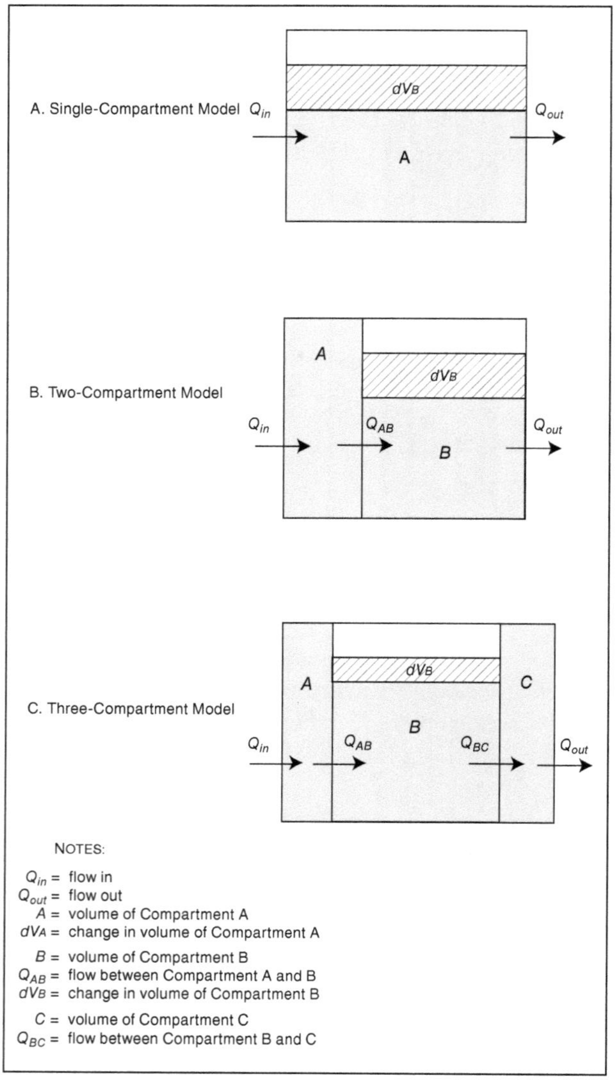

Figure C-8. Schematic of flow-through compartment model

Two-compartment model for flow-through model: In this model, the volume of compartment A is fixed, therefore $\frac{dV_A}{dt} = 0$ and $Q_{in} = Q_{AB}$. Then for compartment A we have

$$\frac{dC_A}{dt} = \frac{Q_{in}(C_{in} - C_A)}{V_A} \qquad \text{(Eq C-18)}$$

The volume in compartment B is variable and is a function of inflow Q_{in} and outflow Q_{out}. Therefore, the change of fluoride concentration is $Q_{in}C_A - Q_{out}C_B$. Substituting these values in Eq C-1 and C-2 will result in

$$\frac{dC_B}{dt} = \frac{Q_{in}(C_A - C_B)}{V_B^{(n)}} \qquad \text{(Eq C-19)}$$

Three-compartment model for flow-through model: The model assumes constant and equal volumes for compartments A and C. Therefore, the inflow to compartment A is the same as the flow from compartment A to B, $Q_{in} = Q_{AB}$. For the same reason, flow from B to C is the same as the outflow from compartment C, $Q_{BC} = Q_{CB}$. Compartment B has a variable volume that is a function of the difference between inflow and outflow. Considering these conditions for compartments A, B, and C, we will have

$$\frac{dC_A}{dt} = \frac{Q_{in}(C_{in} - C_A)}{V_A} \qquad \text{(Eq C-20)}$$

$$\frac{dC_B}{dt} = \frac{Q_{in}(C_A - C_B)}{V_B^{(n)}} \qquad \text{(Eq C-21)}$$

$$\frac{dC_C}{dt} = \frac{Q_{out}(C_B - C_C)}{V_C} \qquad \text{(Eq C-22)}$$

Figure C-9. Schematic representation of three-compartment tank model

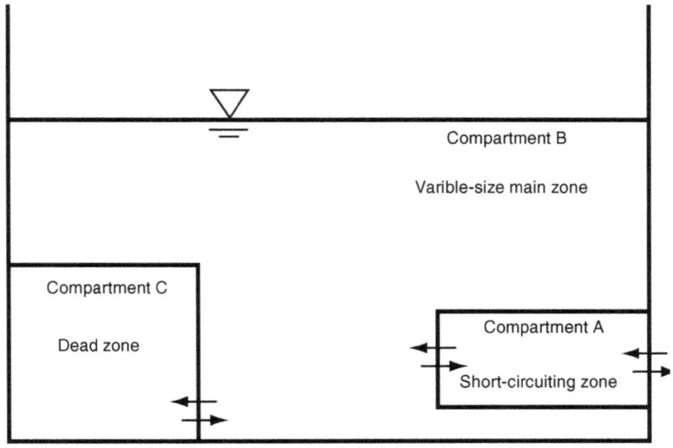

Validation of Compartment Model Assumptions

Water quality models generally represent tanks as continuously stirred tank reactors (CSTR) (Grayman and Clark 1991, 1993), which is useful assumption. However, analysis of field sampling studies and data from laboratory studies of tanks indicate that there is considerable variation in the behavior of tanks and that in many cases the assumption that tanks are completely mixed is not a good representation.

Two related phenomena that lead to the breakdown of the CSTR assumption are short-circuiting between the inlet and outlet of a tank, and the presence of "dead zones" within the tank where mixing with the remainder of the tank is limited. In short-circuiting, a direct flow path between the influent and effluent line results in a last in–first out situation rather than full mixing. These phenomena may be due to design features or may result from environmental factors, such as temperature stratification.

A three-compartment model is suggested as a means of representing a wide range of tank configurations. As illustrated in Figure C-9, the three-compartment model includes (1) compartment A, which represents the volume of the tank near the inlet/outlet where short-circuiting would most likely occur; (2) compartment C, which represents the tank volume where exchange with the main body of the tank is limited (dead zone); and (3) compartment B, the main body of the tank.

These separate compartments are purely conceptual in nature. For example, compartment C may represent several different parts of the tank (e.g., corners in a rectangular basin or the upper portions of tank) where the interchange is limited. In the model, compartments A and C are assumed to be of fixed volume. Compartment B varies in size as the water level in the tank changes. Each of the compartments is assumed to be represented as continuously stirred tank reactors, though at any time the concentration of the constituent of interest may vary between the compartments. A time lag is introduced through a numerical-difference solution technique that operates on an hourly time step.

In this model, all flow is assumed to enter and exit through compartment A. In order to maintain the fixed volume of this compartment, the flow rate between compartments A and B must be the same as flow rate between compartment A and the distribution system. The flow rate between compartments B and C is assumed to be adjustable. The lower the exchange rate, the "deader" the zone.

The control variables are the fixed volumes of compartments A and C and the exchange rate between compartments B and C. If the volumes of compartment A and C are set close to zero,the resulting model acts like a single-compartment CSTR. The values for the three input parameters and the relationship of the compartment volumes to the total volume of the tank result in highly variable behavior. It is important that the sum of the fixed volumes of compartments A and C is always less than the volume of the water in the tank or else the volume of compartment B would become negative and continuity would not be maintained. Similarly, the exchange rate between B and C should be such that the total volume exchanged during a time step is less than the volume of compartment C, otherwise questionable concentrations may result.

In the model, a constituent may be characterized as *conservative* or *nonconservative*. For nonconservative constituents such as chlorine, a first-order exponential decay is assumed. The model may also be used to model the age of water. In this case, the age of water entering the tank must be estimated, and the model then calculates the age in each of the three compartments.

Historical data on the inflow and outflow concentrations for the period being simulated may also be provided by the user so that the model can display the comparison of simulated and historical concentrations. Output from the model includes both graphical and tabular displays. Graphical output includes a plot of predicted concentration over time for the three compartments and a plot of tank flow rates over the same time period. Figure C-10 shows an example of such a plot. A second set of plots for each model run displays the predicted concentrations for compartment A along with the historical data and a similar plot for compartment B, as shown in Figure C-10.

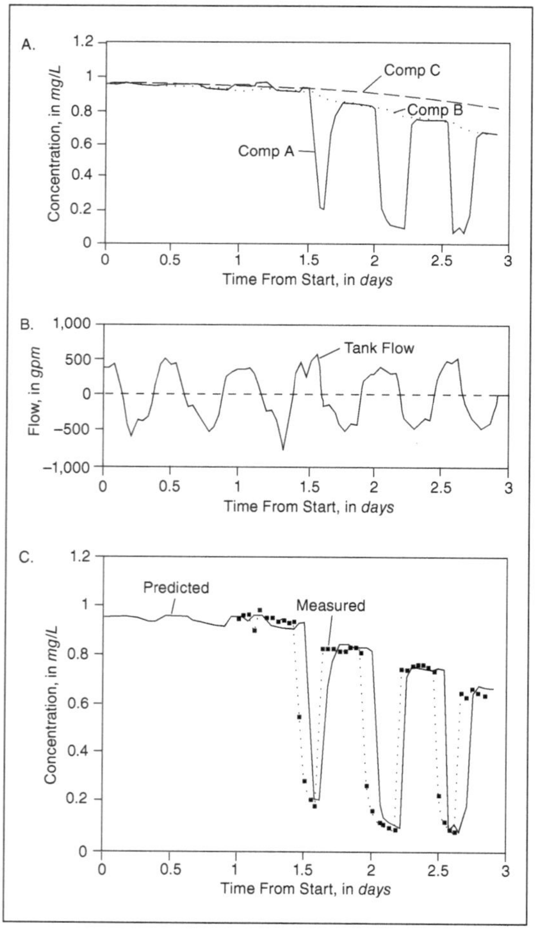

Figure C-10. Sample output from three-compartment tank model

Model Results

The three-compartment model was applied to both the Prospect tanks and the Brushy Plains tank, another tank in the SCCRWA service area. Fluoride was modeled at the Prospect tanks, and fluoride, chlorine, and water age were studied at the Brushy Plains tank. For the Prospect tanks, there was no indication of a dead zone, so only compartments A and B were represented in the model.

As a basis of comparison, a single-compartment model was assumed and the results compared to the concentrations in the effluent line as determined from the field study (Figure C-11). As illustrated, though the predicted concentration curve generally followed the trend of observed concentrations, it did not represent the variations that occurred for an 8-hr to 12-hr period as the tank switched from filling to emptying. As illustrated in Figure C-12, the two-compartment model with compartment A equal to 100,000 ft^3 (2,800 m^3)

Figure C-11. Fluoride concentration at Prospect tank

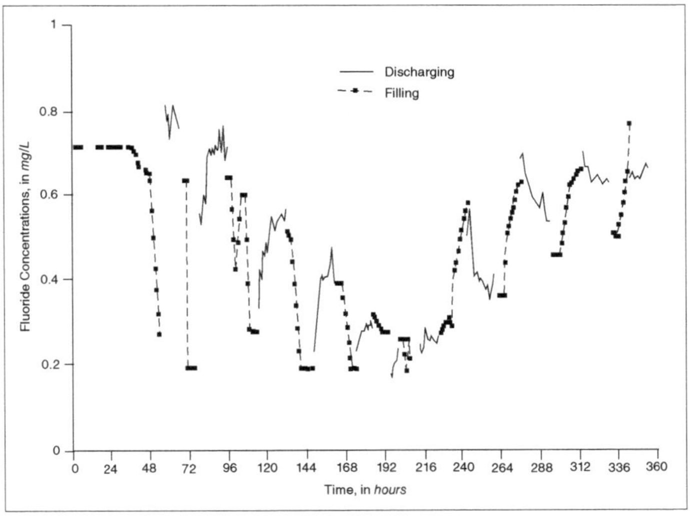

resulted in a significantly improved representation of outflow concentrations.

The dominant phenomenon for the Brushy Plains tank was the apparent dead zone. To model this, compartment C volume was set at 100,000 gal (378,500 L) and an exchange rate between compartments B and C was set at 50 gpm (189 L/min). These parameter values resulted in a good fit when modeling fluoride concentrations in the tank discharge. For chlorine, the decay coefficient also significantly affected model results, as shown in Figure C-13, where coefficients of K = 0 and K = 1 days^{-1}, respectively, were used.

Water age was also simulated for the Brushy Plains tank. Age of water was estimated for the tank inflow and for the initial tank water age by using historical chlorine data. Using these values, the water age was manually calculated and entered into the model. The resulting predicted time patterns of water age within each compartment are shown in Figure C-14. The figure indicates a general increasing trend for water age in the tank, with fluctuation around this trend being most pronounced for compartment A as the tank changes from filling to emptying and vice versa. Because

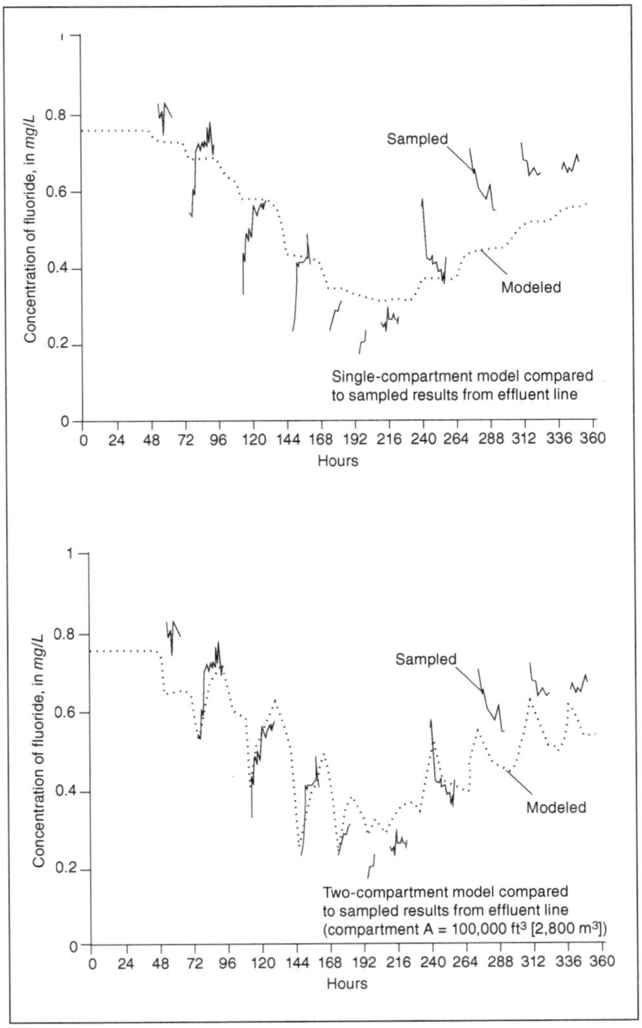

Figure C-12. Comparison of modeled and sampled fluoride concentrations at Prospect tank

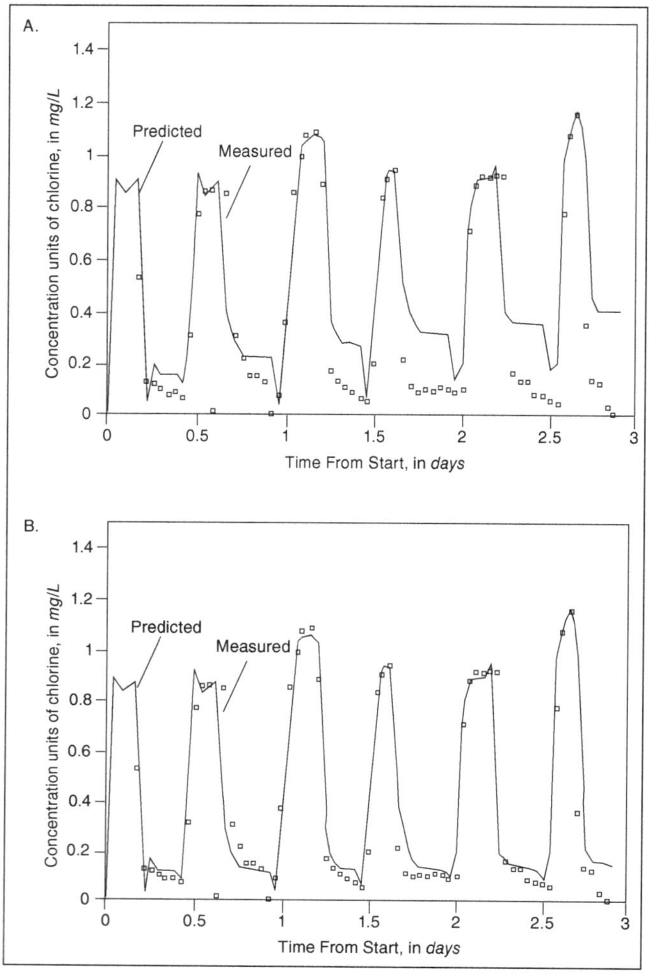

Figure C-13. Comparison of predicted and sampled chlorine concentrations at Brushy Plains tank with (A) no decay and (B) decay

of the reduced exchange between compartments B and C, the age in compartment C is always greater than in compartment B. The relationship between water age in the three compartments is a good indicator of the dynamics occurring within the three-compartment tank model.

Parameterization of the model (i.e., selection of the three parameters) as described to this point is a trial-and-error process. In addition to tank design, environmental parameters such as the relationship between ambient and water temperatures, and conditions within the tank, may affect the model parameters. For the application to the two tanks, several different combinations of parameters were explored, and even though an exhaustive search was not performed, an excellent relationship between simulated and actual values was achieved. It is hoped that

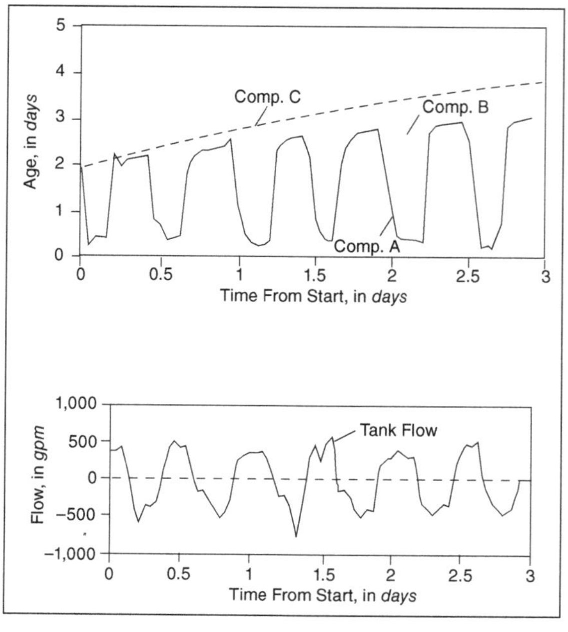

Figure C-14. Simulated water age by compartment at Brushy Plains tank

greater experience with the model will lead to some general guidelines for selection of parameters to represent tanks of differing design or environmental conditions.

A consequence of long residence times is shown in Figure C-13. As can be seen, the discharged water has very low residuals due to the long residence times in the tank (as implied from the previous study).

In the following sections, a more formal approach to compartmentalization is assumed. In addition, a technique that deals with time-varying influent and effluent concentrations of a tracer is presented.

Design Guidelines for Tanks and Storage Reservoirs

Grayman et al. (2004), based on experimental and field studies, developed a set of guidelines to assist water system managers in managing tanks in such a way as to enhance water quality. Mixing and aging are two related phenomena that affect the water quality changes that can occur within distribution system storage tanks and reservoirs. Long residence times depress disinfectant residuals and can promote bacterial regrowth, and uneven mixing can result in zones of older water. For example, Figure C-15 (see appendix E) illustrates the relative chlorine residual values comparing mixed flow to plug flow. For short residence times and low decay coefficients, little loss in chlorine residual in the tank is experienced under both plug-flow and mixed-flow conditions. But as the decay rate and/or residence times increase, the chlorine loss increases much more rapidly under the plug-flow regime than in the mixed-flow regime. As a result, minimizing detention time and avoiding zones of water that remain in the storage facility for long periods of time should be objectives of design and operation of distribution system storage facilities. Grayman et al. suggest the following guidelines:

- *Distribution system reservoirs:* When comparing two tanks of the same size and inflow–outflow characteristics, tanks operated under plug-flow conditions will generally lose more disinfectant than tanks operated under mixed-flow conditions.
- *Baffling in distribution system tank and reservoirs:* Internal baffles are placed in vessels to direct and control flow through the facility. In distribution system tanks and reservoirs where mixed flow is preferable to plug flow, introduction of baffles inhibits mixing and, especially in fill-and-draw operations, can produce both stagnant water and poor mixing zones.
- *Mixing through development of a turbulent jet:* Mixing a fluid requires a source of energy. In distribution system storage facilities, this energy is normally introduced from the facility's inflow. As the water enters the facility, jet flow occurs, the ambient water is entrained into the jet, and circulation patterns are formed, resulting in mixing. For efficient mixing, the jet's path must be long enough to allow the entrainment and mixing process to develop. This means that the inlet jet should not be pointed directly toward nearby impediments such as walls, the reservoir bottom, or deflectors.
- *Stratification in storage facilities:* Temperature differences between the inflow into a storage tank and the ambient water temperature in the tank can lead to a buoyant jet. Positive buoyancy (when the temperature of the inflow is above that of the ambient water temperature) causes the inflow to rise toward the water surface. Negative buoyancy (when the temperature of the inflow is below the ambient water temperature) causes the inflow to sink toward the bottom. An inflow jet with excessive buoyancy (either positive or negative) relative to the momentum in the inflow jet will lead to ineffective mixing and can cause stable stratified conditions within the tank.

- *Tank mixing time:* For a tank operating in the fill-and-draw mode, mixing occurs primarily during the fill cycle. As a result, if the tank is relatively well-mixed at the end of the fill cycle, mixing problems are unlikely. For a wide array of tank and reservoir designs, experimentation has shown that the mixing time primarily depends on the facility's diameter or width, the water height, and the inlet momentum. Because of the highly significant effects on mixing time of the inlet diameter and the amount of water exchanged during the fill cycle, inlet diameters should be sized to ensure that adequate mixing occurs and that the tank is adequately exercised to provide the required mixing.
- *Inlet configurations:* In general, the location and orientation of inlets are secondary in terms of having an impact on mixing, compared with the factors described previously.
- *Standpipes:* In standpipes, the depth exceeds the diameter of the storage facility. Because of the dependence on tank height, standpipes are more susceptible to stratification and incomplete mixing. Existing standpipes should be analyzed for potential problems and monitored more frequently to identify possible stratification that will result in significant variations in disinfectant residuals at different depths in the standpipe.
- *Detention times:* Long detention times can lead to low disinfectant residuals and bacterial regrowth, even in well-mixed tanks. Poor mixing can result in pockets of older water and low disinfectant residuals. Even with good mixing, however, low residuals and the potential for bacterial regrowth can result from excessive detention time in a tank. Detention time can be estimated by dividing the duration of an average fill-and-draw cycle by the fraction of the water that is exchanged during the cycle. The aging of water in distribution system storage facilities can lead to deterioration of the water quality as a result of disinfectant residual loss and bacterial regrowth. Facilities should be operated to ensure that the age of the water is not excessive, taking into account the quality of the water, its reactivity, the type of disinfectant used, and the travel time before and after the water's entry into the storage facility. For example, if a tank is experiencing excess water age, modifications that address mixing only will have little effect on the underlying problem. Figure C-16 (see appendix E) illustrates the effect of reservoir turnover. Various monitoring and modeling tools such as water quality and temperature measurements are available to assess mixing and aging processes in existing or proposed tanks and reservoirs. Using these methods can lead to designs and operational policies that result in improved water quality in the storage facility. Poor mixing in storage facilities can exacerbate aging problems by creating zones of older water. To discourage the formation of zones of stagnant, older water, distribution system storage facilities should be designed and operated to encourage good mixing. This differs significantly from clearwells and contact chambers at treatment plants where plug flow is desired for its ability to produce adequate disinfectant contact time. Tanks and reservoirs can experience aging or mixing problems, or both. These problems can be interrelated or independent.

According to Mahmood et al. (2005), applied computational fluid dynamics (CFD) modeling, water temperature measurements, and chlorine residual measurements can be effective tools in predicting mixing characteristics inside storage tanks. These techniques can be used to determine mixing characteristics within storage tanks, identify tanks with poor mixing characteristics (and thus the greatest potential for water quality deterioration), and help modify designs to improve mixing characteristics. The authors applied these tools to storage tanks in the distribution system for the city of Virginia Beach, Va., which includes three types of storage tanks: elevated tanks, underground tanks, and standpipes. For one of the elevated tanks and a standpipe, the mixing characteristics were evaluated using computational fluid dynamics modeling, temperature measurements, and chlorine residual measurements. For the ground storage tank, the mixing characteristics were evaluated using temperature measurements and chlorine residual measurements. Figure C-17 (see appendix E) shows the results of temperature monitoring for a 1-mil-gal elevated tank. Several observations can be made from Figure C-17:

- A cyclical variation of water temperatures within the tank appears to be related to the fill-and-draw operation. During the fill cycle, the cooler inflow tended to increase the temperature difference between the bottom of the tank and the 19-ft (5.5m) depth. During the drain cycle, a trend of increasing temperature was observed as the warmer water in the upper region of the tank was drawn down and came into contact with the temperature probes.
- There was minimal tendency of the water temperatures to converge (from top to bottom) during the filling cycles. This result suggests that only limited mixing was achieved throughout the full depth of the water during a cycle.
- The observed temperature stratification in this tank appeared to be attributable primarily to the large diameter (48-in. or 1,200-mm) of the inlet–outlet riser, which resulted in low inlet momentum.

Mahmood et al. concluded the following:

- Inlet momentum is a key factor for mixing of water in storage tanks. The higher the inlet momentum, the better the mixing characteristics in the storage tanks. Increasing the flow rate could offer a simple way to increase momentum, but it may not be practical to increase the flow rate into tanks because of limitations of system hydraulics.
- The location and orientation of the inlet pipe relative to the tank walls have a significant effect on mixing characteristics. For example, when the height of a tank is much greater than the diameter or width, the location of the inlet pipe in the horizontal direction at the bottom of the tank will likely result in incomplete mixing of the water in the tank.
- Differences in the temperature between the bulk tank water and the inflow significantly affect the mixing characteristics of a storage tank. Depending on the location of the inlet pipe and tank geometry, the water entering the tank from the buried pipes may be cooler than the bulk water in the tank during the summer or warmer than the bulk water in the tank during the winter. Temperature differential may result in density gradients inside the tank and cause stratification and poor mixing.
- CFD modeling and temperature measurements can be effective tools to predict mixing characteristics in storage tanks. The adequacy of inlet momentum and suitability of inlet pipe orientation to provide sufficient mixing throughout the storage tank as well as the effect of design changes on mixing characteristics can be effectively investigated using CFD modeling. Temperature measurements of storage tanks are relatively simple to conduct and are highly effective in predicting the potential for thermal stratification that results from poor mixing characteristics.

References

American Water Works Association (AWWA). 2008. *AWWA Manual M31, Distribution System Requirements for Fire Protection*. 4th ed. Denver, Colo.: AWWA.

Boulos, P.F., W.M. Grayman, R.W. Bowcock, J.W. Clapp, L.A. Rossman, R.M. Clark, R.A. Deininger, and A.K. Dhingra. 1996. Hydraulic Mixing and Free Chlorine Residual in Reservoirs. *Journal AWWA*, 88(7):48–59.

Characklis, W.G. 1988. *Bacterial Regrowth in Distribution Systems*. Denver, Colo.: AwwaRF and AWWA.

Clark, R.M., W.M. Grayman, J.A. Goodrich, R.A. Deininger, and A.F. Hess. 1991. Field Testing Distribution Water Quality Models. *Journal AWWA*, 83(7):67–75.

Clark, R.M., F. Abdesaken, P.F. Boulos, and R.E. Mau. 1996. Mixing in Distribution System Storage Tanks: Its Effects on Water Quality. *Jour. Environ. Eng.*, 122(9):814–821.

Danckwerts, P.V. 1958. Continuous Flow Systems. *Chemical Engineering Science*, 2(1):1–18.

Germeles, A.E. 1975. Forced Plumes and Mixing of Liquids in Tanks. *Jour. Fluid Mechanics*, 71(3).

Grayman, W.M., R.M. Clark, and J.A. Goodrich. 1991. The Effects of Operation, Design and Location of Storage Tanks on the Water Quality in a Distribution System. In *Proc. of the Water Quality Modeling in Distribution Systems Conf.* Denver, Colo.: AwwaRF and AWWA.

Grayman, W.M., and R.M. Clark. 1991. Water Quality Modeling in a Distribution System. In *Proc. AWWA 1991 Annual Conference*. Denver, Colo.: AWWA.

Grayman, W.M., and R.M. Clark. 1993. Using Computers to Determine the Effect of Storage on Water Quality. *Journal AWWA*, 85(7):67–77.

Grayman, W.M., L.A. Rossman, R.A. Deininger, C.D. Smith, C.N. Arnold, and J.F. Smith. 2004. Mixing and Aging of Water in Distribution System Storage Facilities. *Journal AWWA*, 98(9):70–80.

Kennedy, M.S., S. Sarikelle, S. Moegling, and K. Suravallop. 1991. Mixing Characteristics in Distribution System Storage Reservoirs. In. *Proc. AWWA 1991 Annual Conference*. Denver, Colo.: AWWA.

Kennedy, M.S., S.S. Moegling, and K. Suravallop. 1993. Assessing the Effects of Storage Tank Design. *Journal AWWA*, 85(7):78–88.

Mahmood, F., J.G. Pimblett, N.O. Grace, and W.M. Grayamn. 2005. Evaluation of Water Mixing Characteristics in Distribution System Storage Tanks. *Journal AWWA*, 97(3):74–88.

Skov, K.R., A.F. Hess, and D.B. Smith. 1991. Field Sampling Procedures for Calibration of a Water Distribution System Hydraulic Model. In *Water Quality Modeling in Distribution Systems*. Denver, Colo.: AwwaRF, USEPA, and AWWA.

Appendix D

Conversion Charts

Table D-1. Conversion of US customary units

		Linear Measurement		
fathoms	×	6	=	feet (ft)
feet (ft)	×	12	=	inches (in.)
inches (in.)	×	0.0833	=	feet (ft)
miles (mi)	×	5,280	=	feet (ft)
yards (yd)	×	3	=	feet (ft)
yards (yd)	×	36	=	inches (in.)
		Circular Measurement		
degrees (angle)	×	60	=	minutes (angle)
degrees (angle)	×	0.01745	=	radians
		Area Measurement		
acres	×	43,560	=	square feet (ft^2)
square feet (ft^2)	×	144	=	square inches ($in.^2$)
square inches ($in.^2$)	×	0.00695	=	square feet (ft^2)
square miles (mi^2)	×	640	=	acres
square miles (mi^2)	×	27,880,000	=	square feet (ft^2)
square miles (mi^2)	×	3,098,000	=	square yards (yd^2)
square yards (yd^2)	×	9	=	square feet (ft^2)
		Volume Measurement		
acre-feet (acre-ft)	×	43,560	=	cubic feet (ft^3)
acre-feet (acre-ft)	×	325,851	=	gallons (gal)
barrels (bbl)	×	42	=	gallons (gal)
board foot (fbm)			=	144 square inches × 1 inch
cubic feet (ft^3)	×	1,728	=	cubic inches ($in.^3$)
cubic feet (ft^3)	×	7.48052	=	gallons (gal)
cubic feet (ft^3)	×	29.92	=	quarts (qt)
cubic feet (ft^3)	×	59.84	=	pints (pt)
cubic feet (ft^3)	×	0.000023	=	acre feet (acre-ft)
cubic inches ($in.^3$)	×	0.00433	=	gallons (gal)
cubic inches ($in.^3$)	×	0.00058	=	cubic feet (ft^3)
drops	×	60	=	teaspoons (tsp)
gallons (gal)	×	0.1337	=	cubic feet (ft^3)
gallons (gal)	×	231	=	cubic inches ($in.^3$)
gallons (gal)	×	0.0238	=	barrels (bbl)
gallons (gal)	×	4	=	quarts (qt)

(continues)

Table D-1. Conversion of US customary units (continued)

gallons (gal)	×	8	=	pints (pt)
gallons (gal)	×	128	=	ounces (oz)
gallons, US	×	0.83267	=	gallons, Imperial
gallons (gal)	×	0.00000308	=	acre-feet (acre-ft)
gallons (gal)	×	0.0238	=	barrels (42 gal) (bbl)
gallons, Imperial	×	1.20095	=	gallons, US
pints (pt)	×	2	=	quarts (qt)
quarts (qt)	×	4	=	gallons (gal)
quarts (qt)	×	57.75	=	cubic inches (in.^{3})
		Pressure Measurement		
atmospheres	×	29.92	=	inches of mercury
atmospheres	×	33.90	=	feet of water
atmospheres	×	14.70	=	pounds per square inch (lb/in.^{2})
feet of water	×	0.8826	=	inches of mercury
feet of water	×	0.02950	=	atmospheres
feet of water	×	0.4335	=	pounds per square inch (lb/in.^{2})
feet of water	×	62.43	=	pounds per square foot (lb/ft^2)
feet of water	×	0.8876	=	inches of mercury
inches of mercury	×	1.133	=	feet of water
inches of mercury	×	0.03342	=	atmospheres
inches of mercury	×	0.4912	=	pounds per square inch (lb/in.^{2})
inches of water	×	0.002458	=	atmospheres
inches of water	×	0.07355	=	inches of mercury
inches of water	×	0.03613	=	pounds per square inch (lb/in.^{2})
pounds/square in. (lb/in.^{2})	×	0.01602	=	feet of water
pounds/square foot (lb/ft^2)	×	6,954	=	pounds per square inch (lb/in.^{2})
pounds/square in. (lb/in.^{2})	×	2.307	=	feet of water
pounds/square inch (lb/in.^{2})	×	2.036	=	inches of mercury
pounds/square inch (lb/in.^{2})	×	27.70	=	inches of water
feet suction lift of water	×	0.882	=	inches of mercury
		Weight Measurement		
cubic feet of ice	×	57.2	=	pounds (lb)
cubic feet of water (50°F)	×	62.4	=	pounds of water
cubic inches of water	×	0.036	=	pounds of water
gallons water (50°F)	×	8.3453	=	pounds of water
milligrams/liter (mg/L)	×	0.0584	=	grains per gallon (US) (gpg)
milligrams/liter (mg/L)	×	0.07016	=	grains per gallon (Imp)
milligrams/liter (mg/L)	×	8.345	=	pounds per million gallons (lb/mil gal)
ounces (oz)	×	437.5	=	grains (gr)
parts per million (ppm)	×		=	milligrams per liter (mg/L) (for normal water applications)
grains per gallon (gpg)	×	17.118	=	parts per million (ppm)
grains per gallon (gpg)	×	142.86	=	pounds per million gallons (lb/mil gal)
percent solution	×	10,000	=	milligrams per liter (mg/L)
pounds (lb)	×	16	=	ounces (oz)
pounds (lb)	×	7,000	=	grains (gr)
pounds (lb)	×	0.0004114	=	tons (short)

Table D-1. Conversion of US customary units (continued)

pounds/cubic inch (lb/in.^{3})	×	1,728	=	pounds per cubic foot (lb/ft^3)
pounds of water	×	0.0166032	=	cubic feet (ft^3)
pounds of water	×	2,768	=	cubic inches (in.^{3})
pounds of water	×	0.1198	=	gallons (gal)
tons (short)	×	2,000	=	pounds (lb)
tons (short)	×	0.89287	=	tons (long)
tons (long)	×	2,240	=	pounds (lb)
cubic feet air (@ 60°F and 29.92 in. mercury)	×	0.0763	=	pounds (lb)
		Flow Measurement		
barrels per hour (bbl/hr)	×	0.70	=	gallons per minute (gpm)
acre-feet/minute	×	325.851	=	gallons per minute (gpm)
acre-feet/minute	×	726	=	cubic feet per second (ft^3/sec)
cubic feet/minute (ft^3/min)	×	0.1247	=	gallons per second (gps)
cubic feet/minute (ft^3/min)	×	62.43	=	pounds of water per minute
cubic feet/second (ft^3/sec)	×	448.831	=	gallons per minute (gpm)
cubic feet/second (ft^3/sec)	×	0.646317	=	million gallons per day (mgd)
cubic feet/second (ft^3/sec)	×	1.984	=	acre-feet per day (acre-ft/day)
gallons/minute (gpm)	×	1,440	=	gallons per day (gpd)
gallons/minute (gpm)	×	0.00144	=	million gallons per day (mgd)
gallons/minute (gpm)	×	0.00223	=	cubic feet per second (ft^3/sec)
gallons/minute (gpm)	×	0.1337	=	cubic feet per minute (ft^3/min)
gallons/minute (gpm)	×	8.0208	=	cubic feet per hour (ft^3/hr)
gallons/minute (gpm	×	0.00442	=	acre-feet per day (acre-ft/day)
gallons/minute (gpm)	×	1.43	=	barrels (42 gal) per day (bbl/day)
gallons water/minute	×	6.0086	=	tons of water per 24 hours
million gallons/day (mgd)	×	1.54723	=	cubic feet per second (ft^3/sec)
million gallons/day (mgd)	×	92.82	=	cubic feet per minute (ft^3/min)
million gallons/day (mgd)	×	694.4	=	gallons per minute (gpm)
million gallons/day (mgd)	×	3.07	=	acre-feet per day (acre-ft/day)
pounds of water/minute	×	26.700	=	cubic feet per second (ft^3/sec)
miner's inch			=	flow through an orifice of in.^{2} under a head of 4 to 6 in.
miner's inches (9 gpm)	×	8.98	=	gallons per minute (gpm)
miner's inches (9 gpm)	×	1.2	=	cubic feet per minute (ft^3/min)
miner's inches (11.25 gpm)	×	11.22	=	gallons per minute (gpm)
miner's inches (11.25 gpm)	×	1.5	=	cubic feet per minute (ft^3/min)
		Work Measurement		
British thermal units (Btu)	×	777.5	=	foot-pounds (ft-lb)
British thermal units (Btu)	×	39,270	=	horsepower-hours (hp•hr)
British thermal units (Btu)	×	29,280	=	kilowatt-hours (kW•hr)
foot-pounds (ft-lb)	×	1,286	=	British thermal units (Btu)
foot-pounds (ft-lb)	×	50,500,000	=	horsepower-hours (hp•hr)
foot-pounds (ft-lb)	×	37,660,000	=	kilowatt-hours (kW•hr)
horsepower-hours (hp•hr)	×	2,547	=	British thermal units (Btu)
horsepower-hours (hp•hr)	×	0.7457	=	kilowatt-hours (kW•hr)
kilowatt-hours (kW•hr)	×	3,415	=	British thermal units (Btu)

(continues)

Table D-1. Conversion of US customary units (continued)

kilowatt-hours (kW•hr)	×	1.241	=	horsepower-hours (hp•hr)
Power Measurement				
boiler horsepower	×	33,480	=	British thermal units per hour (Btu/hr)
boiler horsepower	×	9.8	=	kilowatts (kW)
British thermal units/second (Btu/sec)	×	1.0551	=	kilowatts (kW)
British thermal units/minute (Btu/min)	×	12.96	=	foot-pounds per second (ft-lb/sec)
British thermal units/minute (Btu/min)	×	0.02356	=	horsepower (hp)
British thermal units/minute (Btu/min)	×	0.01757	=	kilowatts (kW)
British thermal units/hour (Btu/hr)	×	0.293	=	watts (W)
British thermal units/hour (Btu/hr)	×	12.96	=	foot-pounds per minute (ft-lb/min)
British thermal units/hour (Btu/hr)	×	0.00039	=	horsepower (hp)
foot-pounds per second (ft-lb/sec)	×	771.7	=	British thermal units per minute (Btu/min)
foot-pounds per second (ft-lb/sec)	×	1,818	=	horsepower (hp)
foot-pounds per second (ft-lb/sec)	×	1,356	=	kilowatts (kW)
foot-pounds per minute (ft-lb/min)	×	303,000	=	horsepower (hp)
foot-pounds per minute (ft-lb/min)	×	226,000	=	kilowatts (kW)
horsepower (hp)	×	42.44	=	British thermal units per minute (Btu/min)
horsepower (hp)	×	33,000	=	foot-pounds per minute (ft-lb/min)
horsepower (hp)	×	550	=	foot-pounds per second (ft-lb/sec)
horsepower (hp)	×	1,980,000	=	foot-pounds per hour (ft-lb/hr)
horsepower (hp)	×	0.7457	=	kilowatts (kW)
horsepower (hp)	×	745.7	=	watts (W)
kilowatts (kW)	×	0.9478	=	British thermal units per second (Btu/sec)
kilowatts (kW)	×	56.92	=	British thermal units per minute (Btu/min)
kilowatts (kW)	×	3,413	=	British thermal units per hour (Btu/hr)
kilowatts (kW)	×	44,250	=	foot-pounds per minute (ft-lb/min)
kilowatts (kW)	×	737.6	=	foot-pounds per second (ft-lb/sec)
kilowatts (kW)	×	1.341	=	horsepower (hp)
tons of refrig. (US)	×	288,000	=	British thermal units per 24 hours
watts (W)	×	0.05692	=	British thermal units per minute (Btu/min)
watts (W)	×	0.7376	=	foot-pounds (force) per second (ft-lb/sec)
watts (W)	×	44.26	=	foot-pounds per minute (ft-lb/min)
watts (W)	×	1,341	=	horsepower (hp)
Velocity Measurement				
feet/minute (ft/min)	×	0.01667	=	feet per second (ft/sec)
feet/minute (ft/min)	×	0.01136	=	miles per hour (mph)
feet/second (ft/sec)	×	0.6818	=	miles per hour (mph)
miles/hour (mph)	×	88	=	feet per minute (ft/min)
miles/hour (mph)	×	1.467	=	feet per second (ft/sec)
Miscellaneous				
grade: 1 percent (or 0.01)			=	1 foot per 100 feet

Table D-2. Metric conversions

Linear Measurement				
inch (in.)	×	25.4	=	millimeters (mm)
inch (in.)	×	2.54	=	centimeters (cm)
foot (ft)	×	304.8	=	millimeters (mm)
foot (ft)	×	30.48	=	centimeters (cm)
foot (ft)	×	0.3048	=	meters (m)
yard (yd)	×	0.9144	=	meters (m)
mile (mi)	×	1,609.3	=	meters (m)
mile (mi)	×	1.6093	=	kilometers (km)
millimeter (mm)	×	0.03937	=	inches (in.)
centimeter (cm)	×	0.3937	=	inches (in.)
meter (m)	×	39.3701	=	inches (in.)
meter (m)	×	3.2808	=	feet (ft)
meter (m)	×	1.0936	=	yards (yd)
kilometer (km)	×	0.6214	=	miles (mi)
Area Measurement				
square meter (m^2)	×	10,000	=	square centimeters (cm^2)
hectare (ha)	×	10,000	=	square meters (m^2)
square inch ($in.^2$)	×	6.4516	=	square centimeters (cm^2)
square foot (ft^2)	×	0.092903	=	square meters (m^2)
square yard (yd^2)	×	0.8361	=	square meters (m^2)
acre	×	0.004047	=	square kilometers (km^2)
acre	×	0.4047	=	hectares (ha)
square mile (mi^2)	×	2.59	=	square kilometers (km^2)
square centimeter (cm^2)	×	0.16	=	square inches ($in.^2$)
square meters (m^2)	×	10.7639	=	square feet (ft^2)
square meters (m^2)	×	1.1960	=	square yards (yd^2)
hectare (ha)	×	2.471	=	acres
square kilometer (km^2)	×	247.1054	=	acres
square kilometer (km^2)	×	0.3861	=	square miles (mi^2)
Volume Measurement				
cubic inch ($in.^3$)	×	16.3871	=	cubic centimeters (cm^3)
cubic foot (ft^3)	×	28,317	=	cubic centimeters (cm^3)
cubic foot (ft^3)	×	0.028317	=	cubic meters (m^3)
cubic foot (ft^3)	×	28.317	=	liters (L)
cubic yard (yd^3)	×	0.7646	=	cubic meters (m^3)
acre foot (acre-ft)	×	1233.48	=	cubic meters (m^3)
ounce (US fluid) (oz)	×	0.029573	=	liters (L)
quart (liquid) (qt)	×	946.9	=	milliliters (mL)
quart (liquid) (qt)	×	0.9463	=	liters (L)
gallon (gal)	×	3.7854	=	liters (L)
gallon (gal)	×	0.0037854	=	cubic meters (m^3)
peck (pk)	×	0.881	=	decaliters (dL)
bushel (bu)	×	0.3524	=	hectoliters (hL)
cubic centimeters (cm^3)	×	0.061	=	cubic inches ($in.^3$)
cubic meter (m^3)	×	35.3183	=	cubic feet (ft^3)

(continues)

Table D-2. Metric conversions (continued)

cubic meter (m^3)	×	1.3079	=	cubic yards (yd^3)
cubic meter (m^3)	×	264.2	=	gallons (gal)
cubic meter (m^3)	×	0.000811	=	acre-feet (acre-ft)
liter (L)	×	1.0567	=	quart (liquid) (qt)
liter (L)	×	0.264	=	gallons (gal)
liter (L)	×	0.0353	=	cubic feet (ft^3)
decaliter (dL)	×	2.6417	=	gallons (gal)
decaliter (dL)	×	1.135	=	pecks (pk)
hectoliter (hL)	×	3.531	=	cubic feet (ft^3)
hectoliter (hL)	×	2.84	=	bushels (bu)
hectoliter (hL)	×	0.131	=	cubic yards (yd^3)
hectoliter (hL)	×	26.42	=	gallons (gal)
		Pressure Measurement		
pound/square inch (psi)	×	6.8948	=	kilopascals (kPa)
pound/square inch (psi)	×	0.00689	=	pascals (Pa)
pound/square inch (psi)	×	0.070307	=	kilograms/square centimeter (kg/cm^2)
pound/square foot (lb/ft^2)	×	47.8803	=	pascals (Pa)
pound/square foot (lb/ft^2)	×	0.000488	=	kilograms/square centimeter (kg/cm^2)
pound/square foot (lb/ft^2)	×	4.8824	=	kilograms/square meter (kg/m^2)
inches of mercury	×	3,376.8	=	pascals (Pa)
inches of water	×	248.84	=	pascals (Pa)
bar	×	100,000	=	newtons per square meter
pascals (Pa)	×	1	=	newtons per square meter
pascals (Pa)	×	0.000145	=	pounds/square inch (psi)
kilopascals (kPa)	×	0.145	=	pounds/square inch (psi)
pascals (Pa)	×	0.000296	=	inches of mercury (at 60°F)
kilogram/square centimeter (kg/cm^2)	×	14.22	=	pounds/square inch (psi)
kilogram/square centimeter (kg/cm^2)	×	28.959	=	inches of mercury (at 60°F)
kilogram/square meter (kg/m^2)	×	0.2048	=	pounds per square foot (lb/ft^2)
centimeters of mercury	×	0.4461	=	feet of water
		Weight Measurement		
ounce (oz)	×	28.3495	=	grams (g)
pound (lb)	×	0.045359	=	grams (g)
pound (lb)	×	0.4536	=	kilograms (kg)
ton (short)	×	0.9072	=	megagrams (metric ton)
pounds/cubic foot (lb/ft^3)	×	16.02	=	grams per liter (g/L)
pounds/million gallons (lb/mil gal)	×	0.1198	=	grams per cubic meter (g/m^3)
gram (g)	×	15.4324	=	grains (gr)
gram (g)	×	0.0353	=	ounces (oz)
gram (g)	×	0.0022	=	pounds (lb)
kilograms (kg)	×	2.2046	=	pounds (lb)
kilograms (kg)	×	0.0011	=	tons (short)
megagram (metric ton)	×	1.1023	=	tons (short)
grams/liter (g/L)	×	0.0624	=	pounds per cubic foot (lb/ft^3)

Table D-2. Metric conversions (continued)

grams/cubic meter (g/m^3)	×	8.3454	=	pounds/million gallons (lb/mil gal)
		Flow Rates		
gallons/second (gps)	×	3.785	=	liters per second (L/sec)
gallons/minute (gpm)	×	0.00006308	=	cubic meters per second (m^3/sec)
gallons/minute (gpm)	×	0.06308	=	liters per second (L/sec)
gallons/hour (gph)	×	0.003785	=	cubic meters per hour (m^3/hr)
gallons/day (gpd)	×	0.000003785	=	million liters per day (ML/day)
gallons/day (gpd)	×	0.003785	=	cubic meters per day (m^3/day)
cubic feet/second (ft^3/sec)	×	0.028317	=	cubic meters per second (m^3/sec)
cubic feet/second (ft^3/sec)	×	1,699	=	liters per minute (L/min)
cubic feet/minute (ft^3/min)	×	472	=	cubic centimeters/second (cm^3/sec)
cubic feet/minute (ft^3/min)	×	0.472	=	liters per second (L/sec)
cubic feet/minute (ft^3/min)	×	1.6990	=	cubic meters per hour (m^3/hr)
million gallons/day (mgd)	×	43.8126	=	liters per second (L/sec)
million gallons/day (mgd)	×	0.003785	=	cubic meters per day (m^3/day)
million gallons/day (mgd)	×	0.043813	=	cubic meters per second (m^3/sec)
gallons/square foot (gal/ft^2)	×	40.74	=	liters per square meter (L/m^2)
gallons/acre/day (gal/acre/day)	×	0.0094	=	cubic meters/hectare/day m^3/ha/day)
gallons/square foot/day (gal/ft^2/day)	×	0.0407	=	cubic meters/square meter/day (m^3/m^2/day)
gallons/square foot/day (gal/ft^2/day)	×	0.0283	=	liters/square meter/day (L/m^2/day)
gallons/square foot/minute (gal/ft^2/min)	×	2.444	=	cubic meters/square meter/hour (m^3/m^2/hr) = m/hr
gallons/square foot/minute (gal/ft^2/min)	×	0.679	=	liters/square meter/second (L/m^2/sec)
gallons/square foot/minute (gal/ft^2/min)	×	40.7458	=	liters/square meter/minute (L/m^2/min)
gallons/capita/day (gpcd)	×	3.785	=	liters/day/capita (L/d per capita)
liters/second (L/sec)	×	22,824.5	=	gallons per day (gpd)
liters/second (L/sec)	×	0.0228	=	million gallons per day (mgd)
liters/second (L/sec)	×	15.8508	=	gallons per minute (gpm)
liters/second (L/sec)	×	2.119	=	cubic feet per minute (ft^3/min)
liters/minute (L/min)	×	0.0005886	=	cubic feet per second (ft^3/sec)
cubic centimeters/second (cm^3/sec)	×	0.0021	=	cubic feet per minute (ft^3/min)
cubic meters/second (m^3/sec)	×	35.3147	=	cubic feet per second (ft^3/sec)
cubic meters/second (m^3/sec)	×	22.8245	=	million gallons per day (mgd)
cubic meters/second (m^3/sec)	×	15,850.3	=	gallons per minute (gpm)
cubic meters/hour (m^3/hr)	×	0.5886	=	cubic feet per minute (ft^3/min)
cubic meters/hour (m^3/hr)	×	4.403	=	gallons per minute (gpm)
cubic meters/day (m^3/day)	×	264.1720	=	gallons per day (gpd)
cubic meters/day (m^3/day)	×	0.00026417	=	million gallons per day (mgd)
cubic meters/hectare/day (m^3/ha/day)	×	106.9064	=	gallons per acre per day (gal/acre/day)
cubic meters/square meter/day (m^3/m^2/day)	×	24.5424	=	gallons/square foot/day (gal/ft^2/day)

(continues)

Table D-2. Metric conversions (continued)

liters/square meter/minute (L/m²/min)	×	0.0245	=	gallons/square foot/minute (gal/ft²/min)
liters/square meter/minute (L/m²/min)	×	35.3420	=	gallons/square foot/day (gal/ft²/day)
		Work, Heat, and Energy		
British thermal units (Btu)	×	1.0551	=	kilojoules (kJ)
British thermal units (Btu)	×	0.2520	=	kilogram-calories (kg-cal)
foot-pound (force) (ft-lb)	×	1.3558	=	joules (J)
horsepower-hour (hp•hr)	×	2.6845	=	megajoules (MJ)
watt-second (W-sec)	×	1.000	=	joules (J)
watt-hour (W•hr)	×	3.600	=	kilojoules (kJ)
kilowatt-hour (kW•hr)	×	3,600	=	kilojoules (kJ)
kilowatt-hour (kW•hr)	×	3,600,000	=	joules (J)
British thermal units per pound (Btu/lb)	×	0.5555	=	kilogram-calories per kilogram (kg-cal/kg)
British thermal units per cubic foot (Btu/ft³)	×	8.8987	=	kilogram-calories/cubic meter (kg-cal/m³)
kilojoule (kJ)	×	0.9478	=	British thermal units (Btu)
kilojoule (kJ)	×	0.00027778	=	kilowatt-hours (kW•hr)
kilojoule (kJ)	×	0.2778	=	watt-hours (W•hr)
joule (J)	×	0.7376	=	foot-pounds (ft-lb)
joule (J)	×	1.0000	=	watt-seconds (W-sec)
joule (J)	×	0.2399	=	calories (cal)
megajoule (MJ)	×	0.3725	=	horsepower-hour (hp•hr)
kilogram-calories (kg-cal)	×	3.9685	=	British thermal units (Btu)
kilogram-calories per kilogram (kg-cal/kg)	×	1.8000	=	British thermal units per pound (Btu/lb)
kilogram-calories per liter (kg-cal/L)	×	112.37	=	British thermal units per cubic foot (Btu/ft³)
kilogram-calories/cubic meter (kg-cal/m³)	×	0.1124	=	British thermal units per cubic foot (Btu/ft³)
		Velocity, Acceleration, and Force		
feet per minute (ft/min)	×	18.2880	=	meters per hour (m/hr)
feet per hour (ft/hr)	×	0.3048	=	meters per hour (m/hr)
miles per hour (mph)	×	44.7	=	centimeters per second (cm/sec)
miles per hour (mph)	×	26.82	=	meters per minute (m/min)
miles per hour (mph)	×	1.609	=	kilometers per hour (km/hr)
feet/second/second (ft/sec²)	×	0.3048	=	meters/second/second (m/sec²)
inches/second/second (in./sec²)	×	0.0254	=	meters/second/second (m/sec²)
pound-force (lbf)	×	4.44482	=	newtons (N)
centimeters/second (cm/sec)	×	0.0224	=	miles per hour (mph)
meters/second (m/sec)	×	3.2808	=	feet per second (ft/sec)
meters/minute (m/min)	×	0.0373	=	miles per hour (mph)
meters per hour (m/hr)	×	0.0547	=	feet per minute (ft/min)
meters per hour (m/hr)	×	3.2808	=	feet per hour (ft/hr)
kilometers/second (km/sec)	×	2.2369	=	miles per hour (mph)
kilometers/hour (km/hr)	×	0.0103	=	miles per hour (mph)

Table D-2. Metric conversions (continued)

meters/second/second (m/sec^2)	×	3.2808	=	feet/second/second (ft/sec^2)
meters/second/second (m/sec^2)	×	39.3701	=	inches/second/second (in./sec^2)
newtons (N)	×	0.2248	=	pounds force (lbf)

Table D-3. Celsius/Fahrenheit comparison graph

Decimal Equivalents of Fractions			
Fraction	Decimal	Fraction	Decimal
1/64	0.01563	33/64	0.51563
1/32	0.03125	17/32	0.53125
3/64	0.04688	35/64	0.54688
1/16	0.06250	9/16	0.56250
5/64	0.07813	37/64	0.57813
3/32	0.09375	19/32	0.59375
7/64	0.10938	39/64	0.60938
1/8	0.12500	5/8	0.62500
9/64	0.14063	41/64	0.64063
5/32	0.15625	21/32	0.65625
11/64	0.17188	43/64	0.67188
3/16	0.18750	11/16	0.68750
13/64	0.20313	45/64	0.70313
7/32	0.21875	23/32	0.71875
15/64	0.23438	47/64	0.73438
1/4	0.25000	3/4	0.75000
17/64	0.26563	49/64	0.76563
9/32	0.28125	25/32	0.78125
19/64	0.29688	51/64	0.79688
10/32	0.31250	13/16	0.81250
21/64	0.32813	53/64	0.82813
11/32	0.34375	27/32	0.84375
23/64	0.35938	55/64	0.85938
3/8	0.37500	7/8	0.87500
25/64	0.39063	57/64	0.89063
13/32	0.40625	29/32	0.90625
27/64	0.42188	59/64	0.92188
7/16	0.43750	15/16	0.93750
29/64	0.45313	61/64	0.95313
15/32	0.46875	31/32	0.96875
31/64	0.48438	63/64	0.98438
1/2	0.50000		

Water Equivalents and Data

- 1 US gallon of water weighs 8.345 pounds.
- 1 cubic foot of water equals 7.48 gallons.
- 1 foot head of water develops 0.433 pounds per square inch.
- Pounds per hour times 0.12 equals gallons per hour.
- Grains per gallon times 0.143 equals pounds per 1,000 gallons.
- Parts per million divided by 120 equals pounds per 1,000 gallons.
- 1 grain per gallon equals 17.1 parts per million.
- Estimated flow in gallons per minute equals pipe diameter in inches squared times 20.
- 1 boiler horsepower based on 10 square feet of heating surface requires 4 gallons per hour of feedwater.
- 1 pound of coal will produce 7 to 10 pounds of steam.
- 1 gallon of oil will produce 70 to 120 pounds of steam.
- 1,000 cubic feet of natural gas will produce 600 pounds of steam.
- Saturated salt brine for zeolite regeneration contains 2.48 pounds of salt per gallon or 18.5 pounds per cubic foot.
- Refrigeration tonnage is gallons per minute of cooling water times increased temperature divided by 24.
- Cooling tower makeup is estimated at 1½ gallons per hour per ton of refrigeration.
- 1 ton of refrigeration is 288,000 Btu.

Water Conversions

Water is composed of two gases, hydrogen and oxygen, in the ratio of two volumes of the former to one of the latter. It is never found pure in nature because of the readiness with which it absorbs impurities from the air and soil.

- One foot of water column at 39.1°F = 62.425 pounds on the square foot.
- One foot of water column at 39.1°F = 0.4335 pound on the square inch.
- One foot of water column at 39.1°F = 0.0295 atmospheric pressure.
- One foot of water column at 39.1°F = 0.8826 inch mercury column at 32°F.
- One foot of water column at 39.1°F = 773.3 feet of air column at 32°F and atmospheric pressure.
- One pound pressure per square foot = 0.01602 foot water column at 39.1°F.
- One pound pressure per square foot = 2.307 feet water column at 39.1°F.
- One atmospheric pressure = 29.92 inches mercury column = 33.9 feet water column.
- One inch of mercury column at 32°F = 1.133 feet water column.
- One foot of air column at 32°F and 1 atmospheric pressure = 0.001293 foot water column.

Appendix E

Color Figures

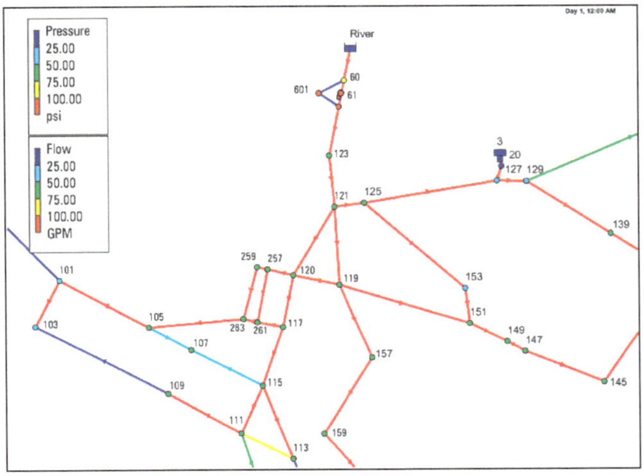

Figure 3-2. EPANET graphical output showing flow and pressure (Panguluri et al. 2005)

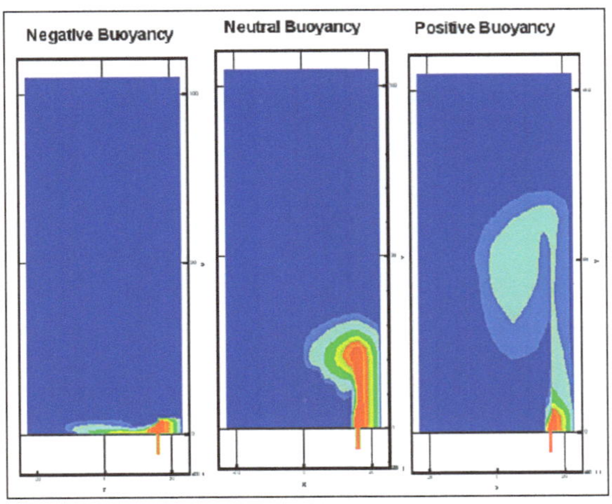

Figure 3-5. Graphical output from a CFD model showing tracer concentration in a tank (Panguluri et al. 2005)

Figure 3-6. A large physical model of a tank located in Bureau of Reclamation laboratory (courtesy of Bates Engineering)

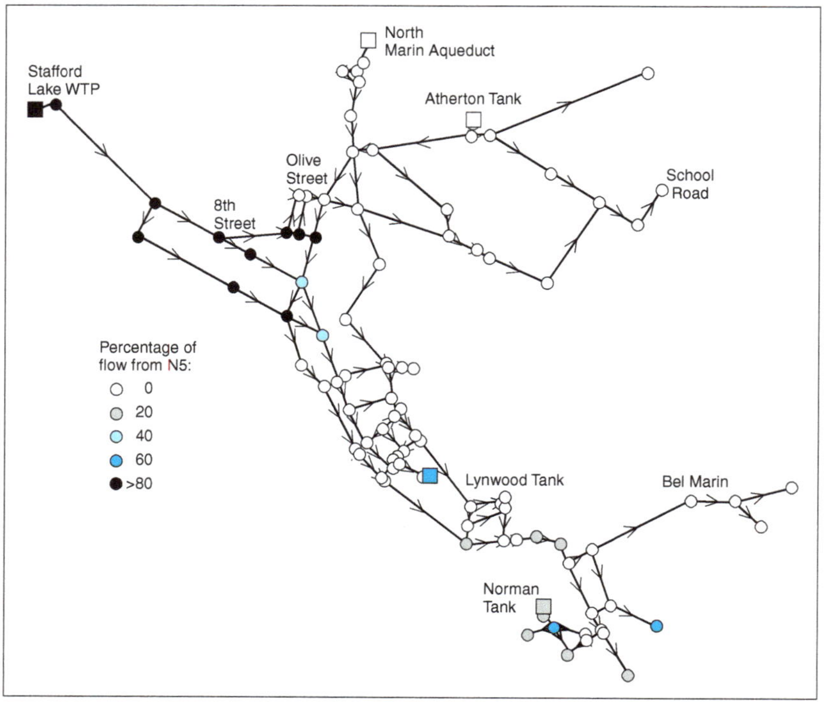

Figure 8-19a. North Marin hydraulic calibrations for 6:00 a.m. showing direction of flow and percentage of flow from N5 (Stafford Lake WTP)

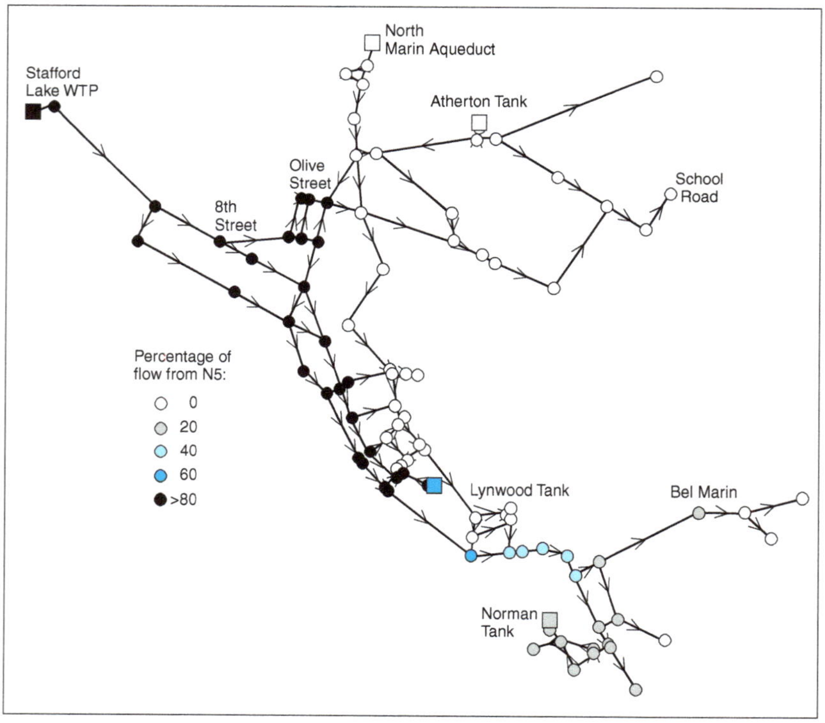

Figure 8-19b. North Marin hydraulic calibrations for 12:00 a.m. showing direction of flow and percentage of flow from N5 (Stafford Lake WTP)

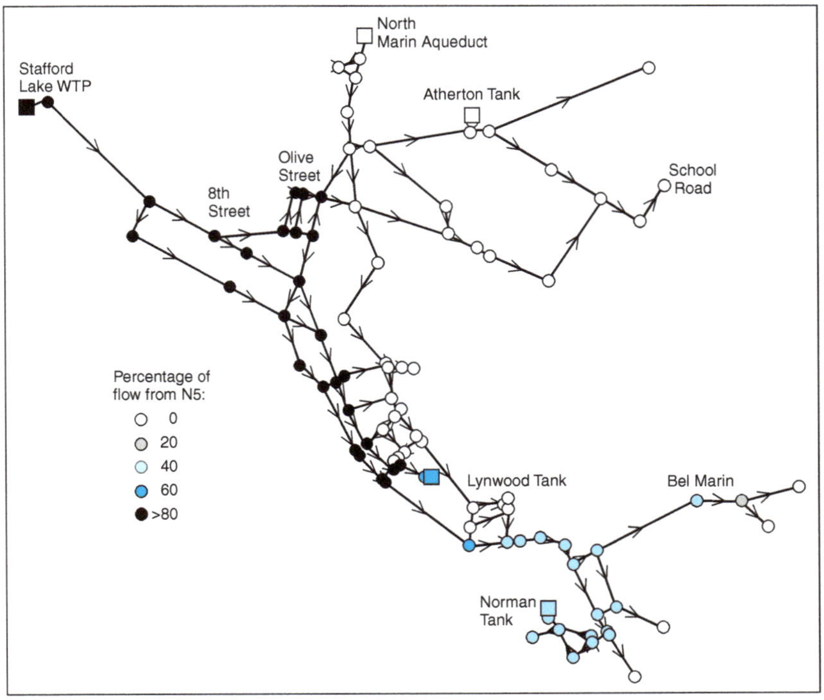

Figure 8-19c. North Marin hydraulic calibrations for 6:00 p.m. showing direction of flow and percentage of flow from N5 (Stafford Lake WTP)

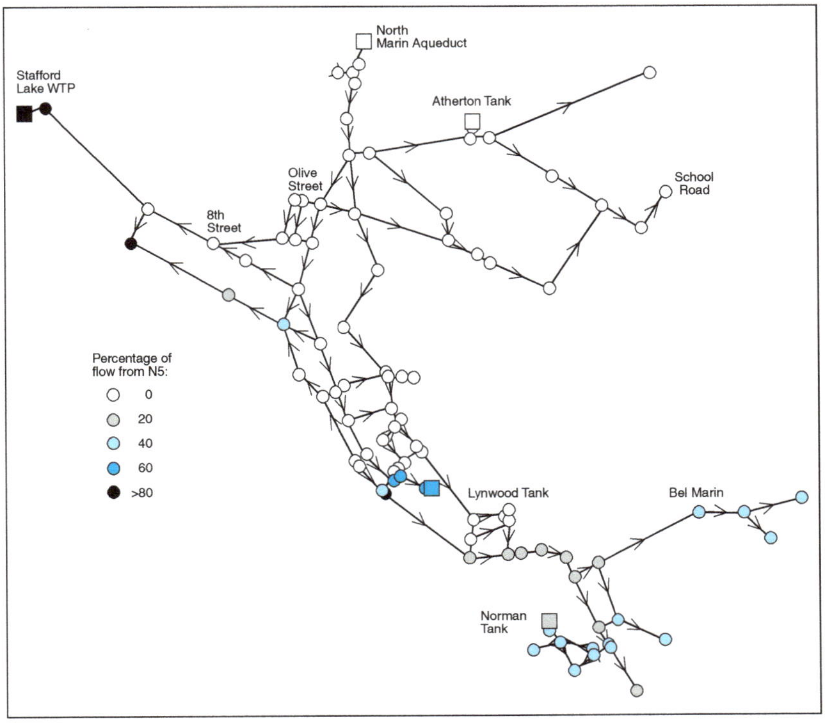

Figure 8-19d. North Marin hydraulic calibrations for 12:00 p.m. showing direction of flow and percentage of flow from N5 (Stafford Lake WTP)

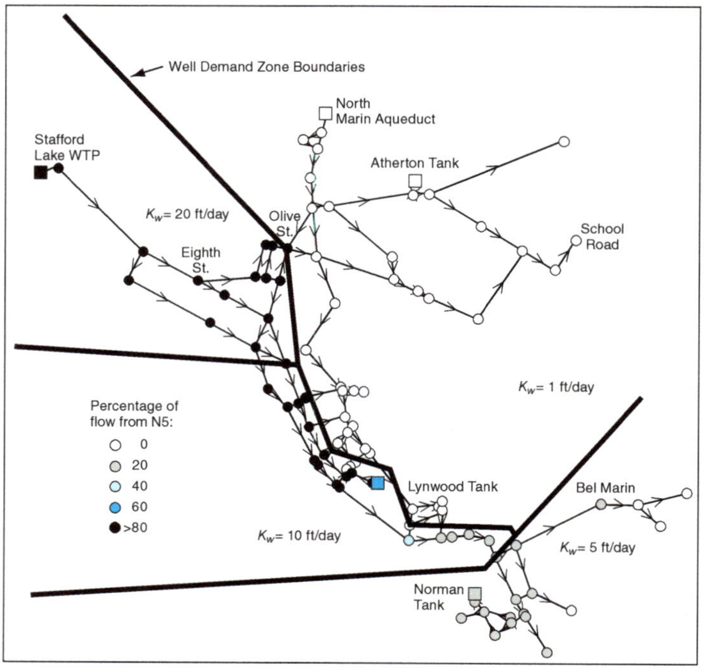

Figure 8-27. North Marin hydraulic calibration for 12:00 a.m. incorporating wall demand by zones

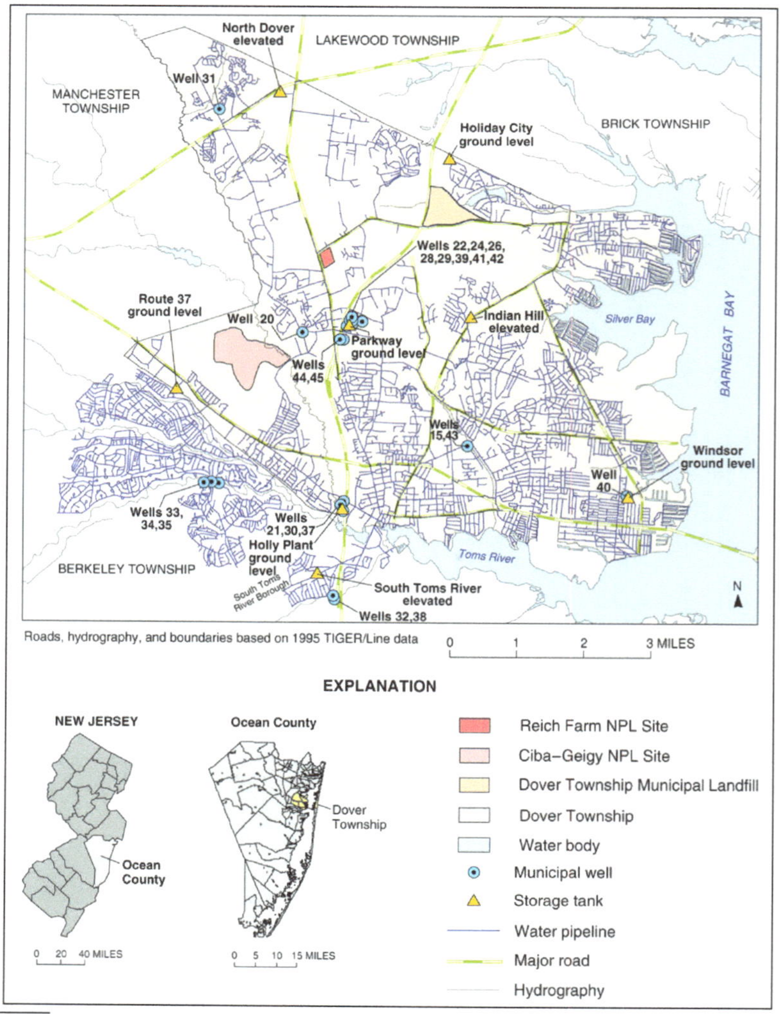

Figure 9-2. Investigation area, Dover Township (Maslia et al. 2001)

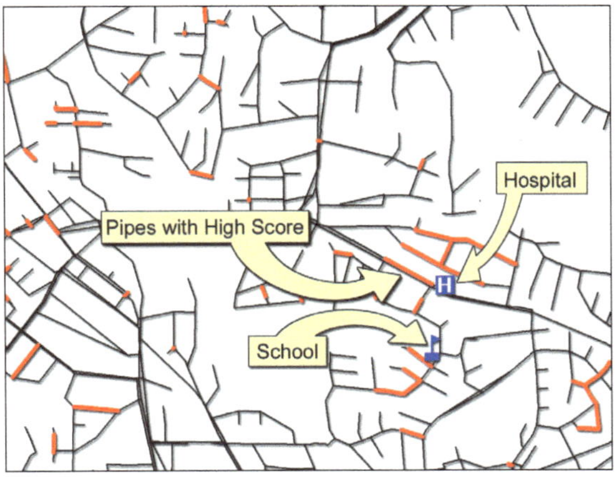

Figure 9-5. Pipes with scores greater than 27 overlaid with critical facility locations (Panguluri et al. 2005)

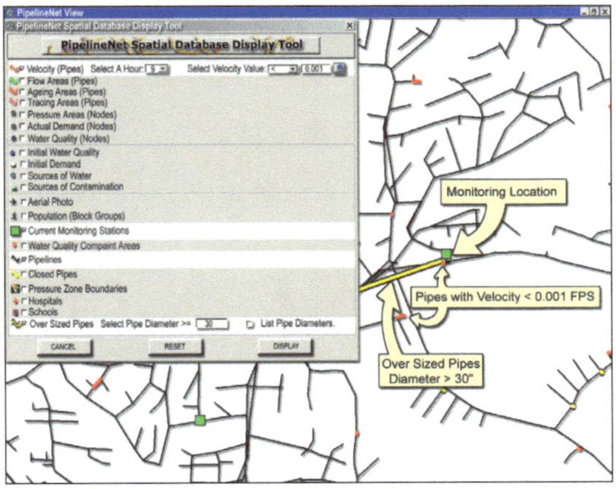

Figure 9-6. Display of low-velocity, oversized pipes, and current monitoring stations using the spatial database tool (Panguluri et al. 2005)

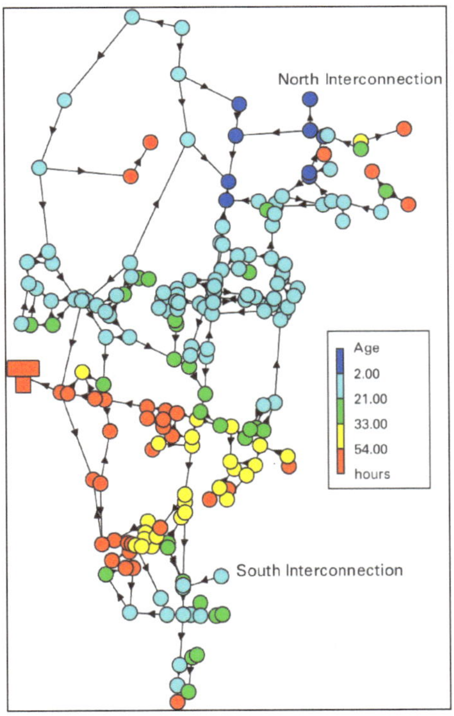

Figure 9-9. Average water age in the distribution system over last 24 hr of a 2-week simulation (Panguluri et al. 2005)

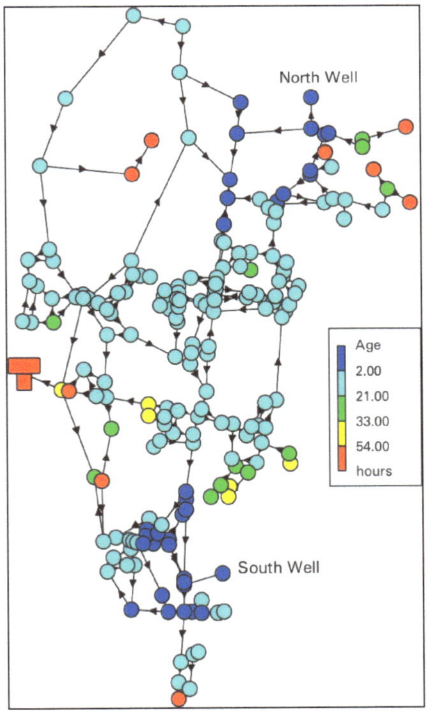

Figure 9-10. Minimum water age in the distribution system over last 24 hr of a 2-week simulation (Panguluri et al. 2005)

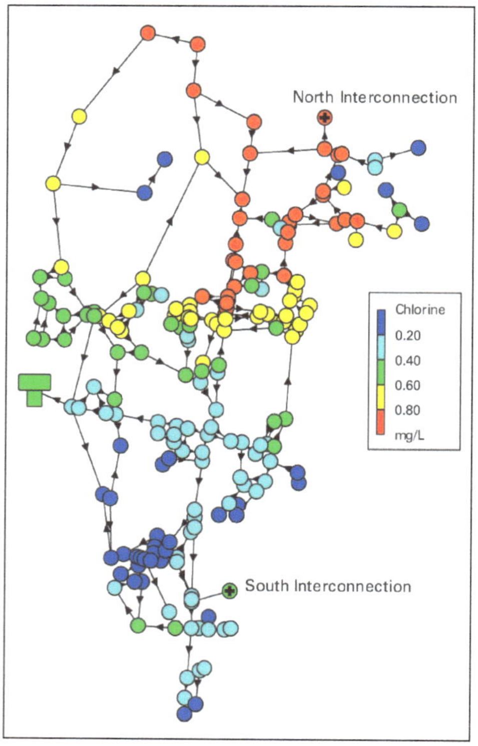

Figure 9-12. Minimum chlorine residual in distribution system over last 24 hr of a 2-week simulation (Panguluri et al. 2005)

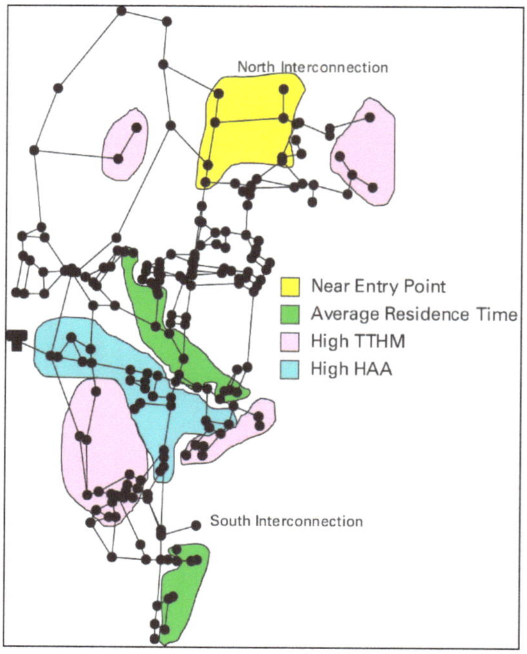

Figure 9-13. Zones representing potential monitoring locations for IDSE based on modeling (Panguluri et al. 2005)

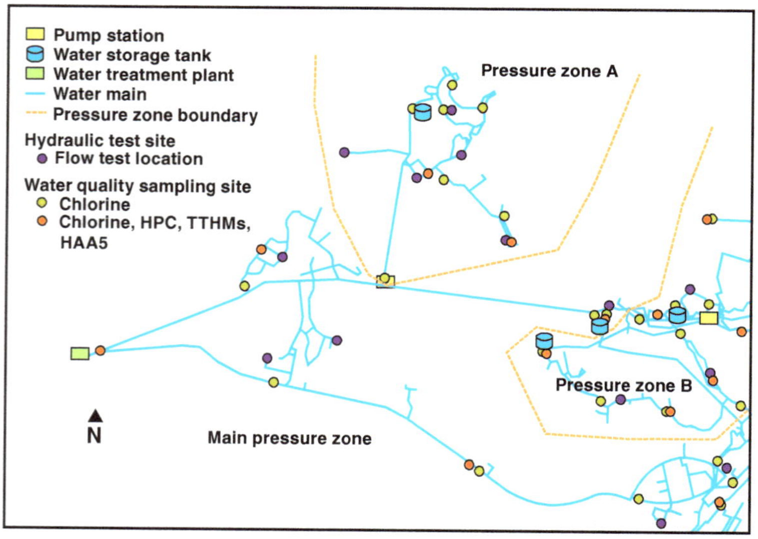

HPC—heterotrophic plate count, HAA5—the sum of five haloacetic acids, TTHMs—total trihalomethanes

Figure 9-14. Hydraulic and water quality field test locations for utility D (Yang et al. 2007)

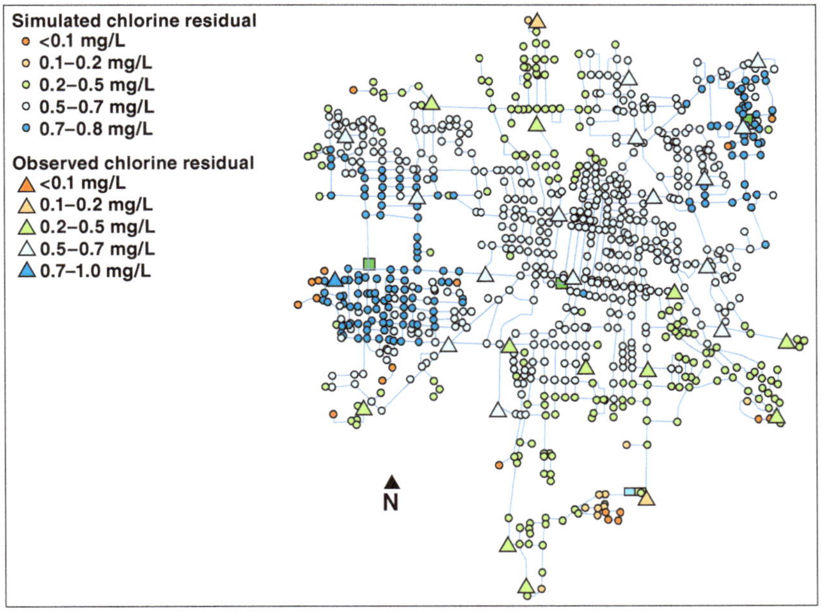

Figure 9-15. Model predicted versus observed chlorine residuals in utility C distribution system (Yang et al. 2007)

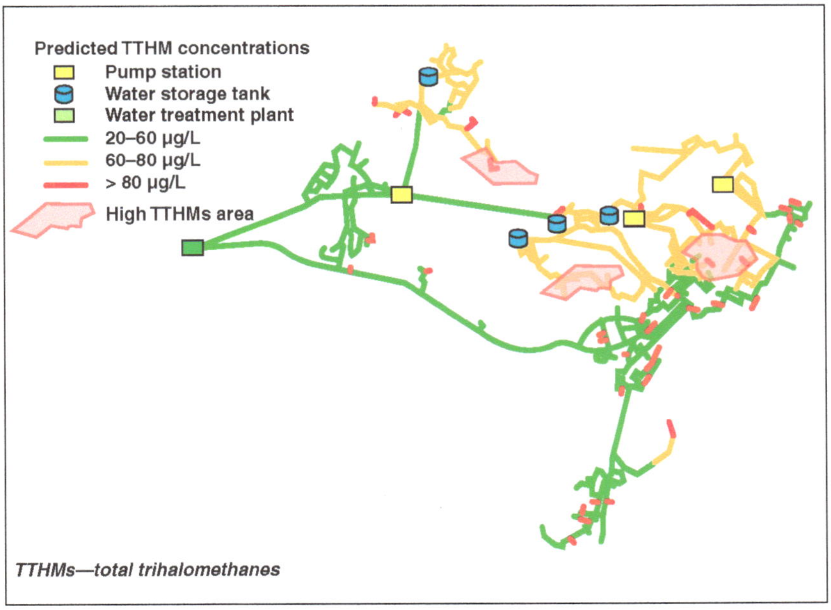

Figure 9-18. Predicted TTHM concentrations in utility D distribution system (Yang et al. 2007)

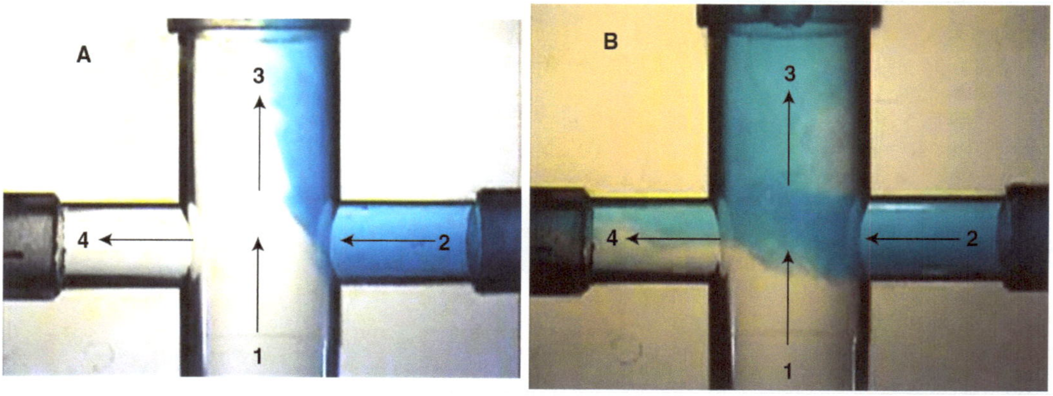

Figure 10-11. Video images of solute mixing in a pipe junction with adjacent inlets and unequal pipe diameters show the mixing behavior at a 1:14 momentum ratio (A: small pipe/large pipe) and at a 4:1 momentum ratio (B: small pipe/large pipe) (Ho and O'Rear 2009)

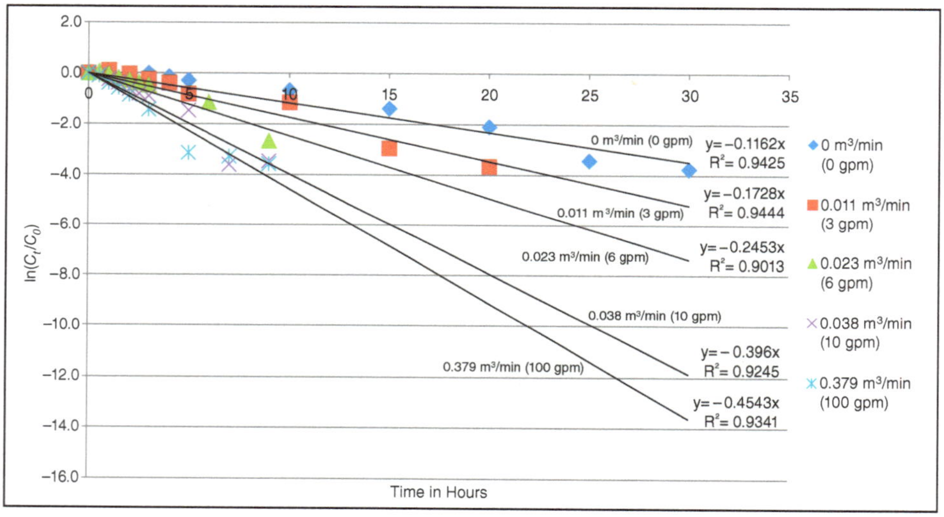

Figure 12-3. For Loop 4 (unlined metallic), ln (C_t/C_0) versus time in hours uncorrected for bulk demand

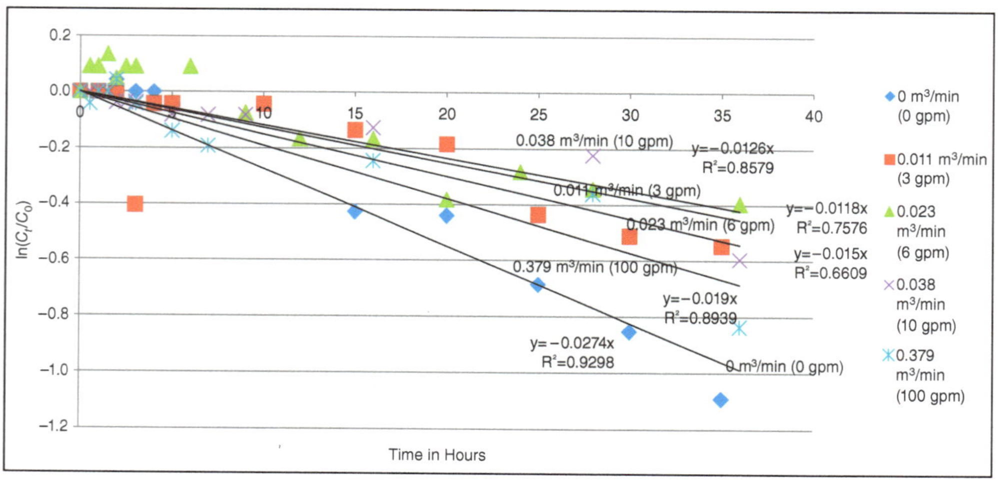

Figure 12-4. For Loop 5 (PVC), ln(C_t/C_0) versus time in hours uncorrected for bulk demand

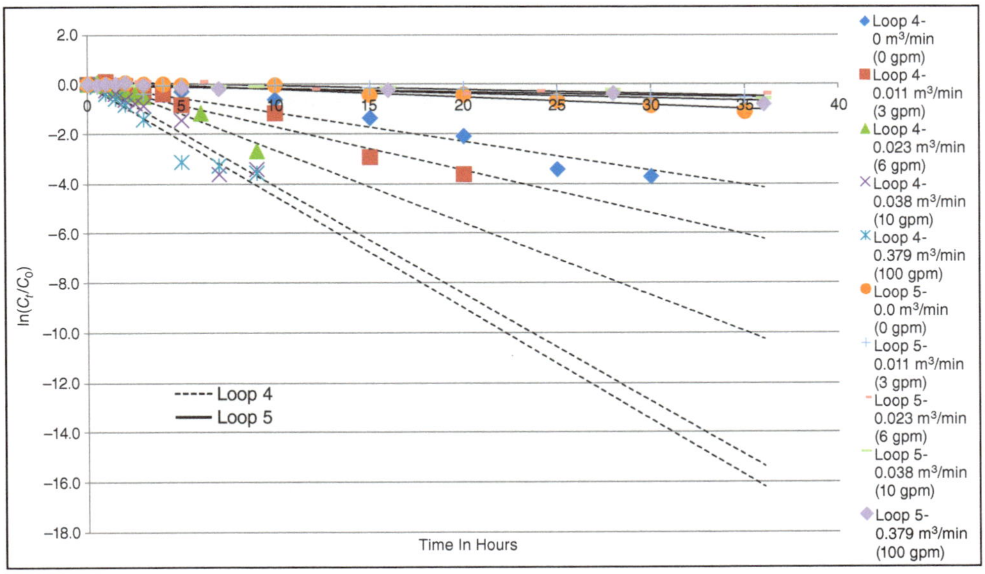

Figure 12-5. For Loops 4 (unlined metallic pipe) and 5 (PVC), ln (C_t/C_0) versus time in hours uncorrected for bulk demand

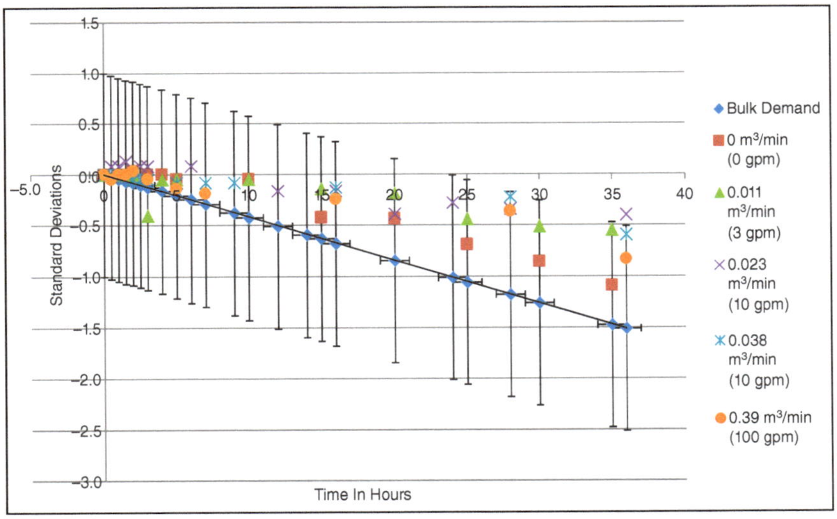

Figure 12-7. Comparison of bulk decay to PVC wall demand values—one standard deviation

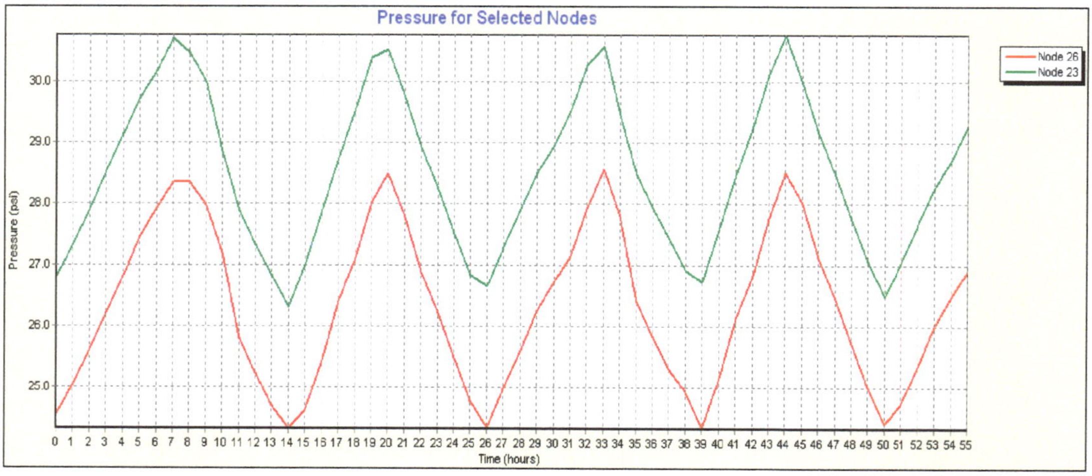

Figure B-5. Minimum pressure versus time for the system

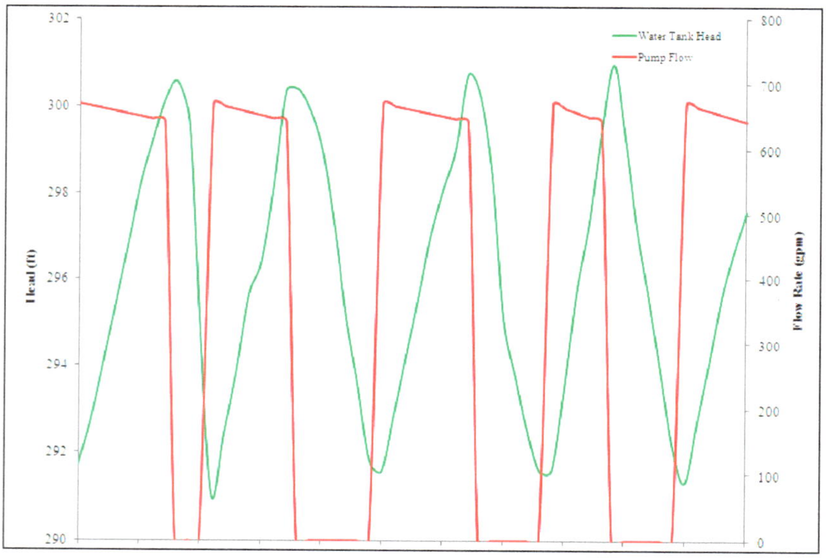

Figure B-8. Comparison of water tank head to pump flow rate

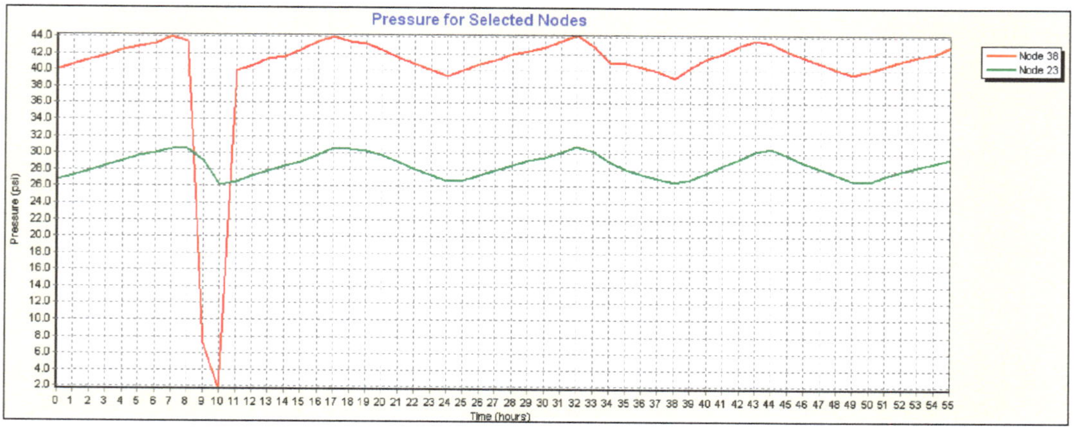

Figure B-9. Pressure versus time for Node 23 and the new subdivision

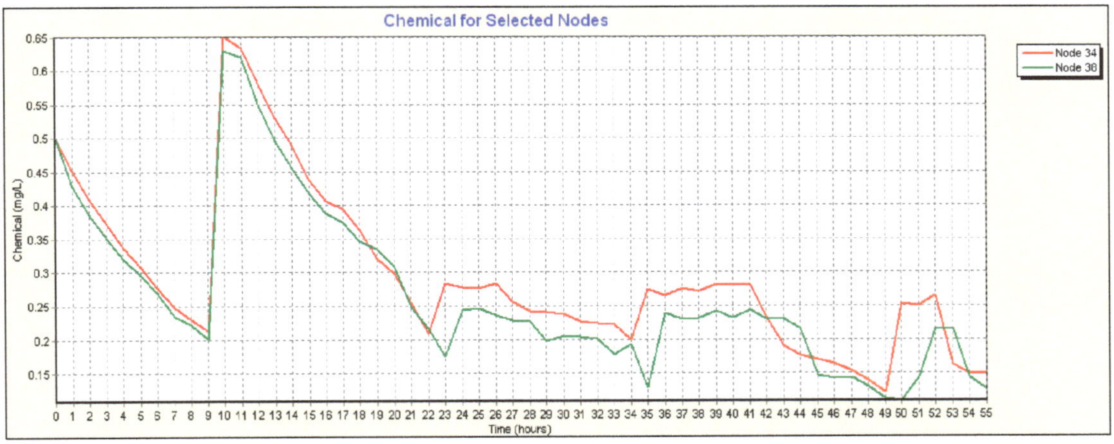

Figure B-10. Chlorine residual versus time for Node 34 and the new subdivision (Node 38)

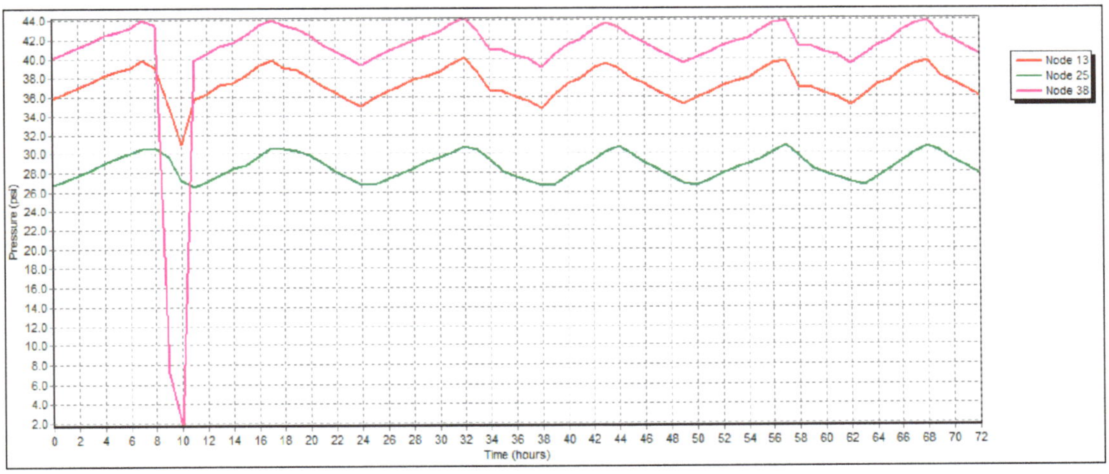

Figure B-16. Pressure nodes 13, 25, and 38

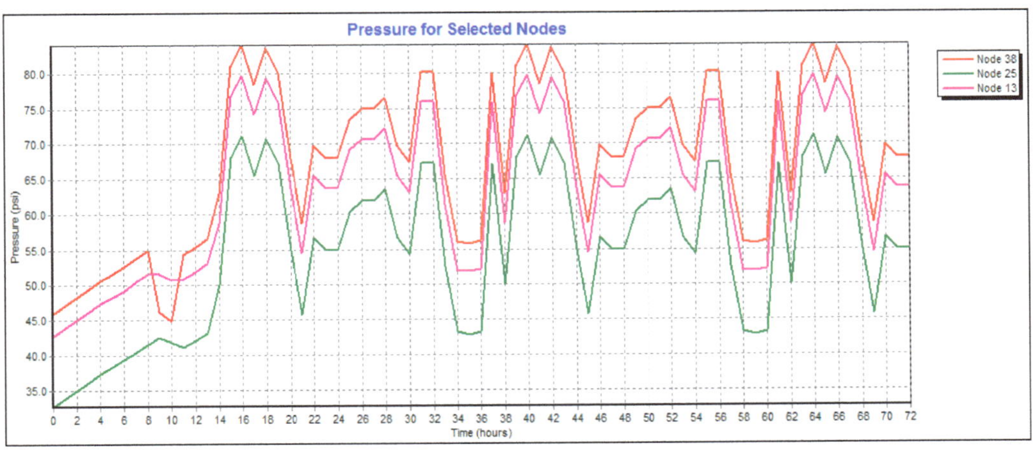

Figure B-17. Time series pressure after modifications in CH/BP WDS (except booster pumps)

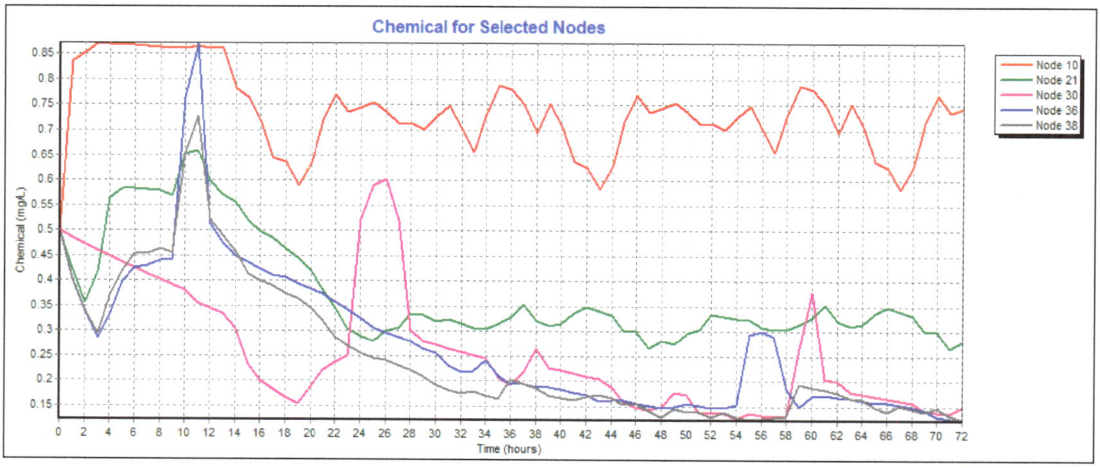

Figure B-18. Time series chlorine residual after pressure improvement in CH/BP WDS

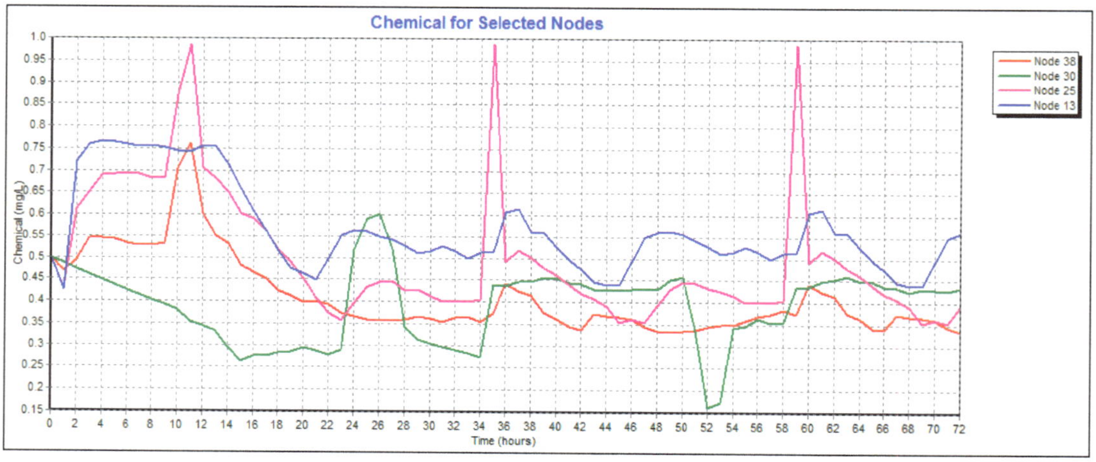

Figure B-19. Time series chlorine residual after booster setpoints in CH/BP WDS

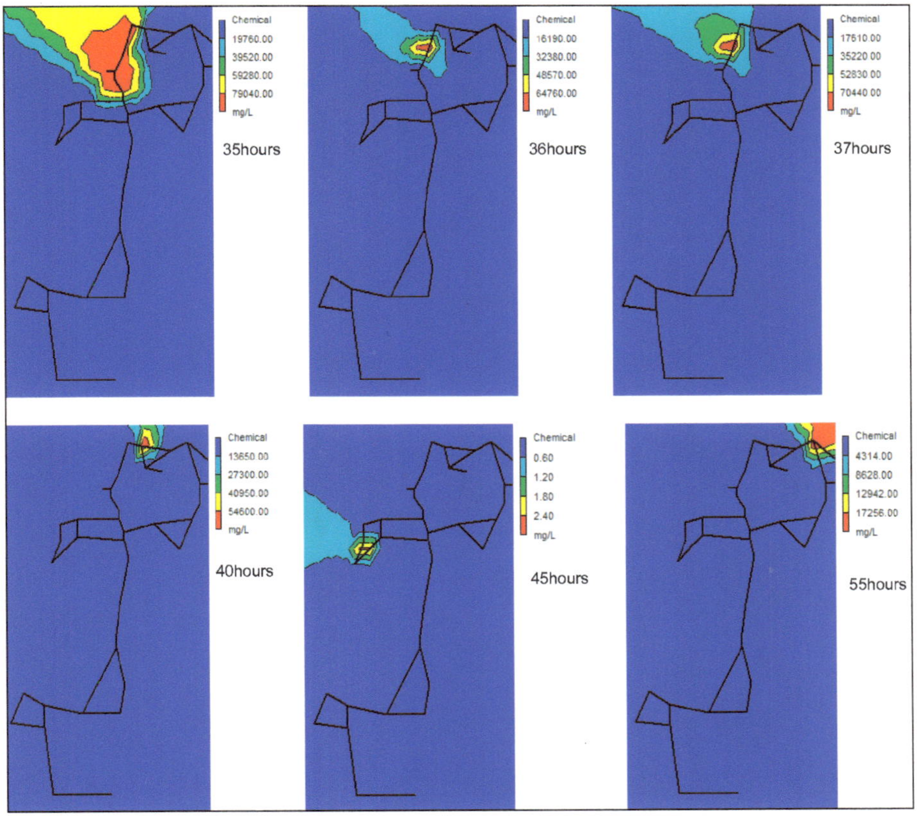

Figure B-21. Contour plots contamination event in tank Node 26

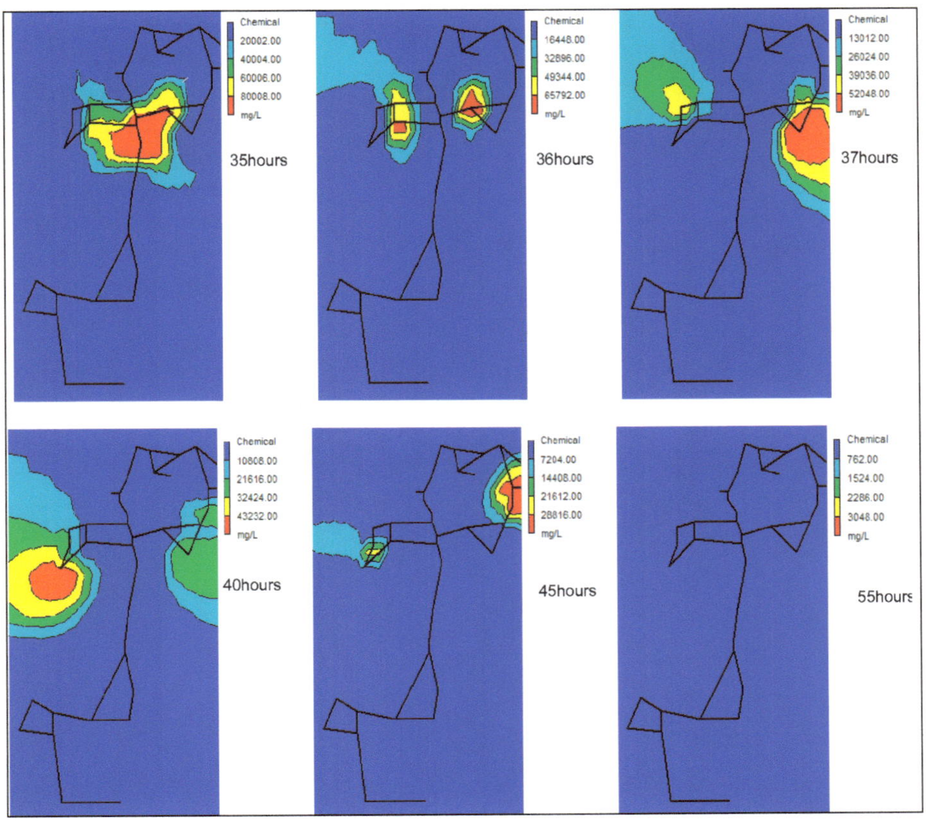

Figure B-22. Contour plots contamination event in tank Node 12

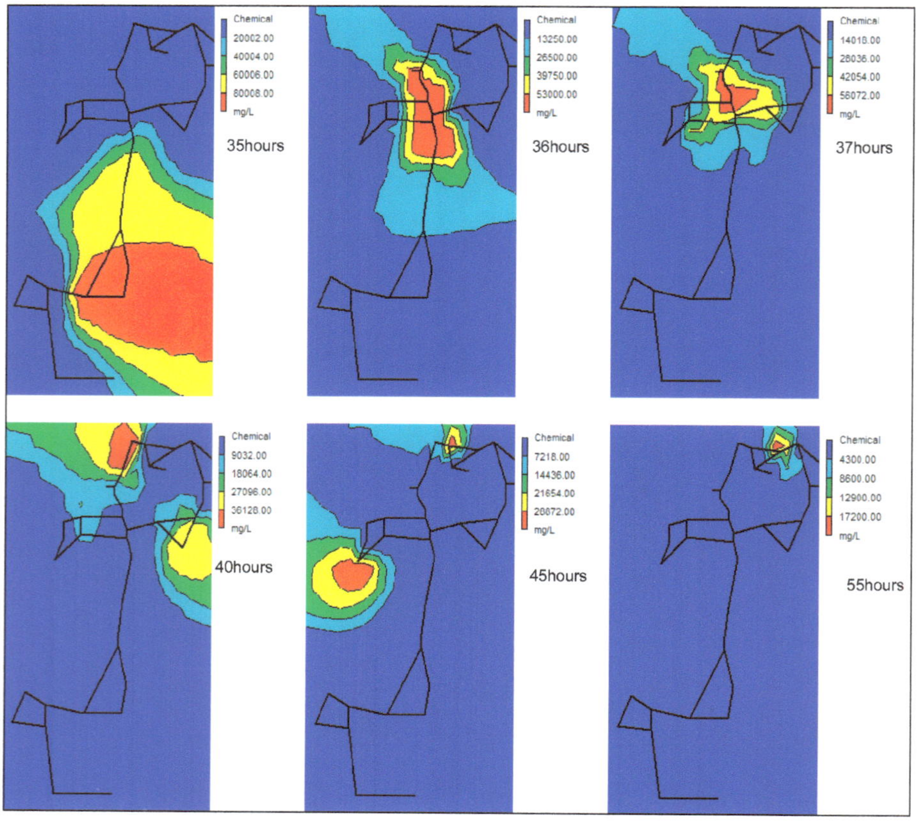

Figure B-23. Contour plots contamination event in tank Node 6

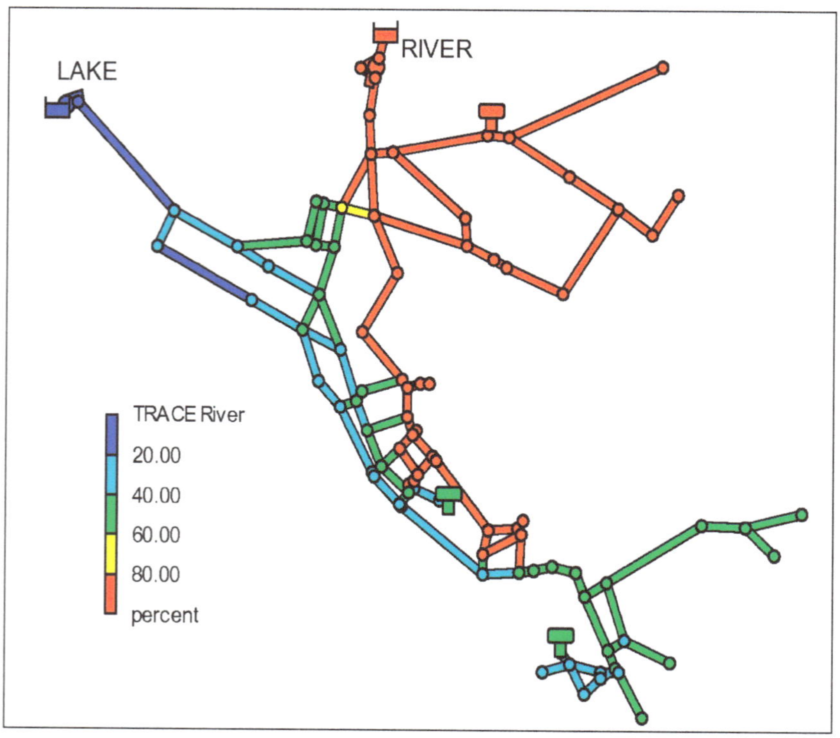

Figure B-24. Example of a two-source water distribution system showing the average percent of water originating from the North Marin Aqueduct

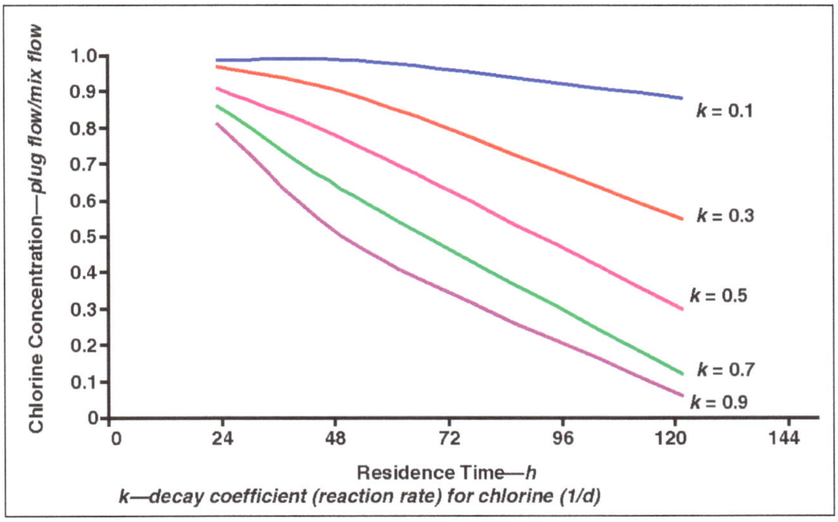

Figure C-15. Ratio of average outflow chlorine residual between plug flow and mixed flow as a function of decay rate and tank residence time (Grayman et al. 2004)

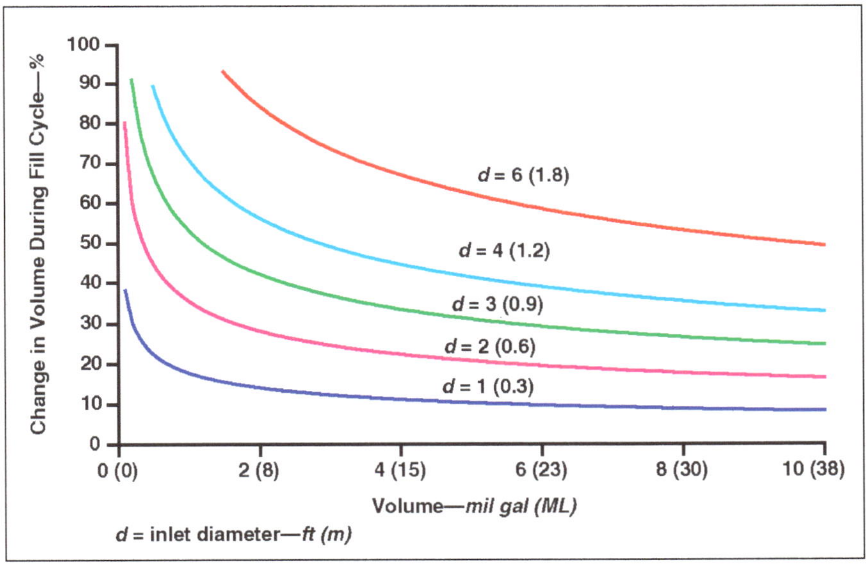

Figure C-16. Reservoir turnover required to assure good mixing during fill cycle (Grayman et al. 2004)

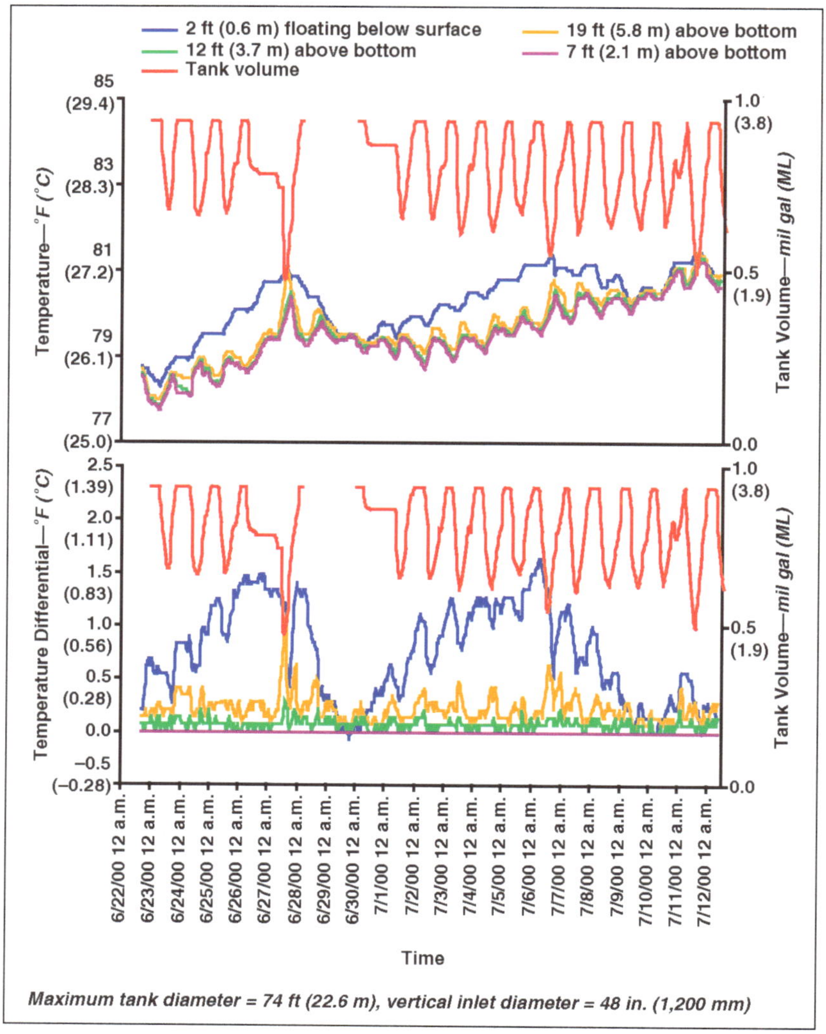

Figure C-17. Temperature profile at various depths for 1-mil-gal (3.8-ML) elevated tank (Mahmood et al. 2005)

Index

Note: *f.* indicates figure; *t.* indicates table

About the Author

Robert M. Clark received a BS Degree in Civil Engineering from Oregon State University (1960), a BS Degree in Mathematics from Portland State University (1961), an MS in Mathematics from Xavier University (1964), an MS in Civil Engineering from Cornell University (1968), and a PhD in Environmental Engineering from the University of Cincinnati (1976). Dr. Clark is a registered engineer in the State of Ohio and worked as an environmental engineer in the US Public Health Service and the US Environmental Protection Agency (USEPA) from 1961 to 2002. He was Director of the USEPA's Water Supply and Water Resources Division in the Office of Research and Development's National Risk Management Research Laboratory for 14 years (1985–1999). In 1999, he was appointed to a Senior Expert position in USEPA with the title Senior Research Engineering Advisor and served on USEPA's Water Protection Task Force in 2001–2002. His research activities with USEPA included initiating and directing USEPA's efforts on water quality modeling in drinking water distribution systems, analyzing and modeling the performance of water treatment unit processes, and developing strategies and techniques for the protection of critical infrastructure.

Dr. Clark retired from USEPA in August 2002 and is now an independent consultant and an Adjunct Professor of Civil and Environmental Engineering at the University of Cincinnati (UC). As a consultant he has worked extensively in the fields of homeland security, modeling water quality fate and transport, and the development of condition assessment techniques for drinking water infrastructure. He was a member of the National Research Council's Committee on "Public Water Distribution Systems: Assessing and Reducing Risks." He is currently a member of the UC/EPA Water Resources Adaptation Program team studying the vulnerability of urban water infrastructure to the impacts of climate change. Dr. Clark is also a member of the *Journal of the American Water Works Association* Peer Review Editorial Board.

Dr. Clark has published more than 380 papers and 5 books and has been professionally active in the American Water Works Association (AWWA) and the American Society of Civil Engineers (ASCE). He is a lifetime member of both organizations. He served as a trustee and chairman of AWWA's Research Division from 1984 to 1988 and was chair of ASCE's Environmental Engineering Division. He has served on AWWA's Total Coliform Rule Review Committee and on numerous Project Advisory Committees for the American Water Works Association Research Foundation (Water Research Foundation). He served on the organizing committee for ASCE's Water Distribution System Symposium held in Cincinnati in August 2006.

Dr. Clark has received numerous awards for his work including:

- American Society of Civil Engineers Environmental and Water Resources Institute's Best Paper Award from the *Journal of Water Resources Planning and Management* (2006).
- ASCE Environmental and Water Resources Institute Lifetime Achievement Award, in recognition of a lifelong and eminent contribution to the environmental and water resources engineering disciplines through practice, research, and public service (2004).
- US Environmental Protection Agency Distinguished Service Career Award for lifetime accomplishments and leadership as a researcher and manager in the field of water supply (2002).
- USEPA Diversity Leadership Award from USEPA's Office of Research and Development for enhancing the careers of ORD staff (1998).
- ASCE Rudolph Hering Medal for the best paper published by the Environmental Engineering Division (1996).
- USEPA Science Achievement Award in Water Quality for research characterizing the factors that cause deterioration of water quality in distribution systems (1995).
- USEPA Gold Medal Award for assisting the city of Milwaukee in providing solutions to the cryptosporidiosis outbreak of 1993 (1994).
- American Water Works Association A.P. Black Award for Outstanding Achievement in Water Supply Research (1993).
- AWWA Publication Award (1990).

www.ingramcontent.com/pod-product-compliance
Lightning Source LLC
LaVergne TN
LVHW071239210926
848486LV00007B/304

* 9 7 8 1 5 8 3 2 1 8 1 6 7 *